Probability and Statistics in Engineering and Management Science

Probability and Statistics in Engineering and Management Science

Second Edition

William W. Hines
Georgia Institute of Technology

Douglas C. Montgomery
Georgia Institute of Technology

John Wiley & Sons
New York   Chichester   Brisbane   Toronto

Library of Congress Cataloging in Publication Data

Hines, William W
   Probability and statistics in engineering
and management science.

   Bibliography: p.
   Includes index.
   1. Engineering—Statistical methods.
I. Montgomery, Douglas C., joint author.
II. Title.
TA340.H55   1980   519.1'02'462   79-26257
ISBN 0-471-04759-7

Printed in the United States of America

10 9 8 7 6 5 4 3 2 1

To Gayle and Martha

Preface

This book has been written for a first course in applied probability and statistics for undergraduate students in engineering, the physical sciences, and management science or operations research. The emphasis is on developing the statistical background required for model formulation and decision making. Knowledge of differential and integral calculus is assumed, and some familiarity with matrix algebra is also required for Chapter 13.

The second edition is a major revision of the original book. The applied probability topics are presented in Chapters 1 to 7 and 16. The material in Chapter 16 on stochastic processes and queueing is new. The engineering statistics topics are presented in Chapters 8 to 15 and 17, including quality control and reliability (Chapter 15) and statistical decision theory (Chapter 17). This material has been extensively reorganized. Point and confidence interval estimation is covered in Chapter 9, and Chapter 10 is devoted exclusively to hypothesis testing. Simple linear regression and correlation is covered in Chapter 12, while multiple regression is presented in Chapter 13. Chapter 14 on the design of experiments has been expanded somewhat.

Each new concept in the book is illustrated by one or more numerical examples. Our experience from the first edition is that the examples are an important part of the text and should be carefully studied. Homework problems play an important role in applied probability and statistics courses. The book contains over 500 exercises, ranging from computational problems to extensions of the basic methodology. Many of the examples and exercises in the second edition are new.

The book contains enough material for either two semesters or three quarters of work. At Georgia Tech, the applied probability topics are covered in one quarter, and the engineering statistics material is covered in two quarters. We have also used the book in a one-quarter (4-credit) intensive course for graduate students without previous statistical training. This course surveys the applied probability topics covered in Chapters 1 to 7 and concentrates on the engineering statistics topics in Chapters 8 to 12.

We would like to express our appreciation to the many students and instructors who have used the first edition of this book, and who have made helpful suggestions for its revision. We also thank Dr. Robert N. Lehrer and Dr. Michael E. Thomas for their continued support and encouragement in preparing the original text and the second edition. We are indebted to Professor E. S. Pearson and the *Biometrika* Trustees, John Wiley & Sons, Prentice-Hall, The American Statistical Association, The Institute of Mathematical Statistics, and the editors of *Biometrics* for permission to use copyrighted material. Dr. Frank B. Alt of the University of Maryland made many useful contributions to the original book that are also incorporated in the second edition. We would also like to thank Dr. Michael P. Deisenroth of Purdue University for providing Table I of the Appendix and Ms. Elizabeth A. Peck of the Coca-Cola Company for preparing the solutions to several of the examples in Chapters 12 and 13. We are grateful to the Office of Naval Research for supporting much of the research that is included in the chapters on regression analysis and quality control. Finally, we thank Mrs. Joyce Williams for typing the final draft of the manuscript.

William W. Hines
Douglas C. Montgomery

Contents

13   Multiple Regression   392

14   Design of Experiments   460

Chapter 1

An Introduction to Probability

1-1 Introduction

Statistics is a science that deals with the analysis of data and the process of making decisions about the system from which the data were obtained. Applications of statistics occur in many fields, including engineering, the physical sciences, business, the health and biological sciences, the social sciences, and education. This book is concerned with statistics as applied in engineering, the physical sciences, and management science.

It is convenient to identify two major branches of statistics: probability and inferential statistics. *Probability* is a methodology that permits the description of random variation in systems. For example, suppose that an engineer is trying to determine the number of trunk lines required to provide an adequate level of service at a telecommunications facility. Calls arrive at this facility in a random fashion, and the call handling time or service time is not constant from call to call. Through the use of probabilistic methods, it is possible to develop an analytic model of this system and determine relatively accurately such decision variables as the number of trunk lines and operators required to provide a specified service level. *Inferential statistics* uses sample data to draw general conclusions about the population from which the sample was taken. For example, suppose that we are interested in testing a manufacturer's claim about the lot average warpwise breaking strength of a specific cloth. A sample of cloth specimens is selected and tested. Techniques of statistical inference can then be used to analyze this sample data and to draw conclusions about the breaking strength of the entire lot of cloth.

Probability theory provides the mathematical foundation and language of statistics. Chapters 1 to 7 present basic probabilistic methods and models, as well as many of the standard probability distributions used widely in engineering. Chapters 8 to 13 present techniques for statistical inference, including descriptive methods for the efficient organization and summarization of data, and analytical methods for decision making. Chapter 15

1

deals with statistical quality control and reliability, and Chapter 16 gives an introduction to stochastic processes and queueing. These are major application areas for probabilistic and statistical models. Chapter 17 surveys statistical decision theory, a unified approach to many of the statistical inference procedures of previous chapters.

This chapter will present the basic elements of probability theory. Topics include elementary definitions, properties of probabilities, and several techniques for problem solution.

1-2 A Review of Sets

To present the basic concepts of probability theory, we will use some ideas from the theory of sets. A *set* is an aggregate or collection of objects. Sets are usually designated by capital letters, A, B, C, and so on. The members of the set A are called the elements of A. In general, when x is an element of A we write $x \in A$ and, if x is not an element of A, we write $x \notin A$. In specifying membership we may resort either to *enumeration* or to a *defining property*. These ideas are illustrated in the following examples. Braces are used to denote a set, and the colon within the braces is short hand for the term "such that."

● **Example 1-1.** The set whose elements are the integers 5, 6, 7, 8 is a finite set with four elements. We could denote this by

$$A = \{5, 6, 7, 8\}$$

Note that $5 \in A$ and $9 \notin A$ are both true, where $\in$ reads "is an element of" and $\notin$ reads "is not an element of."

● **Example 1-2.** If we write $V = \{a, e, i, o, u\}$ we have defined the set of vowels in the English alphabet. We may use a defining property and write this as

$$V = \{* : * \text{ is a vowel in the English alphabet}\}$$

● **Example 1-3.** If we say that A is the set of all real numbers between 0 and 1 inclusive, we might also denote A by a defining property as

$$A = \{x : x \in R, 0 \le x \le 1\}$$

where R is the set of all real numbers.

● **Example 1-4.** The set $B = \{-3, +3\}$ is the same set as

$$B = \{x : x \in R, x^2 = 9\}$$

where R is again the set of real numbers.

The *universal set* is the set of all objects under consideration, and it is generally denoted by U. Another special set is the *null set* or *empty set*, usually denoted by $\emptyset$. To illustrate this concept, consider a set

$$A = \{x : x \in R, x^2 = -1\}$$

The universal set here is R, the set of real numbers. Obviously, set A is empty since there are no real numbers having the defining property $x^2 = -1$. We should point out that $B = \{0\} \neq \emptyset = \{\ \}$.

If two sets are considered, say A and B, we call A a *subset* of B, denoted $A \subset B$, if each element in A is also an element of B. The sets A and B are said to be *equal* ($A = B$) if and only if $A \subset B$ and $B \subset A$. As direct consequences of this we may show that:

1. For any set A, $\emptyset \subset A$.
2. For a given U, then A considered in the context of U satisfies the relation $A \subset U$.
3. For a given set A, $A \subset A$ (a reflexive relation).
4. If $A \subset B$ and $B \subset C$, then $A \subset C$ (a transitive relation).

An interesting consequence of set equality is that the order of element listing is immaterial. To illustrate, let $A = \{a, b, c\}$ and $B = \{c, a, b\}$. Obviously $A = B$ by our definition. Furthermore, when defining properties are used, the sets may be equal although the defining properties are outwardly different. As an example of the second consequence, we let $A = \{x : x \in R, x$ is an even, prime number\}, and $B = \{x : x \in R, x + 3 = 5\}$, and since the integer 2 is the only even prime, $A = B$.

We now consider some operations on sets. Let A and B be any subsets of the universal set U. Then

1. The *complement* of A (with respect to U) is the set made up of the elements of U that do not belong to A. We denote this complementary set as $\bar{A}$. That is,

$$\bar{A} = \{x : x \in U, x \notin A\}$$

2. The *intersection* of A and B is the set of elements that belong to both A *and* B. We denote the intersection as $A \cap B$. In other words,

$$A \cap B = \{x : x \in A \ and \ x \in B\}$$

We should also note that $A \cap B$ is a *set*, and we could give this set some designator such as C.

3. The *union* of A and B is the set of elements that belong to *at least one*

of the sets A and B. If D represents the union, then

$$D = A \cup B = \{x : x \in A \ or \ x \in B \ (\text{or both})\}$$

These operations are illustrated in the following examples.

● **Example 1-5.** Let U be the set of letters in the alphabet, that is, $U = \{* : * \text{ is a letter of the English alphabet}\}$; and let $A = \{** : ** \text{ is a vowel}\}$ and $B = \{*** : *** \text{ is one of the letters } a, b, c\}$. As a consequence of the definitions,

$$\bar{A} = \text{the set of consonants}$$
$$\bar{B} = \{d, e, f, g, \ldots, x, y, z\}$$
$$A \cup B = \{a, b, c, e, i, o, u\}$$
$$A \cap B = \{a\}$$

● **Example 1-6.** If the universal set is defined as $U = \{1, 2, 3, 4, 5, 6, 7\}$, and three subsets, $A = \{1, 2, 3\}$, $B = \{2, 4, 6\}$, $C = \{1, 3, 5, 7\}$, are defined, then we see immediately from the definitions that

$$\bar{A} = \{4, 5, 6, 7\}, \qquad \bar{B} = \{1, 3, 5, 7\}, \qquad \bar{C} = \{2, 4, 6\} = B$$
$$A \cup B = \{1, 2, 3, 4, 6\}, \qquad A \cup C = \{1, 2, 3, 5, 7\}, \qquad B \cup C = U$$
$$A \cap B = \{2\}, \qquad A \cap C = \{1, 3\}, \qquad B \cap C = \emptyset$$

The *Venn diagram* can be used to illustrate certain set operations. A rectangle is drawn to represent the universal set U. A subset A of U is represented by the region within a circle drawn inside the rectangle. Then $\bar{A}$ will be represented by the area of the rectangle outside of the circle, as illustrated in Fig. 1-1. Using this notation, the intersection and union are illustrated in Fig. 1-2.

The operations of intersection and union may be extended in a straightforward manner to accommodate any finite number of sets. In the case of

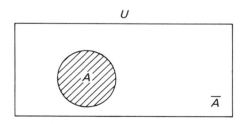

Fig. 1-1. A set in a Venn diagram.

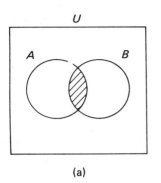
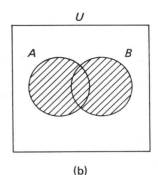

(a)   (b)

Fig. 1-2. The intersection and union of two sets in a Venn diagram. (*a*) The intersection shaded. (*b*) The union shaded.

three sets, say A, B, and C, $A \cup B \cup C$ has the property that $A \cup (B \cup C) = (A \cup B) \cup C$, which property obviously holds since they have identical members. Similarly, we see that $A \cap B \cap C = (A \cap B) \cap C = A \cap (B \cap C)$. A list of several important laws obeyed by sets relative to the operations previously defined is given here.

Identity laws:	$A \cup \emptyset = A$	$A \cap U = A$
	$A \cup U = U$	$A \cap \emptyset = \emptyset$
Idempotent laws:	$A \cup A = A$	$A \cap A = A$
Complement laws:	$A \cup \bar{A} = U$	$A \cap \bar{A} = \emptyset$; $\bar{\bar{A}} = A$
Commutative laws:	$A \cup B = B \cup A$	$A \cap B = B \cap A$
De Morgan's law:	$\overline{A \cup B} = \bar{A} \cap \bar{B}$	$\overline{A \cap B} = \bar{A} \cup \bar{B}$
Associative laws:	$A \cup (B \cup C) = (A \cup B) \cup C$	
	$A \cap (B \cap C) = (A \cap B) \cap C$	
Distributive laws:	$A \cup (B \cap C) = (A \cup B) \cap (A \cup C)$	
	$A \cap (B \cup C) = (A \cap B) \cup (A \cap C)$	

The reader is asked in Exercise 1-1 to illustrate some of these statements with Venn diagrams. Formal proofs are usually more lengthy.

In the case of more than three sets, we use a subscript to generalize. Thus, if n is a positive integer, and $B_1, B_2, \ldots, B_n$ are given sets, then $B_1 \cap B_2 \cap \ldots \cap B_n$ is the set of elements belonging to *all* of the sets, and $B_1 \cup B_2 \cup \ldots \cup B_n$ is the set of elements that belong to *at least one* of the given sets.

If A and B are sets, then the set of all ordered pairs (a, b) such that $a \in A$ and $b \in B$ is called the *Cartesian product set* of A and B. The usual notation is $A \times B$. We thus have

$$A \times B = \{(a, b) : a \in A \text{ and } b \in B\}$$

Let r be a positive integer greater than 1, and let $A_1, \ldots, A_r$ represent sets. Then the Cartesian product set is given by

$$A_1 \times A_2 \times \ldots \times A_r = \{(a_1, a_2, \ldots, a_r) : a_j \in A_j \text{ for } j = 1, 2, \ldots, r\}$$

Frequently, the *number* of elements in a set is of some importance, and we denote by $n(A)$ the number of elements in set A. If the number is *finite*, we say we have a *finite set*. Should the set be infinite, such that the elements can be put into a one-to-one correspondence with the natural numbers, then the set is called a *denumerably infinite set*. A final consideration is the *infinite set*. For example, if $a < b$, then the set $A = \{x : x \in R, a \le x \le b\}$ is infinite, where R is the set of real numbers.

1-3 Experiments and Sample Spaces

We will be concerned with experiments that are called *random experiments*, and although no precise definition is to be given we will note that such experiments have several characteristics in common. First, the experiment is such that it is impossible to state a particular outcome, but we can define the *set* of *all possible* outcomes. Second, the experiment may be repeated indefinitely under unchanged conditions. Finally, the outcomes of repeated experiments occur in a random fashion, or by chance.

The experiment may be considered a *process of observation*, and we will require our experiments to be *idealized*. For example, when a coin is tossed, we will rule out the possibility that it will land on edge.

If we pursue "experiments" theoretically and without ambiguity, we must first agree about the possible outcomes, since *these outcomes define the particular idealized experiment*. The set of all possible outcomes of the experiment is called the *sample space*. We follow convention and use the symbols $\mathscr{E}$ and $\mathscr{S}$ to designate the *experiment* and the *sample space*, respectively.

Several experiments and associated sample spaces are illustrated below. Note that in specifying an experiment, we first specify an operation or procedure to be carried out and then specify the method of observation.

● **Example 1-7**
$\mathscr{E}_1$: Toss a true coin and observe the "up" face.
$\mathscr{S}_1$: $\{H, T\}$.

● **Example 1-8**
$\mathscr{E}_2$: Toss a true coin three times and observe the sequence of heads and tails.
$\mathscr{S}_2$: $\{HHH, HHT, HTH, HTT, THH, THT, TTH, TTT\}$.

- **Example 1-9**

    $\mathscr{E}_3$: Toss a true coin three times and observe the total number of heads.

    $\mathscr{S}_3$: $\{0, 1, 2, 3\}$.

- **Example 1-10**

    $\mathscr{E}_4$: Toss a pair of dice and observe the "up" faces.

    $\mathscr{S}_4$: $\{(1, 1), (1, 2), (1, 3), (1, 4), (1, 5), (1, 6),$

    $(2, 1), (2, 2), (2, 3), (2, 4), (2, 5), (2, 6),$

    $(3, 1), (3, 2), (3, 3), (3, 4), (3, 5), (3, 6),$

    $(4, 1), (4, 2), (4, 3), (4, 4), (4, 5), (4, 6),$

    $(5, 1), (5, 2), (5, 3), (5, 4), (5, 5), (5, 6),$

    $(6, 1), (6, 2), (6, 3), (6, 4), (6, 5), (6, 6)\}$.

- **Example 1-11**

    $\mathscr{E}_5$: An automobile door is assembled with a large number of spot welds. After assembly, each weld is inspected, and the total number of defectives is counted.

    $\mathscr{S}_5$: $\{0, 1, 2, \ldots, K\}$, where $K =$ the total number of welds in the door.

- **Example 1-12**

    $\mathscr{E}_6$: A cathode ray tube is manufactured, put on life test, and aged to failure. The elapsed time (in hours) at failure is recorded.

    $\mathscr{S}_6$: $\{t : t \geq 0\}$.

- **Example 1-13**

    $\mathscr{E}_7$: A lot composed of N motors contains D defectives $(D \leq N)$. The items are selected one by one (without replacement) until the last defective item is found.

    $\mathscr{S}_7$: $\{D, D + 1, \ldots, N\}$.

- **Example 1-14**

    $\mathscr{E}_8$: In a cordwood type module for a digital computer circuit, a diode is pulse welded between two strips of nickel ribbon. Each weld is visually inspected and probed at each end before classification as good, G, or defective, D.

    $\mathscr{S}_8$: $\{GG, GD, DG, DD\}$.

- **Example 1-15**

    $\mathscr{E}_9$: In a particular chemical plant the volume produced per day for a particular product ranges between a minimum value, b, and a maximum value, c, which corresponds to capacity. A day is randomly selected and the amount produced is observed.

    $\mathscr{S}_9$: $\{x : x \in R, b \leq x \leq c\}$.

● **Example 1-16**

$\mathscr{E}_{10}$: An extrusion plant is engaged in making up an order for pieces 20 feet long. Inasmuch as the extrusion process creates scrap at both ends, the extruded bar must exceed 20 feet. Because of costs involved, the amount of scrap is critical. A bar is extruded, trimmed, and finished, and the total length of scrap is measured.

$\mathscr{S}_{10}$: $\{x : x \in R, x > 0\}$.

● **Example 1-17**

$\mathscr{E}_{11}$: In an orbital satellite, the three components of velocity, v_x, v_y, v_z, are monitored from the ground as a function of time. At one minute after launch these are printed for a control unit.

$\mathscr{S}_{11}$: $\{(v_x, v_y, v_z) : v_x, v_y, v_z \text{ are real numbers}\}$.

● **Example 1-18**

$\mathscr{E}_{12}$: In the preceding example, the velocity components are continuously recorded with a very high speed printer for 24 hours of operation.

$\mathscr{S}_{12}$: The space is complicated here as we have all possible *realizations* of the functions $v_x(t)$, $v_y(t)$, $v_z(t)$ for $0 \le t \le 24$ to consider.

All of these examples have the characteristics required of random experiments. With the exception of Example 1-18, the description of the sample space is straightforward, and although repetition is not considered, ideally we could repeat the experiments. To illustrate the phenomena of random occurrence, consider Example 1-7. Obviously, if $\mathscr{E}_1$ is repeated indefinitely, we obtain a sequence of *heads* and *tails*. A pattern emerges as we continue the experiment. Notice that since the coin is true, we should obtain heads approximately one-half of the time. In recognizing the *idealization* in the model, we simply agree on a theoretical possible set of outcomes. In $\mathscr{E}_1$, we ruled out the possibility of having the coin land on edge, and in $\mathscr{E}_6$, where we recorded the elapsed time to failure, the idealized sample space consisted of all nonnegative real numbers.

1-4   Events

An *event*, say A, is associated with the sample space of the experiment $\mathscr{E}$. If the sample space is considered to be the universal set, then event A is simply some subset of $\mathscr{S}$. Note that both $\emptyset$ and $\mathscr{S}$ are subsets of $\mathscr{S}$. As a general rule, a capital letter will denote an event. The following events relate to experiments $\mathscr{E}_1, \mathscr{E}_2, \ldots, \mathscr{E}_{10}$, described in the preceding section. These are provided for illustration only; many other events could have been described for each case.

$\mathscr{E}_1 . A$: The coin toss yields a head $\{H\}$.

$\mathscr{E}_2 . A$: All the coin tosses give the same face $\{HHH, TTT\}$.

$\mathscr{E}_3 . A$: The total number of heads is two $\{2\}$.

$\mathscr{E}_4 . A$: The sum of the "up" faces is seven $\{(1, 6), (2, 5), (3, 4), (4, 3), (6, 1), (5, 2)\}$.

$\mathscr{E}_5 . A$: The number of defective welds does not exceed 5 $\{0, 1, 2, 3, 4, 5\}$.

$\mathscr{E}_6 . A$: The time to failure is greater than 1000 hours $\{t : t > 1000\}$.

$\mathscr{E}_7 . A$: The last defective is not found until the last item is tested $\{N\}$.

$\mathscr{E}_8 . A$: Neither weld is bad $\{GG\}$.

$\mathscr{E}_9 . A$: The volume produced is between $a > b$ and c $\{x : x \in R, b < a < x < c\}$.

$\mathscr{E}_{10} . A$: The scrap does not exceed one foot $\{x : x \in R, 0 \le x \le 1\}$.

Since an event is a set, the intersection, $\cap$, union, $\cup$, and contained, $\subset$, operations are defined for events, and the laws and properties of Section 1-2 hold. If the intersection of two or more events associated with the outcome of a particular experiment is the empty set, the events are said to be *mutually exclusive*. We express this by writing $A \cap B = \emptyset$ in the case of two events, or $A_1 \cap A_2 \cap \ldots \cap A_k = \emptyset$ in the case of k events. The case of $k = 3$ is illustrated in Fig. 1-3. We should emphasize that these multiple events are associated with one experiment.

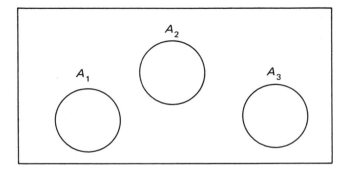

Fig. 1-3. Three mutually exclusive events.

1-5   Probability Definitions

An axiomatic approach is taken to define probability as a *set function* where the elements of the domain are sets and the elements of the range are real numbers. If event A is an element in the domain of this function, we use customary functional notation $P(A)$, $f(A)$, and so on, to designate the corresponding element in the range.

Definition

If an experiment $\mathscr{E}$ has sample space $\mathscr{S}$ and an event A is defined on $\mathscr{S}$, then $P(A)$ is a real number called the probability of event A or the probability of A, and the function $P(\cdot)$ has the following properties:

1. $0 \le P(A) \le 1$ for each event A of $\mathscr{S}$.
2. $P(\mathscr{S}) = 1$.
3. For any finite number k of mutually exclusive events defined on $\mathscr{S}$,

$$P\left(\bigcup_{i=1}^{k} A_i\right) = \sum_{i=1}^{k} P(A_i)$$

4. If $A_1, A_2, A_3, \ldots$ is a denumerable sequence of mutually exclusive events defined on $\mathscr{S}$, then

$$P(A_1 \cup A_2 \cup A_3 \cup \ldots) = P(A_1) + P(A_2) + P(A_3) + \cdots$$

Note that the properties of the definition given do not tell the experimenter *how* to assign probabilities; however, they do restrict the way in which the assignment may be accomplished. In practice, probability is assigned on the basis of (1) estimates obtained from previous experience, (2) an analytical consideration of experimental conditions, or (3) assumption.

In order to illustrate the assignment of probability based on experience, we consider the repetition of the experiment and the relative frequency of the occurrence of the event of interest.

This notion of relative frequency has intuitive appeal, and it involves the conceptual repetition of an experiment and a counting of both the number of repetitions and the number of times the event in question occurs. More precisely, $\mathscr{E}$ is repeated m times and two events are denoted as A and B. We let m_A and m_B be the number of times A and B occur in the m repetitions.

Definition

The value $f_A = m_A/m$ is called the *relative frequency* of event A. It has the following properties:

1. $0 \leq f_A \leq 1$.
2. $f_A = 0$ if and only if A never occurs, and $f_A = 1$ if and only if A occurs on every repetition.
3. If A and B are mutually exclusive events, then $f_{A \cup B} = f_A + f_B$.

The fact that f_A tends to stabilize as m becomes large is a consequence of a regularity, which in turn is a consequence of the requirement of reproducibility of the experiment. That is, as the experiment is repeated, the relative frequency of event A will vary less and less (from repetition to repetition) as the number of repetitions increases. The concept of relative frequency and the tendency toward stability lead to one method for assigning probability. If an experiment $\mathscr{E}$ has sample space $\mathscr{S}$ and an event A is defined, and if the relative frequency, f_A, approaches some number p_A as the number of repetitions increases, then the number p_A is ascribed to A as its probability, that is, as $m \to \infty$,

$$P(A) = \frac{m_A}{m} = p_A \tag{1-1}$$

In practice, something less than infinite replication must obviously be accepted.

Another method for assigning probability to an event where the sample space has a finite number, n, of elements, e_i, and where the probability assigned to an outcome is $p_i = P(E_i)$, where $E_i = \{e_i\}$, and

$$p_i \geq 0 \qquad i = 1, 2, \ldots, n$$

while

$$p_1 + p_2 + \cdots + p_n = 1$$

is

$$P(A) = \sum_{i\,:\,e_i \in A} p_i \tag{1-2}$$

This is simply a statement that the probability of event A is the sum of the probabilities associated with the outcomes making up event A.

● **Example 1-19.** The sample space associated with a particular experiment has only three outcomes, e_1, e_2, and e_3. Outcome e_1 is one-half as likely as e_2, which is three times as likely as e_3. It follows that $p_1 = \frac{1}{2}p_2$, $p_2 = 3p_3$, and since $p_1 + p_2 + p_3 = 1$, $p_1 = \frac{3}{11}$, $p_2 = \frac{6}{11}$, and $p_3 = \frac{2}{11}$. Now suppose event $A = \{e_1, e_2\}$, then $P(A) = \frac{3}{11} + \frac{6}{11} = \frac{9}{11}$.

● **Example 1-20.** Suppose the coin in Example 1.8 is biased so that the outcomes of the sample space $\mathscr{S} = \{HHH, HHT, HTH, HTT, THH, THT, TTH, TTT\}$ have probabilities $p_1 = \frac{1}{27}$, $p_2 = \frac{2}{27}$, $p_3 = \frac{2}{27}$, $p_4 = \frac{4}{27}$, $p_5 = \frac{2}{27}$, $p_6 = \frac{4}{27}$, $p_7 = \frac{4}{27}$, $p_8 = \frac{8}{27}$, where $e_1 = HHH$, $e_2 = HHT$, etc. If we let event A be the event that all tosses yield the same face, then $P(A) = \frac{1}{27} + \frac{8}{27} = \frac{1}{3}$.

In the case where the elementary outcomes are equally likely, $p_1 = p_2 = \ldots = p_n$, and

$$P(A) = \frac{n(A)}{n} \qquad (1\text{-}3)$$

where $n(A)$ represents the number of outcomes contained in A.

● **Example 1-21.** Consider Example 1-8, where a true coin is tossed three times, and consider event A, that all coins show the same face.

$$P(A) = \frac{n(A)}{n} = \frac{2}{8}$$

since there are 8 total outcomes and 2 are favorable to event A. Since the coin was assumed to be true, all 8 possible outcomes are equally likely.

● **Example 1-22.** Assume that the dice in Example 1-10 are true, and consider event A, that the sum of the "up" faces is 7. Using the results of Equation (1-3) we note that there are 36 outcomes of which 6 are favorable to the event in question, so that $P(A) = \frac{1}{6}$.

Note that these examples are extremely simple in two respects: the sample space is of a highly restricted type, and the counting process is easy. Combinatorial methods frequently become necessary as the counting becomes more involved. A number of excellent problems and examples are given by Feller (1968), in which the clever use of counting methods is required to obtain a solution. Basic counting methods are reviewed in Section 1-9.

Some important theorems regarding probability follow.

Theorem 1-1

If $\emptyset$ is the empty set, then $P(\emptyset) = 0$.

Proof

Note that $\mathscr{S} = \mathscr{S} \cup \emptyset$, and $P(\mathscr{S}) = P(\mathscr{S}) + P(\emptyset)$ from property 4; therefore $P(\emptyset) = 0$.

Theorem 1-2

$P(\bar{A}) = 1 - P(A)$.

Proof

Note that $\mathscr{S} = A \cup \bar{A}$, and $P(\mathscr{S}) = P(A) + P(\bar{A})$ from property 4, but from property 2, $P(\mathscr{S}) = 1$; therefore $P(\bar{A}) = 1 - P(A)$.

Theorem 1-3

$P(A \cup B) = P(A) + P(B) - P(A \cap B)$.

Proof

Since $A \cup B = A \cup (B \cap \bar{A})$, where A and $(B \cap \bar{A})$ are mutually exclusive, and $B = (A \cap B) \cup (B \cap \bar{A})$, where $(A \cap B)$ and $(B \cap \bar{A})$ are mutually exclusive, then $P(A \cup B) = P(A) + P(B \cap \bar{A})$, and $P(B) = P(A \cap B) + P(B \cap \bar{A})$. Subtracting, $P(A \cup B) - P(B) = P(A) - P(A \cap B)$, and thus $P(A \cup B) = P(A) + P(B) - P(A \cap B)$.

The Venn diagram shown in Fig. 1-4 is helpful in following the argument of the proof for Theorem 1-3.

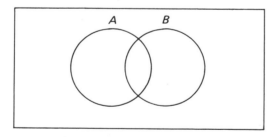

Fig. 1-4. Venn diagram for two events.

Theorem 1-4

$$P(A \cup B \cup C) = P(A) + P(B) + P(C) - P(A \cap B) - P(A \cap C) - P(B \cap C) + P(A \cap B \cap C).$$

Proof

We may write $A \cup B \cup C = (A \cup B) \cup C$ and use Theorem 1-3 since $A \cup B$ is an event. The reader is asked to provide the details in Exercise 1-2.

Theorem 1-5

$$P(A_1 \cup A_2 \cup A_3 \cup \ldots \cup A_k) = \sum_{i=1}^{k} P(A_i) - \sum_{i<j=2}^{k} P(A_i \cap A_j)$$
$$+ \sum_{i<j<r=3}^{k} P(A_i \cap A_j \cap A_r) + \cdots$$
$$+ (-1)^{k-1} P(A_1 \cap A_2 \cap A_3 \cap \ldots \cap A_k).$$

Proof
Refer to Exercise 1-3.

Theorem 1-6

If $A \subset B$, then $P(A) \leq P(B)$.

Proof
If $A \subset B$, then $B = A \cup (\bar{A} \cap B)$ and $P(B) = P(A) + P(\bar{A} \cap B) \geq P(A)$ since $P(\bar{A} \cap B) \geq 0$.

● **Example 1-23.** If A and B are mutually exclusive events, and it is known that $P(A) = .20$ while $P(B) = .30$, we can evaluate several probabilities:

(a) $P(\bar{A}) = 1 - P(A) = .80$.
(b) $P(\bar{B}) = 1 - P(B) = .70$.
(c) $P(A \cup B) = P(A) + P(B) = .2 + .3 = .5$.
(d) $P(A \cap B) = 0$.
(e) $P(\bar{A} \cap \bar{B}) = P(\overline{A \cup B})$, by De Morgan's law
$\qquad = 1 - P(A \cup B)$
$\qquad = 1 - [P(A) + P(B)] = .5$.

● **Example 1-24.** Suppose events A and B are not mutually exclusive and we know that $P(A) = .20$, $P(B) = .30$, and $P(A \cap B) = .10$. Then evaluating the same probabilities as before, we obtain

(a) $P(\bar{A}) = 1 - P(A) = .80$.
(b) $P(\bar{B}) = 1 - P(B) = .70$.
(c) $P(A \cup B) = P(A) + P(B) - P(A \cap B) = .2 + .3 - .1 = .4$.
(d) $P(A \cap B) = .1$.
(e) $P(\bar{A} \cap \bar{B}) = P(\overline{A \cup B}) = 1 - [P(A) + P(B) - P(A \cap B)] = .6$.

1-6   Conditional Probability

As noted in Section 1-4, an event is associated with a sample space and the event is represented by a subset of $\mathscr{S}$. The probabilities discussed in Section 1-5 all relate to the entire sample space. We have used the symbol $P(A)$ to denote the probability of these events; however, we could have used the symbol $P(A|\mathscr{S})$, read as "the probability of A, given sample space $\mathscr{S}$." We will frequently be interested in evaluating the probability of events where the event is *conditioned* on some subset of the sample space.

Some illustrations of this idea should be helpful. Consider a group of 100 persons of whom 40 are college graduates, 20 are self-employed, and 10 are both college graduates and self-employed. Let B represent the set of college graduates and A represent the set of self-employed, so that $A \cap B$ is the set of college graduates who are self-employed. From the group of 100, one person is to be randomly selected. (Each person is given a number from 1 to 100, and 100 chips with the same numbers are agitated, with one being selected by a blindfolded outsider.) Then, $P(A) = .2$, $P(B) = .4$, and $P(A \cap B) = .1$ if the entire sample space is considered. As noted, it may be more instructive to write $P(A|\mathscr{S})$, $P(B|\mathscr{S})$, and $P(A \cap B|\mathscr{S})$ in such a case. Suppose the following event is considered: self-employed *given* that the person is a college graduate $(A|B)$. Obviously the sample space is reduced in that only college graduates are considered (Fig. 1-5). The probability, $P(A|B)$ is thus given by

$$P(A|B) = \frac{P(A \cap B)}{P(B)} = \frac{.1}{.4} = .25$$

The reduced sample space consists of the set of all subsets of $\mathscr{S}$ that belong to B. Of the subsets belonging to B, $A \cap B$ satisfies the condition.

As a second illustration, consider the case where a sample of size two is

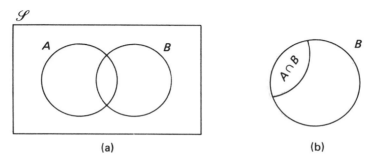

(a)   (b)

Fig. 1-5.   Conditional probability. (*a*) Initial sample space. (*b*) Reduced sample space.

randomly selected from a lot of size ten. It is known that the lot has seven good and three bad items. Let A be the event that the first item selected is good, and B be the event that the second item selected is good. If the items are selected *without replacement*, that is, the first item is not replaced before the second item is selected, then

$$P(A) = \frac{7}{10}$$

and

$$P(B|A) = \frac{6}{9}$$

If the first item is replaced before the second item is selected, the conditional probability $P(B|A) = P(B) = \frac{7}{10}$, and the events A and B resulting from the two selection experiments making $\mathscr{E}$ are said to be *independent*. A formal definition of $P(A|B)$ will be given later, and independence will be discussed in detail. The following examples will help to develop some intuitive feeling for conditional probability.

● **Example 1-25.** Recall Example 1-10 where two dice are tossed, and assume that each die is true. The 36 possible outcomes are enumerated on page 7. If we consider two events,

$$A = \{(d_1, d_2) : d_1 + d_2 = 4\}$$
$$B = \{(d_1, d_2) : d_2 \geq d_1\}$$

where d_1 is the value of the up face of the first die and d_2 is the value of the up face of the second die, then $P(A) = \frac{3}{36}$, $P(B) = \frac{21}{36}$, $P(B|A) = \frac{2}{3}$, and $P(A|B) = \frac{2}{21}$. Note also that $P(A \cap B) = \frac{2}{36}$. The probabilities were obtained from a direct consideration of the sample space and the counting of outcomes. Note that

$$P(A|B) = \frac{P(A \cap B)}{P(B)} \quad \text{and} \quad P(B|A) = \frac{P(A \cap B)}{P(A)}$$

● **Example 1-26.** A coeducational college has three curricula: science, management, and engineering. By sex, the enrollment is as follows:

	Science	Management	Engineering	Total
Male	250	350	200	800
Female	100	50	50	200
Total	350	400	250	1000

Let S_1: The student is male.
    S_2: The student is female.
    C_1: The student is in the science curriculum.
    C_2: The student is in the management curriculum.
    C_3: The student is in the engineering curriculum.

If a student is to be randomly selected from the student body, then we obtain

	C_1	C_2	C_3	Total
S_1	.25	.35	.20	.80
S_2	.10	.05	.05	.20
Total	.35	.40	.25	1.00

where the numbers in the body of the table represent $P(S_i \cap C_j)$ $i = 1, 2$; $j = 1, 2, 3$. Note that $P(S_1) = .80$, and $P(S_2) = .20$. These are sometimes called marginal probabilities. In a like manner, $P(C_1) = .35$, $P(C_2) = .40$, and $P(C_3) = .25$ are marginal probabilities. In order to find $P(C_3|S_2)$, we may work with either the count data or the probabilities. In either case we obtain

$$P(C_3|S_2) = \frac{50}{200} = \frac{.05}{.20} = .25$$

In general,

$$P(C_j|S_i) = \frac{P(C_j \cap S_i)}{P(S_i)} \qquad i = 1, 2 \qquad j = 1, 2, 3$$

We may define the conditional probability of event A given event B as

$$P(A|B) = \frac{P(A \cap B)}{P(B)} \qquad \text{if } P(B) > 0 \tag{1-4}$$

This definition results from the intuitive notion presented in the preceding discussion. The conditional probability $P(\cdot|\cdot)$ satisfies the properties required of probabilities. That is,

1. $0 \le P(A|B) \le 1$.
2. $P(\mathscr{S}|B) = 1$.
3. $P(A_1 \cup A_2 \cup A_3 \cup \ldots|B) = P(A_1|B) + P(A_2|B) + P(A_3|B) + \ldots$,   for $A_1, A_2, A_3, \ldots$, a denumerable sequence of disjoint events.
4. $P\left(\bigcup_{i=1}^{k} A_i|B\right) = \sum_{i=1}^{k} P(A_i|B)$ for $(A_i \cap A_j) = \emptyset$ if $i \ne j$.

In practice we may solve problems by using Equation (1-4) and calculating $P(A \cap B)$ and $P(B)$ with respect to the original sample space (as was illustrated in Example 1-26) or by considering the probability of A with respect to the reduced sample space B (as was illustrated in Example 1-25).

A restatement of Equation (1-4) leads to what is often called the *multiplication rule*, that is,

$$P(A \cap B) = P(B) \cdot P(A|B) \qquad P(B) > 0$$

and

$$P(A \cap B) = P(A) \cdot P(B|A) \qquad P(A) > 0 \qquad (1\text{-}5)$$

The second statement is an obvious consequence of Equation (1-4) with the conditioning on event A rather than event B.

It should be noted that if A and B are *mutually exclusive* as indicated in Fig. 1-6, then $A \cap B = \emptyset$ so that $P(A|B) = 0$ and $P(B|A) = 0$.

In the other extreme, if $B \subset A$ as shown in Fig. 1-7, then $P(A|B) = 1$. In the first case, A and B cannot occur simultaneously, so that the knowledge of the occurrence of B tells us that A does not occur. In the second case, if B occurs, A must occur. There are many cases where the events are totally unrelated, and knowledge of the occurrence of one has no bearing on and

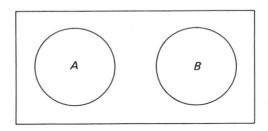

Fig. 1-6.   Mutually exclusive events.

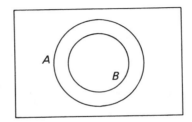

Fig. 1-7.   Event B as a subset of A.

yields no information about the other. Consider, for example, the simple experiment where a true coin is tossed twice. Event A is the event that the first toss results in a "heads," and event B is the event that the second toss results in a "heads." Note that $P(A) = \frac{1}{2}$ since the coin is true, and $P(B|A) = \frac{1}{2}$ since the coin is true and it has no memory. The occurrence of event A did not in any way affect the occurrence of B and, if we wanted to find the probability of A and B occurring, that is, $P(A \cap B)$, we find that

$$P(A \cap B) = P(A) \cdot P(B|A) = \frac{1}{2} \cdot \frac{1}{2} = \frac{1}{4}$$

We may observe that if we had no knowledge about the occurrence or nonoccurrence of A we have $P(B) = P(B|A)$ as in this example.

Informally speaking, two events are considered to be *independent* if the probability of the occurrence of one is not affected by the occurrence or nonoccurrence of the other. This leads to the following definition.

Definition

A and B are independent if and only if

$$P(A \cap B) = P(A) \cdot P(B)$$

An immediate consequence of this definition is the following theorem.

Theorem 1-7

If A and B are independent events, then

$$P(A|B) = P(A) \qquad \text{and} \qquad P(B|A) = P(B) \tag{1-6}$$

Insofar as intuition is concerned, Theorem 1-7 is appealing. The theorem is an immediate consequence of Equation (1-4) and the definition of independence.

The following theorem is sometimes useful. Proof is given here only for the first part.

Theorem 1-8

If A and B are independent events, then

1. A and $\bar{B}$ are independent events.
2. $\bar{A}$ and $\bar{B}$ are independent events.
3. $\bar{A}$ and B are independent events.

Proof: Part 1

$$P(A \cap \bar{B}) = P(A) \cdot P(\bar{B}|A)$$
$$= P(A) \cdot [1 - P(B|A)]$$
$$= P(A) \cdot [1 - P(B)]$$
$$= P(A) \cdot P(\bar{B})$$

In practice, there are many situations where it may not be easy to determine whether two events are independent; however, there are numerous other cases where the requirements may be either justified or approximated from a physical consideration of the experiment. A sampling experiment will serve to illustrate.

● **Example 1-27.** Suppose a random sample of size two is to be selected from a lot of size 100, and it is known that 98 of the 100 items are good. The sample is taken in such a manner that the first item is observed and replaced before the second item is selected. If we let

A: First item observed is good

B: Second item observed is good

and if we want to determine the probability that both items are good, then

$$P(A \cap B) = P(A) \cdot P(B) = \frac{98}{100} \cdot \frac{98}{100} = .9604$$

If the sample is taken "without replacement" so that the first item is not replaced before the second item is selected, then

$$P(A \cap B) = P(A) \cdot P(B|A) = \frac{98}{100} \cdot \frac{97}{99} = .9602$$

The results are obviously very close, and one common practice is to assume the events independent when the *sampling fraction* (sample size/population size) is small, say less than .1.

● **Example 1-28.** The field of *reliability engineering* has developed rapidly since the early 1960s. One type of problem encountered is that of estimating system reliability given subsystem reliabilities. Reliability is defined here as the probability of proper functioning for a stated period of time. Consider the structure of a simple serial system, shown in Fig. 1-8. The system functions if and only if both subsystems function. If the subsystems survive independently, then

$$\text{System reliability} = R_s = R_1 \cdot R_2$$

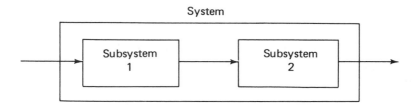

Fig. 1-8. A simple serial system.

where R_1 and R_2 are the reliabilities for subsystems 1 and 2 respectively. For example, if $R_1 = .90$ and $R_2 = .80$, then $R_s = .72$.

Example 1-28 illustrates the need to generalize the concept of independence to more than two events. Suppose the system consisted of three subsystems or perhaps twenty subsystems. What conditions would be required in order to allow the analyst to obtain an estimate of system reliability by obtaining the product of the subsystem reliabilities?

Definition

The k events $A_1, A_2, \ldots, A_k$ are mutually independent if and only if the probability of the intersection of any $2, 3, \ldots, k$ of these sets is the product of their respective probabilities.

Stated more precisely, we require that for $r = 2, 3, \ldots, k$

$$P(A_{i_1} \cap A_{i_2} \cap A_{i_3} \cap \ldots \cap A_{i_r}) = P(A_{i_1}) \cdot P(A_{i_2}) \cdot P(A_{i_3}) \cdot \ldots \cdot P(A_{i_r})$$

$$= \prod_{j=1}^{r} P(A_{i_j})$$

In the case of serial system reliability calculations where mutual independence may reasonably be assumed, the system reliability is a product of subsystem reliabilities

$$R_s = R_1 R_2 \cdots R_k \tag{1-7}$$

In the above definition there are $2^k - k - 1$ conditions to be satisfied. Consider three events A, B, and C. These are independent if and only if $P(A \cap B) = P(A) \cdot P(B)$, $P(A \cap C) = P(A) \cdot P(C)$, $P(B \cap C) = P(B) \cdot P(C)$, and $P(A \cap B \cap C) = P(A) \cdot P(B) \cdot P(C)$. The following example illustrates a case where events are independent in pairs but not in triplets.

● **Example 1-29.** Suppose the sample space, with equally likely outcomes, for a particular experiment is as follows:

$$\mathscr{S} = \{(0, 0, 0), (0, 1, 1), (1, 0, 1), (1, 1, 0)\}$$

Let A_0: First digit is zero. B_1: Second digit is one.
  A_1: First digit is one. C_0: Third digit is zero.
  B_0: Second digit is zero. C_1: Third digit is one.

It follows that

$$P(A_0) = P(A_1) = P(B_0) = P(B_1) = P(C_0) = P(C_1) = \tfrac{1}{2}$$

and it is easily seen that

$$P(A_i \cap B_j) = \tfrac{1}{4} = P(A_i) \cdot P(B_j) \qquad i = 0, 1, j = 0, 1$$

$$P(A_i \cap C_j) = \tfrac{1}{4} = P(A_i) \cdot P(C_j) \qquad i = 0, 1, j = 0, 1$$

$$P(B_i \cap C_j) = \tfrac{1}{4} = P(B_i) \cdot P(C_j) \qquad i = 0, 1, j = 0, 1$$

however, we note that

$$P(A_0 \cap B_0 \cap C_0) = \tfrac{1}{4} \neq P(A_0) \cdot P(B_0) \cdot P(C_0)$$

and

$$P(A_0 \cap B_0 \cap C_1) = 0 \neq P(A_0) \cdot P(B_0) \cdot P(C_1)$$

There are six other triplets to which this could be extended.

The concept of *independent experiments* is introduced to complete this section. If we consider two experiments denoted $\mathscr{E}_1$ and $\mathscr{E}_2$ and let A_1 and A_2 be arbitrary events defined on the respective sample spaces $\mathscr{S}_1$ and $\mathscr{S}_2$ of the two experiments, then the following definition can be given.

Definition

If $P(A_1 \cap A_2) = P(A_1) \cdot P(A_2)$ then $\mathscr{E}_1$ and $\mathscr{E}_2$ are said to be independent experiments.

1-7   Partitions, Total Probability, and Bayes' Theorem

A *partition* of the sample space may be defined as follows.

Definition

If $B_1, \ldots, B_k$ are disjoint subsets of $\mathscr{S}$ (mutually exclusive events), and if $B_1 \cup B_2 \cup \ldots \cup B_k = \mathscr{S}$, then these subsets are said to form a partition of $\mathscr{S}$.

When the experiment is performed, one and only one of the events, B_i, occurs. If we have a partition of $\mathscr{S}$, then only one of the events, B_i, occurs when the experiment is performed.

● **Example 1-30.** A particular binary "word" consists of five "bits," b_1, b_2, b_3, b_4, b_5, where $b_i = 0, 1$; $i = 1, 2, 3, 4, 5$. An experiment consists of transmitting a "word," and it follows that there are thirty-two possible words. If the events are as follows:

$B_1 = \{(0, 0, 0, 0, 0), (0, 0, 0, 0, 1)\}$

$B_2 = \{(0, 0, 0, 1, 0), (0, 0, 0, 1, 1), (0, 0, 1, 0, 0), (0, 0, 1, 0, 1), (0, 0, 1, 1, 0), (0, 0, 1, 1, 1)\}$

$B_3 = \{(0, 1, 0, 0, 0), (0, 1, 0, 0, 1), (0, 1, 0, 1, 0), (0, 1, 0, 1, 1), (0, 1, 1, 0, 0), (0, 1, 1, 0, 1), (0, 1, 1, 1, 0), (0, 1, 1, 1, 1)\}$

$B_4 = \{(1, 0, 0, 0, 0), (1, 0, 0, 0, 1), (1, 0, 0, 1, 0), (1, 0, 0, 1, 1), (1, 0, 1, 0, 0), (1, 0, 1, 0, 1), (1, 0, 1, 1, 0), (1, 0, 1, 1, 1)\}$

$B_5 = \{(1, 1, 0, 0, 0), (1, 1, 0, 0, 1), (1, 1, 0, 1, 1), (1, 1, 1, 0, 0), (1, 1, 1, 0, 1), (1, 1, 1, 1, 0), (1, 1, 0, 1, 0)\}$

$B_6 = \{(1, 1, 1, 1, 1)\}$

then $\mathscr{S}$ is partitioned by the events B_1, B_2, B_3, B_4, B_5, and B_6.

In general, if k events, B_i $(i = 1, 2, \ldots, k)$, form a partition and A is an arbitrary event with respect to $\mathscr{S}$, then we may write

$$A = (A \cap B_1) \cup (A \cap B_2) \cup \ldots \cup (A \cap B_k)$$

so that

$$P(A) = P(A \cap B_1) + P(A \cap B_2) + \cdots + P(A \cap B_k)$$

since the events $(A \cap B_i)$ are pairwise mutually exclusive. (See Fig. 1-9 for $k = 4$.) It does not matter that $A \cap B_i = \emptyset$ for some or all of the i since $P(\emptyset) = 0$.

Using the results of Equation (1-5) we can state the following theorem.

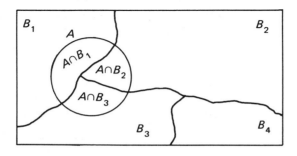

Fig. 1-9. Partition of $\mathscr{S}$.

Theorem 1-9

If $B_1, \ldots, B_k$ represents a partition of $\mathcal{S}$ and A is an arbitrary event on $\mathcal{S}$, then the *total probability* of A is given by

$$P(A) = P(B_1) \cdot P(A|B_1) + P(B_2) \cdot P(A|B_2) + \cdots + P(B_k) \cdot P(A|B_k)$$

The result of Theorem 1-9 is very useful, as there are numerous practical situations in which $P(A)$ cannot be computed directly. However, with the information that B_i has occurred, it is possible to evaluate $P(A|B_i)$ and thus determine $P(A)$ when the values $P(B_i)$ are obtained.

Another important result of the total probability law is known as *Bayes' theorem*.

Theorem 1-10

If $B_1, B_2, \ldots, B_k$ constitute a partition of the sample space $\mathcal{S}$ and A is an arbitrary event on $\mathcal{S}$, then for $r = 1, 2, \ldots, k$

$$P(B_r|A) = \frac{P(B_r) \cdot P(A|B_r)}{\sum\limits_{i=1}^{k} P(B_i) \cdot P(A|B_i)} \qquad (1\text{-}8)$$

Proof

$$P(B_r|A) = \frac{P(B_r \cap A)}{P(A)}$$

$$= \frac{P(B_r) \cdot P(A|B_r)}{\sum\limits_{i=1}^{k} P(B_i) \cdot P(A|B_i)}$$

The numerator is a result of Equation (1-5) and the denominator is a result of Theorem 1-9.

● **Example 1-31.** Three firms supply NPN transistors to a manufacturer of telemetry equipment. All are supposedly made to the same specifications. However, the manufacturer has for several years tested each of two quality parameters on the transistors, and records indicate the following information where a transistor is declared defective if either parameter is out of specification.

Firm	Fraction Defective	Fraction Supplied By
1	.02	.15
2	.01	.80
3	.03	.05

The manufacturer has stopped testing because of the costs involved, and it may be reasonably assumed that the fractions defective and the inventory mix are the same as during the period of record keeping. The director of manufacturing randomly selects a transistor, takes it to the test department, and finds that it is defective. If we let A be the event that an item is defective, and B_i be the event that the item came from firm i ($i = 1, 2, 3$), then we can evaluate $P(B_i|A)$. Suppose, for instance, that we are interested in determining $P(B_3|A)$. Then

$$P(B_3|A) = \frac{P(B_3) \cdot P(A|B_3)}{P(B_1) \cdot P(A|B_1) + P(B_2) \cdot P(A|B_2) + P(B_3) \cdot P(A|B_3)}$$

$$= \frac{(.05)(.03)}{(.15)(.02) + (.80)(.01) + (.05)(.03)} = \frac{3}{25}$$

1-8   Finite Sample Spaces

Experiments that give rise to a finite sample space have already been discussed, and the methods for assigning probabilities to events in such spaces have been presented. We can use Equations (1-1), (1-2), and (1-3) and deal with "equally likely" and "not equally likely" outcomes, respectively. In some situations we will have to resort to the relative frequency concept and successive trials (experimentation) to *estimate* probabilities, as indicated in Equation (1-1), with some finite m. In this section, we deal with equally likely outcomes and Equation (1-3). Note that this equation represents a special case of Equation (1-2), where $p_1 = p_2 = \ldots = p_n = 1/n$.

In order to evaluate probabilities, $P(A) = n(A)/n$, we must be able to determine both n, the number of *outcomes*, and $n(A)$, the *number of outcomes favorable to event A*. If there are n outcomes in $\mathscr{S}$, then there are 2^n possible subsets of $\mathscr{S}$ including the empty set and $\mathscr{S}$.

The requirement for the n outcomes to be equally likely is an important one, and there will be numerous applications where the experiment will specify that one (or more) item is (are) selected at *random* from a population (group) of N items without replacement.

If n represents the sample size ($n \leq N$), and the selection is random, then each possible selection (sample) is equally likely. It will soon be seen that there are $N!/[n!(N - n)!]$ such samples, so the probability of getting a particular sample must be $[n!(N - n)!/N!]$. It should be carefully noted that one sample differs from another if one (or more) item appears in one sample and not the other. The population items must thus be identifiable. In order to illustrate, suppose the population has four chips labeled a, b, c, and d. The sample size is to be two ($n = 2$). The *possible* results of the selection, disregarding order, are: *ab, ac, ad, bc, bd, cd*. If the sample is random the probability of obtaining each sample is $\frac{1}{6}$. The mechanics of selecting random samples vary a great deal, and devices such as *random number tables* and *icosahedron dice* are frequently used.

1-9 Enumeration Methods

This section will review several methods of counting or enumeration. These techniques are frequently useful in problem solution.

1-9.1 Tree Diagram

In simple experiments, a tree diagram may be useful in the enumeration of the sample space. Consider Example 1-8, where a true coin is to be tossed three times. The set of possible outcomes could be found by taking all the paths in the tree diagram shown.

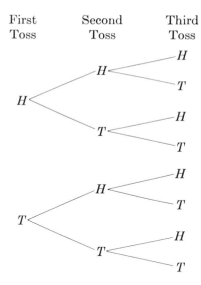

It should be noted that there are 2 outcomes to each trial, 3 trials, and $2^3 = 8$ outcomes {*HHH, HHT, HTH, HTT, THH, THT, TTH, TTT*}.

1-9.2 Multiplication Principle

If sets $A_1, A_2, \ldots, A_k$ have, respectively, $n_1, n_2, \ldots, n_k$ elements, then there are $n_1 \cdot n_2 \cdot \ldots \cdot n_k$ ways to first select an element from A_1, then select an element from $A_2, \ldots$, and finally select an element from A_k.

In the special case where $n_1 = n_2 = \ldots = n_k = n$, there are n^k possible selections. This was the situation encountered in the coin-tossing experiment of Example 1-8.

Suppose we consider some compound experiment $\mathscr{E}$ consisting of k experiments, $\mathscr{E}_1, \mathscr{E}_2, \ldots, \mathscr{E}_k$. If the sample spaces $\mathscr{S}_1, \mathscr{S}_2, \ldots, \mathscr{S}_k$ contain $n_1, n_2, \ldots, n_k$ outcomes respectively, then there are $n_1 \cdot n_2 \cdot \ldots \cdot n_k$ outcomes to $\mathscr{E}$. In addition, if the n_j outcomes of $\mathscr{S}_j$ are equally likely for $j = 1, 2, 3, \ldots, k$, then the $n_1 \cdot n_2 \cdot \ldots \cdot n_k$ outcomes of $\mathscr{E}$ are equally likely.

● **Example 1-32.** Suppose we toss a true coin and cast a true die. Since the coin and the die are true, the two outcomes to $\mathscr{E}_1$, $\mathscr{S}_1 = \{H, T\}$, are equally likely and the six outcomes to $\mathscr{E}_2$, $\mathscr{S}_2 = \{1, 2, 3, 4, 5, 6\}$, are equally likely. Since $n_1 = 2$ and $n_2 = 6$, there are twelve outcomes to the total experiment and the outcomes are equally likely. Due to the simplicity of the experiment in this case, a tree diagram permits an easy and complete enumeration.

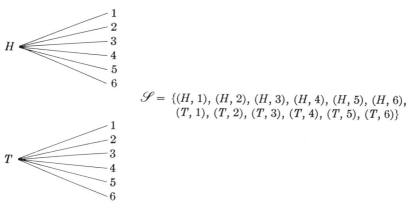

$$\mathscr{S} = \{(H, 1), (H, 2), (H, 3), (H, 4), (H, 5), (H, 6),$$
$$(T, 1), (T, 2), (T, 3), (T, 4), (T, 5), (T, 6)\}$$

● **Example 1-33.** A manufacturing process is operated with very little "in-process inspection." When items are completed they are transported to an inspection area, and four characteristics are inspected, each by a different inspector. The first inspector rates a characteristic according to one of four ratings, the second inspector uses three ratings, and the third and fourth inspectors use two ratings each. Each inspector marks the rating on the item identification tag. There would be a total of $4 \cdot 3 \cdot 2 \cdot 2 = 48$ ways in which the item may be marked.

1-9.3 Permutations

A permutation is an arrangement of distinct objects. One permutation differs from another if the order of arrangement differs or if the content differs. To illustrate, suppose we consider four distinct chips labeled a, b, c, and d. We wish to consider all permutations of these chips taken one at a time. These would be:

$$a$$
$$b$$
$$c$$
$$d$$

If we are to consider all permutations taken two at a time, these would be:

ab	*bc*
ba	*cb*
ac	*bd*
ca	*db*
ad	*cd*
da	*dc*

Note that permutations *ab* and *ba* differ because of a difference in order of the objects, while permutations *ac* and *ab* differ because of content differences. In order to generalize, we consider the case where there are n distinct objects from which we plan to select permutations of r objects ($r \leq n$). The number of such permutations, P_r^n, is given by

$$P_r^n = n(n-1)(n-2)(n-3) \cdot \ldots \cdot (n-r+1)$$
$$= \frac{n!}{(n-r)!} \tag{1-9}$$

This is a result of the fact that there are n ways to select the first object, $(n-1)$ ways to select the second, ..., $[n-(r-1)]$ ways to select the rth and the application of the multiplication principle. Note that $P_n^n = n!$.

1-9.4   Combinations

A combination is an arrangement of distinct objects where one combination differs from another only if the content of the arrangement differs. In the case of the four lettered chips a, b, c, d, the combinations of the chips taken two at a time are

$$ab$$
$$ac$$
$$ad$$
$$bc$$
$$bd$$
$$cd$$

We are interested in determining the number of combinations when there are n distinct objects to be selected r at a time. Since the number of permutations was the number of ways to select r objects from the n *and* permute the r objects, we note that

$$P_r^n = r! \cdot \binom{n}{r}$$

where $\binom{n}{r}$ represents the number of *combinations*. It follows that

$$\binom{n}{r} = P_r^n/r! = \frac{n!}{r!(n-r)!} \tag{1-10}$$

In the illustration with the four chips where $r = 2$, the reader may readily verify that $P_2^4 = 12$ and $\binom{4}{2} = 6$, as we found by complete enumeration.

For present purposes, $\binom{n}{r}$ is defined where n and r are integers such that $0 \le r \le n$; however, the terms $\binom{n}{r}$ may be generally defined for real n and any nonnegative integer, r. In this case we write

$$\binom{n}{r} = \frac{n(n-1)(n-2)\cdot \ldots \cdot(n-r+1)}{r!}$$

The reader will recall the *binomial theorem*:

$$(a+b)^n = \sum_{r=0}^{n} \binom{n}{r} a^r b^{n-r} \tag{1-11}$$

The numbers $\binom{n}{r}$ are thus called binomial coefficients.

Returning briefly to the definition of random sampling from a finite population without replacement, there were N objects with n to be selected. There are thus $\binom{N}{n}$ different samples. If the sampling process is random, each possible sample has probability $1/\binom{N}{n}$ of being the one selected.

Two identities which are helpful in problem solution are

$$\binom{n}{r} = \binom{n}{n-r} \tag{1-12}$$

and

$$\binom{n}{r} = \binom{n-1}{r-1} + \binom{n-1}{r} \tag{1-13}$$

To prove Equation (1-12) we note that

$$\binom{n}{r} = \frac{n!}{r!(n-r)!} = \frac{n!}{(n-r)!r!} = \binom{n}{n-r}$$

To prove Equation (1-13), expand the right-hand side and collect terms.

To verify that a finite collection of n elements has 2^n subsets as indicated earlier, we note that

$$2^n = (1+1)^n = \sum_{r=0}^{n} \binom{n}{r} = \binom{n}{0} + \binom{n}{1} + \cdots + \binom{n}{n}$$

from Equation (1-10). The right side of this relationship gives the total number of subsets since $\binom{n}{0}$ is the number of subsets with 0 elements, $\binom{n}{1}$ is the number with one element, . . . , and $\binom{n}{n}$ is the number with n elements.

● **Example 1-34.** A production lot of size 100 is known to be 5 percent defective. A random sample of 10 items is selected without replacement. In order to determine the probability that there will be no defectives in the sample, we resort to counting both the number of possible samples and the number of samples favorable to event A, where event A is taken to mean that there are no defectives. The number of possible samples is $\binom{100}{10} = \frac{100!}{10!(90)!}$. The number "favorable to A" is $\binom{5}{0} \cdot \binom{95}{10}$, so that

$$P(A) = \frac{\binom{5}{0}\binom{95}{10}}{\binom{100}{10}} = \frac{\frac{5!}{0!5!}\frac{95!}{10!85!}}{\frac{100!}{10!90!}} = .58375$$

To generalize the preceding example, we consider the case where the population has N items of which D belong to some class of interest (such as defective). A random sample of size n is selected without replacement. If A denotes the event of obtaining exactly r items from the class of interest in the sample, then

$$P(A) = \frac{\binom{D}{r}\binom{N-D}{n-r}}{\binom{N}{n}} \qquad r = 0, 1, 2, \ldots, \min(n, D) \qquad (1\text{-}14)$$

1-9.5 Permutations of Like Objects

In the event there are k distinct classes of objects, and the objects within the class are not distinct, the following result is obtained where n_1 is the number in the first class, n_2 is the number in the second class, ..., n_k is the number in the kth class, and $n_1 + n_2 + \cdots + n_k = n$:

$$P^{\,n}_{n_1, n_2, \ldots, n_k} = \frac{n!}{n_1! \cdot n_2! \cdot \ldots \cdot n_k!}$$

1-10 Discrete Sample Spaces

Experiments are frequently encountered where the sample space is not finite but rather contains countably many outcomes. That is, $\mathscr{S} = \{e_1, e_2, e_3, \ldots, e_k, \ldots\}$ where the outcomes may be put in a one-to-one correspondence with the natural numbers. The general definition of probability is still applied; however, the definition resulting in Equation (1-2), which was used in assigning probabilities in the finite sample space, must be modified to the extent that the probabilities assigned the outcomes are p_i ($i = 1, 2, 3, \ldots$), where

$$(a) \quad p_i \geq 0 \text{ for all } i$$

$$(b) \quad \sum_{i=1}^{\infty} p_i = 1$$

Thus, the *discrete sample space* is a more general classification that consists of both *finite sample spaces* and *denumerably infinite sample spaces.*

1-11 Summary

This chapter has introduced the concept of random experiments, the sample space and events, and presented a formal definition of probability. This was followed by methods to assign probability to events. Theorems 1-1 to 1-6 provide important results for dealing with the probability of special events. Conditional probability was defined and illustrated, along with the concept of independent events. In addition, we considered partitions of the sample space, total probability, and Bayes' Theorem. Finite sample spaces with their special properties were discussed, and enumeration methods were reviewed for use in assigning probability to events in the case of equally likely experimental outcomes. The concepts presented in this chapter form an important background for the rest of the book.

1-12 **Exercises**

1-1.   Use Venn diagrams to verify each of the following statements:

Idempotent laws: $A \cup A = A$ $A \cap A = A$

Complement laws: $A \cup \bar{A} = U$ $A = \bar{\bar{A}}$

Commutative laws: $A \cup B = B \cup A$ $A \cap B = B \cap A$

Associative laws: $A \cup (B \cup C) = (A \cup B) \cup C$
$A \cap (B \cap C) = (A \cap B) \cap C$

Distributive laws: $A \cup (B \cap C) = (A \cup B) \cap (A \cup C)$
$A \cap (B \cup C) = (A \cap B) \cup (A \cap C)$

1.2.   Complete the details of the Proof for Theorem 1-4 in the text.

1-3.   Prove Theorem 1-5 and Theorem 1-7.

1-4.   Prove the second and third parts of Theorem 1-8.

1-5.   Television sets are given a final inspection following assembly. Three types of defects are identified as critical, major, and minor defects and are coded A, B, and C, respectively, by a mail-order house. Data are analyzed with the following results.

Sets having only critical defects	3%
Sets having only major defects	2
Sets having only minor defects	10
Sets having only critical *and* major defects	4
Sets having only critical *and* minor defects	5
Sets having only major *and* minor defects	3
Sets having all three types of defects	1

(*a*) What fraction of the sets has no defects?

(*b*) Sets with either critical defects or major defects (or both) get a complete rework. What fraction falls in this category?

1-6.   A single bolt is selected at random from a box of 10,000. Three different kinds of defects, A, B, and C, are known to occur in these particular bolts. Type A defects occur 1 percent of the time, type B defects occur .5 percent of the time, and type C occur .75 percent. In addition, it is estimated that .25 percent have both A and B defects, .30 percent have both A and C, .20 percent have B and C, and .10 percent have all three defects. What is the probability that the sample bolt has at least one of the three types of defects?

1-7.   In a human factors laboratory, the reaction times of human subjects are measured as the elapsed time from the instant a position number is displayed on a digital display until the subject presses a button located at the position indicated. Two subjects are involved, and times are measured in seconds for each subject (t_1, t_2). What is the sample space for this experiment? Present the following events as subsets and mark them on a diagram: $(t_1 + t_2)/2 \le .05$, max $(t_1, t_2) \le .05$, $|t_1 - t_2| \le .02$.

1.8. Consider an experiment to determine the speed of a molecule of a particular gas. Is this a deterministic or nondeterministic experiment? Why? If the experiment is nondeterministic describe the sample space.

1-9. Four men, John, Bill, Sam, and Dave, are hired by the Brown Manufacturing Company. These men are to be assigned at random to either the shipping or receiving departments. Specify a sample space and the event that Bill and Dave will be assigned to the same service.

1-10. Transistors from a box are tested one at a time and marked either defective or nondefective. This is continued until either two defective items are found or five items have been tested. Describe the sample space for this experiment.

1-11. Let the universal set be represented by all the integers 1 to 10. Furthermore, let $A = \{3, 4, 5\}$, $B = \{4, 5, 6\}$, and $C = \{6, 7, 8\}$. Find the following:

(a) $\bar{A} \cap B$
(b) $\bar{A} \cap \bar{B} \cap C$
(c) $\bar{A} \cup B$
(d) $\overline{A \cup (B \cap C)}$
(e) $\overline{\bar{A} \cup B}$
(f) $\overline{A \cup B \cup C}$

1-12. If a set has four elements, how many subsets does it have?

1-13. Three printers do work for the publications office of Georgia Tech. The publications office does not negotiate a contract penalty for late work, and the data below reflect a large amount of experience with these printers.

Printer, i	Fraction of Contracts Held by Printer i	Fraction of Time Delivery More than One Month Late
1	.2	.1
2	.3	.4
3	.5	.2

A department observes that its recruiting booklet is more than a month late, what is the probability that the contract is held by printer 3?

1-14. A chemist is interested in analyzing water samples at a steam electrical plant. The tests will be misleading if either nitrogen or silicone is present in the samples. Automatic devices hold the probability of both elements being present simultaneously to a negligible level. Write a general expression for the probability that the tests have been misleading.

1-15. The XYZ Company replaces the gearbox on its turbine every two years. The probability that the gearbox lasts two years without failure is .80. Two types of failures, say A and B, can occur. Type A failures occur three times as often as do those of type B. What are the probabilities of the two types of failures?

1-16. A political prisoner in Russia is to be exiled to either Siberia or the Ural. The probabilities of being sent to these two places are .8 and .2, respectively. It is also known that if a resident of Siberia is selected at random the probability is .5 that he will be wearing a fur coat, whereas that probability is .7 for a resident of the Ural. Upon arriving in exile, the first person the prisoner sees is not wearing a fur coat. What is the probability he is in Siberia?

1-17. In a new braking device designed to prevent automobile skids, there is a considerable amount of electronic and hydraulic hardware. The entire system may be broken down into three series subsystems that operate independently: an electronics system, a hydraulic system, and a mechanical activator. On a particular braking the reliabilities of these units are approximately, .995, .993, and .994, respectively. Estimate the system reliability.

1-18. Consider the diagram of an electronic system, which shows the probabilities of the system components operating properly. What is the probability that the entire system operates if assembly III and at least one of the components in assembly I and II must operate for the assembly to operate? Assume that the components of each assembly operate independently and that the assemblies operate independently.

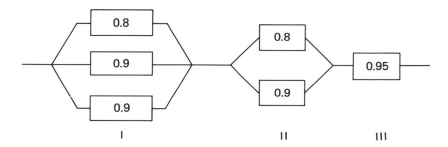

1-19. How is the probability of system operation affected if, in the above problem, the probability of successful operation for the third component in assembly I changes from .9 to .5?

1-20. The quality control manager of a certain company is interested in testing a finished product, which is available in lots of size 50. She would like to scrap the lot if she can be reasonably sure that 10 percent of the items are defective. She decides to select a random sample of 10 items without replacement and scrap the lot if it contains one or more defective items. Does this procedure seem reasonable?

1-21. A trucking firm has a contract to ship a load of goods from city W to city Z. There are no direct routes connecting W to Z, but there are six roads from W to X and five roads from X to Z. How many total routes are there to be considered?

1-22. A state with 800,000 registered vehicles is considering altering its license plates to contain six symbols, the first four being letters and the last two being numbers. Is the proposed scheme feasible?

1-23. The manager of a small plant wishes to determine the number of ways he can assign men to the first shift. He has 15 men who can serve as operators of the production equipment, 8 men who can serve as maintenance personnel, and 4 who can be supervisors. If the shift requires 6 operators, 2 maintenance men, and 1 supervisor, how many ways can the first shift be manned?

1-24. In a large discount house, all inventory items are given an alphabetic code of five letters. It is common practice not to use the same letter twice in a code. How many different codes are possible?

1-25. A production lot has 100 units of which 25 are known to be defective. A random sample of 4 units is selected without replacement. What is the probability that the sample will contain no more than 2 defective units?

1-26. In inspecting incoming lots of merchandise, the following inspection rule is used where the lots contain 200 units. A random sample of 10 items is selected. If there is no more than one defective item in the sample, the lot is accepted. Otherwise it is returned to the vendor. If the fraction defective in the original lot is p', determine the probability of accepting the lot as a function of p'.

1-27. In a plastics plant 12 pipes empty different chemicals into a mixing vat. Each pipe has a five-position gauge that measures the rate of flow into the vat. One day, while experimenting with various mixtures, a solution is obtained which emits a poisonous gas. The settings on the gauges were not recorded. What is the probability of obtaining this same solution when randomly experimenting again?

1-28. Eight equally skilled men and women are applying for two jobs. Because the two new employees must work closely together, their personalities should be compatible. To achieve this, the personnel manager has administered a test and must compare the scores for each possibility. How many comparisons must the manager make?

1-29. By accident, a chemist combined two laboratory substances that yielded a desirable product. Unfortunately, her assistant did not record the names of the ingredients. There are forty substances available in the lab. If the two in question must be located by successive trial-and-error experiments, what is the maximum number of tests that might be made?

1-30. Suppose, in the previous problem, a known catalyst was used in the first accidental reaction. Because of this, the order in which the ingredients are mixed is important. What is the maximum number of tests that might be made?

1-31. A company plans to build five additional warehouses at new locations. Fifteen locations are under consideration. How many total possible choices are there?

1-32. Washing machines can have five kinds of major and five kinds of minor defects. In how many ways can one major and one minor defect occur? In how many ways can two major and two minor defects occur?

1-33. A milling machine may suffer failure at 12 different parts. Given that a nonworking machine has failures at three parts, how many ways may this occur?

1-34. The industrial engineering department of the XYZ Company is performing a work sampling study on eight technicians. The engineer wishes to randomize the order in which he visits the technicians' work areas. In how many ways may he arrange these visits?

1-35. Suppose there are n people in a room. If a list is made of all their birthdays (the specific month and day of the month), what is the probability that two or more persons have the same birthday? Assume there are 365 days in the year and that each day is equally likely to occur for any person's birthday. Let B be the event that two or more persons have the same birthday. Find $P(B)$ and $P(\bar{B})$ for $n = 10, 20, 21, 22, 23, 24, 25, 30, 40, 50$, and 60.

1-36. A hiker leaves point A shown in the figure below, choosing at random one path from AB, AC, AD, and AE. At each subsequent junction she chooses another path at random. What is the probability that she arrives at point X?

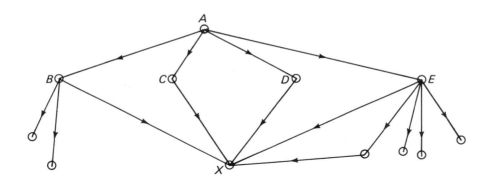

1-37. Two balls are drawn from an urn containing m balls numbered from 1 to m. The first ball is kept if it is numbered 1, and returned to the urn otherwise. What is the probability that the second ball drawn is numbered 2?

1-38. Consider a set of n objects, all of different quality (or value), and assume that it is always possible to know which of a given pair of objects is better. Suppose the objects are presented one at a time and at random to a decision maker. This decision maker either selects an object as the "best" and examines no more objects, or rejects the object *forever* and examines another one. Suppose the decision maker selects the ith object as the "best." What is the probability that this object is actually the best of all n objects, both inspected and uninspected?

1-39. In a certain dice game, players continue to throw two dice until they either win or lose. The player wins on the first throw if the sum of the two upturned faces is either 7 or 11, and loses if the sum is 2, 3, or 12. Otherwise, the sum of the faces becomes the player's "point." The player continues to throw until the first succeeding throw on which he makes his point (in which case he wins), or until he throws a 7 (in which case he loses). What is the probability that the player with the dice will eventually win the game?

Chapter 2

One-Dimensional Random Variables

2-1 Introduction

The objectives of this chapter are to introduce the concept of random variables, to define and illustrate probability distributions and cumulative distribution functions, and to present useful characterizations for random variables.

When describing the sample space of a random experiment, it is not necessary to specify that an individual outcome be a number. In several examples this was not the case, such as in Example 1–8 where a true coin was tossed three times, and $\mathscr{S} = \{HHH, HHT, HTH, HTT, THH, THT, TTH, TTT\}$, or Example 1–14 where probes of welds of diode leads to nickel ribbon yielded a sample space, $\mathscr{S} = \{GG, GD, DG, DD\}$.

In most experimental situations, however, we are interested in numerical outcomes. For example, in the illustration involving coin tossing we might assign some real number x to every element of the sample space. In general, we want to assign a real number x to every outcome, e, of the sample space, $\mathscr{S}$. A functional notation will be used initially, so that $x = X(e)$, where X is the function. The domain of X is $\mathscr{S}$, and the numbers in the range are real numbers. The function X is called a random variable. Figure 2-1 illustrates the nature of this function.

Definition

If $\mathscr{E}$ is an experiment having sample space $\mathscr{S}$, and X is a *function* which assigns a real number $X(e)$ to every outcome $e \in \mathscr{S}$, then $X(e)$ is called a *random variable*.

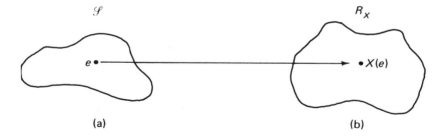

Fig. 2-1.   The concept of a random variable. (a) $\mathscr{S}$: the sample space of $\mathscr{E}$. (b) R_X: the range space of X.

● **Example 2-1.** Consider the coin-tossing experiment discussed in the preceding paragraphs. If X is the number of heads showing, then $X(HHH) = 3$, $X(HHT) = 2$, $X(HTH) = 2$, $X(HTT) = 1$, $X(THH) = 2$, $X(THT) = 1$, $X(TTH) = 1$, and $X(TTT) = 0$. The range space $R_X = \{x : x = 0, 1, 2, 3\}$ in this example (see Fig. 2-2).

The reader should recall that, for all functions and for every element in the domain, there is exactly one value in the range. In the case of the random variable, for every outcome $e \in \mathscr{S}$ there corresponds exactly one value $X(e)$. It should be noted that different values of e may lead to the same x, as was the case where $X(TTH) = 1$, $X(THT) = 1$, and $X(HTT) = 1$ in the preceding example.

Where the outcome in $\mathscr{S}$ is already the numerical characteristic desired, then $X(e) = e$, the identity function. Example 1-12, in which a cathode ray tube was aged to failure, is a good example. Recall that $\mathscr{S} = \{t : t \geq 0\}$. If X is

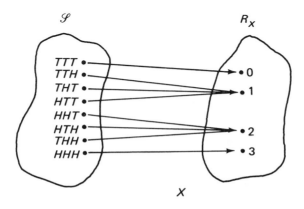

Fig. 2-2.   The number of heads in three coin tosses.

the time to failure, then $X(t) = t$. Some authors call this type of sample space a *numerical-valued phenomenon.*

The range space, R_X, is made up of the possible values of X, and in subsequent work it will not be necessary to indicate the functional nature of X. Here we are concerned with events that are associated with R_X, and the random variable X will induce probabilities onto these events. If we return again to the coin-tossing experiment for illustration and assume the coin to be true, there are eight equally likely outcomes, *HHH, HHT, HTH, HTT, THH, THT, TTH, TTT*, each having probability $\frac{1}{8}$. Now suppose A is the event "exactly two heads" and, as previously, we let X represent the number of heads (see Fig. 2-2). The event that $(X = 2)$ relates to R_X, not $\mathscr{S}$; however, $P(X = 2) = P(A) = \frac{3}{8}$, since $A = \{HHT, HTH, THH\}$ is the equivalent event in $\mathscr{S}$, and probability was defined on events in the sample space. The random variable X induced the probability of $\frac{3}{8}$ to the event $(X = 2)$. Note that parentheses will be used to denote an event in the range of the random variable, and in general we will write $P(X = x)$ rather than $P_X(X = x)$.

In order to generalize this notion, consider the following definition.

Definition

If $\mathscr{S}$ is the sample space of an experiment $\mathscr{E}$ and a random variable, X, with range space R_X, is defined on $\mathscr{S}$, and furthermore if event A is an event in $\mathscr{S}$ while event B is an event in R_X, then A and B are equivalent events if

$$A = \{e \in \mathscr{S} : X(e) \in B\}$$

Figure 2-3 illustrates this concept.

More simply, if event A in $\mathscr{S}$ consists of all outcomes in $\mathscr{S}$ for which $X(e) \in B$, then A and B are equivalent events. Whenever A occurs, B occurs;

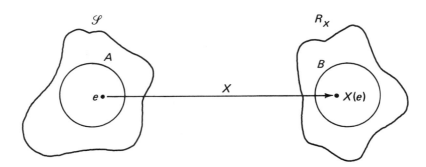

Fig. 2-3. Equivalent events.

and whenever B occurs, A occurs. Note that A and B are associated with different spaces.

Definition

If A is an event in the sample space and B is an event in the range space R_X of the random variable X, then we define the probability of B as

$$P_X(B) = P(A) \qquad \text{where } A = \{e \in \mathscr{S} : X(e) \in B\}$$

With this definition, we may assign probabilities to events in R_X in terms of probabilities defined on events in $\mathscr{S}$, and we will *suppress* the function X, so that $P_X(X = 2) = \frac{3}{8}$ in the familiar coin-tossing example means that there is an event $A = \{HHT, HTH, THH\} = \{e : X(e) = 2\}$ in the sample space with probability $\frac{3}{8}$. In subsequent work, we will not deal with the nature of the function X since we are interested in the values of the range space and their associated probabilities. While outcomes in the sample space may not be real numbers, all elements of the range space of X are real numbers.

The following examples illustrate the sample space-range space relationship, and the concern with the range space rather than the sample space is evident, since numerical results are of interest.

● **Example 2-2.** Consider the tossing of two true dice as described in Example 1-10. (The sample space was described in Chapter 1.) Suppose we define a random variable Y as the sum of the "up" faces. Then $R_Y = \{2, 3, 4, 5, 6, 7, 8, 9, 10, 11, 12\}$, and the probabilities are $(\frac{1}{36}, \frac{2}{36}, \frac{3}{36}, \frac{4}{36}, \frac{5}{36}, \frac{6}{36}, \frac{5}{36}, \frac{4}{36}, \frac{3}{36}, \frac{2}{36}, \frac{1}{36})$, respectively. Table 2-1 shows equivalent events. The reader will recall that there are 36 outcomes which, since the dice are true, are equally likely.

TABLE 2-1 **Equivalent Events**

Some Events in R_Y	Equivalent Events in $\mathscr{S}$	Probability
$Y = 2$	$\{(1, 1)\}$	$\frac{1}{36}$
$Y = 3$	$\{(1, 2), (2, 1)\}$	$\frac{2}{36}$
$Y = 4$	$\{(1, 3), (2, 2), (3, 1)\}$	$\frac{3}{36}$
$Y = 5$	$\{(1, 4), (2, 3), (3, 2), (4, 1)\}$	$\frac{4}{36}$
$Y = 6$	$\{(1, 5), (2, 4), (3, 3), (4, 2), (5, 1)\}$	$\frac{5}{36}$
$Y = 7$	$\{(1, 6), (2, 5), (3, 4), (4, 3), (5, 2), (6, 1)\}$	$\frac{6}{36}$
$Y = 8$	$\{(2, 6), (3, 5), (4, 4), (5, 3), (6, 2)\}$	$\frac{5}{36}$
$Y = 9$	$\{(3, 6), (4, 5), (5, 4), (6, 3)\}$	$\frac{4}{36}$
$Y = 10$	$\{(4, 6), (5, 5), (6, 4)\}$	$\frac{3}{36}$
$Y = 11$	$\{(5, 6), (6, 5)\}$	$\frac{2}{36}$
$Y = 12$	$\{(6, 6)\}$	$\frac{1}{36}$

● **Example 2-3.** One hundred cardiac pacemakers were placed on life test in a saline solution held as close to body temperature as possible. The test is functional, with pacemaker output monitored by a system providing for output signal conversion to digital form for comparison against a design standard. The test was initiated on July 1, 1979. When a pacer output varies from the standard by as much as 2 percent, this is considered a failure and the computer records the date and the time of day (d, t). If X is the random variable "time to failure," then $\mathscr{S} = \{(d, t) : d = \text{date}, \ t = \text{time}\}$ and $R_X = \{x : x \geq 0\}$. The random variable X is the total number of elapsed time units since the module went on test. We will deal directly with X and its probability law. This concept will be discussed in the following sections.

2-2 Discrete Random Variables

If the range space R_X of the random variable X is either finite or countably infinite, then X will be called a *discrete random variable.* In this case, $R_X = \{x_1, x_2, \ldots, x_k, \ldots\}$.

● **Example 2-4.** Suppose that the number of working days in a particular year is 250 and that the records of employees are marked for each day they are absent from work. An experiment consists of randomly selecting a record to observe the days marked absent. The random variable X is defined as the number of days absent, so that $R_X = \{0, 1, 2, \ldots, 250\}$. This is an example of a discrete random variable with a finite number of possible values.

● **Example 2-5.** A Geiger counter is connected to a gas tube in such a way that it will record the background radiation count for a selected time interval $\{0, t\}$. The random variable of interest is the count. If X denotes the random variable, then $R_X = \{0, 1, 2, \ldots, k, \ldots\}$, and we have, at least conceptually, a countably infinite range space (outcomes can be placed in a one-to-one correspondence with the natural numbers) so that the random variable is discrete.

Definition

If X is a discrete random variable, we associate a number $p_X(x_i) = P(X = x_i)$ with each outcome x_i, in R_X for $i = 1, 2, \ldots, n, \ldots$, where the numbers $p_X(x_i)$ satisfy

1. $p_X(x_i) \geq 0$ for all i

2. $\sum_{i=1}^{\infty} p_X(x_i) = 1$

The function p_X is called the *probability function* or *probability law* of the random variable, and the collection of pairs $[(x_i, p_X(x_i)), i = 1, 2, \ldots]$ is called the *probability distribution* of X. The function p_X is usually presented in either *tabular, graphical,* or *mathematical* form as illustrated in the following examples. The subscript, X, will usually be deleted unless needed for clarification.

● **Example 2-6.** In the coin-tossing experiment of Example 1-8, where X = the number of heads, the probability distribution is given in both tabular and graphical form in Fig. 2-4. It will be recalled that $R_X = \{0, 1, 2, 3\}$.

Tabular Presentation   Graphical Presentation

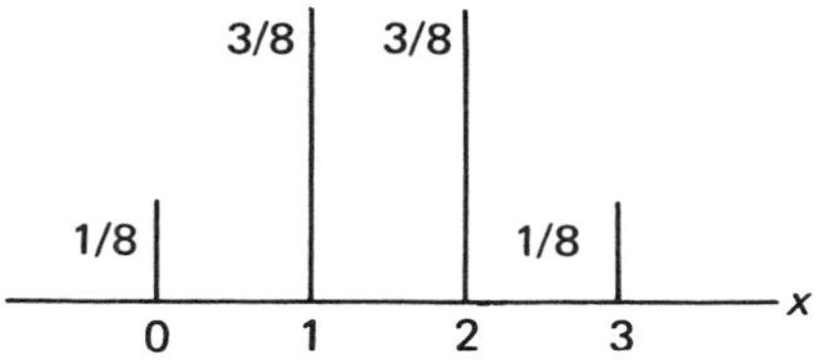

x	p(x)
0	1/8
1	3/8
2	3/8
3	1/8

Fig. 2-4. Probability distribution for coin-tossing experiment.

● **Example 2-7.** Suppose we have a random variable X with a probability distribution given by the relationship

$$p(x) = \binom{n}{x} p^x (1-p)^{n-x} \qquad x = 0, 1, \ldots, n$$
$$= 0 \qquad\qquad\qquad \text{otherwise} \qquad (2\text{-}1)$$

where n is a positive integer and $0 < p < 1$. This relationship is known as the *binomial distribution*, and it will be studied in more detail later. Although it would be possible to display this model in graphical or tabular form for particular n and p by evaluating $p(x)$ for $x = 0, 1, 2, \ldots, n$, this is seldom done in practice.

● **Example 2-8.** Recall the earlier discussion of random sampling from a finite population without replacement. Suppose there are N objects of which D are defective. A random sample of size n is selected without replacement, and if we let X represent the number of defectives in the sample, then

$$p(x) = \frac{\binom{D}{x}\binom{N-D}{n-x}}{\binom{N}{n}} \qquad x = 0, 1, 2, \ldots, \min(n, D)$$
$$= 0 \qquad\qquad\qquad \text{otherwise} \qquad (2\text{-}2)$$

This distribution is known as the *hypergeometric distribution.* In a particular case, suppose $N = 100$ items, $D = 5$ items, and $n = 4$, then

$$p(x) = \frac{\binom{5}{x}\binom{95}{4-x}}{\binom{100}{4}} \qquad x = 0, 1, 2, 3, 4$$

$$= 0 \qquad\qquad \text{otherwise} \qquad\qquad (2\text{-}3)$$

In the event that either tabular or graphical presentation is desired, this would be as shown in Fig. 2-5; however, unless there is some special reason to use these forms, we will use the mathematical relationship.

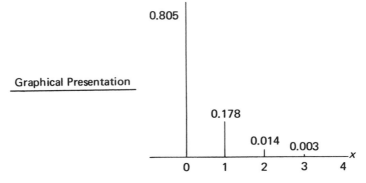

x	$p(x)$
0	$\binom{5}{0}\binom{95}{4} \Big/ \binom{100}{4} \doteq 0.805$
1	$\binom{5}{1}\binom{95}{3} \Big/ \binom{100}{4} \doteq 0.178$
2	$\binom{5}{2}\binom{95}{2} \Big/ \binom{100}{4} \doteq 0.014$
3	$\binom{5}{3}\binom{95}{1} \Big/ \binom{100}{4} \doteq 0.003$
4	$\binom{5}{4}\binom{95}{0} \Big/ \binom{100}{4} \doteq 0.000$

Tabular Presentation

Graphical Presentation

0.805

0.178

0.014   0.003

Fig. 2-5. Some hypergeometric probabilities $N = 100$, $D = 5$, $n = 4$.

● **Example 2-9.** In Example 2-5, where the Geiger counter was prepared for detecting background radiation count, we might use the following relationship that has been experimentally shown to be appropriate:

$$p(x) = e^{-\lambda t}(\lambda t)^x/x! \qquad x = 0, 1, 2, \ldots \qquad \lambda > 0$$
$$= 0 \qquad\qquad\quad \text{otherwise} \qquad\qquad\qquad (2\text{-}4)$$

This is called the *Poisson distribution*, and at a later point it will be derived analytically. The parameter λ is the mean rate in "hits" per unit time, and x is the number of these "hits."

These examples have illustrated some discrete probability distributions and alternate means of presenting the pairs $[(x_i, p(x_i)), i = 1, 2, \ldots]$. In later sections, a number of probability distributions will be developed, each from a set of postulates motivated from considerations of *real world phenomena*.

A general graphical presentation of the discrete distribution from Example 2-8 is given in Fig. 2-6. This geometric interpretation is often useful in developing an intuitive feeling for discrete distributions. There is a close analogy to mechanics if we consider the probability distribution as a mass of one unit distributed over the real line in amounts $p(x_i)$ at points $x_i, i = 1, 2, \ldots, n$.

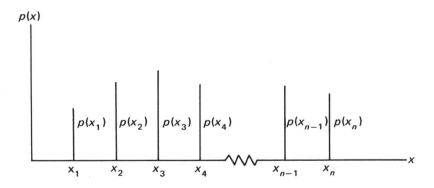

Fig. 2-6.   Geometric interpretation of a probability distribution.

2-3 Continuous Random Variables

If the range space, R_X, of the random variable X is an interval or a collection of intervals, then X will be called a *continuous random variable*.

Definition

For a continuous random variable X, we define

$$P(a \leq X \leq b) = \int_a^b f_X(x) \, dx \tag{2-5}$$

where the function f_X, denoted as the *probability density function* (*pdf*), satisfies the following conditions:

1. $f_X(x) \geq 0$ for all $x \in R_X$ (2-6)

2. $\int_{R_X} f_X(x) \, dx = 1$

The subscript X will usually be deleted unless needed for clarification.

These concepts are illustrated in Fig. 2-7. This definition stipulates the existence of a function f defined on R_X such that

$$P\{e : a \leq X(e) \leq b\} = \int_a^b f(x) \, dx$$

where e is an outcome in the sample space. We are concerned only with R_X and f. It is important to realize that $f(x)$ does not represent the probability of anything, and that only when the function is integrated between two points does it yield a probability.

A result of the definition and Equation (2-5) is that for any specified single value of X, say x_0, $P(X = x_0) = 0$ since

$$\int_{x_0}^{x_0} f(x) \, dx = 0$$

This result may initially seem to contradict intuition; however, if we consider

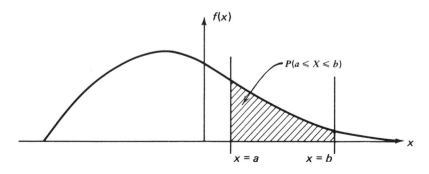

Fig. 2-7. Hypothetical probability density function.

the fact that we allow X to assume all values in some interval, then $P(X = x_0) = 0$ is not equivalent to saying the event $(X = x_0)$ in R_X is impossible. Recall that if $A = \emptyset$, then $P(A) = 0$; however, the fact that $P(X = x_0) = 0$ and that the set $A = \{x : x = x_0\}$ is not empty clearly indicates that the converse is not true.

An immediate result of this is that $P(a \le X \le b) = P(a < X \le b) = P(a < X < b) = P(a \le X < b)$, where X is continuous.

● **Example 2-10.** The time to failure of the cathode ray tube described in Example 1-12 has the following probability density function:

$$f(t) = \lambda e^{-\lambda t} \qquad t \ge 0$$
$$= 0 \qquad \text{otherwise} \qquad (2\text{-}7)$$

where $\lambda > 0$ is a constant known as the failure rate. This probability density function is called the *exponential density*, and experimental evidence has indicated that it is appropriate to describe the time to failure (a real world occurrence) for some types of components. In this example suppose we want to find $P(T \ge 100 \text{ hours})$. This is equivalent to stating $P(100 \le T < \infty)$, and from Equation (2-5)

$$P(T \ge 100) = \int_{100}^{\infty} \lambda e^{-\lambda t}\, dt$$
$$= e^{-100\lambda}$$

We might again employ the concept of conditional probability and determine $P(T \ge 100 \mid T > 99)$, the probability the tube lives at least 100 hours given that it has lived beyond 99 hours. From our earlier work,

$$P(T \ge 100 | T > 99) = \frac{P(T \ge 100 \text{ and } T > 99)}{P(T > 99)}$$

$$= \frac{\int_{100}^{\infty} \lambda e^{-\lambda t}\, dt}{\int_{99}^{\infty} \lambda e^{-\lambda t}\, dt} = \frac{e^{-100\lambda}}{e^{-99\lambda}} = e^{-\lambda}$$

● **Example 2-11.** A random variable X has the probability density function given below and shown graphically in Fig. 2-8.

$$f(x) = x \qquad 0 \le x < 1$$
$$= 2 - x \qquad 1 \le x < 2$$
$$= 0 \qquad \text{otherwise}$$

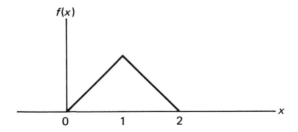

Fig. 2-8.   An example of a density function.

The following probabilities are calculated:

(a) $P\left(-1 < X < \dfrac{1}{2}\right) = \displaystyle\int_{-1}^{0} 0 \, dx + \int_{0}^{1/2} x \, dx = \dfrac{1}{8}$

(b) $P\left(X \le \dfrac{3}{2}\right) = \displaystyle\int_{-\infty}^{0} 0 \, dx + \int_{0}^{1} x \, dx + \int_{1}^{3/2} (2 - x) \, dx$

$$= 0 + \frac{1}{2} + \left(2x - \frac{x^2}{2}\right)_{1}^{3/2}$$

$$= \frac{1}{2} + \frac{3}{8} = \frac{7}{8}$$

(c) $P(X \le 3) = 1$

(d) $P(X \ge 2.5) = 0$

(e) $P\left(\dfrac{1}{4} < X < \dfrac{3}{2}\right) = \displaystyle\int_{1/4}^{1} x \, dx + \int_{1}^{3/2} (2 - x) \, dx$

$$= \frac{15}{32} + \frac{3}{8} = \frac{27}{32}$$

In describing probability density functions, a mathematical model is usually employed. A graphical or geometric presentation may also be instructive. The area under the density function corresponds to probability, and the total area is one. The student familiar with mechanics might consider the probability of one to be distributed over the real line according to f. In Fig. 2-9, the intervals $[a, b]$ and $[b, c]$ are of the same length; however, the probability associated with $[a, b]$ is greater.

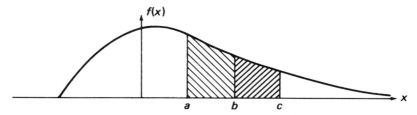

Fig. 2-9. A density function.

2-4 Some Characteristics of Distributions

While a discrete distribution is completely specified by the pairs $[(x_i, p(x_i));$ $i = 1, 2, \ldots, n, \ldots]$, and a probability density function is likewise specified by $[(x, f(x)); x \in R_X]$, it is often convenient to work with some descriptive characteristics of the random variable. In this section we introduce two widely used descriptive measures, as well as a general expression for other similar measures. The first of these is the *first moment* about the origin. This is called the *mean of the random variable* and is denoted by the Greek letter μ, where

$$\mu = \sum_i x_i p(x_i) \qquad \text{for discrete } X$$

$$= \int_{-\infty}^{\infty} xf(x)\, dx \qquad \text{for continuous } X \qquad (2\text{-}8)$$

This measure provides an indication of *central tendency* in the random variable.

● **Example 2-12.** Returning to the coin-tossing experiment where X represents the number of heads and the probability distribution is as shown in Fig. 2-4, the calculation of μ yields

$$\mu = \sum_{i=1}^{4} x_i p(x_i) = 0 \cdot \left(\frac{1}{8}\right) + 1 \cdot \left(\frac{3}{8}\right) + 2 \cdot \left(\frac{3}{8}\right) + 3 \cdot \left(\frac{1}{8}\right) = \frac{3}{2}$$

as indicated in Fig. 2-10. In this particular example, because of symmetry, the value μ could have been easily determined from inspection.

● **Example 2-13.** In Example 2-11, a density f was defined as

$$f(x) = x \qquad 0 \le x \le 1$$
$$= 2 - x \qquad 1 \le x \le 2$$
$$= 0 \qquad \text{otherwise}$$

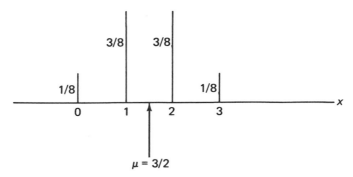

Fig. 2-10. Calculation of the mean.

The mean is determined as follows:

$$\mu = \int_0^1 x \cdot x \, dx + \int_1^2 x \cdot (2 - x) \, dx$$
$$+ \int_{-\infty}^0 x \cdot 0 \, dx + \int_2^\infty x \cdot 0 \, dx = 1$$

Another measure describes the spread or dispersion of the probability associated with elements in R_X. This measure is called the *variance*, denoted by the Greek letter σ^2, and is defined as follows:

$$\sigma^2 = \sum_i (x_i - \mu)^2 p(x_i) \qquad \text{for discrete } X$$
$$= \int_{-\infty}^\infty (x - \mu)^2 f(x) \, dx \quad \text{for continuous } X \qquad (2\text{-}9)$$

This is the second moment about the mean, and it corresponds to the moment of inertia in mechanics. Consider Fig. 2-11, where two hypothetical discrete distributions are shown in graphical form. Note that the mean equals one in

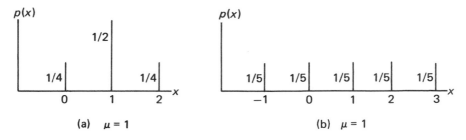

Fig. 2-11. Some hypothetical distributions.

both cases. The variance for the discrete random variable shown in Fig. 2-11*a* is

$$\sigma^2 = (0-1)^2 \cdot \left(\frac{1}{4}\right) + (1-1)^2 \cdot \left(\frac{1}{2}\right) + (2-1)^2 \cdot \left(\frac{1}{4}\right) = \frac{1}{2}$$

and the variance of the discrete random variable shown in Fig. 2-11*b* is

$$\sigma^2 = (-1-1)^2 \cdot \frac{1}{5} + (0-1)^2 \cdot \frac{1}{5} + (1-1)^2 \cdot \frac{1}{5}$$

$$+ (2-1)^2 \cdot \frac{1}{5} + (3-1)^2 \cdot \frac{1}{5} = 2$$

which is four times as great as the variance of the random variable shown in Fig. 2.11*a*.

If the units on the random variable are, say feet, then the units of the mean are the same, but the units on the variance would be feet squared. Another measure of dispersion, called the *standard deviation*, is defined as the positive square root of the variance and denoted by σ, where

$$\sigma = \sqrt{\sigma^2} \tag{2-10}$$

It is noted that the units of σ are the same as those of the random variable, and a small value for σ indicates little dispersion while a large value indicates greater dispersion.

An alternate form of Equation (2-9) is obtained by algebraic manipulation as

$$\sigma^2 = \sum_i x_i^2 p(x_i) - \mu^2 \qquad \text{for discrete } X$$

$$= \int_{-\infty}^{\infty} x^2 f(x)\, dx - \mu^2 \qquad \text{for continuous } X \tag{2-11}$$

This simply indicates that the second moment about the mean is equal to the second moment about the origin less the square of the mean. The reader familiar with engineering mechanics will recognize that the development leading to Equation (2-11) is of the same nature as that leading to the *theorem of moments* in mechanics.

● **Example 2-14.**

(*a*) *Coin tossing—Example 2-6.*
Recall that $\mu = \frac{3}{2}$ from Example 2-12, and

$$\sigma^2 = \left(0-\frac{3}{2}\right)^2 \cdot \frac{1}{8} + \left(1-\frac{3}{2}\right)^2 \cdot \frac{3}{8} + \left(2-\frac{3}{2}\right)^2 \cdot \frac{3}{8} + \left(3-\frac{3}{2}\right)^2 \cdot \frac{1}{8} = \frac{3}{4}$$

Using the alternate form

$$\sigma^2 = \left[0^2 \cdot \frac{1}{8} + 1^2 \cdot \frac{3}{8} + 2^2 \cdot \frac{3}{8} + 3^2 \cdot \frac{1}{8}\right] - \left(\frac{3}{2}\right)^2 = \frac{3}{4}$$

which is only slightly easier.

(b) *Binomial distribution—Example 2-7.*
Recall that $\mu = np$ from Equation (2-8), and

$$\sigma^2 = \sum_{x=0}^{n} (x - np)^2 \binom{n}{x} p^x (1-p)^{n-x}$$

or

$$\sigma^2 = \left[\sum_{x=0}^{n} x^2 \cdot \binom{n}{x} p^x (1-p)^{n-x}\right] - (np)^2$$

which simplifies to

$$\sigma^2 = np(1-p) \qquad\qquad (2\text{-}12)$$

(c) Consider the density function $f(x)$, where

$$f(x) = 2e^{-2x} \qquad x \geq 0$$
$$= 0 \qquad\qquad \text{otherwise}$$

Then

$$\mu = \int_{0}^{\infty} x \cdot 2e^{-2x} \, dx = \frac{1}{2}$$

and

$$\sigma^2 = \int_{0}^{\infty} x^2 \cdot 2e^{-2x} \, dx - \left(\frac{1}{2}\right)^2 = \frac{1}{2} - \frac{1}{4} = \frac{1}{4}$$

(d) Another density is $g(x)$, where

$$g(x) = 8xe^{-4x} \qquad x \geq 0$$
$$= 0 \qquad\qquad \text{otherwise}$$

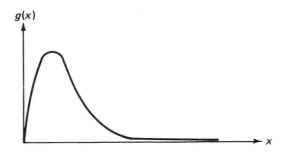

Then

$$\mu = \int_0^\infty x \cdot 8xe^{-4x}\,dx = \frac{1}{2}$$

and

$$\sigma^2 = \int_0^\infty x^2 \cdot 8xe^{-4x}\,dx - \left(\frac{1}{2}\right)^2 = \frac{1}{8}$$

Note that the mean is the same for the densities in parts (c) and (d), with (c) having a variance of twice that of (d).

In the development of the mean and variance, we used the terminology "mean of the random variable" and "variance of the random variable." Some authors use the terminology "mean of the distribution" and "variance of the distribution". Either terminology is acceptable.

In addition to the mean and variance, other moments are also frequently used to describe distributions. That is, the moments of a distribution describe that distribution, measure its properties, and, in certain circumstances, specify it. Moments about the origin are called *origin moments* and are denoted μ_k' for the kth origin moment, where

$$\mu_k' = \sum_i x_i^k\, p(x_i) \qquad \text{for discrete } X$$

$$= \int_{-\infty}^\infty x^k \qquad f(x)\,dx \qquad \text{for continuous } X$$

$$k = 0, 1, 2, \ldots \tag{2-13}$$

Moments about the mean are called *central moments* and are denoted by μ_k where

$$\mu_k = \sum_i (x_i - \mu)^k p(x_i) \qquad \text{for discrete } X$$

$$= \int_{-\infty}^\infty (x - \mu)^k f(x)\,dx \qquad \text{for continuous } X$$

$$k = 0, 1, 2, \ldots \tag{2-14}$$

Note that the mean $\mu = \mu_1'$ and the variance is $\sigma^2 = \mu_2$. Central moments may be expressed in terms of origin moments by the relationship

$$\mu_k = \sum_{j=0}^{k} (-1)^j \binom{k}{j} \mu^j \mu_{k-j}' \qquad k = 0, 1, 2, \ldots \qquad (2\text{-}15)$$

2-5   The Distribution Function

The *distribution function*, or *cumulative distribution function* as it is sometimes called, is an important function in the work of the following chapters.

Definition

The *distribution function* of the random variable X is denoted as F_X and defined as $F_X(x) = P(X \leq x)$ for all real x.

1. If X is discrete, then

$$F_X(x) = \sum_{\substack{i \text{ such that} \\ x_i \leq x}} p(x_i) \qquad (2\text{-}16a)$$

.2. If X is continuous, then

$$F_X(x) = \int_{-\infty}^{x} f(t)\, dt \qquad (2\text{-}16b)$$

The subscript, X, is usually deleted unless needed for clarification.

● **Example 2-15.** In the case of the coin-tossing experiment, the random variable X assumed four values $(0, 1, 2, 3)$ with probabilities $(\frac{1}{8}, \frac{3}{8}, \frac{3}{8}, \frac{1}{8})$. We can state $F(x)$ as follows:

$$
\begin{aligned}
F(x) &= 0 & x &< 0 \\
&= \tfrac{1}{8} & 0 &\leq x < 1 \\
&= \tfrac{4}{8} & 1 &\leq x < 2 \\
&= \tfrac{7}{8} & 2 &\leq x < 3 \\
&= 1 & x &\geq 3
\end{aligned}
$$

A graphical representation is as shown in Fig. 2-12.

● **Example 2-16.** The cathode ray tube described in Example 2-10 had a life with the probability density function given by

$$
\begin{aligned}
f(t) &= \lambda e^{-\lambda t} & t &\geq 0 \\
&= 0 & &\text{otherwise}
\end{aligned}
$$

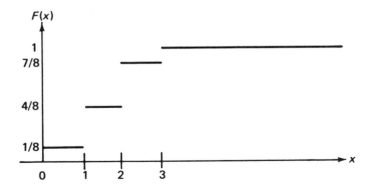

Fig. 2-12. A distribution function.

From this, we can immediately determine $F(t)$ as

$$F(t) = 0 \qquad\qquad\qquad t < 0$$
$$= \int_0^t \lambda e^{-\lambda t}\, ds = 1 - e^{-\lambda t} \qquad t \geq 0$$

This is shown graphically in Fig. 2-13.

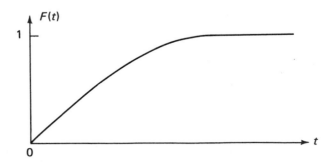

Fig. 2-13. Distribution function—exponentially distributed random variable.

The distribution function has the following properties:

1. $\lim\limits_{x \to \infty} F(x) = 1.$
2. $\lim\limits_{x \to -\infty} F(x) = 0.$

3. The function is nondecreasing; that is, if $b \geq a$, then $F(b) \geq F(a)$.
4. The function is continuous from the right; that is, for all x and $\delta > 0$,

$$\lim_{\delta \to 0} [F(x + \delta) - F(x)] = 0.$$

In the case where the random variable, X, is discrete,

$$p(x_i) = F(x_i) - F(x_{i-1}) \qquad (2\text{-}17)$$

and this relationship leads to a useful result:

$$P(a < X \leq b) = F(b) - F(a) \qquad (2\text{-}18)$$

For a continuous random variable, X, Equation (2-18) holds; however,

$$P(a < X \leq b) = P(a \leq X \leq b) = P(a \leq X < b) = P(a < X < b)$$

It follows from the fundamental theorem of calculus that for all values of x for which F is differentiable,

$$f(x) = \frac{d}{dx} F(x) \qquad (2\text{-}19)$$

● **Example 2-17.** In Example 2-14 we showed that

$$F(x) = 0 \qquad\qquad x < 0$$
$$= 1 - e^{-\lambda x} \qquad x \geq 0$$

For $x \geq 0$, $F'(x) = \lambda e^{-\lambda x}$, which has the density $f(x)$ from which we derived $F(x)$.

2-6   Mixed Distributions

Considerations have thus far been restricted to random variables that are either discrete or continuous, and for most applications this is adequate; however, an occasional situation will arise where the random variable X may assume distinct values $x_1, x_2, \ldots, x_k$ with positive probability and may also assume all of the values in some interval or collection of intervals. For example, consider the distribution of waiting time for a customer arriving at a teller's position in a bank. If there is no one in line, the waiting time is zero, and there is a positive probability of this being the case; but if there is/are one or more customers already in line and/or being waited on, the waiting time will assume some value on an interval. Figure 2-14 shows how the distribution function might appear in the case of waiting time.

A generalization might appear as shown in Fig. 2-15. The properties of distribution functions as stated earlier must be satisfied in this more general case.

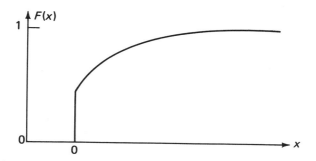

Fig. 2-14. Waiting time distribution function.

Fig. 2-15. A general distribution function.

2-7   Chebyshev's Inequality

In earlier sections of this chapter, it was pointed out that a small variance, σ^2, indicates that large deviations from the mean, μ, are improbable. *Chebyshev's inequality* gives us a means of understanding how the variance measures variability about μ.

Theorem 2-1

Let X be a random variable (discrete or continuous), and let k be some positive number. Then

$$P\{|X - \mu| \geq k\sigma\} \leq \frac{1}{k^2} \tag{2-20}$$

Proof
For continuous X, and a constant $K > 0$, consider

$$\sigma^2 = \int_{-\infty}^{\infty} (x - \mu)^2 \cdot f(x)\, dx = \int_{-\infty}^{\mu - \sqrt{K}} (x - \mu)^2 \cdot f(x)\, dx$$

$$+ \int_{\mu - \sqrt{K}}^{\mu + \sqrt{K}} (x - \mu)^2 \cdot f(x)\, dx + \int_{\mu + \sqrt{K}}^{\infty} (x - \mu)^2 \cdot f(x)\, dx$$

Since

$$\int_{\mu-\sqrt{K}}^{\mu+\sqrt{K}} (x - \mu)^2 \cdot f(x) \, dx \geq 0$$

it follows that

$$\sigma^2 \geq \int_{-\infty}^{\mu-\sqrt{K}} (x - \mu)^2 \cdot f(x) \, dx + \int_{\mu+\sqrt{K}}^{\infty} (x - \mu)^2 \cdot f(x) \, dx$$

Now, $(x - \mu)^2 \geq K$ if and only if $|x - \mu| \geq \sqrt{K}$; therefore,

$$\sigma^2 \geq \int_{-\infty}^{\mu-\sqrt{K}} Kf(x) \, dx + \int_{\mu+\sqrt{K}}^{\infty} Kf(x) \, dx$$

$$\sigma^2 \geq K[P(X \leq \mu - \sqrt{K}) + P(X \geq \mu + \sqrt{K})]$$

and

$$P\{|X - \mu| \geq \sqrt{K}\} \leq \frac{\sigma^2}{K}$$

so that if $k = \sqrt{K}/\sigma$, then

$$P\{|X - \mu| \geq k\sigma\} \leq \frac{1}{k^2}$$

The proof for discrete X is quite similar.
    An alternate form of this inequality

$$P\{|X - \mu| < k\sigma\} \geq 1 - \frac{1}{k^2} \tag{2-21}$$

or

$$P\{\mu - k\sigma < X < \mu + k\sigma\} \geq 1 - \frac{1}{k^2}$$

is often useful.
    The usefulness of Chebyshev's inequality stems from the fact that so little knowledge about the distribution of X is required. Only μ and σ^2 must be known. However, Chebychev's inequality is a weak statement, and this detracts from its usefulness. If the precise form of $f(x)$ or $p(x)$ is known, then a more powerful statement can be made.

● **Example 2-18.** From an analysis of records, a materials control manager estimates that the mean and standard deviation of the "lead time" required in ordering a small valve are 8 days and 1.5 days, respectively. He does not know the distribution of lead time, but he is willing to assume the estimates of the mean and standard deviations to be absolutely correct. The manager would like to determine a time interval such that the probability is at least $\frac{8}{9}$ that the order will be received during that time. That is,

$$1 - \frac{1}{k^2} = \frac{8}{9}$$

so that $k = 3$ and $\mu \pm k\sigma$ gives $8 \pm 3(1.5)$ or [3.5 days to 12.5 days]. It is noted that this interval may very well be too large to be of any value to the manager, in which case he may elect to learn more about the distribution of lead times.

2-8 Summary

This chapter has introduced the idea of random variables. In most engineering and management applications, these are either discrete or continuous; however, Section 2-6 illustrates a more general case. A vast majority of the discrete variables to be considered in this book result from counting processes, whereas the continuous variables are employed to model a variety of measurements. The mean and variance as measures of central tendency and dispersion, and as characterizations of random variables, were presented along with more general moments of higher order. The Chebyshev inequality is presented as a bounding probability that a random variable lies between $\mu - k\sigma$ and $\mu + k\sigma$.

2-9 Exercises

2-1. A poker hand may contain from zero to four aces. If X is the random variable denoting the number of aces, enumerate the range space of X. What are the probabilities associated with each possible value of X?

2-2. A car rental agency has either 0, 1, 2, 3, 4, or 5 cars returned each day, with probabilities $\frac{1}{6}, \frac{1}{6}, \frac{1}{3}, \frac{1}{12}, \frac{1}{6}$, and $\frac{1}{12}$, respectively. Find the mean and the variance of the number of cars returned.

2-3. Let the random variable X denote the sum of the "up" faces of two fair dice. Find the mean and the variance of the probability distribution of X.

2-4. Two regular tetrahedra have their faces numbered 1, 2, 3, and 4, respectively. The two are tossed. Let X be the random variable denoting the sum of their "up" faces. What is the range space of X? What is the mean and variance of the probability distribution of X?

2-5. A random variable X has the probability density function ce^{-x}. Find the proper value of c, assuming $0 \le X < \infty$. Find the mean and the variance of the probability density function of X.

2-6. The probability distribution function that a television tube will fail in t hours is $1 - e^{-ct}$, where c is a parameter dependent on the manufacturer and $t \ge 0$. Find the probability density function of X, the life of the tube.

2-7.   If, in the previous problem, we know that one-half of the tubes failed in 1500 hours, what is the probability that a tube will last as long as 3000 hours?.

2-8.   Show that central moments can be expressed in terms of origin moments by Equation (2-15). *Hint*: See Chapter 3 of Kendall and Stuart (1963).

2-9.   The demand for a product is $-1, 0, +1, +2$ per day with probability $\frac{1}{5}, \frac{1}{10}, \frac{2}{5}, \frac{3}{10}$, respectively. A demand of -1 implies a unit is returned. Find the expected demand and the variance. Sketch the distribution function.

2-10.   The manager of a men's clothing store is concerned over the inventory of suits, which is currently 30 (all sizes). The number of suits sold from now to the end of the season is distributed as

$$f(x) = \frac{e^{-20}20^x}{x!} \qquad x = 0, 1, 2, \ldots$$
$$= 0 \qquad \text{otherwise}$$

Find the probability that he will have suits left over at the season's end.

2-11.   A random variable Y takes on the value $1, 2, 3,$ or 4 with probabilities $(1 + 3k)/4$, $(1 - 2k)/4$, $(1 + 5k)/4$, and $(1 - 6k)/4$, respectively.
(*a*) For what values of k is this a legitimate probability function?
(*b*) Find the cumulative distribution function.
(*c*) Find the mean and variance of Y.

2-12.   Consider the following probability density function:

$$f(x) = kx \qquad\qquad 0 \leq x < 2$$
$$= k(4 - x) \qquad 2 \leq x \leq 4$$
$$= 0 \qquad\qquad \text{otherwise}$$

(*a*) Find the value of k for which f is a probability density function.
(*b*) Find the mean and variance of X.
(*c*) Find the cumulative distribution function.

2-13.   Rework the above problem, except let the probability density function be defined as

$$f(x) = kx \qquad\qquad 0 \leq x < a$$
$$f(x) = k(2a - x) \qquad a \leq x \leq 2a$$
$$= 0 \qquad\qquad \text{otherwise}$$

2-14.   The manager of a job shop does not know the probability distribution of the time required to complete an order. However, from past performance she has been able to estimate the mean and variance as 14 days and 2 (days)^2, respectively. Find an interval such that the probability is .75 that an order is finished during that time.

2-15.   The continuous random variable T has the probability density function $f(t) = kt^2$ for $-1 \leq t \leq 0$. Find the following:
(*a*) The appropriate value of k.
(*b*) The mean and variance of T.
(*c*) The cumulative distribution function.

2-16. The discrete random variable N ($N = 0, 1, \ldots$) has probabilities of occurrence of kr^n ($0 < r < 1$). Find the appropriate value of k.

2-17. The percentage of antiknock additive ($100A$) in a particular gasoline is

$$f(a) = ka^3(1 - a) \quad 0 < a < 1$$
$$= 0 \qquad \text{otherwise}$$

(a) Find the appropriate value of k.
(b) Obtain an expression for the cumulative distribution function.
(c) What is the mean occurrence of additive?

2-18. The postal service requires, on the average, 2 days to deliver a letter across town. The variance is estimated to be .4 (day)^2. If a business executive wants 99 percent of his letters delivered on time, how early should he mail them?

2-19. For the exponential density function of Example 2-10 find:
(a) $P\{t \leq 5\}$
(b) $P\{3 \leq t \leq 8\}$
(c) $P\{1 < t < 3\}$

2-20. Given the cumulative distribution functions below, use Equation (2-18) to develop the probability density functions.
(a) $F(x) = 0 \qquad x < 0$
$\qquad = x^3 \qquad 0 \leq x \leq 1$
$\qquad = 1 \qquad x > 1$

(b) $F(x) = 0 \qquad x < 0$
$\qquad = 1 - \exp\left(\dfrac{-x^2}{2t^2}\right) \qquad x \geq 0 \qquad t > 0$

2-21. Demand for a particular item is characterized by the following discrete distribution:

$$p(d) = k\frac{d^2}{16} \quad d = 1, 2, 3, 4, 5$$
$$= 0 \qquad \text{otherwise}$$

(a) Find the appropriate value of k.
(b) Evaluate the mean and variance of demand.

2-22. Two different real estate developers, A and B, own parcels of land being offered for sale. The probability distributions of selling prices per parcel are shown in the following table.

	Price					
	$1000	$1050	$1100	$1150	$1200	$1350
A	.2	.3	.1	.3	.05	.05
B	.1	.1	.3	.3	.1	.1

Assuming that A and B are operating independently, compute:
(a) The expected selling price of A and of B.
(b) The expected selling price of A given that the B selling price is $1150.
(c) The probability that A and B both have the same selling price.

2-23. Show that the probability function for the sum of values obtained in tossing two dice may be written as

$$p(x_i) = \frac{x_i - 1}{36} \qquad x_i = 2, 3, \ldots, 6$$

$$= \frac{13 - x_i}{36} \qquad x_i = 7, 8, \ldots, 12$$

2-24. Find the mean and variance of the random variable whose probability function is defined in the previous problem.

2-25. Suppose X takes on the values 5 and -5 with probabilities $\frac{1}{2}$. Plot the quantity $P[|X - \mu| \geq k\sqrt{\sigma^2}]$ as a function of k (for $k > 0$). On the same set of axes, plot the same probability determined by Chebyshev's inequality.

2-26. Find the cumulative distribution function associated with

$$f(x) = \frac{x}{t^2} \exp\left(-\frac{x^2}{2t^2}\right) \qquad t > 0, x \geq 0$$

$$= 0 \qquad\qquad \text{otherwise}$$

2-27. Find the cumulative distribution function associated with

$$f(x) = \frac{1}{\sigma\pi} \frac{1}{\{1 + [(x - \mu)^2/\sigma^2]\}} \qquad -\infty < x < \infty, -\infty < \mu < \infty, \sigma > 0$$

2-28. Find the mean and variance of $f(x) = 6x(1 - x), 0 \leq x \leq 1$. Find the cumulative distribution function.

2-29. Consider the probability density function $f(y) = k \sin y, 0 \leq y \leq \pi/2$. What is the appropriate value of k? Find the mean of the distribution.

2-30. Consider the probability density function

$$f(x) = 2(1 - x) \qquad 0 \leq x \leq a$$

$$= 0 \qquad\qquad \text{otherwise}$$

Determine the appropriate value for a.

Chapter 3

Functions of One Random Variable and Expectation

3-1   Introduction

Engineers and management scientists are frequently interested in the behavior of some function, say H, of a random variable X. For example, suppose the circular cross-sectional area of a copper wire is of interest. The relationship, $Y = \pi X^2/4$, where X is the diameter, gives the cross-sectional area. Since X is a random variable, Y also is a random variable, and we would expect to be able to determine the probability distribution of $Y = H(X)$ if the distribution of X is known. The first portion of this chapter will be concerned with problems of this type. This is followed by the concept of expectation, a notion employed extensively throughout the remaining chapters of this book. Approximations are developed for the mean and variance of functions of random variables, and the moment-generating function, a mathematical device for producing moments and describing distributions, is presented with some example illustrations.

3-2   Equivalent Events

Before presenting some specific methods used in determining the probability distribution of a function of a random variable, the concepts involved should be more precisely formulated.

Consider an experiment $\mathscr{E}$ with sample space $\mathscr{S}$. The random variable X is defined on $\mathscr{S}$, assigning values to the outcomes e in $\mathscr{S}$, $X(e) = x$, where the values x are in the range space R_X of X. Now if $Y = H(X)$ is defined so that the values $y = H(x)$ in R_Y, the range space of Y, are real, then Y is a random variable, since for every outcome $e \in \mathscr{S}$, a value y of the random variable Y is determined; that is, $y = H[X(e)]$. This notion is illustrated in Fig. 3-1.

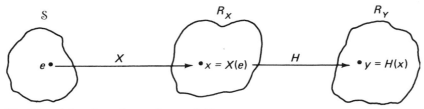

Fig. 3-1.   A function of a random variable.

If C is an event associated with R_Y, the range space of Y, and B is an event in R_X, then B and C are *equivalent events* if they occur together; that is, if $B = \{x \in R_X : H(x) \in C\}$. In addition, if A is an event associated with $\mathscr{S}$ and, furthermore, A and B are equivalent, then A and C are equivalent events.

Definition

If X is a random variable (defined on $\mathscr{S}$) having range space R_X, and if H is a real-valued function, so that $Y = H(X)$ is a random variable with range space R_Y, then for any event $C \subset R_Y$, we define

$$P(C) = P[\{x \in R_X : H(x) \in C\}] \tag{3-1}$$

It is noted that these probabilities relate to probabilities in the sample space. We could write

$$P(C) = P[\{e \in \mathscr{S} : H[X(e)] \in C\}]$$

however, Equation (3-1) indicates the *method* to be used in problem solution. We find the event B in R_X that is equivalent to event C in R_Y; then we find the probability of event B.

● **Example 3-1.** In the case of the cross-sectional area Y of a wire, suppose we know that the diameter has density function:

$$f(x) = 200 \qquad 1.000 \le x \le 1.005$$
$$= 0 \qquad \text{otherwise}$$

In addition, suppose we want to find $P[Y \le (1.01)\pi/4]$. The equivalent event is determined, $P[Y \le (1.01)\pi/4] = P[(\pi/4)X^2 \le (1.01)\pi/4] = P(|X| \le \sqrt{1.01})$. The event $\{x \in R_X : |x| \le \sqrt{1.01}\}$ is in the range space R_X, and since $f(x) = 0$, for all $x < 1.0$, we calculate

$$P(|X| \le \sqrt{1.01}) = P(1.0 \le X \le \sqrt{1.01}) = \int_{1.000}^{\sqrt{1.01}} 200 \, dx$$

$$= 200(\sqrt{1.01} - 1)$$

$$= .9975$$

● **Example 3-2.** In the case of the Geiger counter experiment of Example 2-9, we used the distribution given in Equation (2-3)

$$p(x) = e^{-\lambda t}(\lambda t)^x/x! \qquad x = 0, 1, 2, \ldots$$
$$= 0 \qquad \text{otherwise}$$

Recall that λ, where $\lambda > 0$, represents the mean "hit" rate and t is the time interval for which the counter is operated. Now suppose we wish to find

$$P(Y \le 5)$$

where

$$Y = 2X + 2$$

Proceeding as in the previous example,

$$P(Y \le 5) = P(2X + 2 \le 5) = P\left(X \le \frac{3}{2}\right)$$
$$= [p(0) + p(1)] = [e^{-\lambda t}(\lambda t)^0/0!] + [e^{-\lambda t}(\lambda t)^1/1!]$$
$$= e^{-\lambda t}[1 + (\lambda t)]$$

The event $\{x \in R_X : x \le \frac{3}{2}\}$ is in the range space of X, and we have the function p to work with in that space.

3-3 Functions of a Discrete Random Variable

Suppose that both X and Y are discrete random variables, and let $x_{i_1}, x_{i_2}, \ldots, x_{i_k}, \ldots$, represent the values of X such that $H(x_{i_j}) = y_i$ for some set of index values, $\Omega = \{j : j = 1, 2, \ldots, s_i\}$.

The probability distribution for Y is denoted by $p_Y(y_i)$ and is given by

$$p_Y(y_i) = P(Y = y_i) = \sum_{j \in \Omega} p_X(x_{i_j}) \tag{3-2}$$

For example, in Fig. 3-2 where $s_i = 4$, the probability of y_i is $p_Y(y_i) = p_X(x_{i_1}) + p_X(x_{i_2}) + p_X(x_{i_3}) + p_X(x_{i_4})$.

In the special case where H is such that for each y there is exactly one x, then $p_Y(y_i) = p_X(x_i)$, where $y_i = H(x_i)$. To illustrate these concepts, consider the following examples.

● **Example 3-3.** In the coin-tossing experiment where X represented the number of heads, recall that X assumed four values, 0, 1, 2, 3, with probabilities $\frac{1}{8}, \frac{3}{8}, \frac{3}{8}, \frac{1}{8}$. If $Y = 2X - 1$, then the possible values of Y are $-1, 1, 3, 5$, and $p_Y(-1) = \frac{1}{8}$, $p_Y(1) = \frac{3}{8}$, $p_Y(3) = \frac{3}{8}$, $p_Y(5) = \frac{1}{8}$. In this case, H is such that for each y there is exactly one x.

● **Example 3-4.** X is as in the previous example; however, suppose now that $Y = |X - 2|$, so that the possible values of Y are 0, 1, 2, as indicated in Fig.

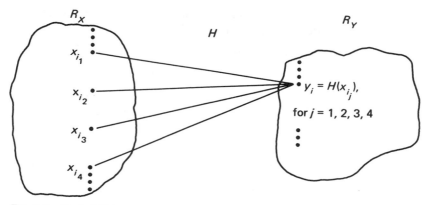

Fig. 3-2. Probabilities in R_Y

3-3. In this case,

$$p_Y(0) = p_X(2) = \frac{3}{8}$$

$$p_Y(1) = p_X(1) + p_X(3) = \frac{4}{8}$$

$$p_Y(2) = p_X(0) = \frac{1}{8}$$

In the event that X is continuous but Y is descrete, the formulation for $p_Y(y_i)$ is as follows:

$$p_Y(y_i) = \int_B f(x)\, dx \qquad (3\text{-}3)$$

where the event B is the event in R_X that is equivalent to the event $(Y = y_i)$ in R_Y.

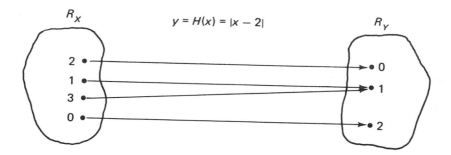

Fig. 3-3. An example function, H.

● **Example 3-5.** Suppose X has a density function given by

$$f(x) = \lambda e^{-\lambda x} \qquad x \geq 0$$
$$= 0 \qquad \text{otherwise}$$

Furthermore, if

$$Y = 0 \qquad \text{for } X \leq 1/\lambda$$
$$= 1 \qquad \text{for } X > 1/\lambda$$

then

$$p_Y(0) = \int_0^{1/\lambda} \lambda e^{-\lambda x}\, dx = -e^{-\lambda x}\Big|_0^{1/\lambda} = 1 - e^{-1} \simeq .6321$$

and

$$p_Y(1) = \int_{1/\lambda}^{\infty} \lambda e^{-\lambda x}\, dx = -e^{-\lambda x}\Big|_{1/\lambda}^{\infty} = e^{-1} \simeq .3679$$

3-4 Continuous Functions of a Continuous Random Variable

If X is a continuous random variable with density function f, and H is also continuous, then $Y = H(X)$ is a continuous random variable. The probability density function for the random variable Y will be denoted by g, and it may be found by performing these three steps.

1. Obtain $G(y) = P(Y \leq y)$ by finding event B in R_X, which is equivalent to the event $(Y \leq y)$ in R_Y.
2. Differentiate $G(y)$ with respect to y to obtain $g(y)$.
3. Find the range space of the new random variable.

These steps are illustrated in the following example.

● **Example 3-6.** Suppose that the random variable X has the following density function:

$$f(x) = x/8 \qquad 0 \leq x \leq 4$$
$$= 0 \qquad \text{otherwise}$$

If $Y = H(X)$ is the random variable for which the density g is desired, and $H(x) = 2x + 8$, as shown in Fig. 3-4, then we proceed according to the steps given above.

(a) $G(y) = P(Y \leq y) = P(2X + 8 \leq y)$
$$= P[X \leq (y-8)/2]$$
$$= \int_0^{(y-8)/2} (x/8)\, dx = \frac{x^2}{16}\Big|_0^{(y-8)/2} = \frac{1}{64}(y^2 - 16y + 64)$$

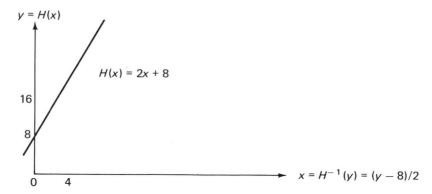

Fig. 3-4.   The function $H(x) = 2x + 8$.

(b) $g(y) = G'(y) = \dfrac{1}{32}(y) - \dfrac{1}{4}$

(c) If $x = 0$, $y = 8$, and if $x = 4$, $y = 16$, so that we have

$$g(y) = (y/32) - \tfrac{1}{4} \qquad 8 \le y \le 16$$
$$= 0 \qquad\qquad\quad \text{otherwise}$$

● **Example 3-7.** Consider the random variable X defined in Example 3-6, and suppose $Y = H(X) = (X - 2)^2$, as shown in Fig. 3-5. Proceeding as in Example 3-6, we find:

(a) $G(y) = P(Y \le y) = P[(X - 2)^2 \le y] = P[-\sqrt{y} \le (X - 2) \le +\sqrt{y}]$

$\qquad = P(2 - \sqrt{y} \le X \le 2 + \sqrt{y})$

$\qquad = \displaystyle\int_{2-\sqrt{y}}^{2+\sqrt{y}} \dfrac{x}{8}\, dx = \dfrac{x^2}{16}\Big|_{2-\sqrt{y}}^{2+\sqrt{y}}$

$\qquad = \dfrac{1}{16}[(4 + 4\sqrt{y} + y) - (4 - 4\sqrt{y} + y)]$

$\qquad = \dfrac{1}{2}\sqrt{y}$

(b) $g(y) = G'(y) = \dfrac{1}{4\sqrt{y}}$

(c) If $x = 2$, $y = 0$, and if $x = 0$ or $x = 4$, $y = 4$; however, g is not defined for $y = 0$. Therefore,

$$g(y) = \dfrac{1}{4\sqrt{y}} \qquad 0 < y \le 4$$
$$= 0 \qquad\qquad \text{otherwise}$$

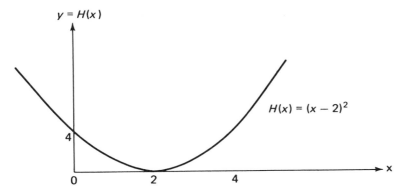

Fig. 3-5.   The function $H(x) = (x-2)^2$.

In Example 3-6, the event in R_X equivalent to $(Y \le y)$ in R_Y was $[X \le (y-8)/2]$; and in Example 3-7, the event in R_X equivalent to $(Y \le y)$ in R_Y was $(2 - \sqrt{y} \le X \le 2 + \sqrt{y})$. In the first example, the function H is a strictly increasing function of x, while in the second example this is not the case.

Theorem 3-1

If X is a continuous random variable with probability density function f that satisfies $f(x) > 0$ for $a < x < b$, and $y = H(x)$ is a continuous strictly increasing or strictly decreasing function of x, then the random variable $Y = H(X)$ has density function

$$g(y) = f(x) \cdot \left| \frac{dx}{dy} \right| \qquad (3\text{-}4)$$

with $x = H^{-1}(y)$ expressed in terms of y. If H is increasing, $g(y) > 0$ if $H(a) < y < H(b)$; and if H is decreasing, $g(y) > 0$ if $H(b) < y < H(a)$.

Proof
(Given only for H increasing. A similar argument holds for H decreasing.)

$$G(y) = P(Y \le y) = P[H(X) \le y]$$
$$= P[X \le H^{-1}(y)]\cdot$$
$$= F[H^{-1}(y)]$$

$$g(y) = G'(y) = \frac{dF(x)}{dx} \cdot \frac{dx}{dy} \qquad \text{by the chain rule}$$

$$= f(x) \cdot \frac{dx}{dy} \qquad \text{where } x = H^{-1}(y)$$

● **Example 3-8.** In Example 3-6, we had

$$f(x) = x/8 \qquad 0 \le x \le 4$$
$$= 0 \qquad \text{otherwise}$$

and $H(x) = 2x + 8$, which is a strictly increasing function. Using Equation (3-4),

$$g(y) = f(x) \cdot \left| \frac{dx}{dy} \right| = \frac{y-8}{16} \cdot \frac{1}{2}$$

since $x = (y-8)/2$. $H(0) = 8$ and $H(4) = 16$; therefore,

$$g(y) = (y/32) - \frac{1}{4} \qquad 8 \le y \le 16$$
$$= 0 \qquad \text{otherwise}$$

3-5   Expectation

If X is a random variable, and $Y = H(X)$ is a function of X, then the *expected value* of $H(X)$ is defined as follows:

$$E[H(X)] = \sum_{\text{all } i} H(x_i) \cdot p(x_i) \qquad \text{for } X \text{ discrete} \tag{3-5}$$

$$E[H(X)] = \int_{-\infty}^{\infty} H(x) \cdot f(x)\, dx \qquad \text{for } X \text{ continuous} \tag{3-6}$$

In the case where X is continuous, we restrict H so that $Y = H(X)$ is a continuous random variable.

The mean and variance, presented earlier, are special applications of Equations (3-5) and (3-6). If $H(X) = X$, we see that

$$E[H(X)] = E(X) = \mu \tag{3-7}$$

Therefore, the expected value of the random variable X is just the mean, μ.

If $H(X) = (X - \mu)^2$, then

$$E[H(X)] = E[(X - \mu)^2] = \sigma^2 \tag{3-8}$$

Thus, the variance of the random variable X may be defined in terms of expectation. Since the variance is utilized extensively, it is customary to introduce a variance *operator* V that is defined in terms of the expected value operator E:

$$V[H(X)] = E\{[H(X) - E(H(X))]^2\} \tag{3-9}$$

Again, in the case where $H(X) = X$,

$$V(X) = E[(X - E(X))^2]$$
$$= E(X^2) - [E(X)]^2 \qquad (3\text{-}10)$$

which is the *variance of X*, denoted by σ^2.

The origin moments and central moments discussed in the previous chapter may also be expressed using the expected value operator as

$$\mu'_k = E(X^k) \qquad (3\text{-}11)$$

and

$$\mu_k = E[(X - E(X))^k]$$

There are two special functions, H, that should be considered at this point. First, suppose $H(X) = aX$, where a is a constant. Then for discrete X,

$$E(aX) = \sum_{\text{all } i} ax_i \cdot p(x_i) = a \sum_{\text{all } i} x_i p(x_i) = aE(X) \qquad (3\text{-}12)$$

and the same result is obtained for continuous X, namely,

$$E(aX) = \int_{-\infty}^{\infty} ax \cdot f(x)\, dx = a \int_{-\infty}^{\infty} xf(x)\, dx = aE(X) \qquad (3\text{-}13)$$

Using Equation (3-9),

$$V(aX) = E\{[aX - aE(X)]^2\} = a^2 V(X) \qquad (3\text{-}14)$$

In the case where $H(X) = a$, a constant, the reader may readily verify that

$$E(a) = a \qquad (3\text{-}15)$$

and

$$V(a) = 0 \qquad (3\text{-}16)$$

● **Example 3-9.** Suppose a contractor is about to bid on a job requiring X days to complete, where X is a random variable denoting the number of days for job completion. Her profit, P, depends on X; that is, $P = H(X)$. The probability distribution of X, $(x, p(x))$, is as follows:

x	$p(x)$
3	$\frac{1}{8}$
4	$\frac{5}{8}$
5	$\frac{2}{8}$

Using the notion of expected value, we calculate the mean and variance of X as follows:

$$E(X) = 3 \cdot \frac{1}{8} + 4 \cdot \frac{5}{8} + 5 \cdot \frac{2}{8} = \frac{33}{8}$$

and

$$V(X) = \left[3^2 \cdot \frac{1}{8} + 4^2 \cdot \frac{5}{8} + 5^2 \cdot \frac{2}{8}\right] - \left(\frac{33}{8}\right)^2 = \frac{23}{64}$$

If the function $H(X)$ is given as:

x	$H(x)$
3	$10,000
4	2,500
5	−7,000

then the expected value of $H(X)$ is

$$E[H(X)] = 10,000\left(\frac{1}{8}\right) + 2500\left(\frac{5}{8}\right) - 7000\left(\frac{2}{8}\right) = \$1062.50$$

and the contractor would view this as the average profit that she would obtain if she bid this job many, many times (actually an infinite number of times), where H remained the same and the random variable X behaved according to the probability function p. The variance of $P = H(X)$ can readily be calculated as

$$V[H(X)] = \left[(10,000)^2 \cdot \frac{1}{8} + (2500)^2 \cdot \frac{5}{8} + (-7000)^2 \cdot \frac{2}{8}\right] - (1062.5)^2$$
$$\simeq \$27.53 \cdot 10^6$$

● **Example 3-10.** A well-known simple inventory problem is "the newsboy problem," described as follows. A newsboy buys papers for 6 cents each and sells them for 10 cents each, and he cannot return unsold papers. Daily demand has the following distribution and each day's demand is independent of the previous day's demand:

Number of customers, x	23	24	25	26	27	28	29	30
Probability, $p(x)$	.01	.04	.10	.10	.25	.25	.15	.10

If the newsboy stocks too many papers, he suffers a loss attributable to the excess supply. If he stocks too few papers, he loses profit because of the excess demand. It seems reasonable for the newsboy to stock some number of papers so as to minimize the *expected* loss. If we let s represent the number of papers stocked, X represent the daily demand, and $L(X, s)$ the newsboy's loss for a particular stock level s, then the loss is simply

$$L(X, s) = .04(X - s) \qquad \text{if } X > s$$
$$= .06(s - X) \qquad \text{if } X \le s$$

and for a given stock level s, the expected loss is:

$$E[L(X, s)] = \sum_{x=23}^{s} .06(s - x) \cdot p(x) + \sum_{x=s+1}^{30} (.04)(x - s) \cdot p(x)$$

and the $E[L(X, s)]$ is evaluated for different values of s.

For $s = 26$:
$$
\begin{aligned}
E[L(X, 26)] &= .06[(26 - 23)(.01) + (26 - 24)(.04) + (26 - 25)(.10) + (26 - 26)(.10)] \\
&\quad + .04[(27 - 26)(.25) + (28 - 26)(.25) + (29 - 26)(.15) \\
&\quad + (30 - 26)(.10)] \\
&= \$.0766
\end{aligned}
$$

For $s = 27$:
$$
\begin{aligned}
E[L(X, 27)] &= .06[(27 - 23)(.01) + (27 - 24)(.04) + (27 - 25)(.10) + (27 - 26)(.10) \\
&\quad + (27 - 27)(.25)] + .04[(28 - 27)(.15) + (29 - 27)(.15) \\
&\quad + (30 - 27)(.10)] \\
&= \$.0616
\end{aligned}
$$

For $s = 28$:
$$
\begin{aligned}
E[L(X, 28)] &= .06[(28 - 23)(.01) + (28 - 24)(.04) + (28 - 25)(.10) + (28 - 26)(.10) \\
&\quad + (28 - 27)(.25) + (28 - 28)(.25)] \\
&\quad + .04[(29 - 28)(.15) + (30 - 28)(.10)] \\
&= \$.0716
\end{aligned}
$$

Thus, the newsboy's policy should be to stock 27 papers if he desires to minimize his expected loss.

● **Example 3-11.** Consider the redundant system shown in the diagram below.

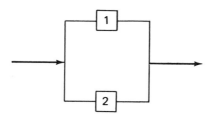

At least one of the units must function, the redundancy is *standby* (meaning that the second unit does not operate until the first fails), switching is perfect, and the system is nonmaintained. It can be shown that under certain conditions when the time to failure for the units of this system has an exponential distribution, then the time to failure for the system has the following probability density function:

$$f(x) = \lambda^2 x e^{-\lambda x} \qquad x > 0, \lambda > 0$$
$$= 0 \qquad \text{otherwise}$$

where λ is the "failure rate" parameter of the component exponential models. The mean time to failure (MTTF) for this system is:

$$E(X) = \int_0^\infty x \cdot \lambda^2 x e^{-\lambda x} \, dx = \frac{2}{\lambda}$$

The terms "mean time to failure" and "expected life" are synonymous.

3-6 Approximations to $E[H(X)]$ and $V[H(X)]$

In cases where $H(X)$ is very complicated, the evaluation of the expectation and variance may be difficult. Often, an approximation to $E[H(X)]$ and $V[H(X)]$ may be obtained by utilizing a Taylor series expansion. To estimate the mean, we expand the function H to two terms of a Taylor series, where the expansion is about $x = \mu$. If $Y = H(X)$, then

$$Y = H(\mu) + (X - \mu)H'(\mu) + \frac{(X-\mu)^2}{2} \cdot H''(\mu) + R$$

where R is the remainder. We use Equations (3-12) through (3-16) to perform the evaluations.

$$E(Y) = E[H(\mu)] + \{E[H'(\mu)X] - E[H'(\mu)\mu]\} + E\left[\frac{1}{2}H''(\mu)(X-\mu)^2\right] + E(R)$$

$$= H(\mu) + \{H'(\mu)\mu - H'(\mu)\mu\} + \frac{1}{2}H''(\mu)V(X) + E(R)$$

$$\approx H(\mu) + \frac{1}{2}H''(\mu)\sigma^2 \tag{3-17}$$

Using only the first two terms and grouping the third into the remainder so that

$$Y = H(\mu) + (X - \mu) \cdot H'(\mu) + R_1$$

where

$$R_1 = R + \frac{(X-\mu)^2}{2} \cdot H''(\mu)$$

then the variance of Y is determined as

$$V(Y) \approx V[H(\mu)] + V[(X - \mu) \cdot H'(\mu)] + V(R_1)$$
$$\approx 0 + V(X) \cdot [H'(\mu)]^2$$
$$\approx [H'(\mu)]^2 \cdot \sigma^2 \tag{3-18}$$

If the variance of X, σ^2, is large and the mean, μ, is small, there may be a rather large error in this approximation. An alternate procedure, known as the Monte Carlo method, may also be used to approximate the mean and variance.

● **Example 3-12.** The surface tension of a liquid is represented by T (dyne/cm), and under certain conditions, $T \approx 2(1 - .005X)^{1.2}$, where X is the liquid temperature in degrees centigrade. If X has probability density function f, where

$$f(x) = 3000x^{-4} \qquad x \geq 10$$
$$= 0 \qquad \text{otherwise}$$

then

$$E(T) = \int_{10}^{\infty} 2(1 - .005x)^{1.2} \cdot 3000x^{-4} \, dx$$

and

$$V(T) = \int_{10}^{\infty} 4(1 - .005x)^{2.4} \cdot 3000x^{-4} \, dx - [E(T)]^2$$

In order to determine these values, it is necessary to evaluate

$$\int_{10}^{\infty} \frac{(1 - .005x)^{1.2}}{x^4} \, dx \qquad \text{and} \qquad \int_{10}^{\infty} \frac{(1 - .005x)^{2.4}}{x^4} \, dx$$

Since the evaluation is difficult, we use the approximations given by Equations (3-17) and (3-18). Note that

$$\mu = E(X) = \int_{10}^{\infty} x \cdot 3000x^{-4} \, dx = -1500x^{-2} \Big|_{10}^{\infty} = 15 \,°\text{C}$$

and

$$\sigma^2 = V(X) = E(X^2) - [E(X)]^2 = \int_{10}^{\infty} x^2 \cdot 3000x^{-4} \, dx - 15^2 = 75(°\text{C})^2$$

Since

$$H(X) = 2(1 - .005X)^{1.2}$$

then

$$H'(X) = -.012(1 - .005X)^{0.2}$$

and

$$H''(X) = .000012(1 - .005X)^{-0.8}$$

so that

$$H(15) = 2[1 - .005(15)]^{1.2} = 1.82$$
$$H'(15) \approx -.012$$

and

$$H''(15) \approx 0$$

Using Equations (3-17) and (3-18),

$$E(T) \simeq H(15) + \frac{1}{2} H''(\mu) \cdot \sigma^2 = 1.82$$

and

$$V(T) \simeq [H'(15)]^2 \cdot \sigma^2 = [-.012]^2 \cdot 75 \simeq .081$$

3-7 The Moment-Generating Function

It is often convenient to utilize a special function in finding the moments of a probability distribution. This special function, called the *moment-generating* function, is defined as follows.

Definition

Given a random variable X, the moment-generating function $M_X(t)$ of its probability distribution is the expected value of e^{tX}. Expressed mathematically,

$$M_X(t) = E(e^{tX}) \tag{3-19}$$

$$= \sum_{\text{all } i} e^{tx_i} p(x_i) \qquad \text{discrete } X \tag{3-20}$$

$$= \int_{-\infty}^{\infty} e^{tx} f(x)\, dx \qquad \text{continuous } X \tag{3-21}$$

For certain probability distributions, the moment-generating function may not exist for all real values of t. However, for the probability distributions treated in this book, the moment-generating function always exists.

Expanding e^{tX} as a power series in t we obtain

$$e^{tX} = 1 + tX + \frac{t^2 X^2}{2!} + \cdots + \frac{t^r X^r}{r!} + \cdots$$

On taking expectation we see that

$$M_X(t) = E[e^{tX}] = 1 + E(X) \cdot t + E(X^2) \cdot \frac{t^2}{2} + \cdots + E(X^r) \cdot \frac{t^r}{r!} + \cdots$$

so that

$$M_X(t) = 1 + \mu_1' \cdot t + \mu_2' \cdot \frac{t^2}{2!} + \cdots + \mu_r' \cdot \frac{t^r}{r!} + \cdots \tag{3-22}$$

Thus, we see that *when $M_X(t)$ is written as a power series in t*, the coefficient of $t^r/r!$ in the expansion is the rth moment about the origin. One procedure, then, for using the moment-generating function would be:

1. Find $M_X(t)$ analytically for the particular distribution.

2. Expand $M_X(t)$ as a power series in t and obtain the coefficient of $t^r/r!$ as the rth origin moment.

The main difficulty in using this procedure is the expansion of $M_X(t)$ as a power series in t.

If we are only interested in the first few moments of the distribution, then the process of determining these moments is usually made easier by noting that the rth derivative of $M_X(t)$, with respect to t evaluated at $t = 0$, is just

$$\frac{d^r}{dt^r} M_X(t)\bigg|_{t=0} = E[X^r e^{tX}]_{t=0} = \mu_r' \tag{3-23}$$

assuming we can interchange the operations of differentiation and expectation. Thus, a second procedure for using the moment-generating function is to

1. Determine $M_X(t)$ analytically for the particular distribution.

2. Find $\mu_r' = \dfrac{d^r}{dt^r} M_X(t)\bigg|_{t=0}$.

Moment-generating functions have many interesting and useful properties which will be used in subsequent chapters; however, for the time being, we will use them only in connection with finding moments of a probability distribution. Perhaps the most important of these properties is that the moment-generating function is unique when it exists, so that if we know the moment-generating function, we immediately know the form of the distribution.

In cases where the moment-generating function does not exist, we may utilize the characteristic function, $C_X(t)$, which is defined to be the expectation of e^{itX}, where $i = \sqrt{-1}$. There are several advantages to using the characteristic function rather than the moment-generating function, but the principal one is that $C_X(t)$ always exists for all t. However, for simplicity, we will use only the moment-generating function.

● **Example 3-13.** Suppose that X has a *binomial distribution*, that is,

$$p(x) = \binom{n}{x} p^x (1-p)^{n-x} \qquad x = 0, 1, 2, \ldots, n$$

$$= 0 \qquad\qquad \text{otherwise}$$

where $0 < p < 1$ and n is a positive integer. The moment-generating function $M_X(t)$ is

$$M_X(t) = \sum_{x=0}^{n} e^{tx} \binom{n}{x} p^x (1-p)^{n-x}$$

$$= \sum_{x=0}^{n} \binom{n}{x} (pe^t)^x (1-p)^{n-x}$$

This last summation is recognized as the binomial expansion of $[pe^t + (1-p)]^n$, so that

$$M_X(t) = [pe^t + (1-p)]^n$$

Taking the derivatives, we obtain

$$M'_X(t) = npe^t[1 + p(e^t - 1)]^{n-1}$$

and

$$M''_X(t) = npe^t(1 - p + npe^t)[1 + p(e^t - 1)]^{n-2}$$

Thus

$$\mu'_1 = \mu = M'_X(t)|_{t=0} = np$$

and

$$\mu'_2 = M''_X(t)|_{t=0} = np(1 - p + np)$$

The second *central moment* may be obtained using $\sigma^2 = \mu'_2 - \mu^2 = np(1-p)$.

● **Example 3-14.** Assume X to have the following distribution:

$$f(x) = \frac{a^b}{\Gamma(b)} x^{b-1} e^{-ax} \qquad 0 \le x < \infty,\ a > 0,\ b > 0$$

$$= 0 \qquad\qquad\qquad \text{otherwise}$$

The moment-generating function is

$$M_X(t) = \int_0^\infty \frac{a^b}{\Gamma(b)} e^{x(t-a)} x^{b-1}\, dx$$

which, if we let $y = x(a - t)$, becomes

$$M_X(t) = \frac{a^b}{\Gamma(b)(a - t)^b} \int_0^\infty e^{-y} y^{b-1}\, dy$$

Since the integral on the right is just $\Gamma(b)$ (as will be shown in Chapter 6), we obtain

$$M_X(t) = \frac{a^b}{(a - t)^b} = \left(1 - \frac{t}{a}\right)^{-b}$$

Now using the power series expansion for

$$\left(1 - \frac{t}{a}\right)^{-b}$$

we find

$$M_X(t) = 1 + b\frac{t}{a} + \frac{b(b+1)}{2!}\left(\frac{t}{a}\right)^2 + \cdots$$

which gives the moments

$$\mu'_1 = \frac{b}{a} \qquad \text{and} \qquad \mu'_2 = \frac{b(b+1)}{a^2}$$

3-8   Summary

This chapter first introduced methods for determining the probability distribution of a random variable that arises as a function of another random variable with known distribution. That is, where $Y = H(X)$, and either X is discrete with known distribution $p(x)$ or X is continuous with known density $f(x)$, methods were presented for obtaining the probability distribution of Y.

The expected value operator was introduced in general terms for $E[H(X)]$, and it was shown that $E(X) = \mu$, the mean, and $E(X - \mu)^2 = \sigma^2$, the variance. The variance operator V was defined as $V(X) = E(X^2) - [E(X)]^2$. Approximations were developed for $E[H(X)]$ and $V[H(X)]$ that are useful when exact methods prove difficult.

The moment-generating function has been presented and illustrated for the moments μ'_r of a probability distribution. It was noted that $E(X^r) = \mu'_r$.

3-9   Exercises

3-1.   A random variable X has probability distribution:

$$p(x) = \frac{x}{6} \qquad x = 1, 2, 3$$
$$= 0 \qquad \text{otherwise}$$

If $Y = (X - 2)^2$,
(a) Find the probability distribution of Y.
(b) Find $E(Y)$ and $V(Y)$.

3-2.   The content of magnesium in an alloy is a random variable, given by the following probability density function,

$$f(x) = \frac{x}{18} \qquad 0 \le x \le 6$$
$$= 0 \qquad \text{otherwise}$$

The profit obtained from this alloy is $P = 10 + 2X$.
(a) Find the probability distribution of P.
(b) What is the expected profit?

3-3.   A manufacturer of color television sets offers a one-year warranty of free replacement if the picture tube fails. He estimates the time to failure, T, to be a random variable with the following probability distribution:

$$f(t) = \frac{1}{4} e^{-t/4} \qquad t > 0$$
$$= 0 \qquad \text{otherwise}$$

(a) What percentage of the sets will he have to service?
(b) If the profit per sale is $200 and the replacement of a picture tube costs $200, find the expected profit of the business.

3-4. Assume that a continuous random variable X has probability distribution

$$f(x) = 2xe^{-x^2} \qquad x \geq 0$$
$$= 0 \qquad \text{otherwise}$$

Find the probability distribution of $Z = X^2$.

3-5. The number of days required to complete a given construction project is denoted by X and considered to be a random variable with the following distribution:

$$p(x) = .2 \qquad x = 10$$
$$= .3 \qquad x = 11$$
$$= .3 \qquad x = 12$$
$$= .1 \qquad x = 13$$
$$= .1 \qquad x = 14$$
$$= 0 \qquad \text{otherwise}$$

The contractors profit is $Y = \$1000(12 - X)$.
(a) Find the probability distribution of Y.
(b) Find $E(X)$, $V(X)$, $E(Y)$, $V(Y)$.

3-6. A used-car salesman finds that he sells either 1, 2, 3, 4, 5, or 6 cars per week with equal frequency. Let X be the number of cars sold.
(a) Find the moment-generating function of X.
(b) Using the moment-generating function find $E(X)$ and $V(X)$.

3-7. The probability function of the random variable X

$$f(x) = \frac{1}{\theta} e^{-(1/\theta)(x-\beta)} \qquad x \geq \beta, \theta > 0$$
$$= 0 \qquad \text{otherwise}$$

is known as the two-parameter exponential distribution. Find the moment-generating function of X. Evaluate $E(X)$ and $V(X)$ using the moment-generating function.

3-8. Let X_1 and X_2 be independent random variables, and $Z = X_1 + X_2$. Suppose $M_{X_1}(t)$, $M_{X_2}(t)$, and $M_Z(t)$ are the moment-generating functions of the three random variables in question. Verify that

$$M_Z(t) = M_{X_1}(t)M_{X_2}(t)$$

3-9. Given that the moment-generating function $M_{X_i}(t) = \exp(\mu_i t + \sigma_i^2 t^2/2)$, find the moment-generating function of $Z = \sum_{i=1}^{n} X_i$, where the X_i are independent random variables.

3-10. The percentage of a certain additive in gasoline determines the selling price. If A is a random variable representing the percentage, then $0 \leq A \leq 1$. If the percentage of A is less than .70 the gasoline is low-test and sells for 92 cents per gallon. If the percentage of A is greater than or equal to .70 the gasoline is high-test and sells for 98 cents per gallon. Find the expected revenue per gallon where $f(a) = 1, 0 \leq a \leq 1$.

3-11. The demand for antifreeze in a season is considered to be a random variable X, with density

$$f(x) = 10^{-6} \qquad 10^6 \le x \le 2 \times 10^6$$
$$= 0 \qquad \text{otherwise}$$

where X is measured in liters. If the manufacturer makes 50 cents profit on each liter she sells in the fall of the year, and if she must carry any excess over to the next year at a cost of 25 cents per liter, find the "optimum" stock level for a particular fall season. *Hint*: This is similar to the problem of Example 3-10.

3-12. The acidity of a certain product, measured on an arbitrary scale, is given by the relationship

$$A = (3 + .05G)^2$$

where G is the amount of one of the constituents having probability distribution

$$f(g) = \frac{1}{32}(5g - 2) \qquad 0 \le g \le 4$$
$$= 0 \qquad \text{otherwise}$$

Evaluate $E(A)$ and $V(A)$ by using the approximations derived in this chapter.

3-13. Suppose that X has probability density function

$$f(x) = 2x \qquad 0 \le x \le 1$$
$$= 0 \qquad \text{otherwise}$$

Find the probability density function of $Y = H(X)$ where $H(x) = 5x + 3$.

3-14. Suppose that X has probability density function

$$f(x) = 2x \qquad 0 \le x \le 1$$
$$= 0 \qquad \text{otherwise}$$

Find the probability density function of $Y = H(X)$ where $H(x) = x^2$.

3-15. Suppose that X has probability density function

$$f(x) = 1 \qquad 1 \le x \le 2$$
$$= 0 \qquad \text{otherwise}$$

Find the probability density function of $Y = H(X)$ where $H(x) = 4 - x^2$.

3-16. Suppose that X has probability density function

$$f(x) = 1 \qquad 1 \le x \le 2$$
$$= 0 \qquad \text{otherwise}$$

Find the probability density function of $Y = H(X)$ where $H(x) = e^x$.

3-17. Suppose that X has probability density function

$$f(x) = e^{-x} \qquad x \ge 0$$
$$= 0 \qquad \text{otherwise}$$

Find the probability density function of $Y = H(X)$ where

$$H(x) = \frac{3}{(1+x)^2}$$

3-18. Show that any continuous probability distribution can be transformed into a very simple rectangular form in which all values of the variable from 0 to 1 are equally probable.

3-19. Consider the probability density function

$$f(x) = k(1-x)^{a-1}x^{b-1} \qquad 0 \leq x \leq 1, a > 0, b > 0$$
$$= 0 \qquad\qquad\qquad \text{otherwise}$$

(a) Evaluate the constant k.
(b) Find the mean.
(c) Find the variance.

3-20. The third moment about the mean is related to the asymmetry, or skewness, of the distribution and is defined to be

$$\mu_3 = E(X - \mu_1')^3$$

Show that $\mu_3 = \mu_3' - 3\mu_2'\mu_1' + 2(\mu_1')^3$. Show that for a symmetric distribution $\mu_3 = 0$.

3-21. Let X be a random variable with probability density function

$$f(x) = ax^2 e^{-bx^2} \qquad x > 0$$
$$= 0 \qquad\qquad \text{otherwise}$$

(a) Evaluate the constant a.
(b) Suppose a new function $Y = 18X^2$ is of interest. Find an approximate value for $E(Y)$ and for $V(Y)$.

3-22. Assume that Y has the density function

$$f(y) = e^{-y} \qquad y > 0$$
$$= 0 \qquad\quad \text{otherwise}$$

Find the approximate value of $E(X)$ and $V(X)$ where

$$X = \sqrt{Y^2 + 36}$$

3-23. The concentration of reactant in a chemical process is a random variable having probability distribution

$$f(r) = 6r(1-r) \qquad 0 \leq r \leq 1$$
$$= 0 \qquad\qquad \text{otherwise}$$

The profit associated with the final product is $P = \$1.00 + \$3.00R$. Find the expected value of P. What is the probability distribution of P?

3-24. Find $E(X)$ where

$$f(x) = \frac{x}{g^2} \exp\left[\frac{-x^2}{2g^2}\right] \qquad x \geq 0, g > 0$$
$$= 0 \qquad\qquad\qquad \text{otherwise}$$

3-25. Try to find $E(X)$ and $V(X)$, where the random variable X has density function

$$f(x) = \frac{1}{\pi}\left[\frac{1}{1+x^2}\right] \qquad -\infty < x < \infty$$

3-26. Let f be a probability density function for which the rth order moment μ_r' exists. Prove that all moments of order less than r also exist.

3-27. A set of constants k_r, called *cumulants*, may be used instead of moments to characterize a probability distribution. If $M_X(t)$ is the moment-generating function of the probability distribution of a random variable X, then the cumulants are defined by the generating function

$$\psi_X(t) = \log M_X(t)$$

Thus, the rth cumulant is given by

$$k_r = \frac{d^r \psi_X(t)}{dt^r}\bigg|_{t=0}$$

Find the cumulants of the distribution whose density function is

$$f(x) = \frac{1}{\sigma\sqrt{2\pi}} \exp\left\{-\frac{1}{2}\left(\frac{x-\mu}{\sigma}\right)^2\right\} \qquad -\infty < x < \infty, \ -\infty < \mu < \infty, \ \sigma > 0$$

Chapter 4

Higher-Dimensional Random Variables

4-1 Introduction

In many situations we must deal with two or more random variables simultaneously. For example, we might select fabricated sheet steel specimens and measure shear strength and weld diameter of spot welds. Thus, both weld shear strength and weld diameter are the random variables of interest.

The objective of this chapter is to formulate *joint probability distributions* for two or more random variables and to present methods for obtaining both *marginal* and *conditional* distributions. Conditional expectation is defined as well as the *regression of the mean*. We also present a definition of *independence* for random variables, and *covariance* and *correlation* are defined. Functions of two or more random variables are presented, and a special case of *linear combinations* is presented with its corresponding moment-generating function. Finally the *law of large* numbers is presented.

The concepts in this chapter require the use of a *random vector*.

Definition

If $\mathscr{S}$ is the sample space associated with an experiment $\mathscr{E}$, and $X_1, X_2, \ldots, X_k$ are functions, each assigning a real number, $X_1(e), X_2(e), \ldots, X_k(e)$, to every outcome e, we call $[X_1, X_2, \ldots, X_k]$ a *k-dimensional random vector* (see Fig. 4-1).

The range space of the random vector $[X_1, X_2, \ldots, X_k]$ is the set of all possible values of the random vector. This may be represented as $R_{X_1 \times X_2 \times \cdots \times X_k}$, where

$$R_{X_1 \times X_2 \times \cdots \times X_k} = \{[x_1, x_2, \ldots, x_k] : x_1 \in R_{X_1}, x_2 \in R_{X_2}, \ldots, x_k \in R_{X_k}\}$$

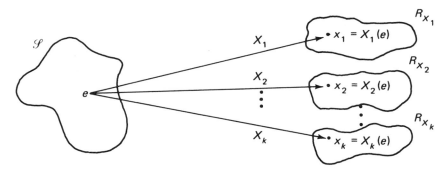

Fig. 4-1. A *k*-dimensional random vector.

This is the Cartesian product of the range space sets. In the case where $k = 2$, that is, where we have a two-dimensional random vector as in the earlier illustrations, $R_{X_1 \times X_2}$ is a subset of the Euclidean plane.

4-2   Two-Dimensional Random Vectors

In most of our considerations here, we will be concerned with two-dimensional random vectors. Sometimes the equivalent term *two-dimensional random variables* will be used.

If the possible values of $[X_1, X_2]$ are either finite or countably infinite in number, then $[X_1, X_2]$ will be a *two-dimensional discrete random vector*. The values of $[X_1, X_2]$ are $[x_{1_i}, x_{2_j}]$, $i = 1, 2, \ldots, n, j = 1, 2, \ldots, m$.

If the possible values of $[X_1, X_2]$ are some uncountable set in the Euclidean plane, then $[X_1, X_2]$ will be a *two-dimensional continuous random vector*. For example, if $a \le x_1 \le b$ and $c \le x_2 \le d$, we would have $R_{X_1 \times X_2} = \{[x_1, x_2]: a \le x_1 \le b, c \le x_2 \le d\}$.

It is also possible for one component to be discrete and the other continuous; however, here we consider only the case where both are discrete or both are continuous.

● **Example 4-1.** Consider the case where weld shear strength and weld diameter are measured. If we let X_1 represent diameter in inches and X_2 represent strength in pounds, and if we know $0 \le x_1 < .25$ inch while $0 \le x_2 \le 2000$ pounds, then the range space for $[X_1, X_2]$ is the set $\{[x_1, x_2]: 0 \le x_1 < .25, 0 \le x_2 \le 2000\}$. This space is shown graphically in Fig. 4-2.

● **Example 4-2.** A small pump is inspected for four quality control characteristics. Each characteristic is classified as good, minor defect (not affecting operation), or major defect (affecting operation). A pump is to be selected and

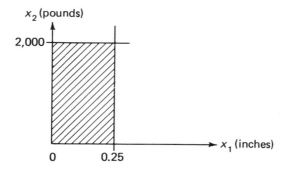

Fig. 4-2. The range space of $[X_1, X_2]$, where X_1 is weld diameter and X_2 is shear strength.

defects counted. If X_1 = the number of minor defects and X_2 = the number of major defects, we know that $x_1 = 0, 1, 2, 3, 4$ and $x_2 = 0, 1, \ldots, 4 - x_1$ because only four characteristics are inspected. The range space for $[X_1, X_2]$ is thus $\{[0, 0], [0, 1], [0, 2], [0, 3], [0, 4], [1, 0], [1, 1], [1, 2], [1, 3], [2, 0], [2, 1], [2, 2], [3, 0], [3, 1], [4, 0]\}$. These possible outcomes are shown in Fig. 4-3.

Definition

Bivariate probability functions:

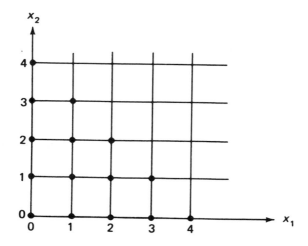

Fig. 4-3. The range space for $[X_1, X_2]$, where X_1 is the number of minor defects and X_2 is the number of major defects. The range space is indicated by heavy dots.

1. *Discrete case.* To each outcome $[x_{1_i}, x_{2_j}]$ of $[X_1, X_2]$, we associate a number

$$p(x_{1_i}, x_{2_j}) = P(X_1 = x_{1_i} \text{ and } X_2 = x_{2_j})$$

where

$$p(x_{1_i}, x_{2_j}) \geq 0 \qquad \text{for all } i, j$$

and

$$\sum_{\text{all } j} \sum_{\text{all } i} p(x_{1_i}, x_{2_j}) = 1 \qquad (4\text{-}1)$$

The values $([x_{1_i}, x_{2_j}], p(x_{1_i}, x_{2_j}))$ for all i, j make up the *probability distribution* of $[X_1, X_2]$.

2. *Continuous case.* If $[X_1, X_2]$ is a continuous random vector with range space R in the Euclidean plane, then f, the *joint density function*, has the following properties:

$$f(x_1, x_2) \geq 0 \qquad \text{for all } (x_1, x_2) \in R$$

and

$$\iint_R f(x_1, x_2) \, dx_1 \, dx_2 = 1$$

A probability statement is then of the form

$$P(a_1 \leq X_1 \leq b_1, a_2 \leq X_2 \leq b_2) = \int_{a_2}^{b_2} \int_{a_1}^{b_1} f(x_1, x_2) \, dx_1 \, dx_2$$

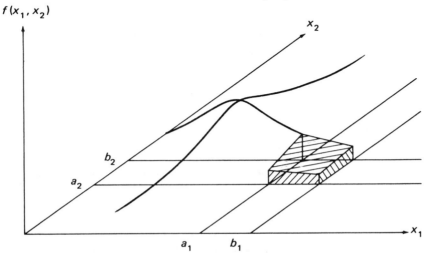

Fig. 4-4. A bivariate density function, where $P(a_1 \leq X_1 \leq b_1, a_2 \leq X_2 \leq b_2)$ is given by the shaded volume.

x_{2_j} \ x_{1_i}	0	1	2	3	4
0	1/30	1/30	2/30	3/30	1/30
1	1/30	1/30	3/30	4/30	
2	1/30	2/30	3/30		
3	1/30	3/30			
4	3/30				

(a)

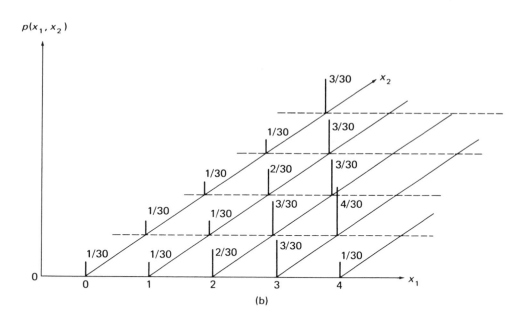

(b)

Fig. 4-5. Tabular and graphical presentation of a bivariate probability distribution. (a) Tabulated values are $p(x_{1_i}, x_{2_j})$. (b) Graphical presentation of discrete bivariate distribution.

See Fig. 4-4. If needed for clarification, a subscript X_1, X_2 may be used with the functions p and f; otherwise it will be deleted.

It should again be noted that $f(x_1, x_2)$ does not represent the probability of anything, and the convention that $f(x_1, x_2) = 0$ for $(x_1, x_2) \notin R$ will be employed so that the second property may be written

$$\int_{-\infty}^{\infty} \int_{-\infty}^{\infty} f(x_1, x_2) \, dx_1 \, dx_2 = 1$$

In the case where $[X_1, X_2]$ is discrete, we might present the probability distribution of $[X_1, X_2]$ in tabular, graphical, or, in some cases, mathematical form. In the case where $[X_1, X_2]$ is continuous, we usually employ a mathematical relationship to present the probability distribution; however, a graphical presentation may occasionally be helpful.

● **Example 4-3.** A hypothetical probability distribution is shown in both tabular and graphical form in Fig. 4-5 for the random variables defined in Example 4-2.

● **Example 4-4.** In the case of the weld diameters represented by X_1 and tensile strength represented by X_2, we might have a uniform distribution as given below:

$$f(x_1, x_2) = \frac{1}{500} \qquad 0 \le x_1 < .25, 0 \le x_2 \le 2000$$

$$= 0 \qquad \text{otherwise}$$

The range space was shown in Fig. 4-2, and if we add another dimension to display graphically $y = f(x_1, x_2)$, then the distribution would appear as in Fig. 4-6. In the univariate case, area corresponded to probability; in the bivariate case, volume under the surface represents the probability.

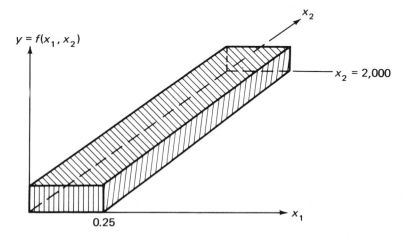

Fig. 4-6. A bivariate uniform density.

For example, suppose we wish to find $P(.1 \leq X_1 \leq .2, 100 \leq X_2 \leq 200)$. This probability would be found by integrating $f(x_1, x_2)$ over the region $.1 \leq x_1 \leq .2$, $100 \leq x_2 \leq 200$. That is,

$$\int_{100}^{200} \int_{.1}^{.2} \frac{1}{500} \, dx_1 \, dx_2 = \frac{1}{50}$$

4-3   Marginal Distributions

Having defined the bivariate probability distribution, sometimes called the *joint probability distribution* (or in the continuous case the *joint density*), a natural question arises as to the distribution of X_1 or X_2 alone. These distributions are called *marginal distributions*. In the discrete case, the marginal distribution of X_1 is

$$p_1(x_{1_i}) = \sum_{\text{all } j} p(x_{1_i}, x_{2_j}) \qquad i = 1, 2, \ldots \tag{4-2}$$

and the marginal distribution of X_2 is

$$p_2(x_{2_j}) = \sum_{\text{all } i} p(x_{1_i}, x_{2_j}) \qquad j = 1, 2, \ldots \tag{4-3}$$

● **Example 4-5.** In Example 4-2 we considered the joint discrete distribution shown in Fig. 4-5. The marginal distributions are shown in Fig. 4-7. We see that $[x_{1_i}, p_1(x_{1_i})]$ is a univariate distribution and it is the distribution of X_1 (the number of minor defects) alone. Likewise $[x_{2_j}, p_2(x_{2_j})]$ is a univariate distribution and it is the distribution of X_2 (the number of major defects) alone.

If $[X_1, X_2]$ is a continuous random vector, the marginal distribution of X_1 is

$$f_1(x_1) = \int_{-\infty}^{\infty} f(x_1, x_2) \, dx_2 \tag{4-4}$$

and the marginal distribution of X_2 is

$$f_2(x_2) = \int_{-\infty}^{\infty} f(x_1, x_2) \, dx_1 \tag{4-5}$$

The function f_1 is the density function for X_1 alone, and the function f_2 is the density function for X_2 alone.

● **Example 4-6.** In Example 4-4, the joint density of $[X_1, X_2]$ was given by

$$f(x_1, x_2) = \frac{1}{500} \qquad 0 \leq x_1 < .25, 0 \leq x_2 \leq 2000$$
$$= 0 \qquad \text{otherwise}$$

$x_2{}_j$ \ $x_1{}_i$	0	1	2	3	4	$p_2(x_2{}_j)$
0	1/30	1/30	2/30	3/30	1/30	8/30
1	1/30	1/30	3/30	4/30		9/30
2	1/30	2/30	3/30			6/30
3	1/30	3/30				4/30
4	3/30					3/30
$p_1(x_1{}_i)$	7/30	7/30	8/30	7/30	1/30	$\Sigma p(x) = 1$

(a)

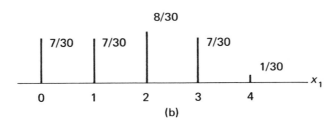

(b)

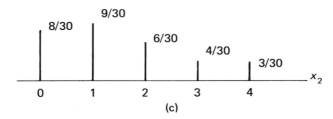

(c)

Fig. 4-7. Marginal distributions for discrete $[X_1, X_2]$. (a) Marginal distributions—tabular form. (b) Marginal distribution $(x_1{}_i, p_1(x_1{}_i))$. (c) Marginal distribution $(x_2{}_j, p_2(x_2{}_j))$.

The marginal distributions of X_1 and X_2 are

$$f_1(x_1) = \int_0^{2000} \frac{1}{500} \, dx_2 = 4 \qquad 0 \le x_1 < .25$$

$$= 0 \qquad \text{otherwise}$$

and

$$f_2(x_2) = \int_0^{.25} \frac{1}{500} \, dx_1 = \frac{1}{2000} \qquad 0 \le x_2 \le 2000$$

$$= 0 \qquad \text{otherwise}$$

These are shown graphically in Fig. 4-8.

The expected value of X_1 (mean of X_1), the expected value of X_2 (mean of X_2), the variance of X_1, and variance of X_2 are determined from the marginal distributions exactly as in the univariate case. Where $[X_1, X_2]$ is *discrete*,

$$E(X_1) = \mu_{X_1} = \sum_{\text{all } i} x_{1_i} p_1(x_{1_i}) = \sum_{\text{all } i} \sum_{\text{all } j} x_{1_i} p(x_{1_i}, x_{2_j}) \tag{4-6}$$

$$V(X_1) = \sigma_1^2 = \sum_{\text{all } i} (x_{1_i} - \mu_{X_1})^2 p_1(x_{1_i})$$

$$= \sum_{\text{all } i} \sum_{\text{all } j} (x_{1_i} - \mu_{X_1})^2 p(x_{1_i}, x_{2_j})$$

$$= \sum_{\text{all } i} x_{1_i}^2 p_1(x_{1_i}) - \mu_{X_1}^2$$

$$= \sum_{\text{all } i} \sum_{\text{all } j} x_{1_i}^2 p(x_{1_i}, x_{2_j}) - \mu_{X_1}^2 \tag{4-7}$$

and

$$E(X_2) = \mu_{X_2} = \sum_{\text{all } j} x_{2_j} p_2(x_{2_j}) = \sum_{\text{all } j} \sum_{\text{all } i} x_{2_j} p(x_{1_i}, x_{2_j}) \tag{4-8}$$

$$V(X_2) = \sigma_2^2 = \sum_{\text{all } j} (x_{2_j} - \mu_{X_2})^2 p_2(x_{2_j})$$

$$= \sum_{\text{all } j} \sum_{\text{all } i} (x_{2_j} - \mu_{X_2})^2 p(x_{1_i}, x_{2_j})$$

$$= \sum_{\text{all } j} x_{2_j}^2 p_2(x_{2_j}) - \mu_{X_2}^2$$

$$= \sum_{\text{all } j} \sum_{\text{all } i} x_{2_j}^2 p(x_{1_i}, x_{2_j}) - \mu_{X_2}^2 \tag{4-9}$$

● **Example 4-7.** In Example 4-5 and Fig. 4-7 marginal distributions for X_1 and X_2 were given. Working with the marginal distribution of X_1 shown in Fig. 4-7b,

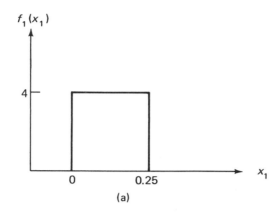

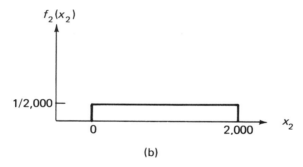

Fig. 4-8. Marginal distributions for bivariate uniform vector $[X_1, X_2]$. (*a*) Marginal distribution of X_1. (*b*) Marginal distribution of X_2.

we may calculate:

$$E(X_1) = \mu_{X_1} = 0 \cdot \frac{7}{30} + 1 \cdot \frac{7}{30} + 2 \cdot \frac{8}{30} + 3 \cdot \frac{7}{30} + 4 \cdot \frac{1}{30} = \frac{8}{5}$$

and

$$V(X_1) = \sigma_1^2 = \left[0^2 \cdot \frac{7}{30} + 1^2 \cdot \frac{7}{30} + 2^2 \cdot \frac{8}{30} + 3^2 \cdot \frac{7}{30} + 4^2 \cdot \frac{1}{30} \right] - \left[\frac{8}{5} \right]^2 = \frac{103}{75}$$

The mean and variance of X_2 could also be determined using the marginal distribution of X_2.

Equations (4–6) to (4–9) show that the mean and variance of X_1 and X_2, respectively, may be determined from the marginal distributions or directly from the joint distribution. In practice, if the marginal distribution has already been determined, it is usually easier to make use of it.

In the case where $[X_1, X_2]$ is *continuous*, then

$$E(X_1) = \mu_{X_1} = \int_{-\infty}^{\infty} x_1 f_1(x_1) \, dx_1 = \int_{-\infty}^{\infty} \int_{-\infty}^{\infty} x_1 f(x_1, x_2) \, dx_2 \, dx_1 \qquad (4\text{-}10)$$

$$\begin{aligned}
V(X_1) = \sigma_1^2 &= \int_{-\infty}^{\infty} (x_1 - \mu_{X_1})^2 f_1(x_1) \, dx_1 \\
&= \int_{-\infty}^{\infty} \int_{-\infty}^{\infty} (x_1 - \mu_{X_1})^2 f(x_1, x_2) \, dx_2 \, dx_1 \\
&= \int_{-\infty}^{\infty} x_1^2 f_1(x_1) \, dx_1 - \mu_{X_1}^2 \\
&= \int_{-\infty}^{\infty} \int_{-\infty}^{\infty} x_1^2 f(x_1, x_2) \, dx_2 \, dx_1 - \mu_{X_1}^2 \qquad (4\text{-}11)
\end{aligned}$$

and

$$E(X_2) = \mu_{X_2} = \int_{-\infty}^{\infty} x_2 f_2(x_2) \, dx_2 = \int_{-\infty}^{\infty} \int_{-\infty}^{\infty} x_2 f(x_1, x_2) \, dx_1 \, dx_2 \qquad (4\text{-}12)$$

$$\begin{aligned}
V(X_2) = \sigma_2^2 &= \int_{-\infty}^{\infty} (x_2 - \mu_{X_2})^2 f_2(x_2) \, dx_2 \\
&= \int_{-\infty}^{\infty} \int_{-\infty}^{\infty} (x_2 - \mu_{X_2})^2 f(x_1, x_2) \, dx_1 \, dx_2 \\
&= \int_{-\infty}^{\infty} x_2^2 f_2(x_2) \, dx_2 - \mu_{X_2}^2 \\
&= \int_{-\infty}^{\infty} \int_{-\infty}^{\infty} x_2^2 f(x_1, x_2) \, dx_1 \, dx_2 - \mu_{X_2}^2 \qquad (4\text{-}13)
\end{aligned}$$

Again, in Equations (4-10) to (4-13), observe that we may use either the marginal densities or the joint density in the calculations.

● **Example 4-8.** In Example 4-4, the joint density of weld diameters, X_1, and shear strength, X_2, was given as

$$f(x_1, x_2) = \frac{1}{500} \qquad 0 \le x_1 < .25, 0 \le x_2 \le 2000$$
$$= 0 \qquad \text{otherwise}$$

and the marginal densities for X_1 and X_2 were given in Example 4-6 as

$$f_1(x_1) = 4 \qquad 0 \le x_1 < .25$$
$$= 0 \qquad \text{otherwise}$$

and

$$f_2(x_2) = \frac{1}{2000} \qquad 0 \le x_2 \le 2000$$
$$= 0 \qquad \text{otherwise}$$

Working with the marginal densities, the mean and variance of X_1 are thus

$$E(X_1) = \mu_{X_1} = \int_0^{.25} x_1 \cdot 4 \, dx_1 = 2(.25)^2 = .125$$

and

$$V(X_1) = \sigma_1^2 = \int_0^{.25} x_1^2 \cdot 4 \, dx_1 - (.125)^2 = \frac{4}{3}(.25)^3 - (.125)^2 \simeq 5.21 \times 10^{-6}$$

4-4 Conditional Distributions

When dealing with two jointly distributed random variables it may be of interest to find the distribution of one of these variables, given a particular value of the other. That is, we may wish to find the distribution of X_1 given that $X_2 = x_2$. This probability distribution would be called the *conditional* distribution of X_1 given that $X_2 = x_2$.

Suppose that the random vector $[X_1, X_2]$ is discrete. From the definition of conditional probability given by Equation (1-4), it is easily seen that the conditional probability distributions are

$$p_{X_2|x_{1_i}}(x_{2_j}) = \frac{p(x_{1_i}, x_{2_j})}{p_1(x_{1_i})} \qquad i = 1, 2, \ldots, j = 1, 2, \ldots \qquad (4\text{-}14)$$

and

$$p_{X_1|x_{2_j}}(x_{1_i}) = \frac{p(x_{1_i}, x_{2_j})}{p_2(x_{2_j})} \qquad i = 1, 2, \ldots, j = 1, 2, \ldots \qquad (4\text{-}15)$$

where $p_1(x_{1_i}) > 0$ and $p_2(x_{2_j}) > 0$.

It should be noted that there are as many conditional distributions of X_2 for given X_1 as there are values x_{1_i}, and there are as many conditional distributions of X_1 for given X_2 as there are values x_{2_j}.

● **Example 4-9.** Consider the counting of minor and major defects of the small pumps in Example 4-2. There will be five conditional distributions of X_2, one for each value of X_1. They are shown in Fig. 4-9. The distribution $p_{X_2|0}(x_{2_j})$, for $X_1 = 0$, is shown in Fig. 4-9a. Figure 4-9b shows the distribution $p_{X_2|1}(x_{2_j})$. Other conditional distributions could likewise be determined for $X_1 = 2, 3$, and 4, respectively. The distribution of X_1 for $X_2 = 3$ is

$$p_{X_1|3}(0) = \frac{1/30}{4/30} = \frac{1}{4}$$

$$p_{X_1|3}(1) = \frac{3/30}{4/30} = \frac{3}{4}$$

$$p_{X_1|3}(x_{1_i}) = 0 \qquad \text{otherwise}$$

x_{2_j}	0	1	2	3	4
$p_{X_2\mid 0}(x_{2_j}) = p(0, x_{2_j})/p_1(0)$	$\dfrac{p(0, 0)}{p_1(0)}$	$\dfrac{p(0, 1)}{p_1(0)}$	$\dfrac{p(0, 2)}{p_1(0)}$	$\dfrac{p(0, 3)}{p_1(0)}$	$\dfrac{p(0, 4)}{p_1(0)}$
Quotient	$\dfrac{1/30}{7/30} = \dfrac{1}{7}$	$\dfrac{1/30}{7/30} = \dfrac{1}{7}$	$\dfrac{1/30}{7/30} = \dfrac{1}{7}$	$\dfrac{1/30}{7/30} = \dfrac{1}{7}$	$\dfrac{3/30}{7/30} = \dfrac{3}{7}$

(a)

x_{2_j}	0	1	2	3	4
$p_{X_2\mid 1}(x_{2_j}) = p(1, x_{2_j})/p_1(1)$	$\dfrac{p(1, 0)}{p_1(1)}$	$\dfrac{p(1, 1)}{p_1(1)}$	$\dfrac{p(1, 2)}{p_1(1)}$	$\dfrac{p(1, 3)}{p_1(1)}$	$\dfrac{p(1, 4)}{p_1(1)}$
Quotient	$\dfrac{1/30}{7/30} = \dfrac{1}{7}$	$\dfrac{1/30}{7/30} = \dfrac{1}{7}$	$\dfrac{2/30}{7/30} = \dfrac{2}{7}$	$\dfrac{3/30}{7/30} = \dfrac{3}{7}$	$\dfrac{0}{7/30} = 0$

(b)

Fig. 4-9. Some examples of conditional distributions.

If $[X_1, X_2]$ is a continuous random vector, the conditional densities are

$$f_{X_2\mid x_1}(x_2) = \frac{f(x_1, x_2)}{f_1(x_1)} \tag{4-16}$$

and

$$f_{X_1\mid x_2}(x_1) = \frac{f(x_1, x_2)}{f_2(x_2)} \tag{4-17}$$

where $f_1(x_1) > 0$ and $f_2(x_2) > 0$.

● **Example 4-10.** Suppose the joint density of $[X_1, X_2]$ is the function f described here and shown in Fig. 4-10:

$$f(x_1, x_2) = x_1^2 + \frac{x_1 x_2}{3} \qquad 0 < x_1 \le 1, 0 \le x_2 \le 2$$
$$= 0 \qquad \text{otherwise}$$

The marginal densities are $f_1(x_1)$ and $f_2(x_2)$. These are determined as

$$f_1(x_1) = \int_0^2 \left(x_1^2 + \frac{x_1 x_2}{3}\right) dx_2 = 2x_1^2 + \frac{2}{3}x_1 \qquad 0 < x_1 \le 1$$
$$= 0 \qquad \text{otherwise}$$

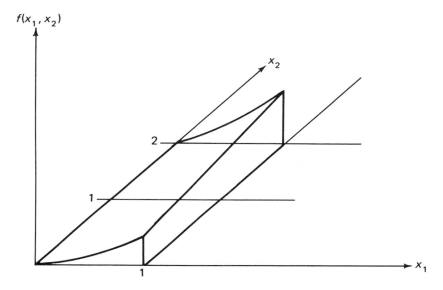

Fig. 4-10.   A bivariate density function.

and

$$f_2(x_2) = \int_0^1 \left(x_1^2 + \frac{x_1 x_2}{3} \right) dx_1 = \frac{1}{3} + \frac{x_2}{6} \qquad 0 \leq x_2 \leq 2$$

$$= 0 \qquad\qquad \text{otherwise}$$

The marginal densities are shown in Fig. 4-11.

The conditional densities may be determined using Equations (4-16) and (4-17) as:

$$f_{X_2|x_1}(x_2) = \frac{x_1^2 + \dfrac{x_1 x_2}{3}}{2x_1^2 + \dfrac{2}{3}x_1} = \frac{1}{2} \cdot \frac{[x_1 + (x_2/3)]}{x_1 + (1/3)} \qquad 0 \leq x_2 \leq 2, 0 < x_1 \leq 1$$

$$= 0 \qquad\qquad\qquad \text{otherwise}$$

and

$$f_{X_1|x_2}(x_1) = \frac{x_1^2 + \dfrac{x_1 x_2}{3}}{\dfrac{1}{3} + \dfrac{x_2}{6}} = \frac{x_1(3x_1 + x_2)}{1 + (x_2/2)} \qquad 0 < x_1 \leq 1, 0 \leq x_2 \leq 2$$

$$= 0, \qquad\qquad\qquad \text{otherwise}$$

Note that for $f_{X_2|x_1}(x_2)$, there are an infinite number of these conditional

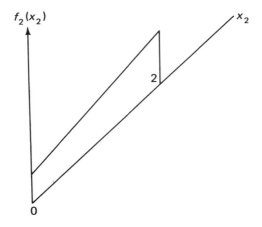

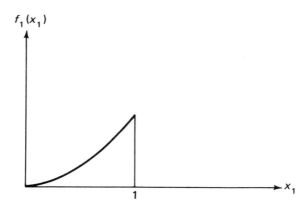

Fig. 4-11.　Some marginal densities.

densities, one for each value $0 < x_1 < 1$. Two of these $f_{X_2|(1/2)}(x_2)$ and $f_{X_2|1}(x_2)$ are shown in Fig. 4-12. Also, for $f_{X_1|x_2}(x_1)$, there are an infinite number of these conditional densities, one for each value $0 \le x_2 \le 2$. Three of these are shown in Fig. 4-13.

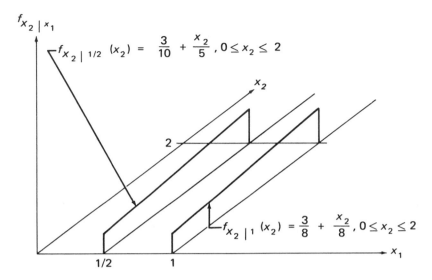

$f_{x_2 \mid x_1}$

$f_{x_2 \mid 1/2} (x_2) = \dfrac{3}{10} + \dfrac{x_2}{5}, 0 \le x_2 \le 2$

$f_{x_2 \mid 1} (x_2) = \dfrac{3}{8} + \dfrac{x_2}{8}, 0 \le x_2 \le 2$

Fig. 4-12. Two conditional densities $f_{x_2 \mid 1/2}$, and $f_{x_2 \mid 1}$.

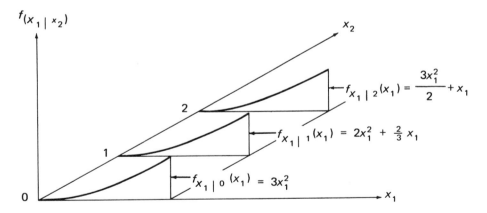

$f(x_1 \mid x_2)$

$f_{x_1 \mid 2}(x_1) = \dfrac{3x_1^2}{2} + x_1$

$f_{x_1 \mid 1}(x_1) = 2x_1^2 + \dfrac{2}{3} x_1$

$f_{x_1 \mid 0} (x_1) = 3x_1^2$

Fig. 4-13. Three conditional densities $f_{x_1 \mid 0}$, $f_{x_1 \mid 1}$, and $f_{x_1 \mid 2}$.

4-5   Conditional Expectation

If $[X_1, X_2]$ is a discrete random vector, the *conditional expections* are

$$E(X_1 \mid x_{2_j}) = \sum_{\text{all } i} x_{1_i} p_{X_1 \mid x_{2_j}}(x_{1_i}) \qquad (4\text{-}18)$$

and

$$E(X_2|x_{1_i}) = \sum_{\text{all } j} x_{2_j} p_{X_2|x_{1_i}}(x_{2_j}) \tag{4-19}$$

Note that there will be an $E(X_1|x_{2_j})$ for each value of x_{2_j}. The value of each $E(X_1|x_{2_j})$ will depend on the value x_{2_j} which is in turn governed by the probability function. Similarly, there will be as many values of $E(X_2|x_{1_i})$ as there are values x_{1_i} and the value of $E(X_2|x_{1_i})$ will depend upon the value x_{1_i} determined by the probability function.

Example 4-11. Consider the probability distribution of the discrete random vector $[X_1, X_2]$, where X_1 represents the number of orders for a large turbine in July and X_2 represents the number of orders in August. The joint distribution as well as the marginal distributions are given in Fig. 4-14. We consider the three conditional distributions, $p_{X_2|0}$, $p_{X_2|1}$, and $p_{X_2|2}$, and the conditional expected values of each:

$p_{X_2|0}(x_{2_j}) = \dfrac{1}{6}, \quad x_{2_j} = 0$	$p_{X_2|1}(x_{2_j}) = \dfrac{1}{10}, \quad x_{2_j} = 0$	$p_{X_2|2}(x_{2_j}) = \dfrac{1}{2}, \quad x_{2_j} = 0$
$= \dfrac{2}{6}, \quad x_{2_j} = 1$	$= \dfrac{5}{10}, \quad x_{2_j} = 1$	$= \dfrac{1}{4}, \quad x_{2_j} = 1$
$= \dfrac{2}{6}, \quad x_{2_j} = 2$	$= \dfrac{3}{10}, \quad x_{2_j} = 2$	$= \dfrac{1}{4}, \quad x_{2_j} = 2$
$= \dfrac{1}{6}, \quad x_{2_j} = 3$	$= \dfrac{1}{10}, \quad x_{2_j} = 3$	$= 0, \quad x_{2_j} = 3$
$= 0, \quad$ otherwise	$= 0, \quad$ otherwise	$= 0, \quad$ otherwise

$$E(X_2|0) = 0 \cdot \frac{1}{6} + 1 \cdot \frac{2}{6} \qquad E(X_2|1) = 0 \cdot \frac{1}{10} + 1 \cdot \frac{5}{10} \qquad E(X_2|2) = 0 \cdot \frac{1}{2} + 1 \cdot \frac{1}{4}$$

$$+ 2 \cdot \frac{2}{6} + 3 \cdot \frac{1}{6} = 1.5 \qquad + 2 \cdot \frac{3}{10} + 3 \cdot \frac{1}{10} = 1.4 \qquad + 2 \cdot \frac{1}{4} + 3 \cdot 0 = .75$$

If $[X_1, X_2]$ is a continuous random vector, the *conditional expectations* are

$$E(X_1|x_2) = \int_{-\infty}^{\infty} x_1 \cdot f_{X_1|x_2}(x_1) \, dx_1 \tag{4-20}$$

and

$$E(X_2|x_1) = \int_{-\infty}^{\infty} x_2 \cdot f_{X_2|x_1}(x_2) \, dx_2 \tag{4-21}$$

and in each case there will be an infinite number of values that the expected value may take. In Equation (4-20), there will be one value of $E(X_1|x_2)$ for

x_{2j} \ x_{1i}	0	1	2	$p_2(x_{2j})$
0	0.05	0.05	0.10	0.2
1	0.10	0.25	0.05	0.4
2	0.10	0.15	0.05	0.3
3	0.05	0.05	0.00	0.1
$p_1(x_{1i})$	0.3	0.5	0.2	

Fig. 4-14. Joint and marginal distributions of $[X_1, X_2]$. Values in body of table are $p(x_{1i}, x_{2j})$.

each value x_2, and in Equation (4-21), there will be one value of $E(X_2|x_1)$ for each value x_1.

● **Example 4-12.** In Example 4-10, we considered a joint density f, where

$$f(x_1, x_2) = x_1^2 + \frac{x_1 x_2}{3} \qquad 0 < x_1 \le 1, 0 \le x_2 \le 2$$

$$= 0 \qquad \text{otherwise}$$

The conditional densities were

$$f_{X_2|x_1}(x_2) = \frac{1}{2} \cdot \frac{x_1 + (x_2/3)}{x_1 + (1/3)} \qquad 0 \le x_2 \le 2, \ 0 < x_1 \le 1$$

and

$$f_{X_1|x_2}(x_1) = \frac{x_1(3x_1 + x_2)}{1 + (x_2/2)} \qquad 0 < x_1 \le 1, 0 \le x_2 \le 2$$

Then, using Equation (4-21), the $E(X_2|x_1)$ is determined as

$$E(X_2|x_1) = \int_0^2 x_2 \cdot \frac{1}{2} \cdot \frac{x_1 + (x_2/3)}{x_1 + (1/3)} \, dx_2$$

$$= \frac{9x_1 + 4}{9x_1 + 3}$$

It should be noted that this is a function of x_1. In the two conditional densities shown in Fig. 4-12, where $x_1 = \frac{1}{2}$ and $x_1 = 1$, the corresponding expected values are $E(X_2|\frac{1}{2}) = \frac{17}{15}$ and $E(X_2|1) = \frac{13}{12}$.

Since $E(X_2|x_1)$ is a function of x_1, and x_1 is a value of the random variable X_1, $E(X_2|X_1)$ is a random variable, and we may consider the expected value of

$E(X_2|X_1)$, that is, $E[E(X_2|X_1)]$. The inner operator is the expectation of X_2 given $X_1 = x_1$, and the outer expectation is with respect to the marginal density of X_1; therefore

$$E[E(X_2|X_1)] = E(X_2) = \mu_{X_2} \qquad (4\text{-}22)$$

and

$$E[E(X_1|X_2)] = E(X_1) = \mu_{X_1} \qquad (4\text{-}23)$$

● **Example 4-13.** In Example 4-12, we found $E(X_2|x_1) = (9x_1 + 4)/(9x_1 + 3)$. Thus

$$E[E(X_2|X_1)] = \int_0^1 \left[\frac{9x_1 + 4}{9x_1 + 3}\right] \cdot \left(2x_1^2 + \frac{2}{3}x_1\right) dx_1$$

$$= \frac{10}{9}$$

Note that this is also $E(X_2)$, since

$$E(X_2) = \int_{-\infty}^{\infty} x_2 f_2(x_2)\, dx_2 = \int_0^2 x_2 \cdot \left(\frac{1}{3} + \frac{x_2}{6}\right) dx_1$$

$$= \frac{10}{9}$$

The *variance operator* may be applied to conditional distributions exactly as in the univariate case.

4-6   Regression of the Mean

It was pointed out that $E(X_2|x_1)$ is a value of $E(X_2|X_1)$ for a particular $X_1 = x_1$, and it is a function of x_1. The graph of this function is called the regression of X_2 on X_1. Alternatively, the function $E(X_1|x_2)$ would be called the regression of X_1 on X_2. This is demonstrated in Fig. 4-15.

● **Example 4-14.** In Example 4-12 we found $E(X_2|x_1)$ for the bivariate density of Example 4-10, that is,

$$f(x_1, x_2) = x_1^2 + \frac{x_1 x_2}{3} \qquad 0 < x_1 \le 1, 0 \le x_2 \le 2$$

$$= 0 \qquad\qquad \text{otherwise}$$

The result was

$$E(X_2|x_1) = \frac{9x_1 + 4}{9x_1 + 3}$$

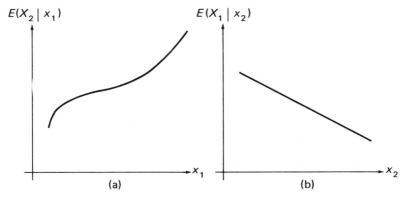

Fig. 4-15. Some regression curves. (a) Regression of X_2 on X_1. (b) Regression of X_1 on X_2.

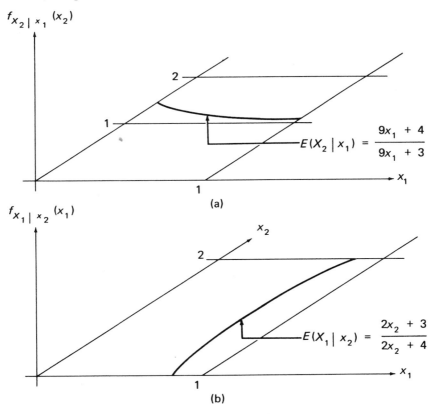

Fig. 4-16. (a) The regression of X_2 on X_1. (b) The regression of X_1 on X_2.

In a like manner, we may find

$$E(X_1|x_2) = \int_0^1 x_1 \cdot \left[\frac{3x_1^2 + x_1 x_2}{1 + (x_2/2)}\right] dx_1$$

$$= \frac{2x_2 + 3}{2x_2 + 4}$$

These regression curves are shown in Fig. 4-16.

Regression will be discussed further in Chapter 12.

4-7 Independence of Random Variables

The notions of independence, and independent random variables, are very useful and important statistical concepts. In Chapter 1, when we were dealing with events, the idea of independent events was introduced and a formal definition presented. We are now concerned with defining *independent random variables*. When the outcome of one variable, say X_1, does not influence the outcome of X_2, and vice versa, we say the random variables X_1 and X_2 are independent.

Definition

1. If $[X_1, X_2]$ is a discrete random vector, then we say that X_1 and X_2 are independent if and only if

$$p(x_{1_i}, x_{2_j}) = p_1(x_{1_i}) \cdot p_2(x_{2_j}) \qquad (4\text{-}24)$$

for all i and j.

2. If $[X_1, X_2]$ is a continuous random vector, then we say that X_1 and X_2 are independent if and only if

$$f(x_1, x_2) = f_1(x_1) \cdot f_2(x_2) \qquad (4\text{-}25)$$

for all x_1 and x_2.

Utilizing this definition, and the properties of conditional probability distributions, we may extend the concept of independence to a theorem.

Theorem 4-1

1. Let $[X_1, X_2]$ be a discrete random vector. Then

$$p_{X_2|x_{1_i}}(x_{2_j}) = p_2(x_{2_j})$$

and

$$p_{X_1|x_{2_j}}(x_{1_i}) = p_1(x_{1_i})$$

for all i and j if and only if X_1 and X_2 are independent.

2. Let $[X_1, X_2]$ be a continuous random vector. Then

$$f_{X_2|x_1}(x_2) = f_2(x_2)$$

and

$$f_{X_1|x_2}(x_1) = f_1(x_1)$$

for all x_1 and x_2 if and only if X_1 and X_2 are independent.

Proof

Refer to Exercise 4-10.

Note that the requirement for the joint distribution to be factorable into the respective marginal distributions is somewhat similar to the requirement that, for independent events, the probability of intersection of events equals the product of the event probabilities.

● **Example 4-15.** A city transit service receives calls from broken-down buses and a wrecker crew must haul the buses in for service. The joint distribution of the number of calls received on Mondays and Tuesdays is given in Fig. 4-17, along with the marginal distributions. The variable X_1 represents the number of calls on Mondays and X_2 represents the number of calls on Tuesdays. A quick inspection will show that X_1 and X_2 are independent, since the joint probabilities are the product of the marginal probabilities.

x_{2_j} \ x_{1_i}	0	1	2	3	4	$p_2(x_{2_j})$
0	0.02	0.04	0.06	0.04	0.04	0.2
1	0.02	0.04	0.06	0.04	0.04	0.2
2	0.01	0.02	0.03	0.02	0.02	0.1
3	0.04	0.08	0.12	0.08	0.08	0.4
4	0.01	0.02	0.03	0.02	0.02	0.1
$p_1(x_{1_i})$	0.1	0.2	0.3	0.2	0.2	

Fig. 4-17.   Joint probabilities for wrecker calls.

4-8   Covariance and Correlation

We have noted that $E(X_1) = \mu_1$ and $V(X_1) = \sigma_1^2$ are the mean and variance of X_1. They may be determined from the marginal distribution of X_1. In a similar manner, μ_2 and σ_2^2 are the mean and variance of X_2. Two measures used in describing the *degree of association* between X_1 and X_2 are the *covariance* of $[X_1, X_2]$ and the *correlation coefficient*.

Definition

If $[X_1, X_2]$ is a two-dimensional random variable, the *covariance*, denoted by σ_{12}, is

$$\text{Cov}(X_1, X_2) = \sigma_{12} = E[(X_1 - E(X_1))(X_2 - E(X_2))] \tag{4-26}$$

and the *correlation coefficient*, denoted by ρ, is

$$\rho = \frac{\text{Cov}(X_1, X_2)}{\sqrt{V(X_1)} \cdot \sqrt{V(X_2)}} = \frac{\sigma_{12}}{\sigma_1 \cdot \sigma_2} \tag{4-27}$$

The correlation coefficient is a dimensionless quantity. By performing the multiplication operations in Equation (4-26) before distributing the outside expected value operator across the resulting quantities, we get an alternate form for the covariance as follows:

$$\text{Cov}(X_1, X_2) = E(X_1 \cdot X_2) - [E(X_1) \cdot E(X_2)] \tag{4-28}$$

Theorem 4-2

If X_1 and X_2 are independent, then $\rho = 0$.

Proof
If X_1 and X_2 are independent,

$$
\begin{aligned}
E(X_1 \cdot X_2) &= \int_{-\infty}^{\infty} \int_{-\infty}^{\infty} x_1 x_2 \cdot f(x_1, x_2)\, dx_1\, dx_2 \\
&= \int_{-\infty}^{\infty} \int_{-\infty}^{\infty} x_1 f_1(x_1) \cdot x_2 f_2(x_2)\, dx_1\, dx_2 \\
&= \left[\int_{-\infty}^{\infty} x_1 f_1(x_1)\, dx_1 \right] \cdot \left[\int_{-\infty}^{\infty} x_2 f_2(x_2)\, dx_2 \right] \\
&= E(X_1) \cdot E(X_2)
\end{aligned}
$$

Thus $\text{Cov}(X_1, X_2) = 0$ from Equation (4-28) and $\rho = 0$ from Equation (4-27). A similar argument would be used for $[X_1, X_2]$ discrete.

The converse of the theorem is not necessarily true, and we may have $\rho = 0$ without the variables being *independent*. In this case they are said to be *uncorrelated*.

Theorem 4-3

The value of ρ will be on the interval $[-1, +1]$, that is,

$$-1 \le \rho \le +1$$

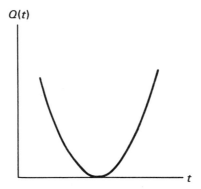

Fig. 4-18. The quadratic $Q(t)$.

Proof

Consider the function Q defined below and illustrated in Fig. 4-18.

$$Q(t) = E[(X_1 - E(X_1)) + t(X_2 - E(X_2))]^2$$
$$= E[X_1 - E(X_1)]^2 + 2tE[(X_1 - E(X_1))(X_2 - E(X_2))] + t^2E[X_2 - E(X_2)]^2$$

Since $Q(t) \geq 0$, the discriminant of $Q(t)$ must be ≤ 0, so that

$$\{2E[(X_1 - E(X_1))(X_2 - E(X_2))]\}^2 - 4E[X_2 - E(X_2)]^2E[X_1 - E(X_1)]^2 \leq 0$$

It follows that

$$4[\text{Cov}\,(X_1, X_2)]^2 - 4V(X_2) \cdot V(X_1) \leq 0$$

so

$$\frac{[\text{Cov}\,(X_1, X_2)]^2}{V(X_1)V(X_2)} \leq 1$$

and

$$-1 \leq \rho \leq +1 \qquad (4\text{-}29)$$

● **Example 4-16.** A continuous random vector $[X_1, X_2]$ has density function f as given below:

$$f(x_1, x_2) = 1 \qquad -x_2 < x_1 < +x_2, 0 < x_2 < 1$$
$$= 0 \qquad \text{otherwise}$$

This function is shown in Fig. 4-19.

The marginal densities are:

$$f_1(x_1) = 1 - x_1 \qquad \text{for } 0 < x_1 < 1$$
$$= 1 + x_1 \qquad \text{for } -1 < x_1 < 0$$
$$= 0 \qquad \text{otherwise}$$

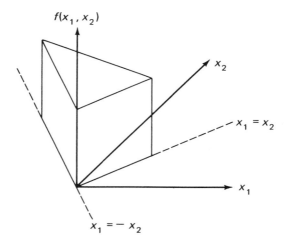

Fig. 4-19. A joint density.

and

$$f_2(x_2) = 2x_2 \qquad 0 < x_2 < 1$$
$$= 0 \qquad \text{otherwise}$$

Since $f(x_1, x_2) \neq f_1(x_1) \cdot f_2(x_2)$, the variables are *not independent*. If we calculate the covariance, we obtain

$$\text{Cov}\,(X_1, X_2) = \int_0^1 \int_{-x_2}^{x_2} x_1 \cdot x_2 \cdot 1 \; dx_1 \; dx_2 - 0 = 0$$

and thus $\rho = 0$, so the variables are *uncorrelated* although they are not independent.

Finally, it is noted that if X_2 is related to X_1 linearly, that is, $X_2 = A + BX_1$, then $\rho^2 = 1$. If $B > 0$, $\rho = +1$; and if $B < 0$, $\rho = -1$.

4-9 The Distribution Function for Two-Dimensional Random Vectors

The distribution function of the random vector $[X_1, X_2]$ is F, where

$$F(x_1, x_2) = P(X_1 \leq x_1, X_2 \leq x_2) \tag{4-30}$$

This is the probability over the shaded region in Fig. 4-20.

If $[X_1, X_2]$ is discrete, then

$$F(x_1, x_2) = \sum_{t_2 = -\infty}^{x_2} \sum_{t_1 = -\infty}^{x_1} p(t_1, t_2) \tag{4-31}$$

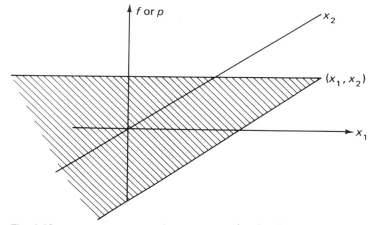

Fig. 4-20. Domain of integration or summation for $F(x_1, x_2)$.

and if $[X_1, X_2]$ is continuous, then

$$F(x_1, x_2) = \int_{-\infty}^{x_2} \int_{-\infty}^{x_1} f(t_1, t_2) \, dt_1 \, dt_2 \qquad (4\text{-}32)$$

Care must be exercised in defining F for the entire Euclidean plane. Once again if needed for clarification, a subscript, X_1, X_2, may be used on the function F as $F_{X_1 X_2}(x_1, x_2)$.

● **Example 4-17.** Suppose X_1 and X_2 have the following density:

$$f(x_1, x_2) = 24x_1 x_2 \qquad x_1 > 0, \, x_2 > 0, \, x_1 + x_2 < 1$$
$$= 0 \qquad \text{otherwise}$$

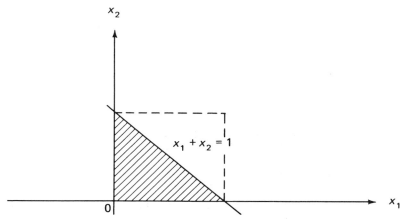

Fig. 4-21. The domain of F.

Looking at the Euclidean plane shown in Fig. 4-21 we see several cases that must be considered:

(a) $x_1 \leq 0$, $F(x_1, x_2) = 0$

(b) $x_2 \leq 0$, $F(x_1, x_2) = 0$

(c) $0 < x_1 < 1$ and $x_1 + x_2 < 1$,

$$F(x_1, x_2) = \int_0^{x_2} \int_0^{x_1} 24t_1 t_2 \, dt_1 \, dt_2$$
$$= 6x_1^2 \cdot x_2^2$$

(d) $0 < x_1 < 1$ and $1 - x_1 \leq x_2 \leq 1$,

$$F(x_1, x_2) = \int_0^{x_2} \int_0^{1-t_2} 24t_1 t_2 \, dt_1 \, dt_2 + \int_0^{1-x_1} \int_{x_1}^{1-t_2} 24t_1 t_2 \, dt_1 \, dt_2$$
$$= 6x_2^2 - 8x_2^3 + 3x_2^4 + 3(1 - x_1)^4 - 8(1 - x_1)^3$$
$$+ 6(1 - x_1)^2(1 - x_1^2)$$

(e) $0 < x_1 < 1$ and $x_2 > 1$,

$$F(x_1, x_2) = \int_0^{x_1} \int_0^{1-t_1} 24t_1 t_2 \, dt_2 \, dt_1$$
$$= 6x_1^2 - 8x_1^3 + 3x_1^4$$

(f) $0 \leq x_2 \leq 1$ and $x_1 \geq 1$,

$$F(x_1, x_2) = 6x_2^2 - 8x_2^3 + 3x_2^4$$

(g) $x_1 \geq 1$ and $x_2 \geq 1$,

$$F(x_1, x_2) = 1$$

The function F has properties analogous to those discussed in the one-dimensional case. We note that when X_1 and X_2 are continuous,

$$\frac{\partial^2 F(x_1, x_2)}{\partial x_1 \, \partial x_2} = f(x_1, x_2)$$

if the derivatives exist.

4-10 Functions of Two Random Variables

Often we will be interested in functions of several random variables; however, at present, this section will concentrate on functions of two random variables, say $Y = H(X_1, X_2)$. Since $X_1 = X_1(e)$ and $X_2 = X_2(e)$, $Y = H[X_1(e), X_2(e)]$ clearly depends on the outcome of the original experiment and, thus, Y is a random variable with range space R_Y.

The problem of finding the distribution of Y is somewhat more involved than in the case of functions of one variable; however, if $[X_1, X_2]$ is discrete, the procedure is straightforward if X_1 and X_2 take on a relatively small number of values.

● **Example 4-18.** If X_1 represents the number of defective units produced by machine No. 1 in one hour and X_2 represents the number of defective units produced by machine No. 2 in the same hour, then the joint distribution might be presented as in Fig. 4-22. Furthermore, suppose the random variable $Y = H(X_1, X_2)$, where $H(x_1, x_2) = 2x_1 + x_2$. It follows that $R_Y = \{0, 1, 2, 3, 4, 5, 6, 7, 8, 9\}$. In order to determine, say $P(Y = 0) = p_Y(0)$, we note that $Y = 0$ if and only if $X_1 = 0$ and $X_2 = 0$, therefore $p_Y(0) = .02$.

We note that $Y = 1$ if and only if $X_1 = 0$ and $X_2 = 1$; therefore $p_Y(1) = .06$. We also note that $Y = 2$ if and only if either $X_1 = 0$, $X_2 = 2$ or $X_1 = 1$, $X_2 = 0$; so $p_Y(2) = .10 + .03 = .13$. Using similar logic, we obtain the rest of the distribution, as follows.

y_i	$p_Y(y_i)$
0	.02
1	.06
2	.13
3	.11
4	.19
5	.15
6	.21
7	.07
8	.05
9	.01

In the case where the random vector is continuous and $H(x_1, x_2)$ is continuous, then $Y = H(X_1, X_2)$ is a continuous, one-dimensional random variable. The general procedure for the determination of the density function of Y is outlined below.

1. We are given $Y = H_1(X_1, X_2)$.
2. Introduce a second random variable $Z = H_2(X_1, X_2)$. The function H_2 is selected for convenience, but we want to be able to solve $y = H_1(x_1, x_2)$ and $z = H_2(x_1, x_2)$ for x_1 and x_2 in terms of y and z.
3. Find $x_1 = G_1(y, z)$, and $x_2 = G_2(y, z)$.
4. Find the following partial derivatives (we assume they exist and are continuous):

$$\frac{\partial x_1}{\partial y} \quad \frac{\partial x_1}{\partial z} \quad \frac{\partial x_2}{\partial y} \quad \frac{\partial x_2}{\partial z}$$

x_{1_i} x_{2_j}	0	1	2	3	$p_2(x_{2_j})$
0	0.02	0.03	0.04	0.01	0.1
1	0.06	0.09	0.12	0.03	0.3
2	0.10	0.15	0.20	0.05	0.5
3	0.02	0.03	0.04	0.01	0.1
$p_1(x_{1_i})$	0.2	0.3	0.4	0.1	

Fig. 4-22. Joint distribution of defectives produced on two machines $p(x_{1_i}, x_{2_j})$.

5. The joint density of $[Y, Z]$, denoted by $\ell(y, z)$, is found as follows:

$$\ell(y, z) = f[G_1(y, z), G_2(y, z)] \cdot |J(y, z)| \qquad (4\text{-}34)$$

where $J(y, z)$, called the *Jacobian* of the transformation, is given by the following determinant:

$$J(y, z) = \begin{vmatrix} \partial x_1/\partial y & \partial x_1/\partial z \\ \partial x_2/\partial y & \partial x_2/\partial z \end{vmatrix} \qquad (4\text{-}35)$$

6. The density of Y, say g, is then found as

$$g(y) = \int_{-\infty}^{\infty} \ell(y, z) \, dz \qquad (4\text{-}36)$$

● **Example 4-19.** Consider the continuous random vector $[X_1, X_2]$ with the following density:

$$f(x_1, x_2) = 4e^{-2(x_1+x_2)} \qquad x_1 > 0, x_2 > 0$$
$$= 0 \qquad \text{otherwise}$$

Suppose we are interested in the distribution of $Y = X_1/X_2$. We will let $y = x_1/x_2$ and choose $z = x_1 + x_2$ so that $x_1 = yz/(1 + y)$ and $x_2 = z/(1 + y)$. It follows that

$$\partial x_1/\partial y = z \left[\frac{1}{(1 + y)^2}\right] \qquad \text{and} \qquad \partial x_1/\partial z = \left[\frac{y}{(1 + y)}\right]$$

$$\partial x_2/\partial y = z \left[\frac{-1}{(1 + y)^2}\right] \qquad \text{and} \qquad \partial x_2/\partial z = \left[\frac{1}{(1 + y)}\right]$$

Therefore,

$$J(y, z) = \begin{vmatrix} \dfrac{z}{(1+y)^2} & \dfrac{y}{(1+y)} \\[3mm] -\dfrac{z}{(1+y)^2} & \dfrac{1}{(1+y)} \end{vmatrix} = \dfrac{z}{(1+y)^3} + \dfrac{zy}{(1+y)^3} = \dfrac{z}{(1+y)^2}$$

and

$$f[G_1(y, z), G_2(y, z)] = 4e^{-2\{[yz/(1+y)]+[z/(1+y)]\}}$$
$$= 4e^{-2z}$$

Thus,

$$\ell(y, z) = [4e^{-2z}] \cdot \dfrac{z}{(1+y)^2}$$

and

$$g(y) = \int_0^\infty 4e^{-2z}[z/(1+y)^2]\, dz$$

$$= \dfrac{1}{(1+y)^2} \qquad y > 0$$

$$= 0 \qquad\qquad \text{otherwise}$$

4-11   Random Vectors of Dimension $n > 2$

Should we have three or more random variables, the random vector will be denoted as $[X_1, X_2, \ldots, X_n]$, and extensions follow from the two-dimensional case. We will assume that the variables are continuous; however, the results may readily be extended to the discrete case substituting the appropriate summation operations for integrals. We assume the existence of a joint density f such that

$$f(x_1, x_2, \ldots, x_n) \geq 0 \tag{4-37}$$

and

$$\int_{-\infty}^{\infty} \int_{-\infty}^{\infty} \cdots \int_{-\infty}^{\infty} f(x_1, x_2, \ldots, x_n)\, dx_n \ldots dx_2\, dx_1 = 1$$

Thus,

$$P(a_1 \leq X_1 \leq b_1, a_2 \leq X_2 \leq b_2, \ldots, a_n \leq X_n \leq b_n)$$

$$= \int_{a_1}^{b_1} \int_{a_2}^{b_2} \cdots \int_{a_n}^{b_n} f(x_1, x_2, \ldots, x_n)\, dx_n \ldots dx_2\, dx_1 \tag{4-38}$$

The marginal densities are determined as follows:

$$f_1(x_1) = \int_{-\infty}^{\infty} \int_{-\infty}^{\infty} \cdots \int_{-\infty}^{\infty} f(x_1, x_2, \ldots, x_n) \, dx_n \ldots dx_2$$

$$f_2(x_2) = \int_{-\infty}^{\infty} \int_{-\infty}^{\infty} \cdots \int_{-\infty}^{\infty} f(x_1, x_2, \ldots, x_n) \, dx_n \ldots dx_3 \, dx_1$$

$$f_n(x_n) = \int_{-\infty}^{\infty} \int_{-\infty}^{\infty} \cdots \int_{-\infty}^{\infty} f(x_1 x_2, \ldots, x_n) \, dx_{n-1} \ldots dx_2 \, dx_1$$

The integration is over all variables having a subscript different from the one for which the marginal density is required.

Definition

The variables $[X_1, \ldots, X_n]$ are independent random variables if and only if for all $[x_1, x_2, \ldots, x_n]$

$$f(x_1, x_2, \ldots, x_n) = f_1(x_1) \cdot f_2(x_2) \cdot \ldots \cdot f_n(x_n) \tag{4-39}$$

The expected value of, say X_1, is

$$E(X_1) = \int_{-\infty}^{\infty} \int_{-\infty}^{\infty} \cdots \int_{-\infty}^{\infty} x_1 \cdot f(x_1, x_2, \ldots, x_n) \, dx_1 \, dx_2 \ldots dx_n \tag{4-40}$$

and the variance is

$$V(X_1) = \int_{-\infty}^{\infty} \int_{-\infty}^{\infty} \cdots \int_{-\infty}^{\infty} (x_1 - \mu_1)^2 \cdot f(x_1, x_2, \ldots, x_n) \, dx_1 \, dx_2 \ldots dx_n \tag{4-41}$$

We recognize these as the mean and variance, respectively, of the marginal distribution of X_1.

In the two-dimensional case considered earlier, geometric interpretations were instructive; however, in dealing with n-dimensional random vectors, the range space is the Euclidean n-space, and graphical presentations are thus not possible. The marginal distributions are, however, in one dimension and the conditional distribution for one variable given values for the other variables is in one dimension. The conditional distribution of X_1 given values $(x_2, x_3, \ldots, x_n)$ is denoted by

$$f_{X_1|x_2, \ldots, x_n}(x_1) = \frac{f(x_1, x_2, \ldots, x_n)}{\displaystyle\int_{-\infty}^{\infty} f(x_1, x_2, \ldots, x_n) \, dx_1} \tag{4-42}$$

and the expected value of X_1 for given $(x_2, \ldots, x_n)$ is

$$E(X_1|x_2, x_3, \ldots, x_n) = \int_{-\infty}^{\infty} x_1 \cdot f_{X_1|x_2, \ldots, x_n}(x_1) \, dx_1 \tag{4-43}$$

The hypothetical graph of $E(X_1|x_2, \ldots, x_n)$ as a function of the vector $[x_2, \ldots, x_n]$ is called the *regression of X_1 on $(X_2, \ldots, X_n)$*.

4-12   Linear Combinations

The consideration of general functions of random variables, say $X_1, X_2, \ldots, X_n$, is beyond the scope of this text. However, there is one particular function of the form $Y = H(X_1, \ldots, X_n)$, where

$$H(X_1, X_2, \ldots, X_n) = a_0 + a_1 X_1 + \cdots + a_n X_n \qquad (4\text{-}44)$$

that is of interest. The a_i are real constants for $i = 0, 1, 2, \ldots, n$. This is called a *linear combination* of the variables $X_1, X_2, \ldots, X_n$. A special situation occurs when $a_0 = 0$ and $a_1 = a_2 = \cdots = a_n = 1$, in which case we have a *sum* $Y = X_1 + X_2 + \cdots + X_n$.

● **Example 4-20.** Four resistors are connected in series as shown in Fig. 4-23. Each resistor has a resistance that is a random variable. The resistance of the assembly may be denoted by Y, where $Y = X_1 + X_2 + X_3 + X_4$.

● **Example 4-21.** Two parts are to be assembled as shown in Fig. 4-24. The clearance can be expressed as $Y = X_1 - X_2$ or $Y = (1)X_1 + (-1)X_2$. Of course, a negative clearance would mean interference. This is a linear combination with $a_0 = 0$, $a_1 = 1$, and $a_2 = -1$.

● **Example 4-22.** A sample of 10 items is randomly selected from the output of a process that manufactures a small shaft used in electric fan motors, and the diameters are to be measured with a value called the sample mean calculated as follows:

$$\bar{X} = \frac{1}{10}(X_1 + X_2 + \cdots + X_{10})$$

The value $\bar{X} = \frac{1}{10}X_1 + \frac{1}{10}X_2 + \cdots + \frac{1}{10}X_{10}$ is a linear combination with $a_0 = 0$ and $a_1 = a_2 = \cdots = a_{10} = \frac{1}{10}$.

Let us next consider how to determine the mean and variance of linear combinations. Consider the sum of two random variables,

$$Y = X_1 + X_2 \qquad (4\text{-}45)$$

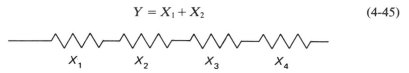

X_1 X_2 X_3 X_4

Fig. 4-23. Resistance in series.

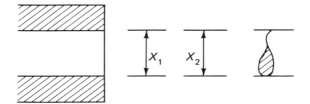

Fig. 4-24. A simple assembly.

The mean of Y or $\mu_Y = E(Y)$ is given as

$$E(Y) = E(X_1) + E(X_2) \tag{4-46}$$

however, the variance calculation is not so obvious.

$$
\begin{aligned}
V(Y) &= E[Y - E(Y)]^2 = E(Y^2) - [E(Y)]^2 \\
&= E[(X_1 + X_2)^2] - [E(X_1 + X_2)]^2 \\
&= E[X_1^2 + 2X_1X_2 + X_2^2] - [E(X_1) + E(X_2)]^2 \\
&= E(X_1^2) + 2E(X_1X_2) + E(X_2^2) - [E(X_1)]^2 - 2E(X_1)\cdot E(X_2) - [E(X_2)]^2 \\
&= \{E(X_1^2) - [E(X_1)]^2\} + \{E(X_2^2) - [E(X_2)]^2\} + 2[E(X_1X_2) - E(X_1)\cdot E(X_2)]
\end{aligned}
$$
$$V(Y) = V(X_1) + V(X_2) + 2\,\mathrm{Cov}\,(X_1, X_2)$$

or

$$\sigma_Y^2 = \sigma_1^2 + \sigma_2^2 + 2\sigma_{12} \tag{4-47}$$

These results generalize to any linear combination

$$Y = a_0 + a_1X_1 + a_2X_2 + \cdots + a_nX_n \tag{4-48}$$

as follows:

$$E(Y) = a_0 + \sum_{i=1}^{n} a_i E(X_i)$$

$$= a_0 + \sum_{i=1}^{n} a_i \mu_i \tag{4-49}$$

where $E(X_i) = \mu_i$, and

$$V(Y) = \sum_{i=1}^{n} a_i^2 V(X_i) + \sum_{i=1}^{n}\sum_{\substack{j=1 \\ i \neq j}}^{n} a_i a_j \,\mathrm{Cov}\,(X_i, X_j) \tag{4-50}$$

or

$$\sigma_Y^2 = \sum_{i=1}^{n} a_i^2 \sigma_i^2 + \sum_{i=1}^{n}\sum_{\substack{j=1 \\ i \neq j}}^{n} a_i a_j \sigma_{ij}$$

If the *variables are independent*, the expression for the variance of Y is greatly simplified, as all the covariance terms are zero. In this situation the variance of Y is simply:

$$V(Y) = \sum_{i=1}^{n} a_i^2 \cdot V(X_i) \tag{4-51}$$

or

$$\sigma_Y^2 = a_1^2 \sigma_1^2 + a_2^2 \sigma_2^2 + \cdots + a_n^2 \sigma_n^2$$

● **Example 4-23.** In Example 4-20, four resistors were connected in series so that $Y = X_1 + X_2 + X_3 + X_4$ was the resistance of the assembly, and X_1 was the resistance of the first resistor, and so on. The mean and variance of Y in terms of the means and variances of the components may be easily calculated. If the resistors are selected randomly for the assembly, it is reasonable to assume that X_1, X_2, X_3, and X_4 are independent so that

$$\mu_Y = \mu_1 + \mu_2 + \mu_3 + \mu_4$$

and

$$\sigma_Y^2 = \sigma_1^2 + \sigma_2^2 + \sigma_3^2 + \sigma_4^2$$

We have said nothing about the distribution of Y; however, given the mean and variance of X_1, X_2, X_3, and X_4, we may readily calculate the mean and the variance of Y since the variables are independent.

● **Example 4-24.** In Example 4-21 where two components were to be assembled, suppose the joint distribution of $[X_1, X_2]$ is

$$f(x_1, x_2) = 8e^{-(2x_1 + 4x_2)} \quad x_1 \ge 0, x_2 \ge 0$$
$$= 0 \quad \text{otherwise}$$

Since $f(x_1, x_2)$ can be easily factored as

$$f(x_1, x_2) = [2e^{-2x_1}] \cdot [4e^{-4x_2}]$$
$$= f_1(x_1) \cdot f_2(x_2)$$

X_1 and X_2 are independent. Furthermore, $E(X_1) = \mu_1 = \frac{1}{2}$, and $E(X_2) = \mu_2 = \frac{1}{4}$. We may calculate the variances

$$V(X_1) = \sigma_1^2 = \int_0^\infty x_1^2 \cdot 2e^{-2x_1} \, dx_1 - \left(\frac{1}{2}\right)^2 = \frac{1}{4}$$

and

$$V(X_2) = \sigma_2^2 = \int_0^\infty x_2^2 \cdot 4e^{-4x_2} \, dx_2 - \left(\frac{1}{4}\right)^2 = \frac{1}{16}$$

We denoted clearance $Y = X_1 - X_2$ or $Y = (1)X_1 + (-1)X_2$ in the example, so

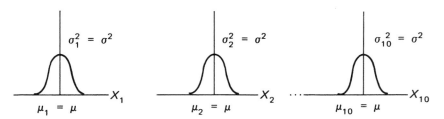

Fig. 4-25. Some identical distributions.

that $E(Y) = \mu_1 - \mu_2 = \frac{1}{2} - \frac{1}{4} = \frac{1}{4}$ and $V(Y) = (1)^2 \cdot \sigma_1^2 + (-1)^2 \cdot \sigma_2^2 = \frac{1}{4} + \frac{1}{16} = \frac{5}{16}$.

● **Example 4-25.** In Example 4-22, we might expect the random variables $X_1, \ldots, X_{10}$ to be independent because of the random sampling process. Furthermore, the distribution for each variable X_i is identical. This is shown in Fig. 4-25. In the earlier example, the linear combination of interest was the *sample mean*

$$\bar{X} = \frac{1}{10} X_1 + \cdots + \frac{1}{10} X_{10}$$

It follows that

$$E(\bar{X}) = \mu_{\bar{X}} = \frac{1}{10} \cdot E(X_1) + \frac{1}{10} \cdot E(X_2) + \cdots + \frac{1}{10} \cdot E(X_{10})$$

$$= \frac{1}{10} \mu + \frac{1}{10} \mu + \cdots + \frac{1}{10} \mu$$

$$= \mu$$

Furthermore,

$$V(\bar{X}) = \sigma_{\bar{X}}^2 = \left(\frac{1}{10}\right)^2 \cdot V(X_1) + \left(\frac{1}{10}\right)^2 \cdot V(X_2) + \cdots + \left(\frac{1}{10}\right)^2 \cdot V(X_{10})$$

$$= \left(\frac{1}{10}\right)^2 \cdot \sigma^2 + \left(\frac{1}{10}\right)^2 \cdot \sigma^2 + \cdots + \left(\frac{1}{10}\right)^2 \cdot \sigma^2$$

$$= \frac{\sigma^2}{10}$$

4-13 Moment-Generating Functions and Linear Combinations

In the case where $Y = aX$, then

$$M_Y(t) = M_X(at) \tag{4-52}$$

For sums of *independent random variables*, $Y = X_1 + X_2 + \cdots + X_n$, then

$$M_Y(t) = M_{X_1}(t) \cdot M_{X_2}(t) \cdot \ldots \cdot M_{X_n}(t) \tag{4-53}$$

and this property has considerable use in statistics. If the linear combination is of the general form $Y = a_0 + a_1X_1 + \cdots + a_nX_n$, and the variables $X_1, \ldots, X_n$ are independent, then

$$M_Y(t) = e^{a_0 t}[M_{X_1}(a_1 t) \cdot M_{X_2}(a_2 t) \cdot \ldots \cdot M_{X_n}(a_n t)]$$

Linear combinations are to be of particular significance in later chapters, and we will discuss them again at greater length.

4-14 The Law of Large Numbers

A special case arises in dealing with sums of independent random variables where each variable may take only two values, 0 and 1. Consider the following formulation. An experiment $\mathscr{E}$ consists of n independent experiments (trials) $\mathscr{E}_j$, $j = 1, 2, \ldots, n$. There are only two outcomes, success, $\{S\}$, and failure, $\{F\}$, to each trial, so that $\mathscr{S}_j = \{S, F\}$. The probabilities

$$P\{S\} = p$$

and

$$P\{F\} = 1 - p = q$$

remain constant for $j = 1, 2, \ldots, n$. We let

$$X_j = \begin{cases} 0, & \text{if the } j\text{th trial results in failure} \\ 1, & \text{if the } j\text{th trial results in success} \end{cases}$$

and

$$Y = X_1 + X_2 + X_3 + \cdots + X_n$$

Thus Y represents the number of successes in n trials; and Y/n is an approximation (or estimator) for the unknown probability p. For convenience, we will let $\hat{p} = Y/n$. Note that this value corresponds to the term f_A used in the relative frequency definition of Chapter 1.

The law of large numbers states that

$$P[|\hat{p} - p| < \epsilon] \geq 1 - \frac{p(1-p)}{n\epsilon^2} \tag{4-54}$$

or equivalently

$$P[|\hat{p} - p| \geq \epsilon] \leq \frac{p(1-p)}{n\epsilon^2} \tag{4-55}$$

To indicate the proof, we note that

$$E(Y) = n \cdot E(X_j) = n[(0 \cdot q) + (1 \cdot p)] = np$$

and

$$V(Y) = nV(X_j) = n[(0^2 \cdot q) + (1^2 \cdot p) - (p)^2] = np(1-p)$$

Since $\hat{p} = (1/n)Y$, we have

$$E(\hat{p}) = \frac{1}{n} \cdot E(Y) = p \qquad (4\text{-}56)$$

and

$$V(\hat{p}) = \left[\frac{1}{n}\right]^2 \cdot V(Y) = \frac{p(1-p)}{n}$$

Using Chebyshev's inequality,

$$P\left[|\hat{p} - p| < k \sqrt{\frac{p(1-p)}{n}}\right] \geq 1 - \frac{1}{k^2} \qquad (4\text{-}57)$$

so that if

$$\epsilon = k \sqrt{\frac{p(1-p)}{n}}$$

then

$$P[|\hat{p} - p| < \epsilon] \geq 1 - \frac{p(1-p)}{n\epsilon^2}$$

Thus for arbitrary $\epsilon > 0$, as $n \to \infty$,

$$P[|\hat{p} - p| < \epsilon] \to 1$$

Equation (4-54) may be rewritten, with an obvious notation, as

$$P[|\hat{p} - p| < \epsilon] \geq 1 - \alpha \qquad (4\text{-}58)$$

We may now fix both ϵ and α in Equation (4-58) and determine the value of n required to satisfy the probability statement as

$$n \geq \frac{p(1-p)}{\epsilon^2 \alpha} \qquad (4\text{-}59)$$

● **Example 4-26.** A production process is operated in such a manner that there is a probability p that each item produced is defective, and p is unknown. A random sample of n items is to be selected to *estimate* p. The estimator to be used is $\hat{p} = Y/n$, where

$$X_j = \begin{cases} 0, & \text{if the } j\text{th item is good} \\ 1, & \text{if the } j\text{th item is defective} \end{cases}$$

and

$$Y = X_1 + X_2 + \cdots + X_n$$

It is desired that the probability be at least .95 that the error, $|\hat{p} - p|$, not exceed .01. In order to determine the required value of n, we note that $\epsilon = .01$, and $\alpha = .05$; however, p is unknown. Equation (4-59) indicates that

$$n \geq \frac{p(1-p)}{(.01)^2 \cdot (.05)}$$

Since p is unknown, the worst possible case must be assumed [note that $p(1-p)$ is maximum when $p = \frac{1}{2}$]. This yields

$$n \geq \frac{(.5)(.5)}{(.01)^2(.05)} = 50,000$$

a very large number indeed.

Example 4-26 demonstrates why the law of large numbers is not used extensively. The requirements of $\epsilon = .01$ and $\alpha = .05$ to give a probability of .95 of the departure $|\hat{p} - p|$ being less than .01 seem reasonable; however, the resulting sample size is very large. In order to resolve problems of this nature we must know the distribution of the random variables involved ($\hat{p}$ in this case). The next three chapters will consider in detail a number of the more frequently encountered distributions.

4-15 Summary

This chapter has presented a number of topics related to jointly distributed random variables and functions of jointly distributed variables. The examples presented illustrate these topics, and the exercises which follow will allow the student to reinforce these concepts.

A great many situations encountered in engineering, science, and management involve situations where several related random variables simultaneously bear on the response being observed. The approach presented in this chapter provides the structure for dealing with several aspects of such problems.

4-16 Exercises

4-1. A refrigerator manufacturer subjects his finished product to a final inspection. Of interest are two categories of defects: scratches or flaws in the porcelain finish, and mechanical defects. The number of each type of defect is a random variable. The results of inspecting 50 refrigerators are shown in the following table, where X represents the occurrence of finish defects and Y represents the occurrence of mechanical defects.
(a) Find the marginal distributions of X and Y.
(b) Find the probability distribution of mechanical defects, given that there are no finish defects.

Y \ X	0	1	2	3	4	5
0	11/50	4/50	2/50	1/50	1/50	1/50
1	8/50	3/50	2/50	1/50	1/50	
2	4/50	3/50	2/50	1/50		
3	3/50	1/50				
4	1/50					

(c) Find the probability distribution of finish defects, given that there are no mechanical defects.

4-2. An inventory manager has accumulated records of demand for her company's product over the last 100 days. The random variable X represents the number of orders received per day and the random variable Y represents the number of units per order. Her data are shown in the following table.

X \ Y	1	2	3	4	5	6	7	8	9
1	10/100	6/100	3/100	2/100	1/100	1/100	1/100	1/100	1/100
2	8/100	5/100	3/100	2/100	1/100	1/100	1/100		
3	8/100	5/100	2/100	1/100	1/100				
4	7/100	4/100	2/100	1/100	1/100				
5	6/100	3/100	1/100	1/100					
6	5/100	3/100	1/100	1/100					

(a) Find the marginal distributions of X and Y.
(b) Find all conditional distributions for Y given X.

4-3. Let X_1 and X_2 be the scores on a general intelligence test and an occupational preference test, respectively. The probability density function of the random variables $[X_1, X_2]$ is given by

$$f(x_1, x_2) = \frac{k}{1000} \qquad 0 \le x_1 \le 100, \, 0 \le x_2 \le 10$$

$$= 0 \qquad \text{otherwise}$$

(a) Find the appropriate value of k.
(b) Find the marginal densities of X_1 and X_2.
(c) Find an expression for the cumulative distribution function $F(x_1, x_2)$.

4-4. Consider a situation in which the surface tension and acidity of a chemical product are measured. These variables are coded such that surface tension is measured on a scale $0 \le X_1 \le 2$, and acidity is measured on a scale $2 \le X_2 \le 4$.

The probability density function of $[X_1, X_2]$ is

$$f(x_1, x_2) = k(6 - x_1 - x_2) \qquad 0 \le x_1 \le 2, 2 \le x_2 \le 4$$
$$= 0 \qquad \text{otherwise}$$

(a) Find the appropriate value of k.
(b) Calculate the probability that $X_1 < 1, X_2 < 3$.
(c) Calculate the probability that $X_1 + X_2 \le 4$.
(d) Find the probability that $X_1 < 1.5$.
(e) Find the marginal densities of both X_1 and X_2.

4-5.   Given the density function

$$f(w, x, y, z) = 16wxyz \qquad 0 \le w \le 1, 0 \le x \le 1, 0 \le y \le 1, 0 \le z \le 1$$
$$= 0 \qquad \text{otherwise}$$

(a) Compute the probability that $W \le \frac{2}{3}$ and $Y \le \frac{1}{2}$.
(b) Compute the probability that $X \le \frac{1}{2}$ and $Z \le \frac{1}{4}$.
(c) Find the marginal density of W.

4-6.   Suppose the joint density of $[X, Y]$ is

$$f(x, y) = \frac{1}{8}(6 - x - y) \qquad 0 \le x \le 2, 2 \le y \le 4$$
$$= 0 \qquad \text{otherwise}$$

Find the conditional densities $f_{X|y}(x)$ and $f_{Y|x}(y)$.

4-7.   For the data in Exercise 4-2 find the expected number of units per order, given that there are three orders per day.

4-8.   Consider the probability distribution of the discrete random vector $[X_1, X_2]$, where X_1 represents the number of orders for aspirin in August at the neighborhood drugstore and X_2 represents the number of orders in September. The joint distribution is shown in the following table.

X_2 \ X_1	51	52	53	54	55
51	0.06	0.05	0.05	0.01	0.01
52	0.07	0.05	0.01	0.01	0.01
53	0.05	0.10	0.10	0.05	0.05
54	0.05	0.02	0.01	0.01	0.03
55	0.05	0.06	0.05	0.01	0.03

(a) Find the marginal distributions.
(b) Find the expected sales in September, given that sales in August were either 51, 52, 53, 54, or 55, respectively.

4-9.   Assume that X_1 and X_2 are coded scores on two intelligence tests, and the probability density function of $[X_1, X_2]$ is given by

$$f(x_1, x_2) = 6x_1^2 x_2 \qquad 0 \le x_1 \le 1, 0 \le x_2 \le 1$$
$$= 0 \qquad \text{otherwise}$$

Find the expected value of the score on test No. 2 given the score on test No. 1. Also, find the expected value of the score on test No. 1 given the score on test No. 2.

4-10.   Let

$$f(x_1, x_2) = 4x_1 x_2 e^{-(x_1^2 + x_2^2)} \qquad 0 \le x_1 < \infty, 0 \le x_2 < \infty$$
$$= 0 \qquad \text{otherwise}$$

(a) Find the marginal distributions of X_1 and X_2.
(b) Find the conditional probability distributions of X_1 and X_2.
(c) Find an expression for the conditional expectations of X_1 and X_2.

4-11.   Assume that $[X, Y]$ is a continuous random vector and that X and Y are independent such that $f(x, y) = g(x)h(y)$. Define a new random variable $Z = XY$. Show that the probability density function of Z, $\ell(z)$, is given by

$$\ell(z) = \int_{-\infty}^{\infty} g(t)h\left(\frac{z}{t}\right)\left|\frac{1}{t}\right| dt$$

Hint: Let $Z = XY$ and $T = X$ and find the Jacobian for the transformation to the joint probability density function of Z and T, say $r(z, t)$. Then integrate $r(z, t)$ with respect to t.

4-12.   Use the result of the previous problem to find the probability density function of the area of a rectangle, $A = S_1 S_2$ where the sides are of random length. Specifically, the sides are independent random variables such that

$$g(s_1) = 2s_1 \qquad 0 \le s_1 \le 1$$
$$= 0 \qquad \text{otherwise}$$

and

$$h(s_2) = \frac{1}{8} s_2 \qquad 0 \le s_2 \le 4$$
$$= 0 \qquad \text{otherwise}$$

Some care must be taken in determining the limits of integration as the variable of integration cannot assume negative values.

4-13.   Assume that $[X, Y]$ is a continuous random vector and that X and Y are independent such that $f(x, y) = g(x)h(y)$. Define a new random variable $Z = X/Y$. Show that the probability density function of Z, $\ell(z)$, is given by

$$\ell(z) = \int_{-\infty}^{\infty} g(uz)h(u)|u| \, du$$

Hint: Let $Z = X/Y$ and $U = Y$, and find the Jacobian for the transformation to the joint probability density function of Z and U, say $p(z, u)$. Then integrate $p(z, u)$ with respect to u.

4-14. Suppose we have a simple electrical circuit in which Ohm's law $V = IR$ holds. We wish to find the probability distribution of resistance given that the probability distributions of voltage (V) and current (I) are known to be:

$$g(v) = e^{-v} \qquad v \geq 0$$
$$= 0 \qquad \text{otherwise}$$
$$h(i) = 3e^{-3i} \qquad i \geq 0$$
$$= 0 \qquad \text{otherwise}$$

Use the results of the previous problem, assuming that V and I are independent random variables.

4-15. Demand for a certain product is a random variable having a mean of 20 units per day and a variance of 9. We define the lead time to be the time that elapses between the placement of an order and its arrival. The lead time for the product is fixed at four days. Find the expected value and the variance of *lead time demand*, assuming demands to be independently distributed.

4-16. Prove in detail the (*a*) and (*b*) parts of Theorem 4-1 (page 104).

4-17. Let X_1 and X_2 be random variables such that $X_2 = A + BX_1$. Show that $\rho^2 = 1$, and that $\rho = -1$ if $B < 0$ and $\rho = +1$ if $B > 0$.

4-18. Let X_1 and X_2 be random variables such that $X_2 = A + BX_1$. Show that the moment-generating function for X_2 is

$$M_{X_2}(t) = e^{At}M_{X_1}(Bt)$$

4-19. Let X_1 and X_2 be distributed according to

$$f(x_1, x_2) = 2 \qquad 0 \leq x_1 \leq x_2 \leq 1$$
$$= 0 \qquad \text{otherwise}$$

Find the correlation coefficient between X_1 and X_2.

4-20. Let X_1 and X_2 be random variables with correlation coefficient ρ_{X_1, X_2}. Suppose we define two new random variables $U = A + BX_1$ and $V = C + DX_2$, where A, B, C, and D are constants. Show that $\rho_{UV} = (BD/|BD|)\rho_{X_1, X_2}$.

4-21. Consider the data shown in Exercise 4-1. Are X and Y independent? Calculate the correlation coefficient.

4-22. A couple wishes to sell their house. The minimum price that they are willing to accept is a random variable, say X, where $s_1 \leq X \leq s_2$. A population of buyers are interested in the house. Let Y, where $p_1 \leq Y \leq p_2$, denote the maximum price they are willing to pay. Y is also a random variable. Assume that the joint distribution of $[X, Y]$ is $f(x, y)$.

(*a*) Under what circumstances will a sale take place?
(*b*) Write an expression for the probability of a sale taking place.
(*c*) Write an expression for the expected price of the transaction.

4-23. Let $[X, Y]$ be uniformly distributed over the semicircle in the following diagram. Thus $f(x, y) = 2/\pi$ if $[x, y]$ is in the semicircle.

(*a*) Find the marginal distributions of X and Y.
(*b*) Find the conditional probability distributions.
(*c*) Find the conditional expectations.

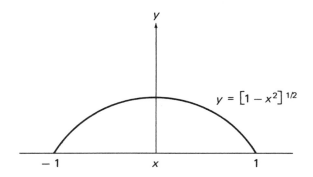

$$y = \left[1 - x^2\right]^{1/2}$$

4-24. Let X and Y be independent random variables. Prove that $E(X|Y) = E(X)$ and that $E(Y|X) = E(Y)$

4-25. Show that, in the continuous case,

$$E[E(X|Y)] = E(X)$$
$$E[E(Y|X)] = E(Y)$$

4-26. Consider the two independent random variables, S and D, whose probability distributions are shown below:

$$f(s) = \frac{1}{30} \qquad 10 \le s \le 40$$
$$= 0 \qquad \text{otherwise}$$

$$g(d) = \frac{1}{20} \qquad 10 \le d \le 30$$
$$= 0 \qquad \text{otherwise}$$

Find the probability distribution of the new random variable

$$W = S + D$$

4-27. If

$$f(x, y) = x + y \qquad 0 < x < 1, 0 < y < 1$$
$$= 0 \qquad \text{otherwise}$$

find the following:

(*a*) $E[X|y]$
(*b*) $E[X]$
(*c*) $E[Y]$

4-28. For the multivariate distribution

$$f(x, y) = \frac{k(1 + x + y)}{(1 + x)^4 (1 + y)^4} \qquad 0 \le x < \infty, 0 \le y < \infty$$

$$= 0 \qquad\qquad\qquad \text{otherwise}$$

(a) Evaluate the constant k.
(b) Find the marginal distribution of X.

4-29. For the multivariate distribution

$$f(x, y) = \frac{k}{(1 + x + y)^n} \qquad x \ge 0, y \ge 0, n > 2$$

$$= 0 \qquad\qquad\qquad \text{otherwise}$$

(a) Evaluate the constant k.
(b) Find $F(x, y)$.

4-30. The manager of a small bank wishes to determine the proportion of the time a particular teller is busy. He decides to observe the teller at n randomly spaced intervals. The estimator of the degree of gainful employment is to be Y/n, where

$$X_i = \begin{cases} 0, \text{ if on the } i\text{th observation, teller is idle} \\ 1, \text{ if on the } i\text{th observation, teller is busy} \end{cases}$$

and $Y = \Sigma_{i=1}^n X_i$. It is desired to estimate p so that the error of the estimate does not exceed .05 with probability .95. Determine the necessary value of n.

4-31. Given the following joint distributions determine whether X and Y are independent.

(a) $g(x, y) = 4xye^{-(x^2+y^2)}$ $\qquad x \ge 0, y \ge 0$
(b) $f(x, y) = 3x^2 y^{-3}$ $\qquad\qquad 0 \le x \le y \le 1$
(c) $f(x, y) = 6(1 + x + y)^{-4}$ $\qquad x \ge 0, y \ge 0$

4-32. Let $f(x, y, z) = h(x)h(y)h(z)$, $x \ge 0$, $y \ge 0$, $z \ge 0$. Determine the probability that a point drawn at random will have a coordinate (x, y, z) that does not satisfy either $x > y > z$ or $x < y < z$.

4-33. Suppose that X and Y are random variables denoting the fraction of a day that a request for merchandise occurs and the receipt of a shipment occurs, respectively. The joint probability density function is

$$f(x, y) = 1 \qquad 0 \le x \le 1, 0 \le y \le 1$$

$$= 0 \qquad \text{otherwise}$$

(a) What is the probability that both the request for merchandise and the receipt of an order occur during the first half of the day?
(b) What is the probability that a request for merchandise occurs after its receipt? Before its receipt?

4-34. Suppose that in the above problem the merchandise is highly perishable and must be requested during the $\frac{1}{4}$-day interval after it arrives. What is the probability that merchandise will not spoil?

4-35.   Let X be a continuous random variable with probability density function $f(x)$. Find a general expression for the new random variable Z, where:

(a) $Z = a + bX$

(b) $Z = 1/X$

(c) $Z = \log_e X$

(d) $Z = e^x$

Chapter 5
Some Important Discrete Distributions

5-1 Introduction

In this chapter we present several discrete probability distributions, developing their analytical form from certain basic assumptions about real world phenomena. We also present some examples of their application. The distributions presented have found extensive application in engineering, operations research, and management science. Four of the distributions, the *binomial*, the *geometric*, the *Pascal*, and the *negative binomial*, stem from a *random process* made up of sequential *Bernoulli trials*. The *hypergeometric distribution*, the *multinomial distribution*, and the *Poisson distribution* will also be presented in this chapter.

5-2 Bernoulli Trials and the Bernoulli Distribution

There are many problems in which the experiment consists of n trials or subexperiments. Here we are concerned with an individual trial that has as its two possible outcomes *success*, $\{S\}$, or *failure*, $\{F\}$. For each trial we thus have:

$\mathcal{E}_j$: Perform an experiment (the jth) and observe the outcome.
$\mathcal{S}_j$: $\{S, F\}$.

For convenience, we will define a random variable $X_j = 1$ if $\mathcal{E}_j$ results in $\{S\}$ and $X_j = 0$ if $\mathcal{E}_j$ results in $\{F\}$. See Fig. 5-1.

The n Bernoulli trials $\mathcal{E}_1, \mathcal{E}_2, \ldots, \mathcal{E}_n$ are called a Bernoulli process if the trials are independent, each trial has only two possible outcomes, say $\{S\}$ or $\{F\}$, and the probability of success remains constant from trial to trial. That is,

$$p(x_1, x_2, \ldots, x_n) = p_1(x_1) \cdot p_2(x_2) \cdot \ldots \cdot p_n(x_n)$$

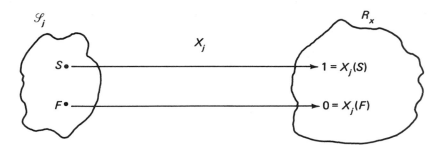

Fig. 5-1. A Bernoulli trial.

and

$$p_j(x_j) = p(x_j) \begin{cases} = p & x_j = 1, j = 1, 2, \ldots, n \\ = (1-p) = q & x_j = 0, j = 1, 2, \ldots, n \\ = 0 & \text{otherwise} \end{cases} \qquad (5\text{-}1)$$

For one trial, the distribution given in Equation (5-1) and Fig. 5-2 is called the Bernoulli distribution.

The mean and variance are calculated as follows:

$$E(X_j) = 0 \cdot q + 1 \cdot p = p$$

and

$$V(X_j) = [(0^2 \cdot q) + (1^2 \cdot p)] - p^2 = p(1-p) \qquad (5\text{-}2)$$

The moment-generating function may be shown to be

$$M_{X_j}(t) = q + pe^t \qquad (5\text{-}3)$$

● **Example 5-1.** Suppose we consider a manufacturing process in which a small steel part is produced by an automatic machine. Furthermore, each part in a

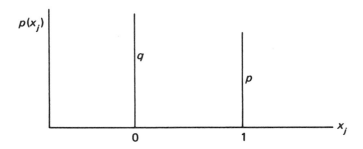

Fig. 5-2. The Bernoulli distribution.

production run of 1000 parts may be classified as defective or good when inspected. We can think of the production of a part as a single trial that results in success (say a defective) or failure (a good item). If we have reason to believe that the machine is just as likely to produce a defective on one run as on another, and if the production of a defective on one run is neither more nor less likely because of the results on the previous runs, then it would be quite reasonable to assume that the production run is a Bernoulli process with 1000 trials. The probability; p, of a defective on one trial is called the *process average fraction defective.*

Note that in the preceeding example the assumption of a Bernoulli process is a *mathematical idealization* of the actual real world situation. Effects of tool wear, machine adjustment, and instrumentation difficulties were ignored. The real world was approximated by a model that did not consider all factors, but nevertheless, the approximation is good enough for useful results to be obtained.

We are going to be primarily concerned with a series of Bernoulli trials. In this case the experiment $\mathscr{E}$ is denoted as $\{(\mathscr{E}_1, \mathscr{E}_2, \ldots, \mathscr{E}_n): \mathscr{E}_j$ are independent Bernoulli trials, $j = 1, 2, \ldots, n\}$. The sample space is

$$\mathscr{S} = \{(x_1, \ldots, x_n): x_i = S \quad \text{or} \quad F, i = 1, \ldots, n\}$$

● **Example 5-2.** Suppose an experiment consists of three Bernoulli trials and the probability of success is p on each trial (see Fig. 5-3). The random variable X

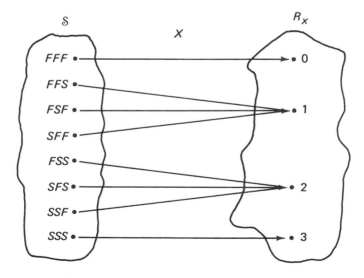

Fig. 5-3. Three Bernoulli trials.

is given by $X = \sum_{j=1}^{3} X_j$. The distribution of X can be determined as follows:

x	$p(x)$
0	$P\{FFF\} = q \cdot q \cdot q = q^3$
1	$P\{FFS\} + P\{FSF\} + P\{SFF\} = 3pq^2$
2	$P\{FSS\} + P\{SFS\} + P\{SSF\} = 3p^2q$
3	$P\{SSS\} = p^3$

5-3   The Binomial Distribution

The random variable X that denotes *the number of successes in n Bernoulli trials has a binomial distribution* given by $p(x)$, where

$$p(x) = \binom{n}{x} p^x (1-p)^{n-x} \qquad x = 0, 1, 2, \ldots, n \qquad (5\text{-}4)$$

$$= 0 \qquad\qquad\qquad \text{otherwise}$$

Example 5-2 clearly illustrates a binomial distribution with $n = 3$. The parameters of the binomial distribution are n and p, where n is a positive integer, and $0 \leq p \leq 1$. A simple derivation is outlined below. Let

$$p(x) = P\{\text{"}x \text{ successes in } n \text{ trials"}\}$$

The probability of the *particular outcome* in $\mathscr{S}$ with Ss for the first x trials and Fs for the last $n - x$ trials is

$$P(\overbrace{SSS \ldots SS}^{x} \overbrace{FF \ldots FF}^{n-x}) = p^x q^{n-x}$$

(where $q = 1 - p$) due to the independence of the trials. There are $\binom{n}{x} = \dfrac{n!}{x!(n-x)!}$ outcomes having exactly x Ss and $(n-x)$ Fs; therefore,

$$p(x) = \binom{n}{x} p^x q^{n-x} \qquad x = 0, 1, 2, \ldots n$$

$$= 0 \qquad\qquad\qquad \text{otherwise}$$

The mean may be determined directly as

$$E(X) = \sum_{x=0}^{n} x \cdot \frac{n!}{x!(n-x)!} p^x q^{n-x}$$

$$= np \sum_{x=1}^{n} \frac{(n-1)!}{(x-1)!(n-x)!} p^{x-1} q^{n-x}$$

and letting $y = x - 1$

$$E(X) = np \sum_{y=0}^{n-1} \frac{(n-1)!}{y!(n-1-y)!} p^y q^{n-1-y}$$

so that

$$E(X) = np \tag{5-5}$$

Using a similar approach we can calculate the variance as

$$V(X) = \sum_{x=0}^{n} \frac{x^2 n!}{x!(n-x)!} p^x q^{n-x} - (np)^2$$

$$= n(n-1)p^2 \sum_{y=0}^{n-2} \frac{(n-2)!}{y!(n-2-y)!} p^y q^{n-2-y} + np - (np)^2$$

so that

$$V(X) = npq \tag{5-6}$$

A much easier approach would have been to consider X as a sum of n independent random variables, each with mean p and variance pq, so that $X = X_1 + X_2 + \cdots + X_n$, then

$$E(X) = p + p + \cdots + p = np$$

and

$$V(X) = pq + pq + \cdots + pq = npq$$

The moment-generating function for the binomial distribution is

$$M_X(t) = (pe^t + q)^n \tag{5-7}$$

● **Example 5-3.** A production process represented schematically by Fig. 5-4 produces thousands of parts per day. On the average, 1 percent of the parts are defective and this average does not vary with time. Every hour, a random sample of 100 parts is selected from a conveyor and several characteristics are observed and measured on each part; however, the inspector classifies the part as either good or defective. If we consider the sampling as $n = 100$ Bernoulli trials with $p = .01$, the total number of defectives in the sample, X, would have a binomial distribution

$$p(x) = \binom{100}{x} (.01)^x (.99)^{100-x} \qquad x = 0, 1, 2, \ldots, 100$$

$$= 0 \qquad\qquad\qquad\qquad \text{otherwise}$$

Suppose the inspector has instructions to stop the process if the sample has more than two defectives. Then, the $P(X > 2) = 1 - P(X \le 2)$, and we may

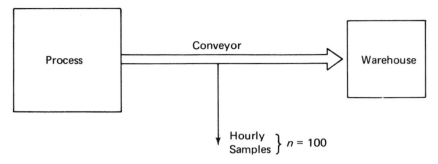

Hourly Samples $\}\ n = 100$

Fig. 5-4.   A sampling situation with attribute measurement.

calculate

$$P(X \leq 2) = \sum_{x=0}^{2} \binom{100}{x} (.01)^x (.99)^{100-x}$$
$$= (.99)^{100} + 100(.01)^1(.99)^{99} + 4950(.01)^2(.99)^{98}$$
$$\approx .92$$

Thus, the probability of the inspector stopping the process is approximately $1 - (.92) = .08$. The mean number of defectives that would be found is $E(X) = np = 100(.01) = 1$, and the variance is $V(X) = npq = .99$.

The cumulative binomial distribution or the distribution function, F, is

$$F(x) = \sum_{k=0}^{x} \binom{n}{k} p^k (1-p)^{n-k} \tag{5-8}$$

This function has been extensively tabulated. For example, see Romig's *50–100 Binomial Tables* (1953) and the *Cumulative Binomial Probability Distribution* (1955).

Another random variable, first noted in the law of large numbers, is frequently of interest. It is the proportion of successes and is denoted by

$$\hat{p} = X/n \tag{5-9}$$

where X has a binomial distribution with parameters n and p. The mean, variance, and moment-generating function are given below:

$$E(\hat{p}) = \frac{1}{n} \cdot E(X) = \frac{1}{n} np = p \tag{5-10}$$

$$V(\hat{p}) = \left(\frac{1}{n}\right)^2 \cdot V(X) = \frac{1}{n^2} \cdot npq = \frac{pq}{n} \tag{5-11}$$

$$M(t) = M_X \left(\frac{t}{n}\right) = (pe^{t/n} + q)^n \tag{5-12}$$

In order to evaluate, say $P(\hat{p} \le p_0)$, where p_0 is some number between 0 and 1, we note that

$$P(\hat{p} \le p_0) = P\left(\frac{X}{n} \le p_0\right) = P(X \le np_0)$$

Since np_0 is possibly not an integer,

$$P(\hat{p} \le p_0) = P(X \le np_0) = \sum_{x=0}^{[[np_0]]} \binom{n}{x} p^x q^{n-x} \qquad (5\text{-}13)$$

where $[[\ \ ]]$ indicates the "greatest integer contained in" function.

- **Example 5-4.** From a flow of product on a conveyor belt between production operations J and $J + 1$, a random sample of 200 units is taken every two hours (see Fig. 5-5).

  Past experience has indicated that if the unit is not properly degreased, the painting operation will not be successful, and, furthermore, on the average 5 percent of the units are not properly degreased. The manufacturing manager has grown accustomed to accepting the 5 percent, but he strongly feels that 6 percent is bad performance and 7 percent is totally unacceptable. He decides to plot the fraction defective in the samples, that is, $\hat{p}$. If the process average stays at 5 percent, he would know that $E(\hat{p}) = .05$. Knowing enough about probability to understand that $\hat{p}$ will vary, he asks the quality control department to determine the $P(\hat{p} > .07 | p = .05)$. This is done as follows:

  $$P(\hat{p} > .07 | p = .05) = 1 - P(\hat{p} \le .07 | p = .05)$$
  $$= 1 - P(X \le 200(.07) | p = .05)$$
  $$= 1 - \sum_{k=0}^{14} \binom{200}{k} (.05)^k (.95)^{200-k}$$
  $$= 1 - .917 = .083$$

- **Example 5-5.** An industrial engineer is concerned about the excessive "avoidable delay" time that one machine operator seems to have. The engineer considers two activities as "avoidable delay time" and "not avoidable delay

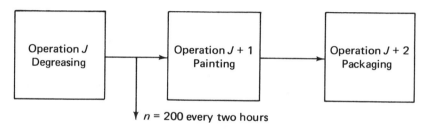

Fig. 5-5. Sequential production operations.

time." She identifies a time-dependent variable as follows:

$$X(t) = 1 \qquad \text{avoidable delay}$$
$$= 0 \qquad \text{otherwise}$$

A hypothetical particular realization of $X(t)$ for two days (960 minutes) is shown in Fig. 5-6.

Rather than have a time study technician continuously analyze this operation, the engineer elects to use "work sampling," and randomly selects n points on the 960-minute span and estimates the fraction of time the "avoidable delay" category exists. She lets $X_i = 1$ if $X(t) = 1$ at the time of the ith observation and $X_i = 0$ if $X(t) = 0$ at the time of the ith observation. The statistic

$$\hat{p} = \frac{\sum_{i=1}^{n} X_i}{n}$$

is to be evaluated. However, $\hat{p}$ is a random variable having a mean equal to p, variance equal to pq/n, and standard deviation equal to $\sqrt{pq/n}$. The procedure outlined is not the best way to go about such a study, but it does illustrate one utilization of the random variable $\hat{p}$.

In summary, analysts must be sure that the phenomenon they are studying may be reasonably considered to be a series of Bernoulli trials in order to use the binomial distribution to describe X, the number of successes in n trials. It is often useful to visualize the graphical presentation of the binomial distribution as shown in Fig. 5-7. The values $p(x)$ increase to a point and then decrease. More precisely, $p(x) > p(x - 1)$ for $x < (n + 1)p$, and $p(x) < p(x - 1)$ for $x > (n + 1)p$. If $(n + 1)p$ is an integer, say m, then $p(m) = p(m - 1)$. There is only one integer such that

$$(n + 1)p - 1 < m \le (n + 1)p$$

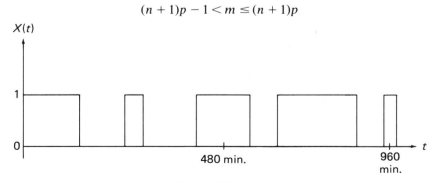

Fig. 5-6.   A hypothetical realization of $X(t)$.

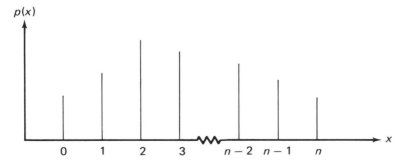

Fig. 5-7.   The binomial distribution.

5-4   The Geometric Distribution

The geometric distribution is also related to a sequence of Bernoulli trials except that the number of trials is not fixed, and, in fact, the random variable of interest, denoted by X, is defined to be the number of trials required to achieve the first success. The sample space and range space for X are illustrated in Fig. 5-8. The range space for X is $R_X = \{1, 2, 3, \ldots\}$, and the distribution of X is given by

$$p(x) = q^{x-1}p \qquad x = 1, 2, \ldots$$
$$= 0 \qquad\qquad \text{otherwise} \qquad\qquad (5\text{-}14)$$

It is easy to verify that this is a probability distribution since

$$\sum_{x=1}^{\infty} pq^{x-1} = p \sum_{k=0}^{\infty} q^k = p \cdot \left[\frac{1}{1-q} \right] = 1$$

and

$$p(x) \geq 0 \qquad \text{for all } x$$

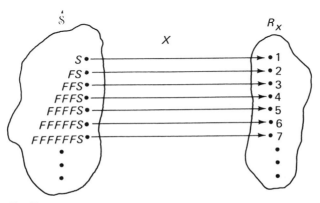

Fig. 5-8.   Sample space and range space for X.

The mean and variance of the geometric distribution are easily found as follows:

$$\mu = E(X) = \sum_{x=1}^{\infty} x \cdot p \cdot q^{x-1} = p \cdot \frac{d}{dq} \sum_{x=1}^{\infty} q^x$$

or

$$\mu = p \frac{d}{dq} \left[\frac{q}{1-q} \right] = \frac{1}{p} \tag{5-15}$$

$$\sigma^2 = V(X) = \sum_{x=1}^{\infty} x^2 \cdot pq^{x-1} - \left(\frac{1}{p}\right)^2 = p \sum_{x=1}^{\infty} x^2 q^{x-1} - \frac{1}{p^2}$$

or

$$\sigma^2 = q/p^2 \tag{5-16}$$

The moment-generating function is

$$M_X(t) = \frac{pe^t}{1 - qe^t} \tag{5-17}$$

● **Example 5-6.** A certain experiment is to be performed until a successful result is obtained. The trials are independent and the cost of performing the experiment is $25,000; however, if a failure results, it costs $5000 to "set up" for the next trial. The experimenter would like to determine the expected cost of the project. If X is the number of trials required to obtain a successful experiment, then the cost function would be

$$C(X) = \$25,000X + \$5000(X - 1)$$
$$= (30,000)X + (-5000)$$

Then

$$E[C(X)] = \$30,000 \cdot E(X) - E(\$5000)$$
$$= \left[30,000 \cdot \frac{1}{p} \right] - 5000$$

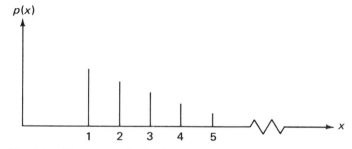

p(x)

Fig. 5-9.   The geometric distribution.

If the probability of success on a single trial is, say, .25, then the $E[C(X)] =$ $\$30,000/.25 - \$5000 = \$115,000$. This may or may not be acceptable to the experimenter. It should also be recognized that it is possible to continue indefinitely without having a successful experiment. Suppose that the experimenter has a maximum of \$500,000. He may wish to find the probability that the experimental work would cost more than this amount, that is,

$$P(C(X) > \$500,000) = P(\$30,000X - \$5000 > \$500,000)$$

$$= P\left(X > \frac{505,000}{30,000}\right)$$

$$= P(X > 16.833)$$

$$= 1 - P(X \le 16)$$

$$= 1 - \sum_{x=1}^{16} .25(.75)^{x-1}$$

$$= 1 - .25 \sum_{x=1}^{16} (.75)^{x-1}$$

$$\simeq .01$$

The experimenter may not be at all willing to run the risk (probability .01) of spending the available \$500,000 without getting a successful run.

The geometric distribution decreases, that is, $p(x) < p(x-1)$ for $x = 2, \ldots$. This is shown graphically in Fig. 5-9.

An interesting and useful property of the geometric distribution is that it has no memory, that is,

$$P(X > x + s | X > s) = P(X > x) \tag{5-18}$$

The geometric distribution is the only discrete distribution having this memoryless property.

5-5 The Pascal Distribution

The *Pascal distribution* also has its basis in Bernoulli trials. It is a logical extension of the geometric distribution. In this case, the random variable X denotes the trial on which the rth success occurs where r is an integer. The distribution of X is

$$p(x) = \binom{x-1}{r-1} p^r q^{x-r} \qquad x = r, r+1, r+2, \ldots \tag{5-19}$$

$$= 0 \qquad\qquad\qquad \text{otherwise}$$

The term $p^r q^{x-r}$ arises from the probability associated with exactly one outcome in $\mathcal{S}$ that has $(x-r)$ Fs (failures) and r Ss (successes). In order for this outcome to occur, there must be $r-1$ successes in the $x-1$ repetitions before the last outcome which is always success. There are thus $\binom{x-1}{r-1}$ arrangements satisfying this condition, and therefore the distribution is as shown in Equation (5-19).

The development thus far has been for integer values of r. If we have arbitrary $r > 0$ and $0 < p < 1$, the distribution of Equation (5-19) is known as the *negative binomial distribution*.

If X has a Pascal distribution, as illustrated in Fig. 5-10, the mean, variance, and moment-generating function are given below:

$$\mu = r/p \tag{5-20}$$

$$\sigma^2 = rq/p^2 \tag{5-21}$$

and

$$M_X(t) = \left(\frac{pe^t}{1 - qe^t}\right)^r \tag{5-22}$$

● **Example 5-7.** The president of a large corporation makes decisions by throwing darts at a board. The center section is marked "yes" and represents a success. The probability of his hitting a "yes" is .6, and this probability remains constant from throw to throw. The president continues to throw until he has three "hits." We denote X as the number of the trial on which he experiences the third hit. The mean is $3/.6 = 5$, meaning that on the average it will take five throws. The president's decision rule is simple. If he gets three hits on or before the fifth throw he decides in favor of the question. The probability that he will decide in favor is therefore

$$P(X \leq 5) = p(3) + p(4) + p(5)$$

$$= \binom{2}{2}(.6)^3(.4)^0 + \binom{3}{2}(.6)^3(.4)^1 + \binom{4}{2}(.6)^3(.4)^2$$

$$= .6636$$

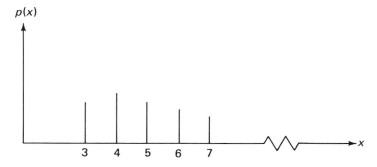

Fig. 5-10.   An example of the Pascal distribution.

5-6   The Multinomial Distribution

An important and useful higher-dimensional random vector has a distribution known as the *multinomial distribution*. Assume an experiment $\mathscr{E}$ with sample space $\mathscr{S}$ is partitioned into k mutually exclusive events, say $B_1, B_2, \ldots, B_k$. We consider n independent repetitions of $\mathscr{E}$ and let $p_i = P(B_i)$ be constant from trial to trial, for $i = 1, 2, \ldots, k$. If $k = 2$, we have Bernoulli trials as described earlier. The random vector, $[X_1, X_2, \ldots, X_k]$, has the following distribution where X_i is the number of times B_i occurs in the n repetitions of $\mathscr{E}$, $i = 1, 2, \ldots, k$.

$$p(x_1, x_2, \ldots, x_k) = \left[\frac{n!}{x_1! x_2! \ldots x_k!}\right] p_1^{x_1} p_2^{x_2} \ldots p_k^{x_k} \tag{5-23}$$

for $x_1 = 0, 1, 2, \ldots$; $x_2 = 0, 1, 2, \ldots$; $\ldots$; $x_k = 0, 1, 2, \ldots$; and where $\Sigma_{i=1}^k x_i = n$.

It should be noted that $X_1, X_2, \ldots, X_k$ are not independent random variables since $\Sigma_{i=1}^k X_i = n$ for any n repetitions.

The mean and variance of X_i, a particular component, are

$$E(X_i) = np_i \tag{5-24}$$

and

$$V(X_i) = np_i(1 - p_i) \tag{5-25}$$

● **Example 5-8.** Mechanical pencils are manufactured by a process involving a large amount of labor in the assembly operations. This is highly repetitive work and incentive pay is involved. Final inspection has revealed that 85 percent of the product is good, 10 percent is defective but may be reworked, and 5 percent is defective and must be scrapped. These percentages remain constant over time. A random sample of twenty items is selected, and if we let

$X_1 =$ number of good items
$X_2 =$ number of defective but reworkable items
$X_3 =$ number of items to be scrapped

then

$$p(x_1, x_2, x_3) = \frac{(20)!}{x_1! x_2! x_3!} (.85)^{x_1} (.10)^{x_2} (.05)^{x_3}$$

Suppose we want to evaluate this probability function for $x_1 = 18$, $x_2 = 2$, and $x_3 = 0$ (we must have $x_1 + x_2 + x_3 = 20$); then

$$p(18, 2, 0) = \frac{(20)!}{(18)! 2! 0!} (.85)^{18} (.10)^2 (.05)^0$$

$$= 190(.85)^{18}(.01)$$

$$\simeq .105$$

5-7 The Hypergeometric Distribution

In an earlier section an example presented the hypergeometric distribution. We will now formally develop this distribution and further illustrate its application. Suppose there is some finite population with N items. Some number $D(D \leq N)$ of the items fall into a class of interest. The particular class will, of course, depend on the situation under consideration. It might be defectives (vs. nondefectives) in the case of a production lot, or persons with blue eyes (vs. not blue-eyed) in a classroom with N students. A random sample of size n is selected *without replacement*, and the random variable of interest, X, is the number of items in the sample that belong to the class of interest. The distribution of X is given below:

$$p(x) = \frac{\binom{D}{x}\binom{N-D}{n-x}}{\binom{N}{n}} \qquad x = 0, 1, 2, \ldots, \min(n, D)$$

$$= 0 \qquad\qquad \text{otherwise} \qquad\qquad (5\text{-}26)$$

The mean and variance of X are given respectively by

$$E(X) = n \cdot \left[\frac{D}{N}\right] \qquad\qquad (5\text{-}27)$$

and

$$V(X) = n \cdot \left[\frac{D}{N}\right] \cdot \left[1 - \frac{D}{N}\right] \cdot \left[\frac{N-n}{N-1}\right] \qquad\qquad (5\text{-}28)$$

Extensive tables of the distribution are given in Lieberman and Owen (1961).

● **Example 5-9.** In a receiving inspection department, lots of a pump shaft are periodically received. The lots contain 100 units and the following *acceptance sampling plan* is used. A random sample of 10 units is selected without replacement. The lot is accepted if the sample has no more than one defective. Suppose a lot is received that is $p'(100)$ percent defective. What is the probability that it will be accepted?

$$P(\text{accept lot}) = P(X \leq 1) = \frac{\sum_{x=0}^{1} \binom{100p'}{x}\binom{100[1-p']}{10-x}}{\binom{100}{10}}$$

$$= \frac{\binom{100p'}{0}\binom{100[1-p']}{10} + \binom{100p'}{1}\binom{100[1-p']}{9}}{\binom{100}{10}}$$

Obviously the probability of accepting the lot is a function of the lot quality, p'. If $p' = .05$, then

$$P(\text{accept lot}) = \frac{\binom{5}{0}\binom{95}{10} + \binom{5}{1}\binom{95}{9}}{\binom{100}{10}} = .923$$

5-8   The Poisson Distribution

One of the most useful discrete distributions is the *Poisson distribution*. The Poisson distribution may be developed in two ways, and both are instructive insofar as they indicate the circumstances where this random variable may be expected to represent the outcome of a random experiment. The first development involves the definition of a *Poisson process*. The second development shows the Poisson distribution to be *a limiting form of the binomial distribution*.

In developing the Poisson process, we initially consider some arbitrary, time-oriented occurrences, often called "arrivals" or "births" (see Fig. 5-11). The random variable of interest, say X_t, is the number of arrivals that occur on the interval $[0, t]$. The range space $R_{X_t} = \{0, 1, 2, \ldots\}$. In developing the distribution of X_t it is necessary to make some assumptions, the plausibility of which is supported by considerable empirical evidence.

The first assumption is that the number of arrivals during *nonoverlapping* time intervals are *independent* random variables. Second, we make the assumption that there exists a positive quantity λ such that for any small time interval, Δt, the following *postulates* are satisfied:

1. *The probability that exactly one arrival will occur in an interval of width Δt is approximately $\lambda \cdot \Delta t$.* The approximation is in the sense that the probability is $\lambda \cdot \Delta t + o_1(\Delta t)$ and $[o_1(\Delta t)/\Delta t] \to 0$ as $\Delta t \to 0$.

2. *The probability that exactly zero arrivals will occur in the interval is approximately $1 - \lambda \cdot \Delta t$.* Again this is in the sense that it is equal to $1 - [\lambda \cdot \Delta t] + o_2(\Delta t)$ and $[o_2(\Delta t)/\Delta t] \to 0$ as $\Delta t \to 0$.

3. *The probability that two or more arrivals occur in the interval is equal to a quantity $o_3(\Delta t)$, where $[o_3(\Delta t)/\Delta t] \to 0$ as $\Delta t \to 0$.*

The parameter λ is sometimes called the mean arrival rate or mean

Fig. 5-11.   The time axis.

occurrence rate. In the development to follow, we let

$$p(x) = P(X_t = x) = p_x(t) \qquad x = 0, 1, 2, \ldots \tag{5-29}$$

We thus fix time at t and obtain

$$p_0(t + \Delta t) \simeq [1 - \lambda \cdot \Delta t] \cdot p_0(t)$$

so that

$$\frac{p_0(t + \Delta t) - p_0(t)}{\Delta t} \simeq -\lambda p_0(t)$$

and

$$\lim_{\Delta t \to 0} \left[\frac{p_0(t + \Delta t) - p_0(t)}{\Delta t} \right] = p_0'(t) = -\lambda p_0(t) \tag{5-30}$$

For $x > 0$

$$p_x(t + \Delta t) \simeq \lambda \cdot \Delta t p_{x-1}(t) + [1 - \lambda \cdot \Delta t] \cdot p_x(t)$$

so that

$$\frac{p_x(t + \Delta t) - p_x(t)}{\Delta t} \simeq \lambda \cdot p_{x-1}(t) - \lambda \cdot p_x(t)$$

and

$$\lim_{\Delta t \to 0} \left[\frac{p_x(t + \Delta t) - p_x(t)}{\Delta t} \right] = p_x'(t) = \lambda p_{x-1}(t) - \lambda p_x(t) \tag{5-31}$$

Summarizing, we have a system of differential equations:

$$p_0'(t) = -\lambda p_0(t) \tag{5-32a}$$

and

$$p_x'(t) = \lambda p_{x-1}(t) - \lambda p_x(t) \qquad x = 1, 2, \ldots \tag{5-32b}$$

The solution to these equations is

$$p_x(t) = (\lambda t)^x e^{-(\lambda t)} / x! \qquad x = 0, 1, 2, \ldots \tag{5-33}$$

Thus for fixed t we let $\alpha = \lambda t$ and obtain the Poisson distribution as

$$p(x) = \frac{\alpha^x e^{-\alpha}}{x!} \qquad x = 0, 1, 2, \ldots \tag{5-34}$$

$$= 0 \qquad \text{otherwise}$$

Note that this distribution was developed as a *consequence* of certain assumptions; thus, when the assumptions hold or approximately hold, the Poisson distribution is an appropriate model. There are many real world phenomena for which the Poisson model is appropriate.

To show how the Poisson distribution may also be developed as a limiting form of the binomial distribution with $\alpha = np$, we return to the binomial distribution

$$p(x) = \frac{n!}{x!(n-x)!} p^x (1-p)^{n-x} \qquad x = 0, 1, 2, \ldots, n$$

If we let $np = \alpha$, so that $p = \alpha/n$ and $1 - p = 1 - \alpha/n = \dfrac{n-\alpha}{n}$, and if we then replace terms involving p by the corresponding terms involving α, we obtain:

$$p(x) = \frac{n(n-1)(n-2)\cdots(n-x+1)}{x!} \left[\frac{\alpha}{n}\right]^x \left[\frac{n-\alpha}{n}\right]^{n-x}$$

$$= \frac{\alpha^x}{x!} \left[(1)\left(1 - \frac{1}{n}\right)\left(1 - \frac{2}{n}\right)\cdots\left(1 - \frac{x-1}{n}\right)\right]\left(1 - \frac{\alpha}{n}\right)^n \left(1 - \frac{\alpha}{n}\right)^{-x} \quad (5\text{-}35)$$

In letting $n \to \infty$ and $p \to 0$ in such a way that $np = \alpha$ remains fixed, the terms $\left(1 - \frac{1}{n}\right), \left(1 - \frac{2}{n}\right), \ldots, \left(1 - \frac{x-1}{n}\right)$ all approach 1 as does $\left(1 - \frac{\alpha}{n}\right)^{-x}$. Now we know that $\left(1 - \frac{\alpha}{n}\right)^n \to e^{-\alpha}$ as $n \to \infty$; thus, the limiting form of Equation (5-35) is $p(x) = (\alpha^x/x!) \cdot e^{-\alpha}$, which is the Poisson distribution.

The *mean* of the random variable $X = X_t$ is α and the variance is also α, as seen below.

$$E(X) = \sum_{x=0}^{\infty} \frac{x e^{-\alpha} \alpha^x}{x!} = \sum_{x=1}^{\infty} \frac{e^{-\alpha} \alpha^x}{(x-1)!}$$

$$= \alpha e^{-\alpha} \left[1 + \frac{\alpha}{1!} + \frac{\alpha^2}{2!} + \cdots\right]$$

$$= \alpha e^{-\alpha} \cdot e^{\alpha}$$

$$= \alpha \qquad (5\text{-}36)$$

Similarly,

$$E(X^2) = \sum_{x=0}^{\infty} \frac{x^2 \cdot e^{-\alpha} \alpha^x}{x!} = \alpha^2 + \alpha$$

so that

$$V(X) = E(X^2) - [E(X)]^2$$

$$= \alpha \qquad (5\text{-}37)$$

The moment-generating function may be shown to be

$$M_X(t) = e^{\alpha(e^t - 1)} \qquad (5\text{-}38)$$

The utility of this generating function is illustrated in the proof of the following theorem.

Theorem 5-1

If $X_1, X_2, \ldots, X_k$ are independently distributed random variables, each having a Poisson distribution with parameter α_i, $i = 1, 2, \ldots, k$, and $Y = X_1 + X_2 + \cdots + X_k$, then Y has a Poisson distribution with parameter

$$\alpha = \alpha_1 + \alpha_2 + \cdots + \alpha_k$$

Proof

The moment-generating function of X_i is

$$M_{X_i}(t) = e^{\alpha_i(e^t - 1)} .$$

and, since $M_Y(t) = M_{X_1}(t) \cdot M_{X_2}(t) \cdot \ldots \cdot M_{X_k}(t)$, then

$$M_Y(t) = e^{(\alpha_1 + \alpha_2 + \cdots + \alpha_k)(e^t - 1)}$$

which is recognized as the moment-generating function of a Poisson random variable with parameter $\alpha = \alpha_1 + \alpha_2 + \cdots + \alpha_k$.

This reproductive property of the Poisson distribution is highly useful. Simply, it states that sums of independent Poisson random variables are distributed according to the Poisson distribution.

Extensive tables for the Poisson distribution are available in Molina (1942). A brief tabulation is given in Table I of the Appendix.

● **Example 5-10.** Suppose a retailer determines that the number of orders for a certain home appliance in a particular period has a Poisson distribution with parameter α. She would like to determine the stock level K for the beginning of the period so that there will be a probability of at least .95 of supplying all customers who order the appliance during the period. She does not wish to back order merchandise or resupply the warehouse during the period. If X represents the number of orders, the dealer wishes to determine K such that

$$P(X \le K) \ge .95$$

or

$$P(X > K) \le .05$$

so that

$$\sum_{x=K+1}^{\infty} e^{-\alpha}(\alpha)^x / x! \le .05$$

The solution may be determined directly from tables of the Poisson distribution.

● **Example 5-11.** The attainable sensitivity for electronic amplifiers and apparatus is limited by noise or spontaneous current fluctuations. In vacuum tubes, one noise source is shot noise due to the random emission of electrons from the heated cathode. If the potential difference between the anode and cathode is so great that all electrons emitted by the cathode have such high velocity that there is no spare charge (accumulation of electrons between the cathode and anode), and if an event or occurrence or arrival is considered to be an emission of an electron from the cathode, then, as Davenport and Root (1958) have shown, the number of electrons, X, emitted from the cathode in time t has a Poisson distribution given by:

$$p(x) = (\lambda t)^x e^{-(\lambda t)}/x! \qquad x = 0, 1, 2, \ldots$$
$$= 0 \qquad \text{otherwise}$$

The parameter λ is the mean rate of emission of electrons from the cathode.

5-9   Some Approximations

It is often useful to approximate one distribution with another, particularly when the approximation is easier to manipulate. The two approximations considered in this section are:

1. The binomial approximation to the hypergeometric distribution.
2. The Poisson approximation to the binomial distribution.

For the hypergeometric distribution, if the *sampling fraction* n/N is small, say less than .1, then the binomial distribution with $p = D/N$ and n provides a good approximation. The smaller the ratio n/N, the better the approximation.

● **Example 5-12.** A production lot of 200 units has 8 defectives. A random sample of 10 units is selected, and we want to find the probability that the sample will contain exactly 1 defective.

$$P(X = 1) = \frac{\binom{8}{1}\binom{192}{9}}{\binom{200}{10}}$$

Since $n/N = \frac{10}{200} = .05$ is small, we let $p = \frac{8}{200} = .04$ and use the binomial approximation

$$p(1) \simeq \binom{10}{1}(.04)^1(.96)^9 \simeq .28$$

In the case of the Poisson approximation to the binomial, we indicated earlier that for large n and small p, the approximation is satisfactory. In utilizing this approximation we let $\alpha = np$. In general, p should be less than .1

TABLE 5-1 Summary of Discrete Distributions

Distribution	Parameters	Probability Function: $p(x)$	Mean	Variance	Moment-Generating Function
Bernoulli	$0 \leq p \leq 1$	$p(x) = p^x q^{1-x} \quad x = 0, 1$ $= 0 \quad$ otherwise	p	pq	$pe^t + q$
Binomial	$n = 1, 2, \ldots$ $0 \leq p \leq 1$	$p(x) = \binom{n}{x} p^x q^{n-x} \quad x = 0, 1, 2, \ldots, n$ $= 0 \quad$ otherwise	np	npq	$(pe^t + q)^n$
Geometric	$0 < p < 1$	$p(x) = pq^{x-1} \quad x = 1, 2, \ldots$ $= 0 \quad$ otherwise	$1/p$	q/p^2	$pe^t/(1 - qe^t)$
Pascal (Neg. Binomial)	$0 < p < 1$ $r = 1, 2, \ldots$ $(r > 0)$	$p(x) = \binom{x-1}{r-1} p^r q^{x-r} \quad x = r, r+1, r+2, \ldots$ $= 0 \quad$ otherwise	r/p	rq/p^2	$\left[\dfrac{pe^t}{1 - qe^t}\right]^r$
Hypergeometric	$N = 1, 2, \ldots, N$ $n = 1, 2, \ldots, N$ $D = 1, 2, \ldots, N$	$p(x) = \dfrac{\binom{D}{x}\binom{N-D}{n-x}}{\binom{N}{n}} \quad x = 0, 1, 2, \ldots, \min(n, D)$ $= 0 \quad$ otherwise	$n\left[\dfrac{D}{N}\right]$	$n\left[\dfrac{D}{N}\right]\left[1 - \dfrac{D}{N}\right]\left[\dfrac{N-n}{N-1}\right]$	See Kendall and Stuart (1963)
Poisson	$\alpha > 0$	$p(x) = e^{-\alpha}(\alpha)^x/x! \quad x = 0, 1, 2, \ldots$ $= 0 \quad$ otherwise	α	α	$e^{\alpha(e^t - 1)}$

in order to apply the approximation. The smaller p and the larger n, the better the approximation.

● **Example 5-13.** The probability that a particular rivet in the wing surface of a new aircraft is defective is .001. There are 4000 rivets in the wing. What is the probability that not more than 6 defective rivets will be installed?

$$P(X \leq 6) = \sum_{x=0}^{6} \binom{4000}{x} (.001)^x (.999)^{4000-x}$$

Using the Poisson approximation,

$$\alpha = 4000(.001) = 4$$

and

$$P(X \leq 6) = \sum_{x=0}^{6} e^{-4}(4)^x/x! = .889$$

5-10 Summary

The distributions presented in this chapter have wide use in engineering, scientific, and management applications. The selection of a specific discrete distribution will depend on how well the assumptions underlying the distribution are met by the phenomenon to be modeled. The distributions presented here were selected because of their wide applicability.

A summary of these distributions is presented in Table 5-1.

5-11 Exercises

5-1. An experiment consists of four independent Bernoulli trials with probability of success p on each trial. The random variable X is the number of successes. Enumerate the probability distribution of X.

5-2. Six independent space missions to the moon are planned. The estimated probability of success on each mission is .95. What is the probability that at least five of the planned missions will be successful?

5-3. The XYZ Company has planned sales presentations to a dozen important customers. The probability of receiving an order as a result of such a presentation is estimated to be .5. What is the probability of receiving four or more orders as the result of the meetings?

5-4. A stockbroker calls her 20 most important customers every morning. If the probability is one in three of making a transaction as the result of such a call, what are the chances of her handling 10 or more transactions?

5-5. A production process that manufactures transistors operates, on the average, at 2 percent fraction defective. Every two hours a random sample of size 50 is taken from the process. If the sample contains more than two defectives the process must be stopped. Determine the probability that the process will be stopped by the sampling scheme.

5-6. Find the mean and variance of the binomial distribution using the moment-generating function; see Equation (5-7).

5-7. A production process manufacturing turn-indicator dash lights is known to produce lights that are 1 percent defective. If this value remains unchanged, and a sample of 100 such lights is randomly selected, find $P(\hat{p} \leq .03)$, where $\hat{p}$ is the sample fraction defective.

5-8. Suppose a random sample of size 200 is taken from a process that is .07 fraction defective. What is the probability that $\hat{p}$ will exceed the true fraction defective by one standard deviation? By two standard deviations? By three standard deviations?

5-9. Five missiles have been built for use in national defense. The probability of a successful firing is, on any one test, .95. Assuming independent firings, what is the probability that the first failure occurs on the fifth firing?

5-10. A real estate agent estimates his probability of selling a house to be .25. He has to see four clients today. If he is successful on the first three calls what is the probability that his fourth call is unsuccessful?

5-11. Suppose five independent identical laboratory experiments are to be undertaken. Each experiment is extremely sensitive to enviromental conditions, and there is only a probability p that it will be completed successfully. Plot, as a function of p, the probability that the fifth experiment is the first failure. Find mathematically the value of p that maximizes the probability of the fifth trial being the first unsuccessful experiment.

5-12. The XYZ Company plans to visit potential customers until a substantial sale is made. Each sales presentation costs $1000. It cost $3000 to travel to the next customer and set up a new presentation.

(a) What is the expected cost of making a sale if the probability of making a sale after any presentation is .10?

(b) If the expected profit at each sale is $15,000, should the trips be undertaken?

(c) If the budget for advertising is only $100,000, what is the probability that this sum will be spent without getting an order?

5-13. Find the mean and variance of the geometric distribution using the moment-generating function.

5-14. A submarine's probability of sinking an enemy vessel with any one firing of its torpedoes is .8. If the firings are independent, determine the probability of a sinking within the first two firings. Within the first three.

5-15. In a southern area, the probability that a thunderstorm will occur on any day during the spring is .05. Assuming independence, what is the probability that the first thunderstorm occurs on April 5? Assume spring begins on March 1.

5-16. A potential customer enters an automobile dealership every hour. The probability of a salesperson concluding a transaction is .25. She is determined to keep working until she has sold three cars. What is the probability that she will have to work exactly eight hours? More than eight hours?

5-17.   A personnel manager is interviewing potential employees in order to fill two jobs. The probability of an interviewee having the necessary qualifications and accepting an offer is .8. What is the probability that exactly four people must be interviewed? What is the probability that fewer than four people must be interviewed?

5-18.   Show that the moment-generating function of the Pascal random variable is as given by Equation (5-22). Use it to determine the mean and variance of the Pascal distribution.

5-19.   The probability that an experiment has a successful outcome is .75. The experiment is to be repeated until five successful outcomes have occurred. What is the expected number of repetitions required? What is the variance?

5-20.   A military commander wishes to destroy an enemy bridge. Each flight of planes he sends out has a probability of .8 of scoring a direct hit on the bridge. It takes four direct hits to completely destroy the bridge. If he can mount seven assaults before the bridge becomes tactically unimportant, what is the probability that the bridge will be destroyed?

5-21.   Three companies X, Y, and Z have probabilities of obtaining an order for a particular type of merchandise of .4, .3, and .3, respectively. Three orders are to be awarded independently. What is the probability that one company receives all the orders?

5-22.   Four companies are interviewing five college students for positions after graduation. Assuming all five receive offers from each company, and the probabilities of the companies hiring a new employee are equal, what is the probability that one company gets all of the new employees? None of them?

5-23.   We are interested in the weight of bags of feed. Specifically, we need to know if any of the four events below has occurred:

$$T_1 = (x \le 10) \qquad p(T_1) = .2$$
$$T_2 = (10 < x \le 11) \qquad p(T_2) = .2$$
$$T_3 = (11 < x \le 11.5) \qquad p(T_3) = .2$$
$$T_4 = (11.5 < x) \qquad p(T_4) = .4$$

If 10 bags are selected at random what is the probability of 4 being less than or equal to 10 pounds, 1 being greater than 10 but less than or equal to 11 pounds, and 2 being greater than 11.5 pounds?

5-24.   In the above problem what is the probability that all 10 bags weigh more than 11.5 pounds? What is the probability that 5 bags weigh more than 11.5 pounds and the remaining 5 weigh less than 10 pounds?

5-25.   A lot of 25 color television tubes is subjected to an acceptance testing procedure. The procedure consists of drawing five tubes at random, without replacement, and testing them. If two or fewer tubes fail, the remaining ones are accepted. Otherwise the lot is rejected. Assume the lot contains four defective tubes.

(a) What is the exact probability of lot acceptance?

(b) What is the probability of lot acceptance computed from the binomial distribution with $p = \frac{4}{25}$?

5-26. Suppose that in Exercise 5-25 the lot size had been 100. Would the binomial approximation be satisfactory in this case?

5-27. A purchaser receives small lots ($N = 25$) of a high-precision device. She wishes to reject the lot 95 percent of the time if it contains as many as seven defectives. Suppose she decides that the presence of one defective in the sample is sufficient to cause rejection. How large should her sample size be?

5-28. Show that the moment-generating function of the Poisson random variable is as given by Equation (5-38).

5-29. The number of automobiles passing through a particular intersection per hour is estimated to be 25. Find the probability that fewer than 10 vehicles pass through during any one-hour interval. Assume that the number of vehicles follows a Poisson distribution.

5-30. Calls arrive at a telephone switchboard such that the number of calls per hour follows a Poisson distribution with mean 10. The current equipment can handle up to 20 calls without becoming overloaded. What is the probability of such an overload occurring?

5-31. The number of red blood cells per square unit visible under a microscope follows a Poisson distribution with mean 4. Find the probability that more than 5 such blood cells are visible to the observer.

5-32. Let X_t be the number of vehicles passing through an intersection during a length of time t. The random variable X_t is Poisson distributed with parameter αt. Suppose an automatic counter has been installed to count the number of passing vehicles. However, this counter is not functioning properly, and each passing vehicle has a probability p of not being counted. Let Y_t be the number of vehicles counted during t. Find the probability distribution of Y_t.

5-33. A large insurance company has discovered that .2 percent of the U.S. population is injured as a result of a particular type of accident. This company has 15,000 policyholders carrying coverage against such an accident. What is the probability that three or fewer claims will be filed against those policies next year? Five or more claims?

5-34. Maintenance crews arrive at a tool crib requesting a particular spare part according to a Poisson distribution with parameter $\alpha = 2$. Three of these spare parts are normally kept on hand. If more than three orders occur, the crews must journey a considerable distance to central stores.

(a) On a given day, what is the probability that such a journey must be made?

(b) What is the expected demand per day for spare parts?

(c) How many spare parts must be carried if the tool crib is to service all incoming crews 90 percent of the time?

(d) What is the expected number of crews serviced daily at the tool crib?

(e) What is the expected number of crews making the journey to central stores?

5-35. A loom experiences one yarn breakage approximately every 10 hours. A particular style of cloth is being produced that will take 25 hours on this loom. If three or more breaks are required to render the product unsatisfactory, find the probability that this style of cloth is finished with acceptable quality.

5-36. The number of people boarding a bus at each stop follows a Poisson distribution with parameter α. The bus company is surveying its usages for scheduling purposes, and has installed an automatic counter on each bus. However, if more than 10 people board at any one stop the counter cannot record the excess and merely registers 10. If X is the number of riders recorded, find the probability distribution of X.

5-37. A mathematics textbook has 200 pages on which typographical errors in the equations could occur. If there are in fact five errors randomly dispersed among these 200 pages, what is the probability that a random sample of 50 pages will contain at least one error? How large must the random sample be to assure that at least three errors will be found with 90 percent probability?

5-38. The probability of a vehicle having an accident at a particular intersection is .0001. Suppose that 10,000 vehicles per day travel through this intersection. What is the probability of no accidents occurring? What is the probability of two or more accidents?

5-39. If the probability of being involved in an auto accident is .01 during any year, what is the probability of having two or more accidents during any 10-year driving period?

5-40. Suppose that the number of accidents to employees working on high-explosive shells over a period of time (say five weeks) is taken to follow a Poisson distribution with parameter $\alpha = 2$.

   (a) Find the probability of 1, 2, 3, 4, or 5 accidents.

   (b) The Poisson distribution has been freely applied in the area of industrial accidents. However, it frequently provides a poor "fit" to actual historical data. Why might this be true? *Hint*: See Kendall and Stuart (1963), pp. 128–30.

Chapter 6

Some Important Continuous Distributions

6-1 Introduction

We will now study several important continuous probability distributions. They are the uniform, exponential, gamma, and Weibull distributions. In Chapter 7 the normal distribution, and several other probability distributions closely related to it, will be presented. The normal distribution is perhaps the most important of all continuous distributions. The reason for postponing its study is one of convenience. It is important enough to warrant a separate chapter.

It has been noted that the range space for a continuous random variable X consists of an interval or a set of intervals. This was illustrated in an earlier chapter, and it was observed that an idealization is involved. For example, if we are measuring the time to failure for an electronic component or the time to process an order through an information system, the measurement devices used are such that there are only a finite number of possible outcomes; however, we will idealize and assume that time may take *any* value on some interval.

6-2 The Uniform Distribution

The uniform density function is defined as

$$f(x) = \frac{1}{\beta - \alpha} \qquad \alpha \le x \le \beta$$
$$= 0 \qquad\qquad \text{otherwise} \qquad\qquad (6\text{-}1)$$

where α and β are real constants with $\alpha < \beta$. The density function is shown in Fig. 6–1. Since a uniformly distributed random variable has a probability density function that is constant over some interval of definition, the constant

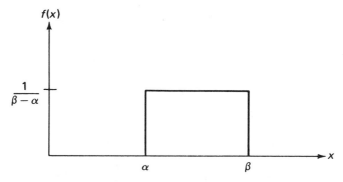

Fig. 6-1.   A uniform density.

must be the reciprocal of the length of the interval in order to satisfy the requirement that

$$\int_{-\infty}^{\infty} f(x)\, dx = 1$$

A uniformly distributed random variable represents the continuous analog to equally likely outcomes in the sense that for any subinterval $[a, b]$, where $\alpha \leq a < b \leq \beta$, the $P(a \leq X \leq b)$ is the same for all subintervals of the same length.

$$P(a \leq X \leq b) = \int_{a}^{b} \frac{dx}{\beta - \alpha} = \frac{(b - a)}{(\beta - \alpha)}$$

The statement that we *choose a point at random on* $[\alpha, \beta]$ simply means that the value chosen, say Y, is uniformly distributed on $[\alpha, \beta]$.

The *mean* and *variance* of the uniform distribution are

$$E(X) = \int_{\alpha}^{\beta} \frac{x\, dx}{\beta - \alpha} = \frac{1}{2(\beta - \alpha)} x^2 \Big|_{\alpha}^{\beta} = \frac{(\beta + \alpha)}{2} \tag{6-2}$$

and

$$V(X) = \int_{\alpha}^{\beta} \frac{x^2\, dx}{\beta - \alpha} - \left[\frac{(\beta + \alpha)}{2}\right]^2$$

$$= \frac{(\beta - \alpha)^2}{12} \tag{6-3}$$

The moment-generating function $M_X(t)$ is found as follows:

$$M_X(t) = E(e^{tX}) = \int_{\alpha}^{\beta} e^{tx} \cdot \frac{1}{\beta - \alpha}\, dx = \frac{1}{t(\beta - \alpha)} e^{tx} \Big|_{\alpha}^{\beta}$$

$$= \frac{e^{t\beta} - e^{t\alpha}}{t(\beta - \alpha)} \qquad \text{for } t \neq 0 \tag{6-4}$$

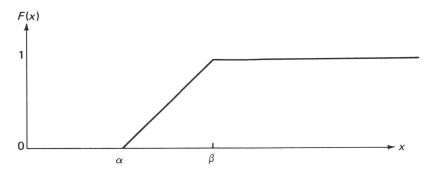

Fig. 6-2. Distribution function for the uniform random variable.

For a uniformly distributed random variable, the distribution function $F(x) = P(X \leq x)$ is given by Equation (6–5) below, and its graph is shown in Fig. 6–2.

$$
\begin{aligned}
F(x) &= 0 & x < \alpha \\
&= \int_{\alpha}^{x} \frac{dx}{\beta - \alpha} = \frac{x - \alpha}{\beta - \alpha} & \alpha \leq x < \beta \\
&= 1 & x \geq \beta
\end{aligned}
\qquad (6\text{-}5)
$$

● **Example 6-1.** A point is chosen at random on the interval $[0, 10]$. Suppose we wish to find the probability that the point lies between $\frac{3}{2}$ and $\frac{7}{2}$. The density of the random variable X is $f(x) = \frac{1}{10}$, $0 \leq x \leq 10$; and $f(x) = 0$, otherwise. Hence, $P(\frac{3}{2} \leq X \leq \frac{7}{2}) = \frac{2}{10}$.

● **Example 6-2.** Numbers of the form $NN.N$ are "rounded off" to the nearest integer. The round-off procedure is such that if the decimal part is less than .5, the round is "down" by simply dropping the decimal part; however, if the decimal part is greater than .5, the round is up, that is, the new number is $[[NN.N]] + 1$ where $[[\ ]]$ is the "greatest integer contained in" function. If the decimal part is exactly .5, a coin is tossed to determine which way to round. The round-off error, X, is defined as the difference between the number before rounding and the number after rounding. These errors are commonly distributed according to the uniform distribution on the interval $[-.5, +.5]$. That is,

$$
\begin{aligned}
f(x) &= 1 & -.5 \leq x \leq +.5 \\
&= 0 & \text{otherwise}
\end{aligned}
$$

● **Example 6-3.** A well-known digital computer simulation language is GPSS (General-Purpose System Simulator). One of the special features of this language is a simple automatic procedure for using the uniform distribution. The user declares a mean and a modifier (e.g., 500, 100). The compiler immediately creates a routine to produce realizations of a random variable X uniformly distributed on [400, 600]. In this language, the uniform distribution is frequently used as an approximation to many other distributions, the exact form of which may be unknown to the user.

6-3   The Exponential Distribution

The exponential distribution has density function

$$f(x) = \lambda e^{-\lambda x} \qquad x \geq 0$$
$$= 0 \qquad \text{otherwise} \qquad (6\text{-}6)$$

where parameter λ is a real, positive constant. A graph of the exponential density is shown in Fig. 6-3.

The exponential density is closely related to the Poisson distribution, and an explanation of this relationship should help the reader develop an understanding of the kinds of situations for which the exponential density is appropriate.

In developing the Poisson distribution from the Poisson postulates and the Poisson process, we fixed time at some value t, and we developed the distribution of the *number of occurrences in the interval* $[0, t]$. We noted this random variable by X, and the distribution was

$$p(x) = e^{-\lambda t}(\lambda t)^x/x! \qquad x = 0, 1, 2, \ldots$$
$$= 0 \qquad \text{otherwise} \qquad (6\text{-}7)$$

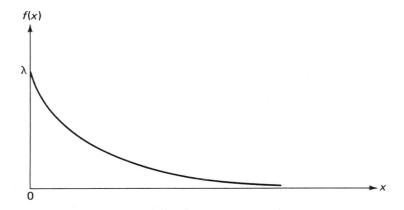

Fig. 6-3.   Exponential density function.

Now consider $p(0)$, which is the probability of no occurrences on $[0, t]$. This is equal to

$$p(0) = e^{-\lambda t} \tag{6-8}$$

Recall that we originally fixed time at t. Another interpretation of $p(0) = e^{-\lambda t}$ is that this is the probability that the time to the first occurrence is greater than t. Considering this time as a random variable T, we note that

$$p(0) = P(T > t) = e^{-\lambda t} \qquad t \geq 0 \tag{6-9}$$

Thus, if we now let time vary and consider the random variable T as the time to occurrence, then

$$F(t) = P(T \leq t) = 1 - e^{-\lambda t} \qquad t \geq 0 \tag{6-10}$$

And, since $f(t) = F'(t)$, we see that the density is

$$f(t) = \lambda e^{-\lambda t} \qquad t \geq 0$$
$$= 0 \qquad \text{otherwise} \tag{6-11}$$

This is the density of Equation (6-6). Thus, the relationship between the exponential and Poisson distributions may be stated as follows: if the number of occurrences has a Poisson distribution as shown in Equation (6-7), then the time between occurrences has an exponential distribution as shown in Equation (6-11). For example, if the number of orders for a certain item received per week has a Poisson distribution, then the time between orders would have an exponential distribution. One variable is discrete (the count) and the other (time) is continuous.

In order to verify that f is a density function, we note that $f(x) \geq 0$ for all x and

$$\int_0^\infty \lambda e^{-\lambda x}\, dx = -e^{-\lambda x}\Big|_0^\infty = 1$$

The *mean* and *variance* of the exponential distribution are

$$E(X) = \int_0^\infty x\lambda e^{-\lambda x}\, dx = -xe^{-\lambda x}\Big|_0^\infty + \int_0^\infty e^{-\lambda x}\, dx = 1/\lambda \tag{6-12}$$

and

$$V(X) = \int_0^\infty x^2 \lambda e^{-\lambda x}\, dx - (1/\lambda)^2 = \left[-x^2 e^{-\lambda x}\Big|_0^\infty + 2\int_0^\infty xe^{-\lambda x}\, dx \right] - (1/\lambda)^2 = 1/\lambda^2 \tag{6-13}$$

The standard deviation is $1/\lambda$, and thus the mean and the standard deviation are equal.

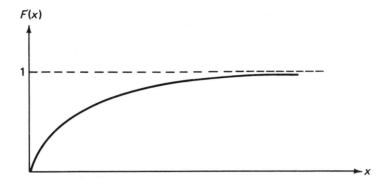

$F(x)$

1

x

Fig. 6-4. Distribution function for exponential density.

The moment-generating function is

$$M_X(t) = \left(1 - \frac{t}{\lambda}\right)^{-1} \tag{6-14}$$

provided $t < \lambda$.

The distribution function F can be obtained by integrating Equation (6-6) as follows:

$$F(x) = 0 \qquad\qquad\qquad\qquad x < 0 \tag{6-15}$$

$$= \int_0^x \lambda e^{-\lambda t}\, dt = 1 - e^{-\lambda x} \qquad x \geq 0$$

Figure 6-4 depicts the distribution function of Equation (6-15).

● **Example 6-4.** An electronic component is known to have a useful life represented by an exponential density with failure rate 10^{-5} failures per hour (that is, $\lambda = 10^{-5}$). The mean time to failure, $E(X)$, is thus 10^5 hours. Suppose we want to determine the fraction of such components that would fail before the mean life or expected life.

$$P\left(T \leq \frac{1}{\lambda}\right) = \int_0^{1/\lambda} \lambda e^{-\lambda x}\, dx = -e^{-\lambda x}\Big|_0^{1/\lambda} = 1 - e^{-1}$$

$$= .63212$$

This result holds for àny value of λ greater than zero. In our example, 63.212 percent of the items would fail before 10^5 hours (see Fig. 6-5).

● **Example 6-5.** Suppose a designer is to make a decision between two manufacturing processes for the manufacture of a certain component. Process A costs C dollars per unit to manufacture a component. Process B costs $k \cdot C$ dollars

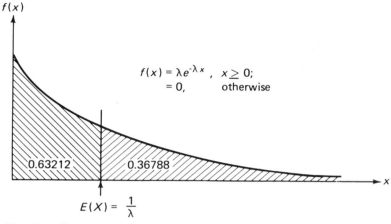

Fig. 6-5. The mean of an exponential distribution.

per unit to manufacture a component, where $k > 1$. Components have an exponential time to failure density with failure rate 200^{-1} failures per hour for process A, while components from process B have a failure rate of 300^{-1} failures per hour. The mean lives are thus 200 hours and 300 hours, respectively, for the two processes. Because of a warranty clause, if a component lasts less than 400 hours, the manufacturer must pay a penalty of K dollars. Let X be the time to failure of each component. Thus

$$C_A = C \qquad \text{if } X \geq 400$$
$$ = C + K \qquad \text{if } X < 400$$

and

$$C_B = kC \qquad \text{if } X \geq 400$$
$$ = kC + K \qquad \text{if } X < 400$$

The expected costs are

$$E(C_A) = (C + K) \int_0^{400} 200^{-1} e^{-200^{-1}x} \, dx + C \int_{400}^{\infty} 200^{-1} e^{-200^{-1}x} \, dx$$
$$= (C + K)\left[-e^{-200^{-1}x} \Big|_0^{400}\right] + C\left[-e^{-200^{-1}x} \Big|_{400}^{\infty}\right]$$
$$= (C + K)[1 - e^{-2}] + C[e^{-2}]$$
$$= C + K(1 - e^{-2})$$

and

$$E(C_B) = (kC + K) \int_0^{400} 300^{-1} e^{-300^{-1}x} \, dx + kC \int_{400}^{\infty} 300^{-1} e^{300^{-1}x} \, dx$$
$$= (kC + K)[1 - e^{-4/3}] + kC[e^{-4/3}]$$
$$= kC + K(1 - e^{-4/3}).$$

Therefore, if $k < 1 - K/C(e^{-2} - e^{-4/3})$, then the ratio $[E(C_A)/E(C_B)] > 1$, and it is likely that the designer would select Process B.

The exponential distribution has an interesting and unique memoryless property for continuous variables; that is,

$$P(X > x + s \mid X > x) = \frac{P(X > x + s)}{P(X > x)}$$

$$= \frac{e^{-\lambda(x+s)}}{e^{-\lambda x}} = e^{-\lambda s}$$

so that

$$P(X > x + s \mid X > x) = P(X > s) \qquad (6\text{-}16)$$

For example, if a cathode ray tube has an exponential time to failure distribution and at time x it is observed to be still functioning, then the *remaining* life has the same exponential failure distribution as the tube had at time zero.

6-4   The Gamma Distribution

A function used in the definition of a gamma distribution is the gamma function defined by

$$\Gamma(n) = \int_0^\infty x^{n-1}e^{-x}\, dx \qquad \text{for } n > 0 \qquad (6\text{-}17)$$

It can be shown that when $n > 0$,

$$\lim_{k\to\infty} \int_0^k x^{n-1}e^{-x}\, dx$$

exists. An important recursive relationship which may easily be shown on integrating Equation (6-17) by parts is

$$\Gamma(n) = (n - 1)\Gamma(n - 1) \qquad (6\text{-}18)$$

If *n is a positive integer*, then

$$\Gamma(n) = (n - 1)! \qquad (6\text{-}19)$$

since $\Gamma(1) = \int_0^\infty e^{-x}\, dx = 1$. Thus, the gamma function is a generalization of the factorial. The student is asked in Exercise 6-17 to verify that

$$\Gamma\!\left(\frac{1}{2}\right) = \int_0^\infty x^{-1/2}e^{-x}\, dx = \sqrt{\pi} \qquad (6\text{-}20)$$

With the use of the gamma function, we are now able to introduce the gamma probability distribution as:

$$f(x) = \frac{\lambda}{\Gamma(r)}(\lambda x)^{r-1}e^{-\lambda x} \qquad x > 0$$

$$= 0 \qquad\qquad\qquad \text{otherwise} \qquad (6\text{-}21)$$

The parameters are $r > 0$ and $\lambda > 0$. The parameter r is usually called the *shape parameter*, and λ is called the *scale parameter*. Figure 6-6 shows several gamma distributions, for $\lambda = 1$ and various r. It should be noted that $f(x) \geq 0$ for all x, and

$$\int_{-\infty}^{\infty} f(x)\, dx = \int_{0}^{\infty} \frac{\lambda}{\Gamma(r)} (\lambda x)^{r-1} e^{-\lambda x}\, dx$$

$$= \frac{1}{\Gamma(r)} \int_{0}^{\infty} y^{r-1} e^{-y}\, dy = \frac{1}{\Gamma(r)} \cdot \Gamma(r)$$

$$= 1$$

There is a close relationship between the exponential distribution and the gamma distribution. Namely, if $r = 1$ the gamma distribution reduces to the exponential distribution. This follows from the general definition that, *if the random variable X is the sum of r independent, exponentially distributed random variables, each with parameter λ, then X has a gamma density with parameters r and λ*. That is to say, if

$$X = X_1 + X_2 + X_3 + \cdots + X_r \tag{6-22}$$

where

$$g(x_j) = \lambda e^{-\lambda x_j} \qquad x_j \geq 0$$
$$= 0 \qquad \text{otherwise}$$

and the X_j are mutually independent, then X has the density given in Equation (6-21). In many applications of the gamma distribution which we will consider, r will be a positive integer, and we may use this knowledge to good advantage in developing the distribution function.

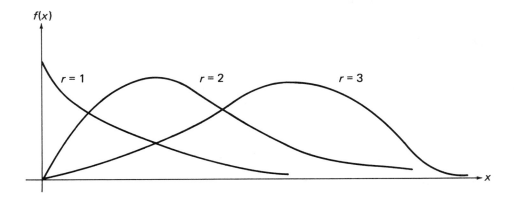

Fig. 6-6. Gamma distribution for $\lambda = 1$.

We may show that the *mean* and *variance* of the gamma distribution are

$$E(X) = r/\lambda \tag{6-23}$$

and

$$V(X) = r/\lambda^2 \tag{6-24}$$

Equations (6-23) and (6-24) represent the mean and variance regardless of whether or not r is an integer; however, when r is an integer and the interpretation given in Equation (6-22) is made, it is obvious that

$$E(X) = \sum_{j=1}^{r} E(X_j) = r \cdot 1/\lambda = r/\lambda$$

and

$$V(X) = \sum_{j=1}^{r} V(X_j) = r \cdot 1/\lambda^2 = r/\lambda^2$$

from a direct application of the expected value and variance operators to the sum of independent random variables.

The moment-generating function for the gamma distribution is

$$M_X(t) = \left(1 - \frac{t}{\lambda}\right)^{-r} \tag{6-25}$$

Recalling that the moment-generating function for the exponential distribution was $[1 - (t/\lambda)]^{-1}$, this result is expected, since

$$M_{(X_1+X_2+\cdots+X_r)}(t) = \prod_{j=1}^{r} M_{X_j}(t) = \left[\left(1 - \frac{t}{\lambda}\right)^{-1}\right]^r \tag{6-26}$$

The distribution function, F, is

$$F(x) = 1 - \int_x^\infty \frac{\lambda}{\Gamma(r)} (\lambda t)^{r-1} e^{-\lambda t} \, dt \qquad x > 0$$
$$= 0 \qquad x \le 0 \tag{6-27}$$

If r is a positive integer, then Equation (6-27) may be integrated by parts giving

$$F(x) = 1 - \sum_{k=0}^{r-1} e^{-\lambda x} (\lambda x)^k / k! \qquad x > 0 \tag{6-28}$$

which is the sum of Poisson terms with mean λx. Thus, tables of the cumulative Poisson may be used to evaluate the distribution function of the gamma.

● **Example 6-6.** A redundant system operates as shown in Fig. 6-7. Initially unit 1 is on line, while unit 2 and unit 3 are on standby. When unit 1 fails, the

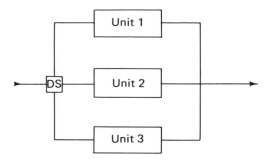

Fig. 6-7. A standby redundant system.

decision switch (DS) switches unit 2 on until it fails and then unit 3 is switched on. The decision switch is assumed to be perfect, so that the system life X may be represented as the sum of the subsystem lives, $X = X_1 + X_2 + X_3$. If the subsystem lives are independent of one another, and if the subsystems each have a life X_j, $j = 1, 2, 3$, having density $g(x_j) = (1/100)\, e^{-x_j/100}$, $x_j \geq 0$, then X will have a gamma density with $r = 3$ and $\lambda = .01$. That is,

$$f(x) = \frac{.01}{2!}(.01x)^2 e^{-.01x} \qquad x > 0$$
$$= 0 \qquad\qquad\qquad \text{otherwise}$$

The probability that the system will operate at least x hours is denoted by $R(x)$ and is called the *reliability function*. Here,

$$R(x) = 1 - F(x) = \sum_{k=0}^{2} e^{-.01x}(.01x)^k/k!$$
$$= e^{-.01x}[1 + (.01x) + (.01x)^2/2]$$

● **Example 6-7.** For a gamma distribution with $\lambda = \frac{1}{2}$ and $r = \nu/2$ where ν is a positive integer, the following single-parameter family of *chi-square distributions with ν degrees of freedom* results:

$$f(\chi^2) = \frac{1}{2^{\nu/2}\Gamma(\nu/2)}(\chi^2)^{(\nu/2)-1}e^{-x^2/2} \qquad \chi^2 > 0$$
$$= 0 \qquad\qquad\qquad\qquad \text{otherwise}$$

This distribution will be discussed further in Chapter 8.

6-5 The Weibull Distribution

The Weibull (1951) distribution has been widely applied to many random phenomena. The principal utility of the Weibull distribution is that it affords an excellent approximation to the probability law of many random variables.

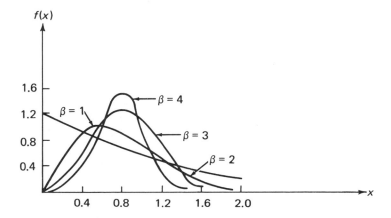

Fig. 6-8. Weibull densities for $\gamma = 0$, $\delta = 1$, and $\beta = 1, 2, 3, 4$.

One important area of application has been as a model for time to failure in electrical and mechanical components and systems. This is discussed in Chapter 15. The density function is given as follows:

$$f(x) = \frac{\beta}{\delta}\left(\frac{x - \gamma}{\delta}\right)^{\beta-1} \exp\left[-\left(\frac{x - \gamma}{\delta}\right)^{\beta}\right] \qquad x \ge \gamma$$
$$= 0 \qquad\qquad\qquad \text{otherwise} \qquad (6\text{-}29)$$

Its parameters are γ, $(-\infty < \gamma < \infty)$ the location parameter, $\delta > 0$ the scale parameter, and $\beta > 0$, the shape parameter. By appropriate selection of these parameters, this density function will closely approximate many observational phenomena.

Figure 6-8 shows some Weibull densities for $\gamma = 0$, $\delta = 1$, and $\beta = 1, 2, 3, 4$. Note that when $\gamma = 0$ and $\beta = 1$, the Weibull distribution reduces to an exponential density with $\lambda = 1/\gamma$.

The *mean* and *variance* of the Weibull distribution can be shown to be

$$E(X) = \gamma + \delta\Gamma\left(1 + \frac{1}{\beta}\right) \qquad (6\text{-}30)$$

$$V(X) = \delta^2\left[\Gamma\left(1 + \frac{2}{\beta}\right) - \Gamma\left(1 + \frac{1}{\beta}\right)^2\right] \qquad (6\text{-}31)$$

The distribution function has the relatively simple form

$$F(x) = 1 - \exp{-\left(\frac{x - \gamma}{\delta}\right)^{\beta}} \qquad x \ge \gamma \qquad (6\text{-}32)$$

● **Example 6-8.** The time to failure distribution for electronic subassemblies is known to have a Weibull density with $\gamma = 0$, $\beta = \frac{1}{2}$, and $\delta = 100$. The fraction

TABLE 6-1 Summary of Continuous Distributions

Density	Parameters	Density Function: $f(x)$	Mean	Variance	Moment-Generating Function
Uniform	α, β $\beta > \alpha$	$f(x) = \dfrac{1}{\beta - \alpha} \qquad \alpha \le x \le \beta$ $\qquad\quad = 0 \qquad\quad$ otherwise	$(\alpha + \beta)/2$	$(\beta - \alpha)^2/12$	$\dfrac{e^{t\beta} - e^{t\alpha}}{t(\beta - \alpha)}$
Exponential	$\lambda > 0$	$f(x) = \lambda e^{-\lambda x} \qquad x > 0$ $\qquad\quad = 0 \qquad\quad$ otherwise	$1/\lambda$	$1/\lambda^2$	$(1 - t/\lambda)^{-1}$
Gamma	$r > 0$ $\lambda > 0$	$f(x) = \dfrac{\lambda}{\Gamma(r)} (\lambda x)^{r-1} e^{-\lambda x} \qquad x > 0$ $\qquad\quad = 0 \qquad\qquad\qquad$ otherwise	r/λ	r/λ^2	$(1 - t/\lambda)^{-r}$.
Weibull	$-\infty < \gamma < \infty$ $\delta > 0$ $\beta > 0$	$f(x) = \dfrac{\beta}{\delta} \left(\dfrac{x - \gamma}{\delta}\right)^{\beta-1} \exp\left[-\left(\dfrac{x - \gamma}{\delta}\right)^\beta\right] \quad x \ge \gamma$ $\qquad\quad = 0 \qquad\qquad\qquad\qquad\qquad$ otherwise	$\gamma + \delta \cdot \Gamma\left(\dfrac{1}{\beta} + 1\right)$	$\delta^2 \left\{ \Gamma\left(\dfrac{2}{\beta} + 1\right) - \left[\Gamma\left(\dfrac{1}{\beta} + 1\right)\right]^2 \right\}$	

expected to survive to, say, 400 hours is thus

$$1 - F(400) = e^{-\sqrt{400/100}} = .1353$$

The mean time to failure is

$$E(X) = 0 + 100(2) = 200 \text{ hours}$$

● **Example 6-9.** Berrettoni (1964) presented a number of applications of the Weibull distribution. The following are examples of natural processes having a probability law closely approximated by the Weibull distribution. The random variable is denoted by X in the examples.

(*a*) Corrosion resistance of magnesium alloy plates.

   X: Corrosion weight loss 10^2 mg/(cm^2)(day) when magnesium alloy plates are immersed in an inhibited aqueous 20 percent solution of MgBr$_2$.

(*b*) Return goods classified by number of weeks after shipment.

   X: Length of the period (10^{-1} weeks) until a customer returns the defective product after shipment.

(*c*) Number of down-times per shift.

   X: Number of down-times per shift times 10^{-1} occurring in a continuous automatic and complicated assembly line.

(*d*) Leakage failure in dry-cell batteries.

   X: Age (years) when leakage starts.

(*e*) Reliability of capacitors.

   X: Life (hours) of 3.3 μF, 50-volt, solid tantalum capacitors operating at an ambient temperature of 125 °C, where the rated catalogue voltage is 33 volts.

6-6   Summary

This chapter has presented four widely used density functions for continuous random variables. The *uniform, exponential, gamma,* and *Weibull* distributions were presented along with underlying assumptions and example applications. Table 6-1 presents a summary of these distributions.

6-7   Exercises

6-1.   A point is chosen at random on the line segment [0, 4]. What is the probability that it lies between $\frac{1}{2}$ and $1\frac{3}{4}$? Between $2\frac{1}{4}$ and $3\frac{3}{8}$?

6-2.   The opening price of a particular stock is uniformly distributed on the interval [$35\frac{3}{4}$, $44\frac{1}{4}$]. What is the probability that, on any day, the opening price is less than 40? Between 40 and 42?

6-3. The random variable X is uniformly distributed on the interval $[0, 2]$. Find the distribution of the random variable $Y = 5 + 2X$.

6-4. A real estate broker charges a fixed fee of $50 plus a 6 percent commission on the landowners' profit. If this profit is uniformly distributed between $0 and $2000, find the probability distribution of the broker's total fees.

6-5. Verify that the moment-generating function for the uniform density is as given by Equation (6-4). Use it to generate the mean and variance.

6-6. Let X be uniformly distributed and symmetric about zero with variance 1. Find the appropriate values for α and β.

6-7. Show how the uniform density function can be used to generate variates from the empirical probability distribution described below:

y	p(y)
1	.3
2	.2
3	.4
4	.1

6-8. The random variable X is uniformly distributed over the interval $[0, 4]$. What is the probability that the roots of $y^2 + 4Xy + X + 1 = 0$ are real?

6-9. Verify that the moment-generating function of the exponential distribution is as given by Equation (6-14). Use it to generate the mean and variance.

6-10. The engine and drive shaft of a new car are guaranteed for one year. The mean lives of engines and drive shafts are estimated to be three years, and the time to failure has an exponential density. The realized profit on a new car is $1000. Including costs of parts and labor the dealer must pay $250 to repair each failure. What is the expected profit per car?

6-11. For the data in the above problem, what percentage of cars will experience failure in the engine and drive shaft during the first six months of use?

6-12. Let the length of time a machine will operate be an exponentially distributed random variable with probability density function $f(t) = \theta e^{-\theta t}$, $t \geq 0$. Suppose an operator for this machine must be hired for a predetermined and fixed length of time, say Y. She is paid d dollars per time period during this interval. The net profit from operating this machine, exclusive of labor costs, is r dollars per time period. Find the value of Y which maximizes the total profit obtained.

6-13. The time to failure of a television tube is estimated to be exponentially distributed with mean three years. A company offers insurance on these tubes for the first year of usage. On what percentage of policies will they have to pay a claim?

6-14. Is there an exponential density that satisfies the following condition?

$$P\{X \leq 2\} = \frac{2}{3} P\{X \leq 3\}$$

If so, find the value of λ.

6-15. Two manufacturing processes are under consideration. The per-unit cost for process I is C, while for process II it is $3C$. Products from both processes have exponential time to failure densities with mean rates of 25^{-1} failures per hour and 35^{-1} failures per hour from I and II, respectively. If a product fails before 15 hours it must be replaced at a cost of Z dollars. Which process would you recommend?

6-16. A transistor has an exponential time to failure distribution with mean time to failure of 20,000 hours. The transistor has already lasted 20,000 hours in a particular application. What is the probability that the transistor fails by 30,000 hours?

6-17. Show that $\Gamma(\tfrac{1}{2}) = \sqrt{\pi}$.

6-18. Prove the gamma function properties given by Equations (6-18) and (6-19).

6-19. A ferry boat will take its customers across a river when 10 cars are aboard. Experience shows that cars arrive at the ferry boat independently and at a mean rate of 7 per hour. Find the probability that the time between consecutive trips will be 1 hour.

6-20. A box of candy contains 24 bars. The demand for these candy bars is exponentially distributed with a mean rate of 6 per hour. What is the probability that a box of candy bars opened at 8:00 A.M. will be empty by noon?

6-21. Show that the moment-generating function of the gamma distribution is as given by Equation (6-25).

6-22. The life of an electronic system is $Y = X_1 + X_2 + X_3 + X_4$, the sum of the subsystem component lives. The subsystems are independent, each having exponential failure densities with a mean time between failures of 4 hours. What is the probability that the system will operate at least 24 hours?

6-23. The replenishment time for a certain product is known to be gamma distributed with mean 40 and variance 400. Find the probability that an order is received within the first 20 days after it is ordered. Within the first 60 days.

6-24. Suppose a gamma distributed random variable is defined over the interval $u \leq X < \infty$ with density function

$$f(x) = \frac{\lambda^r}{\Gamma(r)}(x - u)^{r-1}e^{-\lambda(x-u)} \qquad x \geq u, \lambda \geq 0, r > 0$$

$$= 0 \qquad \text{otherwise}$$

Find the mean of this *three-parameter* gamma distribution.

6-25. The beta probability distribution is defined by

$$f(x) = \frac{\Gamma(\lambda + r)}{\Gamma(\lambda)\Gamma(r)}x^{\lambda-1}(1 - x)^{r-1} \qquad 0 \leq x \leq 1, \lambda > 0, r > 0$$

$$= 0 \qquad \text{otherwise}$$

(*a*) Graph the distribution for $\lambda > 1, r > 1$.
(*b*) Graph the distribution for $\lambda < 1, r < 1$.
(*c*) Graph the distribution for $\lambda < 1, r \geq 1$.
(*d*) Graph the distribution for $\lambda \geq 1, r < 1$.
(*e*) Graph the distribution for $\lambda = r$.

6-26. Show that when $\lambda = r = 1$ the beta distribution reduces to the uniform distribution.

6-27. Show that when $\lambda = 2$, $r = 1$ or $\lambda = 1$, $r = 2$ the beta distribution reduces to a triangular probability distribution.

6-28. Show that if $\lambda = r = 2$ the beta distribution reduces to a parabolic probability distribution. Graph the density function.

6-29. Find the mean and variance of the beta distribution.

6-30. Find the mean and variance of the Weibull distribution.

6-31. The diameter of steel shafts is Weibull distributed with parameters $\gamma = 1.0$ inches, $\beta = 2$, and $\delta = .5$. Find the probability that a randomly selected shaft will not exceed 1.5 inches in diameter.

6-32. The time to failure of a certain transistor is known to be Weibull distributed with parameters $\gamma = 0$, $\beta = \frac{1}{3}$, and $\delta = 400$. Find the fraction expected to survive 600 hours.

6-33. The time to leakage failure in a certain type of dry-cell battery is expected to have a Weibull distribution with parameters $\gamma = 0$, $\beta = \frac{1}{2}$, and $\delta = 400$. What is the probability that a battery will survive beyond 800 hours of use?

6-34. Graph the Weibull distribution with $\gamma = 0$, $\delta = 1$, and $\beta = 1, 2, 3,$ and 4.

6-35. The time to failure density for a small computer system has a Weibull density with $\gamma = 0$, $\beta = \frac{1}{4}$, and $\delta = 200$.
(*a*) What fraction of these units will survive to 1000 hours?
(*b*) What is the mean time to failure?

6-36. A manufacturer of a commercial television monitor guarantees the picture tube for one year (8760 hours). The monitors are used in airport terminals for flight schedules, and they are in continuous use with power on. The mean life of the tubes is 20,000 hours, and they follow an exponential time to failure density. It costs the manufacturer $300 to make, sell, and deliver a monitor that will be sold for $400. It costs $150 to replace a failed tube, including materials and labor. The manufacturer has no replacement obligation beyond the first replacement. What is the manufacturer's expected profit?

6-37. The lead time for orders of diodes from a certain manufacturer is known to have a gamma distribution with a mean of 20 days and a standard deviation of 10 days. Determine the probability of receiving an order within 15 days of the placement date.

Chapter 7

The Normal Distribution

7-1 Introduction

In this chapter we consider the normal distribution. This distribution is very important in both the theory and application of statistics. We also discuss the lognormal and bivariate normal distributions.

The normal distribution was first studied in the eighteenth century when the patterns in errors of measurement were observed to follow a symmetrical, bell-shaped distribution. It was first presented in mathematical form in 1733 by DeMoivre, who derived it as a limiting form of the binomial distribution. The distribution was also known to Laplace no later than 1775. Through historical error, it has been attributed to Gauss, whose first published reference to it appeared in 1809, and the term *Gaussian distribution* is frequently employed. Various attempts were made during the eighteenth and nineteenth centuries to establish this distribution as the underlying probability law for all continuous variates; thus, the name *normal* came to be applied.

7-2 The Normal Distribution

The normal distribution is in many respects the cornerstone of statistics. A random variable X is said to have a normal distribution with mean $\mu(-\infty < \mu < \infty)$ and variance $\sigma^2 > 0$ if it has the density function

$$f(x) = \frac{1}{\sigma\sqrt{2\pi}} e^{-(1/2)[(x-\mu)/\sigma]^2} \qquad -\infty < x < \infty \qquad (7\text{-}1)$$

The distribution is illustrated graphically in Fig. 7-1. The normal distribution is used so extensively that the shorthand notation $X \sim N(\mu, \sigma^2)$ is often used to indicate that the random variable X is normally distributed with mean μ and variance σ^2.

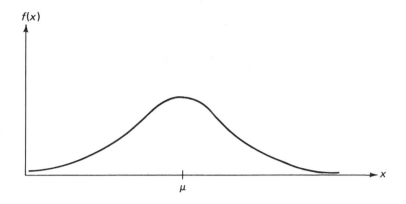

Fig. 7-1.   The normal distribution.

The normal distribution has several important properties:

1. $\int_{-\infty}^{\infty} f(x)\, dx = 1$
2. $f(x) \geq 0$ for all x } required of all density functions.   (7-2)
3. $\lim\limits_{x \to \infty} f(x) = 0$ and $\lim\limits_{x \to -\infty} f(x) = 0$
4. $f[(x + \mu)] = f[-(x - \mu)]$ The density is symmetric about μ.
5. The maximum value of f occurs at $x = \mu$.
6. The points of inflection of f are at $x = \mu \pm \sigma$.

Property (a) may be demonstrated as follows. Let $y = (x - \mu)/\sigma$ in Equation (7-1) and denote the integral as I. That is,

$$I = \frac{1}{\sqrt{2\pi}} \int_{-\infty}^{\infty} e^{-(1/2)y^2}\, dy$$

Our proof that $\int_{-\infty}^{\infty} f(x)\, dx = 1$ will consist of showing that $I^2 = 1$, and then inferring that $I = 1$ since f must be everywhere positive. Defining a second normally distributed variable, Z, we have:

$$I^2 = \frac{1}{\sqrt{2\pi}} \int_{-\infty}^{\infty} e^{-(1/2)y^2}\, dy \; \frac{1}{\sqrt{2\pi}} \int_{-\infty}^{\infty} e^{-(1/2)z^2}\, dz$$

$$= \frac{1}{2\pi} \int_{-\infty}^{\infty} \int_{-\infty}^{\infty} e^{-(1/2)(y^2 + z^2)}\, dy\, dz$$

On changing to polar coordinates with the transformation of variables $y =$

$r \sin \theta$ and $z = r \cos \theta$, the integral becomes

$$I^2 = \frac{1}{2\pi} \int_0^\infty \int_0^{2\pi} r e^{-(1/2)r^2} \, d\theta \, dr$$

$$= \int_0^\infty r e^{-(1/2)r^2} \, dr = 1$$

completing the proof.

The mean of the normal distribution may be determined easily. Since

$$E(X) = \int_{-\infty}^\infty \frac{x}{\sigma\sqrt{2\pi}} e^{-(1/2)[(x-\mu)/\sigma]^2} \, dx$$

and if we let $z = (x - \mu)/\sigma$, we obtain

$$E(X) = \int_{-\infty}^\infty \frac{1}{\sqrt{2\pi}} (\mu + \sigma z) e^{-z^2/2} \, dz$$

$$= \mu \int_{-\infty}^\infty \frac{1}{\sqrt{2\pi}} e^{-z^2/2} \, dz + \sigma \int_{-\infty}^\infty \frac{1}{\sqrt{2\pi}} z e^{-z^2/2} \, dz$$

Since the integrand of the first integral is that of a normal density with $\mu = 0$ and $\sigma^2 = 1$, the value of the first integral is one. The second integral has value zero, that is,

$$\int_{-\infty}^\infty \frac{1}{\sqrt{2\pi}} z e^{-z^2/2} \, dz = -\frac{1}{\sqrt{2\pi}} e^{-z^2/2} \Big|_{-\infty}^\infty = 0$$

and thus

$$E(X) = \mu[1] + \sigma[0]$$

$$= \mu \tag{7-3}$$

To find the variance we must evaluate

$$V(X) = E[(X - \mu)^2] = \int_{-\infty}^\infty (x - \mu)^2 \frac{1}{\sigma\sqrt{2\pi}} e^{-(1/2)[(x-\mu)/\sigma]^2} \, dx$$

and letting $z = (x - \mu)/\sigma$ we obtain

$$V(X) = \int_{-\infty}^\infty \sigma^2 z^2 \frac{1}{\sqrt{2\pi}} e^{-z^2/2} \, dz = \sigma^2 \left[\int_{-\infty}^\infty \frac{z^2}{\sqrt{2\pi}} e^{-z^2/2} \, dz \right]$$

$$= \sigma^2 \left[\frac{-ze^{-z^2/2}}{\sqrt{2\pi}} \Big|_{-\infty}^\infty + \int_{-\infty}^\infty \frac{1}{\sqrt{2\pi}} e^{-z^2/2} \, dz \right]$$

$$= \sigma^2[0 + 1]$$

so that

$$V(X) = \sigma^2 \tag{7-4}$$

In summary the mean and variance of the normal density given in Equation (7-1) are μ and σ^2, respectively.

The *moment-generating function* for the normal distribution can be shown to be

$$M_X(t) = e^{(t\mu + \sigma^2 t^2/2)} \tag{7-5}$$

For the development of Equation (7-5), see Exercise 7-14.

The distribution function F is

$$F(x) = P(X \le x) = \int_{-\infty}^{x} \frac{1}{\sigma\sqrt{2\pi}} e^{-(1/2)[(u-\mu)/\sigma]^2} \, du \tag{7-6}$$

It is impossible to evaluate this integral without resorting to numerical methods, and even then the evaluation would have to be accomplished for each pair (μ, σ^2). However, a simple transformation of variables, $z = (x - \mu)/\sigma$, allows the evaluation to be independent of μ and σ. That is,

$$F(x) = P(X \le x) = P\left(Z \le \frac{x-\mu}{\sigma}\right) = \int_{-\infty}^{(x-\mu)/\sigma} \frac{1}{\sqrt{2\pi}} e^{-z^2/2} \, dz$$

$$= \int_{-\infty}^{(x-\mu)/\sigma} \varphi(z) \, dz = \Phi\left(\frac{x-\mu}{\sigma}\right) \tag{7-7}$$

The density

$$\varphi(z) = \frac{1}{\sqrt{2\pi}} e^{-z^2/2} \qquad -\infty < z < \infty$$

has mean 0 and variance 1, that is, $Z \sim N(0, 1)$, and we say that Z has a *standard normal distribution*. A graph of the probability density function is shown in Fig. 7-2. The corresponding distribution function is Φ, where

$$\Phi(z) = \int_{-\infty}^{z} \frac{1}{\sqrt{2\pi}} e^{-u^2/2} \, du \tag{7-8}$$

and this function has been well tabulated. A table of the integral in Equation (7-8) has been provided in Table II of the Appendix.

The procedure used in solving problems is quite simple. For example, suppose we know that $X \sim N(100, 4)$ and we wish to evaluate $F(104)$, that is, $P(X \le 104)$. Since $F(x) = \Phi[(x - \mu)/\sigma]$, we determine that $\Phi[(104 - 100)/2] = \Phi(2) = .9772$ from Table II in the Appendix.

Note the relationship between the argument of F and the argument of Φ, which measures the departure of x from the mean, μ, in standard deviation (σ)

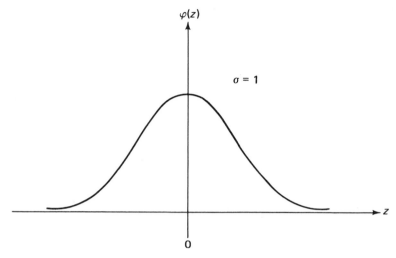

Fig. 7-2.   The standard normal distribution.

units. For example, in the case just considered, $F(104) = \Phi(2)$, which indicates that 104 is *two* standard deviations ($\sigma = 2$) above the mean. In general, $x = \mu + \sigma z$. In solving problems, we sometimes need to use the symmetry property of φ in addition to the tables. It is helpful to make a sketch if there is any confusion in determining exactly which probabilities are required, since the area under the curve and over the interval of interest is the probability that the random variable will lie on the interval.

● **Example 7-1.** The breaking strength (in newtons) of a synthetic fabric is denoted by X, and it is distributed as $N(800, 144)$. The purchaser of the fabric requires the fabric to have a strength of at least 772 N. A fabric sample is randomly selected and tested. To find $P(X \geq 772)$, we first calculate

$$P(X < 772) = P\left(\frac{X - \mu}{\sigma} < \frac{772 - 800}{12}\right)$$
$$= P(Z < -2.33)$$
$$= \Phi(-2.33) = .01$$

Hence, the desired probability, $P(X \geq 772)$, equals .99. Figure 7-3 shows the calculated probability relative to both X and Z. We have chosen to work with the random variable Z because its distribution function is tabulated.

● **Example 7-2.** The time required to repair an automatic loading machine in a complex food-packaging operation of a production process is X minutes.

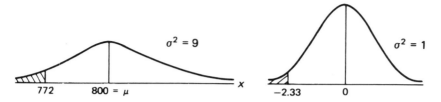

Fig. 7-3. $P(X < 772)$ where $X \sim N(800, 144)$.

Studies have shown that the approximation $X \sim N(120, 16)$ is quite good. A sketch is shown in Fig. 7-4. If the process is down for more than 125 minutes, all equipment must be cleaned, with the loss of food in process. The total cost of food loss and cleaning associated with the long down-time is \$10,000. In order to determine the probability of this occurrence, we proceed as follows:

$$P(X > 125) = P\left(Z > \frac{125 - 120}{4}\right) = P(Z > 1.25)$$
$$= 1 - \Phi(1.25)$$
$$= 1 - .8944$$
$$= .1056$$

Thus, given a breakdown of the packaging machine, the expected cost is $E(C) = .1056(10{,}000 + C_{R_1}) + .8944(C_{R_1})$, where C is the total cost and C_{R_1} is the repair cost. Simplified, $E(C) = C_{R_1} + 1056$. Suppose the management can reduce the mean of the service time distribution to 115 minutes by adding more

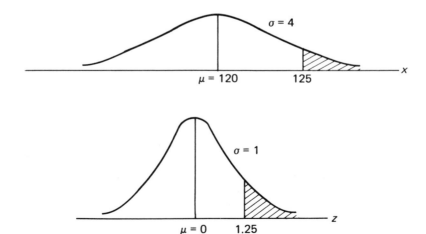

Fig. 7-4. $P(X > 125)$ where $X \sim N(120, 16)$.

maintenance personnel. The new cost for repair will be $C_{R_2} > C_{R_1}$; however,

$$P(X > 125) = P\left(Z > \frac{125 - 115}{4}\right) = P(Z > 2.5)$$

$$= 1 - \Phi(2.5)$$

$$= 1 - .9938$$

$$= .0062$$

so that the new expected cost would be $C_{R_2} + 62$; and one would logically make the decision to add to the maintenance crew if

$$C_{R_2} + 62 < C_{R_1} + 1056$$

or

$$(C_{R_2} - C_{R_1}) < \$994$$

It is assumed that the frequency of breakdowns remains unchanged.

- **Example 7-3.** The pitch diameter of the thread on a fitting is normally distributed with mean .4008 centimeter and standard deviation .0004 centimeter. The design specifications are .4000 ± .0010 centimeter. This is illustrated in Fig. 7-5. We desire to determine what fraction of product is within tolerance. Using the approach employed previously,

$$P(.399 \le X \le .401) = P\left(\frac{.399 - .4008}{.0004} \le Z \le \frac{.4010 - .4008}{.0004}\right)$$

$$= P(-4.5 \le Z \le .5)$$

$$= \Phi(.5) - \Phi(-4.5)$$

$$= .6915 - .0000$$

$$= .6915$$

As process engineers study the results of such calculations, they decide to replace a worn cutting tool and adjust the machine producing the fittings so that the new mean falls directly at the nominal value (the designers' most desired value) of .4000. Then,

$$P(.3990 \le X \le .4010) = P\left(\frac{.3990 - .4}{.0004} \le Z \le \frac{.4010 - .4}{.0004}\right)$$

$$= P(-2.5 \le Z \le +2.5)$$

$$= \Phi(2.5) - \Phi(-2.5)$$

$$= .9938 - .0062$$

$$= .9876$$

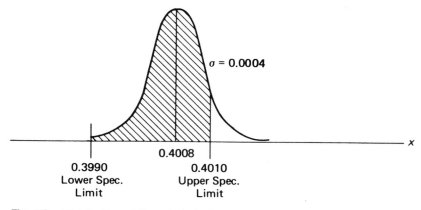

σ = 0.0004

0.4008

0.3990
Lower Spec.
Limit

0.4010
Upper Spec.
Limit

Fig. 7-5. Distribution of thread pitch diameters.

We see that with the adjustments, 98.76 percent of the fittings will be within tolerance. The distribution of adjusted, machine pitch diameters is shown in Fig. 7-6.

● **Example 7-4.** Another type of problem involving the use of tables of the normal distribution sometimes arises. Suppose, for example, that $X \sim N(50, 4)$. Furthermore, suppose we want to determine a value of X, say x, such that $P(X > x) = .025$. Then,

$$P(X > x) = P\left(Z > \frac{x - 50}{2}\right) = .025$$

or

$$P\left(Z \leq \frac{x - 50}{2}\right) = .975$$

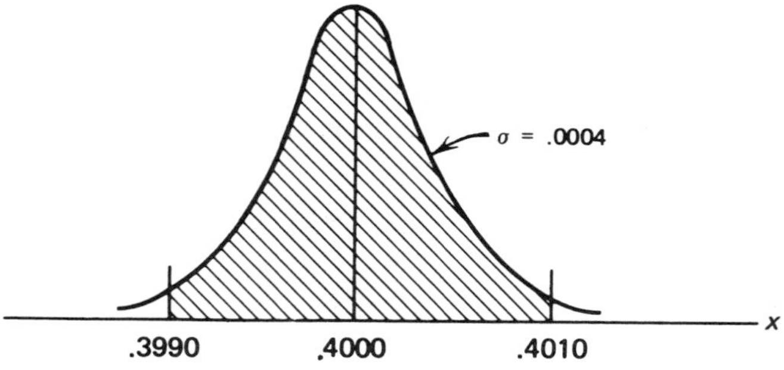

σ = .0004

.3990   .4000   .4010

Fig. 7-6. Distribution of adjusted machine pitch diameters.

so that

$$\frac{x-50}{2} = 1.96$$

and

$$x = 50 + 2(1.96) = 53.92$$

There are several symmetric intervals that arise frequently. Their probabilities are:

$$
\begin{aligned}
P(\mu - 1.00\sigma \; &\leq X \leq \mu + 1.00\sigma) \; = .6826 \\
P(\mu - 1.645\sigma &\leq X \leq \mu + 1.645\sigma) = .90 \\
P(\mu - 1.96\sigma \; &\leq X \leq \mu + 1.96\sigma) \; = .95 \\
P(\mu - 2.57\sigma \; &\leq X \leq \mu + 2.57\sigma) \; = .99 \\
P(\mu - 3.00\sigma \; &\leq X \leq \mu + 3.00\sigma) \; = .9978
\end{aligned}
\tag{7-9}
$$

7-3   The Reproductive Property of the Normal Distribution

Suppose we have n independent, normal random variables $X_1, X_2, \ldots, X_n$, where $X_i \sim N(\mu_i, \sigma_i^2)$, for $i = 1, 2, \ldots, n$. It was shown earlier that if

$$Y = X_1 + X_2 + \cdots + X_n \tag{7-10}$$

then

$$E(Y) = \mu_Y = \sum_{i=1}^{n} \mu_i \tag{7-11}$$

and

$$V(Y) = \sigma_Y^2 = \sum_{i=1}^{n} \sigma_i^2$$

Using moment-generating functions, we see that

$$
\begin{aligned}
M_Y(t) &= M_{X_1}(t) \cdot M_{X_2}(t) \cdot \ldots \cdot M_{X_n}(t) \\
&= [e^{\mu_1 t + \sigma_1^2 t^2/2}] \cdot [e^{\mu_2 t + \sigma_2^2 t^2/2}] \cdot \ldots \cdot [e^{\mu_n t + \sigma_n^2 t^2/2}]
\end{aligned}
\tag{7-12}
$$

Therefore,

$$M_Y(t) = e^{[(\mu_1 + \mu_2 + \cdots + \mu_n)t + (\sigma_1^2 + \sigma_2^2 + \cdots + \sigma_n^2)t^2/2]} \tag{7-13}$$

which is the moment-generating function of a normally distributed random variable with mean $\mu_1 + \mu_2 + \cdots + \mu_n$ and variance $\sigma_1^2 + \sigma_2^2 + \cdots + \sigma_n^2$. Therefore, by the uniqueness property of the moment-generating function, we see that Y is normal with mean μ_Y and variance σ_Y^2.

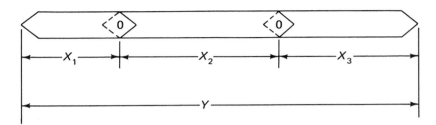

Fig. 7-7.   A linkage assembly.

● **Example 7-5.** An assembly consists of three linkage components as shown in Fig. 7-7. The properties X_1, X_2, and X_3 are given below with means in centimeters and variance in square centimeters.

$$X_1 \sim N(12, .02)$$
$$X_2 \sim N(24, .03)$$
$$X_3 \sim N(18, .04)$$

Links are produced by different machines and operators, so we have reason to assume that X_1, X_2, and X_3 are independent. Suppose we want to determine $P(53.8 \leq Y \leq 54.2)$. Since $Y = X_1 + X_2 + X_3$, Y is distributed normally with mean $\mu_Y = 12 + 24 + 18 = 54$ and variance $\sigma^2 = \sigma_1^2 + \sigma_2^2 + \sigma_3^2 = .02 + .03 + .04 = .09$. Thus,

$$P(53.8 \leq Y \leq 54.2) = P\left(\frac{53.8 - 54}{.3} \leq Z \leq \frac{54.2 - 54}{.3}\right)$$

$$= P\left(-\frac{2}{3} \leq Z \leq +\frac{2}{3}\right)$$

$$= \Phi(.67) - \Phi(-.67)$$

$$= .749 - .251$$

$$= .498$$

These results can be generalized to linear combinations of independent normal variables. Linear combinations of the form

$$Y = a_0 + a_1 X_1 + \cdots + a_n X_n \tag{7-14}$$

were presented earlier and we found that $\mu_Y = a_0 + \sum_{i=1}^{n} a_i \mu_i$. When the variables are independent, $\sigma_Y^2 = \sum_{i=1}^{n} a_i^2 \sigma_i^2$. Again, if $X_1, \ldots, X_n$ are independent and normally distributed, then $Y \sim N(\mu_Y, \sigma_Y^2)$.

● **Example 7-6.** A shaft is to be assembled into a bearing as shown in Fig. 7-8. The clearance is $Y = X_1 - X_2$. Suppose

$$X_1 \sim N(1.500, .0016)$$

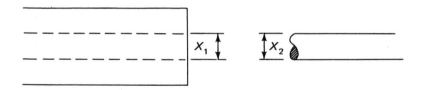

Fig. 7-8.   An assembly.

and
$$X_2 \sim N(1.480, .0009)$$

Then,
$$\mu_Y = a_1\mu_1 + a_2\mu_2$$
$$= (1)(1.500) + (-1)(1.480)$$
$$= .02$$

and
$$\sigma_Y^2 = a_1^2\sigma_1^2 + a_2^2\sigma_2^2$$
$$= (1)^2(.0016) + (-1)^2(.0009)$$
$$= .0025$$

so that
$$\sigma_Y = .05$$

When the parts are assembled, there will be interference if $Y < 0$, so
$$P(\text{interference}) = P(Y < 0) = P\left(Z < \frac{0 - .02}{.05}\right)$$
$$= \Phi(-.4) = .3446$$

This indicates that 34.46 percent of all assemblies attempted would meet with failure. If the designer feels that the *nominal clearance* $\mu_Y = .02$ is as large as it can be made for the assembly, then the only way to reduce the 34.46 percent figure is to reduce the variance of the distributions. In many cases, this can be accomplished by the overhaul of production equipment, better training of production operators, and so on.

7-4   The Central Limit Theorem

If a random variable Y is the *sum of n independent random variables* which satisfy certain general conditions, then for sufficiently large n, Y is approximately normally distributed. We state this as a theorem.

Theorem 7-1 (Central Limit Theorem).

If $X_1, X_2, \ldots, X_n$ is a sequence of n independent random variables with $E(X_i) = \mu_i$ and $V(X_i) = \sigma_i^2$ (both finite) and $Y = X_1 + X_2 + \cdots + X_n$, then under some general conditions

$$Z_n = \frac{Y - \sum_{i=1}^{n} \mu_i}{\sqrt{\sum_{i=1}^{n} \sigma_i^2}} \tag{7-15}$$

has an approximate $N(0, 1)$ distribution as n approaches infinity. If F_n is the distribution function of Z_n, then

$$\lim_{n \to \infty} \frac{F_n(z)}{\Phi(z)} = 1 \tag{7-16}$$

The "general conditions" mentioned in the theorem are informally summarized as follows: The terms X_i, taken individually, contribute a negligible amount to the variance of the sum, and it is not likely that a single term makes a large contribution to the sum.

The proof of this theorem, as well as a rigorous discussion of the necessary assumptions, is beyond the scope of this presentation. There are, however, several observations that should be made. The fact that Y is approximately normally distributed when the X_i terms may have *essentially any distribution* is the basic underlying reason for the importance of the normal distribution. In numerous applications, the random variable being considered may be represented as the sum of n independent random variables, some of which may be measurement error, some due to physical considerations, and so on, and thus the normal distribution provides a good approximation.

A special case of the central limit theorem arises when each of the components has the same distribution.

Theorem 7-2

If $X_1, X_2, \ldots, X_n$ is a sequence of n independent, identically distributed random variables with $E(X_i) = \mu$ and $V(X_i) = \sigma^2$, and $Y = X_1 + X_2 + \cdots + X_n$, then

$$Z_n = \frac{Y - n\mu}{\sigma \sqrt{n}} \tag{7-17}$$

has an approximate $N(0, 1)$ distribution in the sense that if F_n is the distribution function of Z_n, then

$$\lim_{n \to \infty} \frac{F_n(z)}{\Phi(z)} = 1$$

Under the restriction that $M_X(t)$ exists for real t, a straightforward proof may be presented for this form of the central limit theorem. Many mathematical statistics texts present such a proof.

The question immediately encountered in practice is: "How large must n be to get reasonable results using the normal distribution to approximate the distribution of Y?" This is not an easy question to answer since the answer depends on the characteristics of the distribution of the X_i terms as well as the meaning of "reasonable results." From a practical standpoint, some very crude rules of thumb can be given where the distribution of the X_i terms falls into one of three arbitrarily selected groups as follows:

1. Well-behaved—The distribution of X_i does not radically depart from the normal distribution. There is a bell-shaped density that is nearly symmetric. For this case, practitioners in quality control and other areas of application have found that n should be at least 4. That is, $n \geq 4$.

2. Fairly behaved—The distribution of X_i has no prominent mode, and it appears much as a uniform density. In this case $n \geq 12$ is a commonly used rule.

3. Ill-behaved—The distribution has most of its measure in the tails, as in Fig. 7-9. In this case it is most difficult to say; however, in many practical applications, $n \geq 100$ should be satisfactory.

• **Example 7-7.** Small parts are packaged 250 to the crate. Part weights are independent random variables with mean .5 pound and standard deviation .10 pound. Twenty crates are loaded to a pallet. Suppose we wish to find the probability that the parts on a pallet will exceed 2510 pounds in weight. (Neglect both pallet and crate weight). Let

$$Y = X_1 + X_2 + \cdots + X_{5000}$$

represent the total weight of the parts, so that

$$\mu_Y = 5000(.5) = 2500$$

$$\sigma_Y^2 = 5000(.01) = 50$$

and

$$\sigma_Y = \sqrt{50} = 7.071$$

Then

$$P(Y > 2510) = P\left(Z > \frac{2510 - 2500}{7.071}\right)$$

$$1 - \Phi(1.41) = .07929$$

Note that we did not know the distribution of the individual part weights.

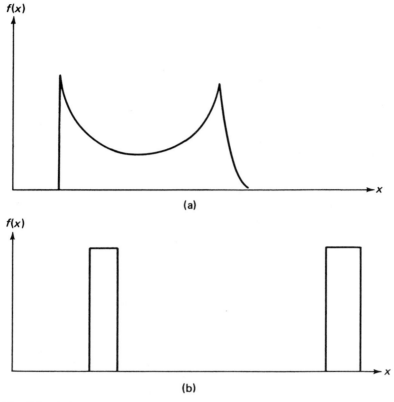

(a)

(b)

Fig. 7-9. Ill-behaved distributions.

● **Example 7-8.** In a construction project, a network of major activities has been constructed to serve as the basis for planning and scheduling. On a *critical path* there are 16 activities. The means and variances are given in Table 7-1.

TABLE 7-1   **Activity Mean Times and Variances (in Weeks and Weeks²)**

Activity	Mean	Variance	Activity	Mean	Variance
1	2.7	1.0	9	3.1	1.2
2	3.2	1.3	10	4.2	.8
3	4.6	1.0	11	3.6	1.6
4	2.1	1.2	12	.5	.2
5	3.6	.8	13	2.1	.6
6	5.2	2.1	14	1.5	.7
7	7.1	1.9	15	1.2	.4
8	1.5	.5	16	2.8	.7

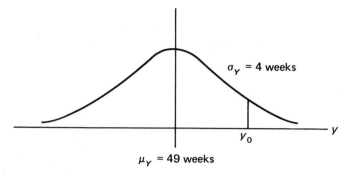

Fig. 7-10.   Distribution of project times.

The activity times may be considered independent and the project time is the sum of the activity times on the critical path, that is, $Y = X_1 + X_2 + \cdots + X_{16}$, where Y is the project time and X_i is the time for the ith activity. Although the distributions of X_i are unknown, the distributions are fairly to well-behaved. The contractor would like to know (1) the expected completion time, and (2) a project time corresponding to a probability of .90 of having the project completed. Calculating μ_Y and σ_Y^2, we obtain

$$\mu_Y = 49 \text{ weeks}$$
$$\sigma_Y^2 = 16 \text{ weeks}^2$$

The expected completion time for the project is thus 49 weeks. In determining the time y_0 such that the probability .9 of having the project completed by that time, Fig. 7-10 may be helpful.

We may calculate

$$P(Y \le y_0) = .90$$

or

$$P\left(Z \le \frac{y_0 - 49}{4}\right) = .90$$

so that

$$\frac{y_0 - 49}{4} = 1.282$$

and

$$y_0 = 49 + 1.282(4)$$
$$= 54.128 \text{ weeks}$$

7-5   The Normal Approximation to the Binomial Distribution

In Chapter 5, the binomial approximation to the hypergeometric distribution was presented as was the Poisson approximation to the binomial distribution.

In this section we consider the normal approximation to the binomial distribution. Since the binomial is a discrete probability distribution, this may seem to go against intuition; however, a limiting process is involved, keeping p of the binomial distribution fixed and letting $n \to \infty$. The approximation is known as the DeMoivre—Laplace approximation.

We recall the binomial distribution as

$$p(x) = \frac{n!}{x!(n-x)!} p^x q^{n-x} \qquad x = 0, 1, 2, \ldots, n$$
$$= 0 \qquad\qquad\qquad\qquad \text{otherwise}$$

Stirling's approximation to $n!$ is

$$n! \simeq (2\pi)^{1/2} e^{-n} n^{n+(1/2)} \tag{7-18}$$

The error

$$\frac{|n! - (2\pi)^{1/2} e^{-n} n^{n+(1/2)}|}{n!} \to 0 \tag{7-19}$$

as $n \to \infty$. Using Stirling's formula to approximate the terms involving $n!$ in the binomial model, we find that, for large n,

$$P(X = x) \simeq \frac{1}{(\sqrt{np(1-p)})\sqrt{2\pi}} e^{(1/2)[(x-np)/\sqrt{np(1-p)}]^2} \tag{7-20}$$

so that

$$P(X \le x) \simeq \Phi\left(\frac{x - np}{\sqrt{npq}}\right) = \int_{-\infty}^{(x-np)/\sqrt{npq}} \frac{1}{\sqrt{2\pi}} e^{-z^2/2} \, dz \tag{7-21}$$

Thus, the quantity $(X - np)/\sqrt{npq}$ *approximately* has a $N(0, 1)$ distribution. If p is close to $\frac{1}{2}$ and $n > 10$, the approximation is fairly good; however, for other values of p, the value of n must be larger. In general, experience indicates that the approximation is fairly good as long as $np > 5$ for $p \le \frac{1}{2}$ or when $nq > 5$ when $p > \frac{1}{2}$.

● **Example 7-9.** In sampling from a production process that produces items of which 20 percent are defective, a random sample of 100 items is selected each hour of each production shift. The number of defectives in a sample is denoted by X. To find, say, $P(X \le 15)$ we may use the normal approximation as follows:

$$P(X \le 15) = P\left(Z \le \frac{15 - 100 \cdot (.2)}{\sqrt{100(.2)(.8)}}\right)$$
$$= P(Z \le -1.25) = \Phi(-1.25) = .1056$$

Since the binomial distribution is discrete and the normal distribution is

TABLE 7-2 Continuity Corrections

Quantity Desired from Binomial Distribution	With Continuity Correction	In Terms of the Distribution Function Φ
$P(X = x)$	$P(x - \frac{1}{2} \le X \le x + \frac{1}{2})$	$\Phi\left(\dfrac{x + \frac{1}{2} - np}{\sqrt{npq}}\right) - \Phi\left(\dfrac{x - \frac{1}{2} - np}{\sqrt{npq}}\right)$
$P(X \le x)$	$P(X \le x + \frac{1}{2})$	$\Phi\left(\dfrac{x + \frac{1}{2} - np}{\sqrt{npq}}\right)$
$P(X < x)$ $= P(X \le x - 1)$	$P(X \le x - 1 + \frac{1}{2})$	$\Phi\left(\dfrac{x - \frac{1}{2} - np}{\sqrt{npq}}\right)$
$P(X \ge -x)$	$P(X \ge x - \frac{1}{2})$	$1 - \Phi\left(\dfrac{x - \frac{1}{2} - np}{\sqrt{npq}}\right)$
$P(X > x)$ $= P(X \ge x + 1)$	$P(X \ge x + 1 - \frac{1}{2})$	$1 - \Phi\left(\dfrac{x + \frac{1}{2} - np}{\sqrt{npq}}\right)$
$P(a \le X \le b)$	$P(a - \frac{1}{2} \le X \le b + \frac{1}{2})$	$\Phi\left(\dfrac{b + \frac{1}{2} - np}{\sqrt{npq}}\right) - \Phi\left(\dfrac{a - \frac{1}{2} - np}{\sqrt{npq}}\right)$

continuous, it is common practice to use *half-interval correction* or *continuity correction*. In fact, this is a necessity in calculating $P(X = x)$. A usual procedure is to go a half-unit on either side of the integer x, depending on the interval of interest. Several cases are shown in Table 7-2.

● **Example 7-10.** Using the data of Example 7-9, where we had $n = 100$ and $p = .2$, we evaluate $P(X = 15)$, $P(X \le 15)$, $P(X < 18)$, $P(X \ge 22)$ and $P(18 < X < 21)$.

(a) $P(X = 15) = P(14.5 \le X \le 15.5) = \Phi\left(\dfrac{15.5 - 20}{4}\right) - \Phi\left(\dfrac{14.5 - 20}{4}\right)$

$$= \Phi(-1.13) - \Phi(-1.38)$$
$$\approx .045.$$

(b) $P(X \le 15) = \Phi\left(\dfrac{15.5 - 20}{4}\right) \approx .1292.$

(c) $P(X < 18) = P(X \le 17) = \Phi\left(\dfrac{17.5 - 20}{4}\right) \approx .266.$

(d) $P(X \ge 22) = 1 - \Phi\left(\dfrac{21.5 - 20}{4}\right) \approx .3541.$

(e) $P(18 < X < 21) = P(19 \le X \le 20)$

$$= \Phi\left(\dfrac{20.5 - 20}{4}\right) - \Phi\left(\dfrac{18.5 - 20}{4}\right)$$
$$\approx .5517 - .3520 = .1997.$$

As discussed in Chapter 5, the variable $\hat{p} = X/n$ is often of interest where X has a binomial distribution with parameters p and n. Interest in this quantity stems primarily from sampling applications, where a random sample of n observations is made, with each observation classified success or failure, and X is the number of successes in the sample. The quantity $\hat{p}$ is simply the sample fraction of successes. Recall that we showed

$$E(\hat{p}) = p \qquad (7\text{-}22)$$

and

$$V(\hat{p}) = \frac{pq}{n}$$

In addition to the DeMoivre–Laplace approximation, note that the quantity

$$Z = \left(\frac{\hat{p} - p}{\sqrt{\dfrac{pq}{n}}} \right) \qquad (7\text{-}23)$$

has an approximate $N(0, 1)$ distribution. This result has proved useful in many applications including those in the areas of quality control, work measurement, reliability engineering, and economics. The results are much more useful than those from the law of large numbers.

● **Example 7-11.** Instead of timing the activity of a maintenance mechanic over the period of a week to determine the fraction of his time spent in an activity classification called "secondary but necessary," a technician elects to use a *work sampling* study, randomly picking 400 time points over the week, taking a flash observation at each, and classifying the activity of the maintenance mechanic. The value X will represent the number of times the mechanic was involved in a "secondary but necessary" activity and $\hat{p} = X/400$. If the true fraction of time that he is involved in this activity is .2, we determine the probability that $\hat{p}$, the estimated fraction, falls between .1 and .3. That is,

$$P(.1 \le \hat{p} \le .3) = \Phi\left(\frac{.3 - .2}{\sqrt{\dfrac{.16}{400}}} \right) - \Phi\left(\frac{.1 - .2}{\sqrt{\dfrac{.16}{400}}} \right)$$

$$= \Phi(5) - \Phi(-5)$$

$$\approx 1.0000$$

7-6 The Lognormal Distribution[1]

The lognormal distribution in its simplest form is the density function of a variable whose logarithm follows the normal probability law. It is the belief of some that the lognormal distribution is as fundamental as the normal dis-

[1]Section 7-6 may be omitted without breaking the continuity of this presentation.

tribution. It arises from the combination of random terms by a multiplicative process.

The lognormal distribution has been applied in a wide variety of fields, including the physical sciences, life sciences, social sciences, and engineering. In engineering applications, the lognormal distribution has been used to describe "time to failure" in reliability engineering and "time to repair" in maintainability engineering.

We consider a random variable X with range space $R_X = \{x : 0 < x < \infty\}$, where $Y = \log_e X$ is normally distributed with mean μ_Y and variance σ_Y^2, that is,

$$E(Y) = \mu_Y \quad \text{and} \quad V(Y) = \sigma_Y^2$$

The density function of X, say f, is

$$f(x) = \frac{1}{x\sigma_Y\sqrt{2\pi}} e^{-(1/2)[(\log x - \mu_Y)/\sigma_Y]^2} \qquad x > 0 \qquad (7\text{-}24)$$
$$= 0 \qquad \qquad \qquad \text{otherwise}$$

A sketch is shown in Fig. 7-11. The mode, a measure of central tendency, corresponds to the value of x for which $f(x)$ is maximum, and the median, another measure of central tendency, corresponds to the value of x for which $P(X \le x) = P(X > x)$. These measures will be discussed in the next chapter.

The *mode* of the distribution occurs at

$$x = e^{\mu_Y - \sigma_Y^2} \qquad (7\text{-}25)$$

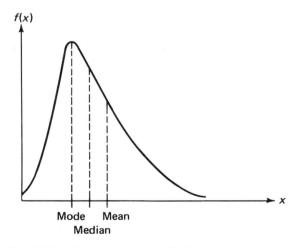

Fig. 7-11.   The lognormal distribution.

and the median at

$$x = e^{\mu_Y} \tag{7-26}$$

and the *mean* and *variance* are

$$E(X) = \mu_X = e^{\mu_Y + (1/2)\sigma_Y^2} \tag{7-27}$$

and

$$V(X) = \sigma_X^2 = e^{2\mu_Y + \sigma_Y^2}(e^{\sigma_Y^2} - 1) \tag{7-28}$$

In general, the kth origin moment is given by

$$\mu_k' = e^{k\mu_Y + (1/2)k^2\sigma_Y^2} \tag{7-29}$$

The third and fourth moments about the mean are

$$\mu_3 = (\mu_X)^3(c^6 + 3c^4) \tag{7-30}$$

and

$$\mu_4 = (\mu_X)^4(c^{12} + 6c^{10} + 15c^8 + 3c^4) \tag{7-31}$$

where $c = (e^{\sigma_Y^2} - 1)$. These moments are applied in two descriptive measures. These measures are the *coefficient of skewness*, ζ_1, and the *coefficient of kurtosis*, ζ_2, defined as

$$\zeta_1 = \frac{\mu_3}{\sigma^3} \tag{7-32}$$

and

$$\zeta_2 = \frac{\mu_4}{\sigma^4} - 3 \tag{7-33}$$

The coefficient of skewness measures departure from symmetry, and the coefficient of kurtosis measures departure from the peakedness displayed by the normal density. For the lognormal distribution,

$$\zeta_1 = c^3 + 3c \tag{7-34}$$

and

$$\zeta_2 = c^8 + 6c^6 + 15c^4 + 16c^2 \tag{7-35}$$

It is clear that $\zeta_1 > 0$ indicates positive skewness, which is obvious from the graph shown in Fig. 7-11. The exponential and gamma distributions presented in the previous chapter also have positive skewness. The large kurtosis for the lognormal distribution ($\zeta_2 > 0$) indicates more peakedness than occurs with the normal distribution.

Where the normal distribution had additive reproductive properties, the lognormal distribution has multiplicative reproductive properties. The results of these properties are as follows:

1. If X has a lognormal distribution with parameters μ_Y and σ_Y^2 and a, b, and d are constants such that $b = e^d$, then $W = bX^a$ has a lognormal distribution with parameters $(d + a\mu_Y)$ and $(a\sigma_Y)^2$.

2. If X_1 and X_2 are independent lognormal variates, with parameters $(\mu_{Y_1}, \sigma_{Y_1}^2)$ and $(\mu_{Y_2}, \sigma_{Y_2}^2)$, respectively, then $W = X_1 \cdot X_2$ has a lognormal distribution with parameters $[(\mu_{Y_1} + \mu_{Y_2}), (\sigma_{Y_1}^2 + \sigma_{Y_2}^2)]$.

3. If $X_1, X_2, \ldots, X_n$ is a sequence of n independent lognormal variates, with parameters $(\mu_{Y_j}, \sigma_{Y_j}^2)$, $j = 1, 2, \ldots, n$, respectively, and $\{a_j\}$ is a sequence of constants while $b = e^d$ is a single constant, and if $\Sigma_{j=1}^n a_j \mu_{Y_j}$ and $\Sigma_{j=1}^n a_j^2 \sigma_{Y_j}^2$ both converge, then the product

$$W = b \prod_{j=1}^{n} X_j^{a_j}$$

has a lognormal distribution with parameters

$$\left(d + \sum_{j=1}^{n} a_j \mu_{Y_j} \right) \quad \text{and} \quad \left(\sum_{j=1}^{n} a_j^2 \sigma_{Y_j}^2 \right)$$

respectively.

4. If X_j $(j = 1, 2, \ldots, n)$ are independent lognormal variates, each with the same parameters (μ_Y, σ_Y^2), then the geometric mean

$$\left(\prod_{j=1}^{n} X_j \right)^{1/n}$$

has a lognormal distribution with parameters μ_Y and σ_Y^2/n.

● **Example 7-12.** The random variable $Y = \log_e X$ has a $N(10, 4)$ distribution so that X has a lognormal distribution with mean and variance of

$$E(X) = e^{10 + (1/2)4} = e^{12} \simeq 162{,}754.79$$

and

$$V(X) = e^{[2(10)+4]}(e^4 - 1)$$
$$= e^{24}(e^4 - 1) \simeq 54.598e^{24}$$

respectively. The mode and median are:

$$\text{mode} = e^6 \simeq 403.43$$

and

$$\text{median} = e^{10} \simeq 22{,}026$$

In order to determine a specific probability, say $P(X \leq 1000)$, we use the transform and determine $P(\log_e X \leq \log_e 1000) = P(Y \leq \log_e 1000)$. The results follow:

$$P(Y \leq \log_e 1000) = P\left(Z \leq \frac{\log_e 1000 - 10}{2}\right)$$

$$= \Phi(-1.55) = .0606$$

● **Example 7-13.** Suppose

$$Y_1 = \log_e X_1 \sim N(4, 1)$$
$$Y_2 = \log_e X_2 \sim N(3, .5)$$
$$Y_3 = \log_e X_3 \sim N(2, .4)$$
$$Y_4 = \log_e X_4 \sim N(1, .01)$$

and furthermore suppose X_1, X_2, X_3, and X_4 are independent random variables. The random variable W defined as follows represents a critical performance variable on a telemetry system:

$$W = e^{1.5}[X_1^{2.5} X_2^2 X_3^7 X_4^{3.1}]$$

By reproductive property 3, W will have a lognormal distribution with parameters

$$1.5 + (2.5 \cdot 4 + .2 \cdot 3 + .7 \cdot 2 + 3.1 \cdot 1) = 16.6$$

and

$$[(2.5)^2 \cdot 1 + (.2)^2 \cdot .5 + (.7)^2 \cdot .4 + (3.1)^2 \cdot (.01)] = 6.562$$

respectively. That is to say, $\log_e W \sim N(16.6, 6.562)$. If the specifications on W are, say, 20,000 to 600,000, we could determine the probability that W would fall within specifications as follows:

$P(20{,}000 \leq W \leq 600 \cdot 10^3)$

$$= P[\log_e(20{,}000) \leq \log_e W \leq \log_e(600 \cdot 10^3)]$$

$$= \Phi\left(\frac{\log_e 600 \cdot 10^3 - 16.6}{\sqrt{6.526}}\right) - \Phi\left(\frac{\log_e 20{,}000 - 16.6}{\sqrt{6.526}}\right)$$

$$\approx \Phi(.51) - \Phi(-2.69) = .6950 - .0036$$

$$= .6914$$

7-7 The Bivariate Normal Distribution

Up to this point, all of the continuous random variables have been of one dimension. A very important two-dimensional probability law which is a generalization of the one-dimensional normal probability law is called the *bivariate normal distribution*. If $[X_1, X_2]$ is a bivariate normal random vector, then the joint density function of $[X_1, X_2]$ is

$$f(x_1, x_2) = \frac{1}{2\pi\sigma_1\sigma_2\sqrt{1-\rho^2}} \exp\left\{ -\frac{1}{2(1-\rho^2)}\left[\left(\frac{x_1-\mu_1}{\sigma_1}\right)^2 \right.\right.$$
$$\left.\left. -2\rho\left(\frac{x_1-\mu_1}{\sigma_1}\right)\left(\frac{x_2-\mu_2}{\sigma_2}\right) + \left(\frac{x_2-\mu_2}{\sigma_2}\right)^2 \right]\right\} \qquad (7\text{-}36)$$

for $-\infty < x_1 < \infty$ and $-\infty < x_2 < \infty$.

The joint probability $P(a_1 \le X_1 \le b_1, a_2 \le X_2 \le b_2)$ is defined as

$$\int_{a_2}^{b_2} \int_{a_1}^{b_1} f(x_1, x_2)\, dx_1\, dx_2 \qquad (7\text{-}37)$$

and is represented by the volume under the surface and over the region $\{(x_1, x_2): a_1 \le x_1 \le b_1, a_2 \le x_2 \le b_2\}$ as shown in Fig. 7-12. Owen (1962) has provided a table of probabilities. The bivariate normal density has five parameters. These are μ_1, μ_2, σ_1, σ_2, and ρ, the correlation coefficient between X_1 and X_2, such that $-\infty < \mu_1 < \infty$, $-\infty < \mu_2 < \infty$, $\sigma_1 > 0$, $\sigma_2 > 0$, and $-1 < \rho < 1$.

The marginal densities f_1 and f_2 are given respectively as

$$f_1(x_1) = \int_{-\infty}^{\infty} f(x_1, x_2)\, dx_2 = \frac{1}{\sigma_1\sqrt{2\pi}}\, e^{-(1/2)[(x_1-\mu_1)/\sigma_1]^2} \qquad (7\text{-}38)$$

for $-\infty < x_1 < \infty$;

$$f_2(x_2) = \int_{-\infty}^{\infty} f(x_1, x_2)\, dx_1 = \frac{1}{\sigma_2\sqrt{2\pi}}\, e^{-(1/2)[(x_2-\mu_2)/\sigma_2]^2} \qquad (7\text{-}39)$$

for $-\infty < x_2 < \infty$.

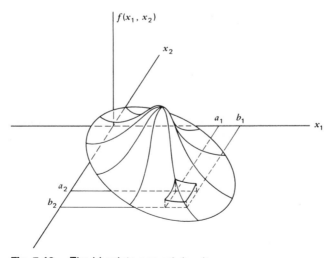

Fig. 7-12.   The bivariate normal density.

We note that these marginal densities are normal, that is,

$$X_1 \sim N(\mu_1, \sigma_1^2) \qquad (7\text{-}40)$$

and

$$X_2 \sim N(\mu_2, \sigma_2^2)$$

so that

$$E(X_1) = \mu_1$$
$$E(X_2) = \mu_2$$
$$V(X_1) = \sigma_1^2$$
$$V(X_2) = \sigma_2^2 \qquad (7\text{-}41)$$

The parameter ρ is represented by the ratio of the covariance to $[\sigma_1 \cdot \sigma_2]$. The covariance is

$$\sigma_{12} = \int_{-\infty}^{\infty} \int_{-\infty}^{\infty} (x_1 - \mu_1)(x_2 - \mu_2) f(x_1, x_2) \, dx_1 \, dx_2$$

Thus

$$\rho = \frac{\sigma_{12}}{[\sigma_1 \cdot \sigma_2]} \qquad (7\text{-}42)$$

The conditional distributions $f_{X_2 \mid x_1}(x_2)$ and $f_{X_1 \mid x_2}(x_1)$ are also important. These conditional densities are normal, as shown here:

$$f_{X_2 \mid x_1} = \frac{f(x_1, x_2)}{f_1(x_1)}$$

$$= \frac{1}{\sigma_2 \sqrt{2\pi} \sqrt{1 - \rho^2}} \, e^{-(1/2)\{x_2 - [\mu_2 + \rho(\sigma_2/\sigma_1)(x_1 - \mu_1)]/\sigma_2 \sqrt{1 - \rho^2}\}^2} \qquad (7\text{-}43)$$

for $-\infty < x_2 < \infty$, and

$$f_{X_1 \mid x_2}(x_1) = \frac{f(x_1, x_2)}{f_2(x_2)}$$

$$= \frac{1}{\sigma_1 \sqrt{2\pi} \sqrt{1 - \rho^2}} \, e^{-(1/2)\{x_1 - [\mu_1 + \rho(\sigma_1/\sigma_2)(x_2 - \mu_2)]/\sigma_1 \sqrt{1 - \rho^2}\}^2} \qquad (7\text{-}44)$$

for $-\infty < x_1 < \infty$. Figure 7-13 illustrates some of these conditional densities. We first consider the distribution $f_{X_2 \mid x_1}$. The mean and variance are

$$E(X_2 \mid x_1) = \mu_2 + \rho(\sigma_2/\sigma_1)(x_1 - \mu_1) \qquad (7\text{-}45)$$

and

$$V(X_2 \mid x_1) = \sigma_2^2(1 - \rho^2) \qquad (7\text{-}46)$$

Furthermore, $f_{X_2 \mid x_1}$ is normal, that is,

$$X_2 \mid x_1 \sim N[\mu_2 + \rho(\sigma_2/\sigma_1)(x_1 - \mu_1), \ \sigma_2^2(1 - \rho^2)]. \qquad (7\text{-}47)$$

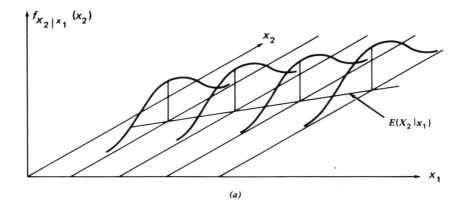

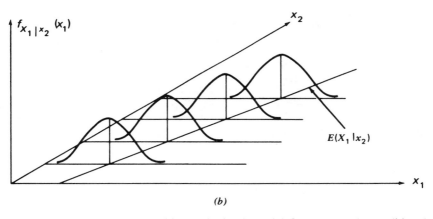

Fig. 7-13. Some typical conditional distributions. (a) Some example conditional distributions of X_2 for a few values of x_1. (b) Some example conditional distributions X_1 for a few values of x_2.

The locus of expected values of X_2 for given x_1 as shown in Equation (7-45) is called the *regression of X_2 on X_1*, and it is linear. Also, the *variance in the conditional distributions is constant* for all x_1.

In the case of the distribution $f_{X_1 \mid x_2}$, the results are similar. That is,

$$E(X_1 \mid x_2) = \mu_1 + \rho(\sigma_1/\sigma_2)(x_2 - \mu_2) \tag{7-48}$$

$$V(X_1 \mid x_2) = \sigma_1^2(1 - \rho^2) \tag{7-49}$$

and

$$X_1 \mid x_2 \sim N[\mu_1 + \rho(\sigma_1/\sigma_2)(x_2 - \mu_2), \sigma_1^2(1 - \rho^2)] \tag{7-50}$$

In the bivariate normal distribution we observe that if $\rho = 0$, the joint density may be factored into the product of the marginal densities and thus X_1 and X_2 are independent. Thus, for a bivariate normal density, zero correlation and independence are equivalent. If planes parallel to the x_1, x_2 plane are passed through the surface shown in Fig. 7-12, the contours cut from the bivariate normal surface are ellipses. The student may wish to show this property.

● **Example 7-14.** In an attempt to substitute a nondestructive testing procedure for a destructive test, an extensive study was made of shear strength, X_2, and weld diameter, X_1, of spot welds, with the following findings.

(a) $[X_1, X_2] \sim$ bivariate normal

(b) $\mu_1 = .20$ inch
$\mu_2 = 1100$ pounds
$\sigma_1^2 = .02$ inch^2
$\sigma_2^2 = 525$ pounds^2
$\rho = .9$

The regression of X_2 on X_1 is thus

$$E(X_2 \mid x_1) = \mu_2 + \rho(\sigma_2/\sigma_1)(x_1 - \mu_1)$$
$$= 1100 + .9\left(\frac{\sqrt{525}}{\sqrt{.02}}\right)(x_1 - .2)$$
$$= (146.7)x_1 + 1070.65$$

and the variance is

$$V(X_2 \mid x_1) = \sigma_2^2(1 - \rho^2)$$
$$= 525(.19) = 99.75$$

In studying these results, the manager of manufacturing notes that since $\rho = .9$, that is, close to 1, weld diameter is highly correlated with shear strength. The specification on shear strength calls for a value greater than 1080. If a weld has diameter .18, he asks: "What is the probability that the strength specification will be met?" The process engineer notes that $E(X_2 \mid .18) = 1097.05$; therefore,

$$P(X_2 \geq 1080) = P\left(Z \geq \frac{1080 - 1097.05}{\sqrt{99.75}}\right)$$
$$= 1 - \Phi(-1.71) = .9564$$

and he recommends a policy such that if the weld diameter is not less than .18, the weld will be classified as satisfactory.

• **Example 7-15.** In developing an admissions policy for a large university, the office of student testing and evaluation has noted that X_1, the combined score on the college board examinations, and X_2, the student grade point average at the end of the freshman year, have a bivariate normal distribution. A grade point of 4.0 corresponds to A. A study indicates that

$$\mu_1 = 1300$$
$$\mu_2 = 2.3$$
$$\sigma_1^2 = 6400$$
$$\sigma_2^2 = .25$$
$$\rho = .6$$

Any student with a grade point average less than 1.5 is automatically dropped at the end of the freshman year; however, an average of 2.0 is considered to be satisfactory.

An applicant takes the college board exams, receives a combined score of 900, and is not accepted. An irate parent argues that the student will do satisfactory work and, specifically, will have better than a 2.0 grade point average at the end of the freshman year. Considering only the probabilistic aspects of the problem, the director of admissions wants to determine $P(X_2 \geq 2.0 \mid x_1 = 900)$. Noting that

$$E(X_2 \mid 900) = 2.3 + (.06)\left(\frac{.5}{80}\right)(900 - 1300)$$

$$= .8$$

and

$$V(X_2 \mid 900) = .16$$

the director proceeds to calculate

$$1 - \Phi\left(\frac{2.0 - .8}{.4}\right) = .0013$$

which predicts only a very slim chance of the parent's claim being valid.

7-8   Summary

This chapter has presented the normal distribution with a number of example applications. The *normal* distribution, the related *standard normal*, and the *lognormal* distributions are univariate, while the *bivariate normal* presents the joint density of two related normal random variables.

The normal distribution forms the basis on which a great deal of the work in statistical inference rests. The wide application of the normal distribution makes it particularly important.

7-9 Exercises

7-1. Let $X \sim N(10, 25)$. Find $P(X \le 15)$, $P(X \ge 12)$, $P(9 \le X \le 20)$, and $P(5 \le X \le 15)$.

7-2. Let $X \sim N(5, 9)$. Find $P(X \le 3)$, $P(X \ge 7)$, and $P(2 \le X \le 11)$.

7-3. Let $X \sim N(5, 4)$. Find b such that $P(X > b) = .25$.

7-4. Let $X \sim N(5, 9)$. Find the values a and b such that $P(a < X < b) = .80$ where the interval a to b is symmetric about the mean.

7-5. Let $X \sim N(10, 9)$. Find a such that $P(X \le a) = .30$ and b such that $P(X \ge b) = .20$.

7-6. The life of a particular type of dry-cell battery is normally distributed with mean 500 days and standard deviation 50 days. What fraction of these batteries would be expected to survive beyond 580 days? What fraction would be expected to fail before 450 days?

7-7. The personnel manager of a large company requires job applicants to take a certain test and achieve a score of 500. If the test scores are normally distributed with mean 485 and standard deviation 20, what percentage of the applicants pass the test?

7-8. A manufacturing process produces bolts that must be between 1.2 and 1.25 inches in diameter. The diameter is known to be normally distributed with mean 1.21 and standard deviation .02. What percentage of the bolts are outside the specifications?

7-9. In Exercise 7-8, what could be done to reduce the percentage of defective bolts produced? Assume that the process variability must remain unchanged.

7-10. If in Exercise 7-8 the process standard deviation could be reduced to a new value σ, find the value of σ that results in 10 percent defective bolts. Assume the process mean remains unchanged.

7-11. The bacteria content of applesauce must be between 65 and 75 to be acceptable. Long experience indicates that the bacteria content is normally distributed with mean 70 and standard deviation 3. What percentage of the cans are not acceptable?

7-12. If in Exercise 7-11 the standard deviation changes to 8.25, what percentage of the cans are not acceptable?

7-13. A certain type of light bulb has an output known to be normally distributed with mean 2500 end footcandles and standard deviation 75 end footcandles. Determine a lower specification limit such that only 5 percent of the manufactured bulbs will be defective.

7-14. Show that the moment-generating function for the normal distribution is as given by Equation (7-5). Use it to generate the mean and variance.

7-15. An assembly consists of three components placed side by side. The length of each component is normally distributed with mean 2 inches and standard deviation .2 inch. Specifications require that all assemblies are between 5.7 and 6.3 inches long. How many assemblies will pass these requirements?

7-16. A shaft whose diameter is $N(1.000, .0025)$ is to be inserted into a collar whose diameter is $N(1.040, .0016)$. Determine the probability of interference.

7-17. Find the mean and variance of the linear combination

$$Y = X_1 + 2X_2 + X_3 + X_4$$

where $X_1 \sim N(2, 3)$, $X_2 \sim N(2, 4)$, $X_3 \sim N(4, 4)$, and $X_4 \sim N(4, 2)$. What is the probability that $10 \le Y \le 14$?

7-18. The price being asked for a certain security is distributed normally with mean \$50.00 and standard deviation \$5.00. Buyers are willing to pay an amount that is also normally distributed with mean \$45.00 and standard deviation \$2.50. What is the probability that a transaction will take place?

7-19. The specifications for a capacitor are that its life must be between 1000 and 5000 hours. The life is known to be normally distributed with mean 3000 hours. The profit realized from each capacitor is \$9.00; however, a failed unit must be replaced at a cost of \$3.00 to the company. Two manufacturing processes can produce capacitors having satisfactory mean lives. The standard deviation for process A is 1000 hours and for process B it is 500 hours. However, process A manufacturing costs are only half those for B. What value of process manufacturing cost is critical, so far as dictating the use of process A or B?

7-20. Suppose that in Example 7-6 the variance of the bearing diameter could be reduced to .0004. What effect would this have on the probability of interference?

7-21. Suppose that $X \sim N(3, 4)$. Find a number C such that $P(X \ge C) = 2P(X < C)$.

7-22. The diameter of a ball bearing is a normally distributed random variable with mean μ and standard deviation 1. Specifications for the diameter are $6 \le X \le 8$, and a ball bearing within these limits yields a profit of C dollars. However, if $X < 6$ then the profit is $-R_1$ dollars or if $X > 8$ the profit is $-R_2$ dollars. Find the value of μ that maximizes the expected profit.

7-23. In the preceding exercise, find the optimum value of μ if $R_1 = R_2 = R$.

7-24. Use the results of Exercise 7-22 with $C = \$8.00$, $R_1 = \$2.00$, and $R_2 = \$4.00$. What is the value of μ that maximizes the expected profit?

7-25. A particular type of transistor is to be used in an application for 50 hours. Two brands of transistors are available, one having a life distribution $N(40, 36)$ and the other having a life distribution $N(48, 9)$. Which transistor is to be preferred in this application?

7-26. Prove that $E(Z_n) = 0$ and $V(Z_n) = 1$, where Z_n is as defined in Theorem 7-1.

7-27. Prove that $E(Z_n) = 0$ and $V(Z_n) = 1$, where Z_n is as defined in Theorem 7-2.

7-28. Let X_i $(i = 1, 2, \ldots, n)$ be independently and identically distributed random variables with mean μ and variance σ^2. The quantity

$$\bar{X} = \frac{1}{n}(X_1 + X_2 + \cdots + X_n) = \frac{1}{n}\sum_{i=1}^{n} X_i$$

is normally distributed with mean μ and variance σ^2/n. Show that $E(\bar{X}) = \mu$ and $V(\bar{X}) = \sigma^2/n$.

7-29. Specifications on a light bulb are that the life be greater than 125 end foot-candles. This life is known to be normally distributed. The only data available to the manufacturer concern the *average* life of test bulbs, which are normally distributed with mean 130 and standard deviation 4. These averages are obtained for groups of three light bulbs. Based on this information, what fraction of the bulbs produced can be expected to fall outside of specifications?

7-30. One hundred small bolts are packed in a box. Each bolt weighs 1 ounce with standard deviation .01 ounce. Find the probability that a box weighs more than 102 ounces.

7-31. An automatic machine is used to fill boxes with soap powder. Specifications require that the boxes weigh between 11.8 and 12.2 ounces. The only data available about machine performance concern the average content of groups of 9 boxes. It is known that the average content is 11.9 ounces with standard deviation .05 ounce. What fraction of the boxes produced are defective? Where should the mean be located in order to minimize this fraction defective? Assume the weight is normally distributed.

7-32. A bus travels between two cities, but visits eight intermediate cities on the route. The mean and standard deviation of the travel times are as follows:

City Pairs	Mean Time (hours)	Standard Deviation (hours)
1-2	2	.5
2-3	3	.5
3-4	2	.25
4-5	4	1.0
5-6	6	.75
6-7	4	.5
7-8	2	.25

What is the probability that the bus completes its journey in 24 hours?

7-33. A production process produces items, of which 10 percent are defective. A random sample of 200 items is selected every day and the number of defective items, say X, is counted. Using the normal approximation to the binomial find:

(a) $P(X \leq 20)$
(b) $P(X = 15)$
(c) $P(15 \leq X \leq 25)$
(d) $P(X = 25)$.

7-34. Suppose we are engaged in a work-sampling study, as described in Example 7-11. Let the true fraction of the time the employee is engaged in useful work be .85. With $n = 200$ observations, find $P(.80 \leq \hat{p} \leq .90)$.

7-35. In a work-sampling study it is often desired to find the number of necessary observations. Given that $p = .1$, find the necessary n such that $P(.05 \leq \hat{p} \leq .15) = .95$.

7-36. The brightness of light bulbs is normally distributed with mean 2500 foot-candles and standard deviation 50 footcandles. The bulbs are tested and all those brighter than 2600 footcandles are placed in a special high-quality lot. What is the probability distribution of the remaining bulbs? What is their expected brightness?

7-37. The random variable $Y = \log_e X$ has a $N(50, 25)$ distribution. Find the mean, variance, mode, and median of X.

7-38. Suppose independent random variables Y_1, Y_2, Y_3 are such that

$$Y_1 = \log_e X_1 \sim N(4, 1)$$
$$Y_2 = \log_e X_2 \sim N(3, 1)$$
$$Y_3 = \log_e X_3 \sim N(2, .5)$$

Find the mean and variance of $W = e^2 X_1^2 X_2^{1.5} X_3^{1.28}$. Determine a set of specifications, L and R, such that

$$P(L \le W \le R) = .90$$

7-39. Show that the density function for a lognormally distributed random variable X is given by

$$f(x) = \frac{1}{\sigma x \sqrt{2\pi}} \exp\left[-\frac{1}{2\sigma^2} (\log_e x - \mu)^2 \right] \qquad x > 0, \sigma > 0, -\infty < \mu < \infty$$
$$= 0 \qquad\qquad\qquad\qquad\qquad \text{otherwise}$$

where $Y = \log_e X \sim N(\mu, \sigma^2)$.

7-40. Consider the bivariate normal density

$$f(x_1, x_2) = \Delta \exp -\left\{ \frac{1}{2(1 - \rho^2)} \left[\frac{x_1^2}{\sigma_1^2} - \frac{2\rho x_1 x_2}{\sigma_1 \sigma_2} + \frac{x_2^2}{\sigma_2^2} \right] \right\} \qquad -\infty < x_1 < \infty, -\infty < x_2 < \infty$$

where Δ is chosen so that f is a probability distribution. Are the random variables X_1 and X_2 independent? Define two new random variables:

$$Y_1 = \frac{1}{(1 - \rho^2)^{1/2}} \left(\frac{X_1}{\sigma_1} - \frac{\rho X_2}{\sigma_2} \right)$$
$$Y_2 = \frac{X_2}{\sigma_2}$$

Show that the two new random variables are independent.

7-41. The life of a tube (X_1) and the filament diameter (X_2) are distributed as a bivariate normal with the following parameters

$$\mu_1 = 2000 \text{ hours}$$
$$\mu_2 = .10 \text{ inch}$$
$$\sigma_1^2 = 2500 \text{ hours}^2$$
$$\sigma_2^2 = .01 \text{ inch}^2$$
$$\rho = .87$$

The quality control manager wishes to determine the life of each tube by measuring the filament diameter. If a filament diameter is .098, what is the probability that the tube will last 1950 hours?

7-42. A college professor has noticed that grades on each of two quizzes have a bivariate normal distribution with the following parameters:

$$\mu_1 = 75$$
$$\mu_2 = 83$$
$$\sigma_1^2 = 25$$
$$\sigma_2^2 = 16$$
$$\rho = .8$$

If a student receives a grade of 80 on the first quiz, what is the probability that she will do better on the second one? How is the answer affected by making $\rho = -.8$?

7-43. Consider the surface $y = f(x_1, x_2)$, where f is the bivariate normal density function.
(a) Prove that $y = $ constant cuts the surface in an ellipse.
(b) Prove that $y = $ constant with $\rho = 0$ and $\sigma_1^2 = \sigma_2^2$ cuts the surface as a circle.

7-44. Let X_1 and X_2 be independent random variables each following a normal density with mean zero and variance σ^2. Find the distribution of

$$R = \sqrt{X_1^2 + X_2^2}$$

The resulting distribution is known as the *Rayleigh* distribution and is frequently used to model the distribution of radial error in a plane. *Hint:* Let $x_1 = r \cos \theta$ and $x_2 = r \sin \theta$. Obtain the joint probability distribution of R and θ, then integrate out θ.

7-45. Using a similar method to that in Exercise 7-44, obtain the distribution of

$$R = \sqrt{X_1^2 + X_2^2 + \cdots + X_n^2}$$

where $X_i \sim N(0, \sigma^2)$ and independent.

7-46. Let the independent random variables $X_i \sim N(0, \sigma^2)$ for $i = 1, 2$. Find the probability distribution of

$$C = \frac{X_1}{X_2}$$

C follows the *Cauchy* distribution. Try to compute $E(C)$.

7-47. Let $X \sim N(0, 1)$. Find the probability distribution of $Y = X^2$. Y is said to follow the *chi-square* distribution with one degree of freedom. It is an important distribution in statistical methodology.

7-48. Let the independent random variables $X_i \sim N(0, 1)$ for $i = 1, 2, \ldots, n$. Show that $Y = \sum_{i=1}^{n} X_i^2$ follows a chi-square distribution with n degrees of freedom.

7-49. Let $X \sim N(0, 1)$. Define a new random variable $Y = |X|$. Then, find the probability distribution of Y. This is often called the *half-normal* distribution.

Chapter 8

Sampling and Descriptive Statistics

In this chapter we discuss sampling and several methods for the presentation and summarization of sample data. We also present several probability distributions useful in the analysis of sample data.

8-1 The Population and the Sample

Statistics is the science of drawing conclusions about a population based on an analysis of sample data from that population. For example, suppose that a soft drink beverage bottler wishes to investigate the bursting strength of glass bottles in a shipment of 10,000 twenty-four bottle cases recently supplied by the bottle vendor. Since the shipment contains a very large number of bottles, it is not economically practical to test each individual bottle. Furthermore, if the test for bursting strength is destructive, testing each bottle would result in loss of the shipment. The logical approach is to select a subset of the bottles in the shipment and test their bursting strength. Then, one hopes that the conclusions about the subset of bottles actually tested can be extended to the entire shipment.

In this example, we may think of the shipment of bottles as a *population*. This terminology has evolved from statistical studies of sociological and economic factors. In general, the term population refers to the universe or totality of the observations with which we are concerned. The size of the population is the number of observations or elements that it contains. This population of soft drink bottles is a finite population of size 240,000. A population may also be of infinite size. In some cases, a finite population may be so large that we can assume its size to be infinite.

A subset of observations selected from a population is called a *sample*. Thus, in the bottle-testing example, we would use a sample of bottles to help draw conclusions about the bursting strength of the bottles in the population.

Sometimes it is necessary to select a sample from a population that exists only conceptually. For example, suppose that a civil engineer makes up several concrete specimens for compression testing. We can view these specimens as a sample from a conceptually infinite population consisting of all possible specimens of this particular type of concrete. There are many different ways to take samples from a population. Furthermore, the conclusions that we can draw about the population will depend on how the sample is selected. Generally speaking, we want the sample to be *representative* of the population. Most of the statistical techniques that we will study assume that the sample is a random sample.

To define a random sample, let X be a random variable with probability distribution $f(x)$. Then the set of n observations $X_1, X_2, \ldots, X_n$, taken on the random variable X, and having numerical outcomes $x_1, x_2, \ldots, x_n$, is called a *random sample* if the observations are obtained by observing X independently under unchanging conditions for n times. Note that the observations $X_1, X_2, \ldots, X_n$ in a random sample are independent random variables with the same probability distribution $f(x)$. That is, the marginal distributions of $X_1, X_2, \ldots, X_n$ are $f(x_1), f(x_2), \ldots, f(x_n)$, respectively, and by independence the joint probability distribution of the random sample is

$$g(x_1, x_2, \ldots, x_n) = f(x_1) \cdot f(x_2) \cdot \ldots \cdot f(x_n) \qquad (8\text{-}1)$$

Thus, in the bottle-bursting strength example, if the random variable X (bursting strength) in the population of bottles is normally distributed, then we would expect each of the observations on bursting strength $X_1, X_2, \ldots, X_n$ in a random sample of n bottles to have exactly the same normal distribution.

It is not always easy to obtain a random sample. This is particularly true when the analyst must rely on judgment in selecting the elements of the sample. Sometimes we may use tables of uniform random numbers, such as those in Table VIII of the Appendix, to generate the sample. To use this approach for a finite population, assign a number to each element of the population and then choose the random sample as those elements that correspond to selected entries in the random number table. This method is particularly useful when the size of the population is small.

● **Example 8-1.** Suppose we wish to take a random sample of 5 batches of raw material out of 25 available batches. We may number the batches with the integers 1 to 25. Now in Table VIII of the Appendix arbitrarily choose a row and column as a starting point. Read down the chosen column, which should have 2 digits, until 5 acceptable numbers are found (an acceptable number lies between 1 and 25). To illustrate, suppose the above process gives us a sequence of numbers that reads 37, 48, 55, **02**, **17**, 61, 70, 43, **21**, 82, 73, **13**, 60, **25**. The bold numbers specify which batches of raw material are to be chosen as the random sample.

8-2 Graphical Presentation of Data

Graphical and tabular methods are usually helpful in the presentation and analysis of data. For example, the data in Table 8-1 represent the bursting strengths in pounds per square inch (psi) of 100 glass, nonreturnable, 1 liter soft drink bottles. These observations were obtained by testing each of a sample of 100 bottles until failure occurred. The data were recorded in the order in which the bottles were tested and, in this form, they do not convey very much information about the bursting strength of the bottles. Note that questions such as "What is the average strength of the bottles?" or "What percentage of the bottles should burst below 230 psi?" are not easy to answer if the data are expressed in this format.

A frequency distribution is helpful in analyzing data. To construct a frequency distribution, we must divide the range of the data into intervals, which are usually called *class intervals*. If possible, the class intervals should be of equal width, to enhance the visual information in the frequency distribution. Some judgment must be used in selecting the number of class intervals in order to give a reasonable display. The number of class intervals used depends on the number of observations and the amount of scatter or dispersion in the data. Choosing the number of class intervals approximately equal to the square root of the sample size often works well in practice.

A frequency distribution for the data in Table 8-1 is shown in Table 8-2. The third column of Table 8-2 contains a relative frequency distribution. The relative frequencies are found by dividing the observed frequency in each class interval by the total number of observations. The fourth column of Table 8-2 expresses the relative frequencies on a cumulative basis. Frequency distributions are often easier to interpret than tables of data. For example, from Table 8-2 it is very easy to see that most of the bottles burst between 230 and 290 psi, and that 13 percent of the bottles burst below 230 psi.

It is also helpful to present the frequency distribution in graphical form, as shown in Fig. 8-1. Such a display is called a *histogram*. The histogram

TABLE 8-1 **Bursting Strength in Pounds per Square Inch for Glass, 1-Liter, Nonreturnable Soft Drink Bottles**

265	197	346	280	265	200	221	265	261	278
205	286	317	242	254	235	176	262	248	250
263	274	242	260	281	246	248	271	260	265
307	243	258	321	294	328	263	245	274	270
220	231	276	228	223	296	231	301	337	298
268	267	300	250	260	276	334	280	250	257
260	281	208	299	308	264	280	274	278	210
234	265	187	258	235	269	265	253	254	280
299	214	264	267	283	235	272	287	274	269
215	318	271	293	277	290	283	258	275	251

TABLE 8-2 **Frequency Distribution for the Bursting Strength Data in Table 8-1**

Class Interval (psi)	Frequency	Relative Frequency	Cumulative Relative Frequency
$170 \le x < 190$	2	.02	.02
$190 \le x < 210$	4	.04	.06
$210 \le x < 230$	7	.07	.13
$230 \le x < 250$	13	.13	.26
$250 \le x < 270$	32	.32	.58
$270 \le x < 290$	24	.24	.82
$290 \le x < 310$	11	.11	.93
$310 \le x < 330$	4	.04	.97
$330 \le x < 350$	3	.03	1.00
	100	1.00	

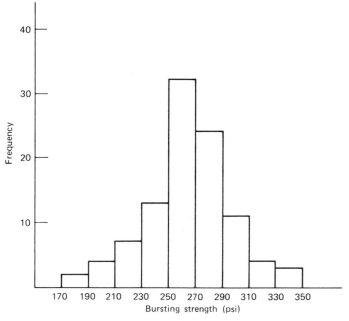

Fig. 8-1. Histogram of bursting strength for 100 glass, 1-liter, nonreturnable soft drink bottles.

provides a visual impression of the shape of the distribution, as well as information about the scatter or dispersion of the data.

In passing from the original data to either a frequency distribution or a histogram, a certain amount of information has been lost in that we no longer have the individual observations. However, this information loss is small compared to the conciseness and ease of interpretation gained in using the frequency distribution and histogram.

8-3 Numerical Description of Data

Just as graphical or tabular displays can improve the display of sample data, numerical descriptions can also be of value. In this section, we present several important numerical measures of sample data.

8-3.1 Measures of Central Tendency

The most common measure of *central tendency*, or location, of the data is the ordinary arithmetic mean, usually called the *sample mean*. If the observations in a sample of size n are $X_1, X_2, \ldots, X_n$, then the sample mean is

$$\bar{X} = \frac{X_1 + X_2 + \cdots + X_n}{n}$$

$$= \frac{\sum\limits_{i=1}^{n} X_i}{n} \qquad (8\text{-}2)$$

For the bottle-bursting strength data in Table 8-1, the sample mean is

$$\bar{x} = \frac{\sum\limits_{i=1}^{100} x_i}{100} = \frac{26{,}406}{100} = 264.06$$

From examination of Fig. 8-1, it seems that the sample mean 264.06 psi is a "typical" value of bursting strength, since it occurs near the middle of the data where the observations are concentrated. However, this impression can be misleading. Suppose that the histogram looked like Fig. 8-2. The mean of this data is still a measure of central tendency, but it does not necessarily imply that most of the observations are concentrated around it. In general, if we think of the observations as having unit mass, the sample mean is just the center of mass of the data.

Another measure of central tendency is the *median*, or the point at which the sample is divided into two equal halves. Let $X_{(1)}, X_{(2)}, \ldots, X_{(n)}$ denote a sample arranged in increasing order of magnitude (that is, $X_{(1)}$ denotes the smallest observation, $X_{(2)}$ denotes the second smallest observation, $\ldots$, and

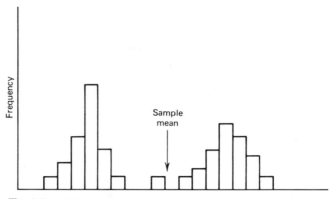

Fig. 8-2.   A histogram.

$X_{(n)}$ denotes the largest observation). Then the median is defined mathematically as

$$M = \begin{cases} X_{([n+1]/2)}, & n \text{ odd} \\ \dfrac{X_{(n/2)} + X_{([n/2]+1)}}{2}, & n \text{ even} \end{cases} \tag{8-3}$$

The median has the advantage that it is not influenced very much by extreme values. For example, suppose that the sample observations are

$$1, 3, 4, 2, 7, 6, \text{ and } 8$$

The sample mean is 4.43, and the sample median is 4. Both quantities give a reasonable measure of the central tendency of the data. Now suppose that the next-to-last observation is changed, so that the data are

$$1, 3, 4, 2, 7, 2519, \text{ and } 8$$

For these data, the sample mean is 363.43. Clearly, in this case the sample mean does not tell us very much about the central tendency of most of the data. The median, however, is still 4, and this is probably a much more meaningful measure of central tendency for the majority of the observations.

The *mode* is the observation that occurs most frequently in the sample. For example, the mode of the sample data

$$2, 4, 6, 2, 5, 6, 2, 9, 4, 5, 2, \text{ and } 1$$

is 2, since it occurs four times, and no other value occurs as often. There may be more than one mode.

If the data are symmetric, then the mean and median coincide. If, in addition, the data have only one mode (we say the data are *unimodal*), then the mean, median, and mode all coincide. If the data are not symmetric, then the median is nearer to the mean than to the mode.

The distribution of the sample mean is well-known and relatively easy to work with. Furthermore, the sample mean is more stable than the sample median, in the sense that it does not vary as much from sample to sample. Consequently, many analytical statistical techniques use the sample mean. However, the median and mode are often helpful descriptive measures.

8-3.2 Measures of Dispersion

The central tendency of the data does not necessarily provide enough information to adequately describe the sample. For example, consider the bursting strengths obtained from two samples of six bottles each.

| *Sample 1:* | 230 | 250 | 245 | 258 | 265 | 240 |
| *Sample 2:* | 190 | 228 | 305 | 240 | 265 | 260 |

The mean of both samples is 248 psi. However, note that the scatter or dispersion of Sample 2 is much greater than that of Sample 1. In this section, we define two widely used measures of dispersion.

The most important measure of dispersion is the *sample variance.* If $X_1, X_2, \ldots, X_n$ is a sample of n observations, then the sample variance is

$$S^2 = \frac{\sum\limits_{i=1}^{n} (X_i - \bar{X})^2}{n-1} \tag{8-4}$$

A more efficient computational formula for the sample variance is obtained as follows:

$$S^2 = \frac{\sum\limits_{i=1}^{n} (X_i - \bar{X})^2}{n-1}$$

$$= \frac{\sum\limits_{i=1}^{n} (X_i^2 + \bar{X}^2 - 2\bar{X} X_i)}{n-1}$$

$$= \frac{\sum\limits_{i=1}^{n} X_i^2 + n\bar{X}^2 - 2\bar{X} \sum\limits_{i=1}^{n} X_i}{n-1}$$

and since $\bar{X} = (1/n) \sum_{i=1}^{n} X_i$, this last equation reduces to

$$S^2 = \frac{\sum\limits_{i=1}^{n} X_i^2 - \left(\sum\limits_{i=1}^{n} X_i\right)^2 / n}{n-1} \tag{8-5}$$

The sample standard deviation S is the positive square root of the sample variance. The units of measurement for S are the same as the units of measurement of the variable, and the units of measurement of S^2 are the square of the units employed for the variable.

● **Example 8-2.** Compute the sample variance and sample standard deviation of the bottle-bursting strength data in Table 8-1. Note that

$$\sum_{i=1}^{100} x_i^2 = 7,074,258.00 \quad \text{and} \quad \sum_{i=1}^{100} x_i = 26,406$$

Consequently, the sample variance is

$$s^2 = \frac{\sum_{i=1}^{100} x_i^2 - \left(\sum_{i=1}^{100} x_i\right)^2 / 100}{99} = \frac{7,074,258.00 - \dfrac{(26,406)^2}{100}}{99} = 1025.15 \text{ psi}^2$$

and the sample standard deviation is

$$s = \sqrt{1025.15} = 32.02 \text{ psi}$$

Another useful measure of dispersion is the *sample range*,

$$R = \max(X_i) - \min(X_i) \tag{8-6}$$

The sample range is very simple to compute, but it ignores all of the information in the sample between the smallest and largest observations. For small sample sizes, say $n \leq 10$, this information loss in some situations is not too serious. The range has found widespread application in statistical quality control, where sample sizes of 4 or 5 are common and computational simplicity is a major consideration. We will discuss the use of the range in statistical quality-control problems in Chapter 15.

● **Example 8-3.** Calculate the ranges of the two samples of bottle-bursting strength given on page 209. For the first sample, we find that

$$R_1 = 265 - 230 = 35$$

while for the second sample,

$$R_2 = 305 - 190 = 115$$

Note that the range of the second sample is much larger than the range of the first, confirming our initial impression that the second sample has greater variablility than the first.

Occasionally, it is desirable to express variation as a fraction of the mean. A measure of relative variation called the *sample coefficient of variation* is defined as

$$CV = \frac{S}{\bar{X}} \qquad (8\text{-}7)$$

The coefficient of variation is useful when comparing the variability of two or more data sets that differ considerably in the magnitude of the observations. For example, the coefficient of variation might be useful in comparing the variability of daily electricity usage within samples of single-family residences in Atlanta, Georgia, and Butte, Montana, during July.

8-3.3 Grouped Data

If the data are in a frequency distribution, it is necessary to modify the computing formulas for the measures of central tendency and dispersion given in Sections 8-3.1 and 8-3.2. Suppose that for each of p distinct values of X, say $X_1, X_2, \ldots, X_p$, the observed frequency is f_j. Then the sample mean and sample variance may be computed as

$$\bar{X} = \frac{\sum\limits_{j=1}^{p} f_j X_j}{\sum\limits_{j=1}^{p} f_j} = \frac{\sum\limits_{j=1}^{p} f_j X_j}{n} \qquad (8\text{-}8)$$

and

$$S^2 = \frac{\sum\limits_{j=1}^{p} f_j X_j^2 - \left(\sum\limits_{j=1}^{p} f_j X_j\right)^2 \Big/ n}{n-1} \qquad (8\text{-}9)$$

respectively.

In many frequency distributions, we can no longer determine the individual observations, but only the class intervals to which they belong. For example, see the frequency distribution of bottle-bursting strengths in Fig. 8-1. In such cases, we can approximate the sample mean and sample variance. This requires that we assume that the observations are concentrated at the center of the class interval. If m_j denotes the midpoint of the jth class interval and there are c class intervals, then the sample mean and sample variance are approximately

$$\bar{X} \simeq \frac{\sum\limits_{j=1}^{c} f_j m_j}{\sum\limits_{j=1}^{c} f_j} = \frac{\sum\limits_{j=1}^{c} f_j m_j}{n} \qquad (8\text{-}10)$$

and

$$S^2 \simeq \frac{\sum\limits_{j=1}^{c} f_j m_j^2 - \left(\sum\limits_{j=1}^{c} f_j m_j\right)^2 \Big/ n}{n-1} \qquad (8\text{-}11)$$

• **Example 8-4.** To illustrate the use of Equations (8-10) and (8-11), we compute the mean and variance of bursting strength for the data in the frequency distribution of Table 8-2. Note that there are $c = 9$ class intervals, and that $m_1 = 180$, $f_1 = 2$, $m_2 = 200$, $f_2 = 4$, $m_3 = 220$, $f_3 = 7$, $m_4 = 240$, $f_4 = 13$, $m_5 = 260$, $f_5 = 32$, $m_6 = 280$, $f_6 = 24$, $m_7 = 300$, $f_7 = 11$, $m_8 = 320$, $f_8 = 4$, $m_9 = 340$, and $f_9 = 3$. Thus

$$\bar{x} \simeq \frac{\sum\limits_{i=1}^{9} f_j m_j}{n} = \frac{26{,}460}{100} = 264.60 \text{ psi}$$

and

$$s^2 \simeq \frac{\sum\limits_{i=1}^{9} f_i m_i^2 - \left(\sum\limits_{i=1}^{9} f_i m_i\right)^2 \Big/ 100}{99} = \frac{7{,}091{,}900 - \dfrac{(26{,}460)^2}{100}}{99} = 914.99 \text{ psi}^2$$

When the data are grouped in class intervals, it is also possible to approximate the median and mode. The median is approximately

$$M \simeq L_M + \left(\frac{\dfrac{n+1}{2} - T}{f_M}\right)\Delta \qquad (8\text{-}12)$$

where L_M is the lower limit of the class interval containing the median (called the median class), f_M is the frequency in the median class, T is the total of all frequencies in the class intervals preceeding the median class, and Δ is the width of the median class. The mode, say MO, is approximately

$$MO \simeq L_{MO} + \left(\frac{a}{a+b}\right)\Delta \qquad (8\text{-}13)$$

where L_{MO} is the lower limit of the modal class (the class interval with the greatest frequency), a is the absolute value of the difference in frequency between the modal class and the preceeding class, b is the absolute value of the difference in frequency between the modal class and the following class, and Δ is the width of the modal class.

Finally, if the data are grouped and coded about an arbitrary origin, say m_0, such that $d_j = (m_j - m_0)/\Delta$, where Δ is the class interval width and m_j is the midpoint of the jth class interval, $j = 1, 2, \ldots, c$, then the sample mean and sample variance are approximately

$$\bar{X} \simeq \Delta\left(\frac{\sum\limits_{i=1}^{c} f_j d_j}{n}\right) + m_0 \qquad (8\text{-}14)$$

and

$$S^2 \simeq \Delta^2 \left(\frac{\sum_{j=1}^{c} f_j d_j^2 - \left(\sum_{j=1}^{c} f_j d_j \right)^2 / n}{n-1} \right) \tag{8-15}$$

These equations are occasionally useful in field experiments where considerable data must be collected.

8-4 Statistics and Sampling Distributions

A *statistic* is any function of the observations in a random sample that does not depend on unknown parameters. For example, if $X_1, X_2, \ldots, X_n$ is a random sample of size n, then the sample mean $\bar{X}$, the sample variance S^2, and the sample standard deviation S are all statistics. Note that since a statistic is a function of the data from a random sample, it is a random variable. That is, if we were to take two different random samples from a population and compute the sample means, we would expect the observed values of the sample means $\bar{x}_1$ and $\bar{x}_2$ to be different.

The process of drawing conclusions about populations based on sample data makes considerable use of statistics. The procedures require that we understand the probabilistic behavior of certain statistics. In general, we call the probability distribution of a statistic a *sampling distribution*. There are several important sampling distributions that will be used extensively in subsequent chapters. In this section, we define and briefly illustrate these sampling distributions.

Consider determining the sampling distribution of the sample mean $\bar{X}$. Suppose that a random sample of size n is taken from a normal population with mean μ and variance σ^2. Now each observation in this random sample is a normally and independently distributed random variable with mean μ and variance σ^2. Consequently, the reproductive property of the normal distribution (see Section 7-3) implies that the sampling distribution of $\bar{X}$ is normal, with mean μ and variance σ^2/n. Furthermore, if the distribution of the population is unknown, the central limit theorem (Section 7-4) implies that for moderate to large samples the sampling distribution of $\bar{X}$ is approximately normal, with mean μ and variance σ^2/n. Therefore, we would conclude that if $\bar{X}$ is the mean of a random sample of size n taken from a population with mean μ and variance σ^2, then the sampling distribution of $\bar{X}$ is normal with mean μ and variance σ^2/n if the population is normally distributed, and the distribution of $\bar{X}$ is approximately normal even if the population follows some nonnormal distribution if the conditions of the central limit theorem are satisfied. It is fairly standard practice to call the standard deviation of a sampling distribution the *standard error* of the statistic. Thus, the standard error of the sample mean $\bar{X}$ is $\sigma/\sqrt{n}$.

Many other useful sampling distributions can be defined in terms of normal random variables. The chi-square distribution is defined below.

Theorem 8-1

Let $Z_1, Z_2, \ldots, Z_k$ be normally and independently distributed random variables, with mean $\mu = 0$ and variance $\sigma^2 = 1$. Then the random variable

$$X^2 = Z_1^2 + Z_2^2 + \cdots + Z_k^2$$

has the probability density function

$$f_{X^2}(u) = \frac{1}{2^{k/2}\Gamma\left(\frac{k}{2}\right)} u^{(k/2)-1} e^{-u/2} \qquad u > 0 \tag{8-16}$$

$$= 0 \qquad \text{otherwise}$$

and is said to follow the chi-square distribution with k degrees of freedom, abbreviated χ_k^2.

For the proof of Theorem 8-1, see Exercises 7-54 and 7-55.

The mean and variance of the χ_k^2 distribution are

$$\mu = k \tag{8-17a}$$

and

$$\sigma^2 = 2k \tag{8-17b}$$

Several chi-square distributions are shown in Fig. 8-3. Note that the chi-square random variable is nonnegative, and that the probability distribution is skewed to the right. However, as k increases, the distribution becomes more

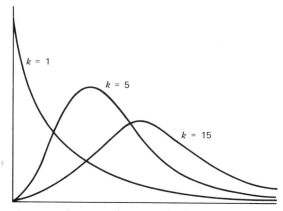

Fig. 8-3. Several chi-square distributions.

symmetric. As $k \to \infty$, the limiting form of the chi-square distribution is the normal distribution.

The percentage points of the χ_k^2 distribution are given in Table III of the Appendix. Define $\chi_{\alpha,k}^2$ as the percentage point or value of the chi-square random variable with k degrees of freedom such that the probability that χ_k^2 exceeds this value is α. That is,

$$P\{X_k^2 \ge \chi_{\alpha,k}^2\} = \int_{\chi_{\alpha,k}^2}^{\infty} f_{X^2}(u)\,du = \alpha$$

This probability is shown as the shaded area in Fig. 8-4. To illustrate the use of Table III, note that

$$P\{X^2 \ge \chi_{0.05,10}^2\} = P\{X^2 \ge 18.31\} = .05$$

That is, the 5 percent point of chi-square with 10 degrees of freedom is $\chi_{.05,10}^2 = 18.31$.

Like the normal distribution, the chi-square distribution has an important reproductive property.

Theorem 8-2 (Additivity Theorem of Chi-Square)

Let $X_1^2, X_2^2, \ldots, X_p^2$ be independent chi-square random variables with $k_1, k_2, \ldots, k_p$ degrees of freedom, respectively. Then the quantity

$$Y = X_1^2 + X_2^2 + \cdots + X_p^2$$

follows the chi-square distribution with degrees of freedom equal to

$$k = \sum_{i=1}^{p} k_i$$

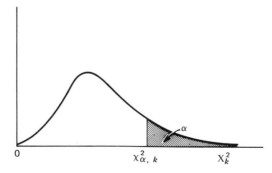

Fig. 8-4. Percentage point $\chi_{\alpha,k}^2$ of the chi-square distribution.

Proof

Note that each chi-square random variable X_i^2 can be written as the sum of the squares of k_i standard normal random variables, say

$$X_i^2 = \sum_{j=1}^{k_i} Z_{ij}^2 \qquad i = 1, 2, \ldots, p$$

Therefore,

$$Y = \sum_{i=1}^{p} X_i^2 = \sum_{i=1}^{p} \sum_{j=1}^{k_i} Z_{ij}^2$$

and since all the random variables Z_{ij} are independent because the X_i^2 are independent, Y is just the sum of the squares of $k = \sum_{i=1}^{p} k_i$ independent standard normal random variables. From Theorem 8-1, it follows that Y is a chi-square random variable with k degrees of freedom.

As an example of a statistic that follows the chi-square distribution, suppose that $X_1, X_2, \ldots, X_n$ is a random sample from a normal population, with mean μ and variance σ^2. The function of the sample variance

$$\frac{(n-1)S^2}{\sigma^2}$$

is distributed as χ_{n-1}^2. We will use this statistic in Chapters 9 and 10.

To illustrate heuristically why the sampling distribution of this particular statistic is chi-square, note that

$$\frac{(n-1)S^2}{\sigma^2} = \frac{\sum_{i=1}^{n} (X_i - \bar{X})^2}{\sigma^2} \tag{8-18}$$

If $\bar{X}$ in Equation (8-18) were replaced by μ, then the distribution of

$$\frac{\sum_{i=1}^{n} (X_i - \mu)^2}{\sigma^2}$$

is χ_n^2, because each term $(X_i - \mu)/\sigma$ is an independent standard normal random variable. Now consider

$$\sum_{i=1}^{n} (X_i - \mu)^2 = \sum_{i=1}^{n} [(X_i - \bar{X}) + (\bar{X} - \mu)]^2$$

$$= \sum_{i=1}^{n} (X_i - \bar{X})^2 + \sum_{i=1}^{n} (\bar{X} - \mu)^2 + 2(\bar{X} - \mu) \sum_{i=1}^{n} (X_i - \bar{X})$$

$$= \sum_{i=1}^{n} (X_i - \bar{X})^2 + n(\bar{X} - \mu)^2$$

Therefore,

$$\frac{\sum_{i=1}^{n} (X_i - \mu)^2}{\sigma^2} = \frac{\sum_{i=1}^{n} (X_i - \bar{X})^2}{\sigma^2} + \frac{(\bar{X} - \mu)^2}{\sigma^2/n}$$

or

$$\frac{\sum_{i=1}^{n} (X_i - \mu)^2}{\sigma^2} = \frac{(n-1)S^2}{\sigma^2} + \frac{(\bar{X} - \mu)^2}{\sigma^2/n} \tag{8-19}$$

Since $\bar{X}$ is normally distributed with mean μ and variance σ^2/n, the quantity $(\bar{X} - \mu)^2/(\sigma^2/n)$ is distributed as χ_1^2. Furthermore, it can be shown that the random variables $\bar{X}$ and S^2 are independent. Therefore, since $\Sigma_{i=1}^{n} (X_i - \mu)^2/\sigma^2$ is distributed as χ_n^2, it seems logical to use the additivity property of chi-square (Theorem 8-2) and conclude that the distribution of $(n-1)S^2/\sigma^2$ is χ_{n-1}^2.

Another important sampling distribution is the t distribution.

Theorem 8-3

Let $Z \sim N(0, 1)$ and V be a chi-square random variable with k degrees of freedom. If Z and V are independent, then the random variable

$$T = \frac{Z}{\sqrt{V/k}}$$

has the probability density function

$$f(t) = \frac{\Gamma[(k+1)/2]}{\sqrt{\pi k}\,\Gamma(k/2)} \cdot \frac{1}{[(t^2/k) + 1]^{(k+1)/2}} \qquad -\infty < t < \infty \tag{8-20}$$

and is said to follow the t distribution with k degrees of freedom, abbreviated t_k.

Proof

Since Z and V are independent, their joint density function is

$$f(z, v) = \frac{(v)^{(k/2)-1}}{\sqrt{2\pi}\,2^{k/2}\Gamma\left(\frac{k}{2}\right)} e^{-(z^2+v)/2} \qquad -\infty < z < \infty, 0 < v < \infty$$

Using the method of Section 4-10 we define a new random variable $U = V$. Thus, the inverse solutions of

$$t = \frac{z}{\sqrt{v/k}}$$

and

$$u = v$$

are

$$z = t\sqrt{\frac{u}{k}}$$

and

$$v = u$$

The Jacobian is

$$J = \begin{vmatrix} \dfrac{t}{2\sqrt{uk}} & \dfrac{u}{k} \\ 1 & 0 \end{vmatrix} = -\sqrt{\dfrac{u}{k}}$$

Thus,

$$|J| = \sqrt{\dfrac{u}{k}}$$

and

$$g(t, u) = \frac{\sqrt{u}}{\sqrt{2\pi k}\, 2^{k/2}\Gamma\left(\dfrac{k}{2}\right)}\, u^{(k/2)-1} e^{-[(u/k)t^2 + u]/2} \tag{8-21}$$

Now, since $V > 0$ we must require that $U > 0$, and since $-\infty < Z < \infty$, then $-\infty < t < \infty$. On rearranging Equation (8-21) we have

$$g(t, u) = \frac{1}{\sqrt{2\pi k}\, 2^{k/2}\Gamma\left(\dfrac{k}{2}\right)}\, u^{(k-1)/2} e^{-(u/2)[(t^2/k)+1]} \qquad 0 < u < \infty, -\infty < t < \infty$$

and since $f(t) = \displaystyle\int_0^\infty g(t, u)\, du$, we obtain

$$f(t) = \frac{1}{\sqrt{2\pi k}\, 2^{k/2}\Gamma\left(\dfrac{k}{2}\right)} \int_0^\infty u^{(k-1)/2} e^{-(u/2)[(t^2/k)+1]}\, du$$

$$= \frac{\Gamma[(k+1)/2]}{\sqrt{\pi k}\,\Gamma\left(\dfrac{k}{2}\right)} \cdot \frac{1}{[(t^2/k)+1]^{(k+1)/2}} \qquad -\infty < t < \infty$$

Primarily because of historical usage, many authors make no distinction between the random variable T and t. The mean and variance of the t distribution are $\mu = 0$ and $\sigma^2 = k/(k-2)$ for $k > 2$, respectively. Several t distributions are shown in Fig. 8-5. The general appearance of the t distribution is similar to the standard normal distribution, in that both distributions are symmetric and unimodal. However, the t distribution has heavier tails than the normal; that is, it exhibits greater variability. As the number of degrees of freedom $k \to \infty$, the limiting form of the t distribution is the standard normal distribution.

The percentage points of the t distribution are given in Table IV of the Appendix. Let $t_{\alpha, k}$ be the percentage point or value of the t random variable

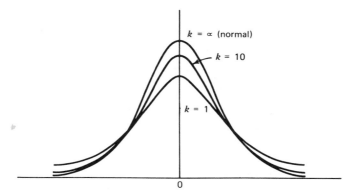

Fig. 8-5. Several t distributions.

with k degrees of freedom such that

$$P\{t \geq t_{\alpha, k}\} = \int_{t_{\alpha, k}}^{\infty} f(t) \, dt = \alpha$$

This percentage point is illustrated in Fig. 8-6. Note that since the t distribution is symmetric about zero, we may find $t_{1-\alpha, k} = -t_{\alpha, k}$. This relationship is useful, since Table IV gives only *upper-tail* percentage points; that is, values of $t_{\alpha, k}$ for $\alpha \leq .50$. To illustrate the use of the table, note that

$$P\{t \geq t_{.05, 10}\} = P\{t \geq 1.813\} = .05$$

That is, the upper 5 percentage point of the t distribution with 10 degrees of freedom is $t_{.05, 10} = 1.813$. Similarly, the lower-tail point $t_{.95, 10} = -t_{.05, 10} = -1.813$.

As an example of a random variable that follows the t distribution, suppose that $X_1, X_2, \ldots, X_n$ is a random sample from a normal distribution with mean μ and variance σ^2, and let $\bar{X}$ and S^2 denote the sample mean and variance.

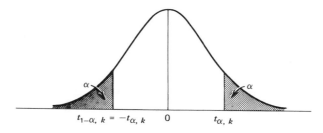

Fig. 8-6. Percentage points of the t distribution.

Consider the statistic

$$\frac{\bar{X} - \mu}{S/\sqrt{n}} \qquad (8\text{-}22)$$

Dividing both the numerator and denominator of Equation (8-22) by σ, we obtain

$$\frac{\dfrac{\bar{X} - \mu}{\sigma}}{S/(\sigma\sqrt{n})} = \frac{\dfrac{\bar{X} - \mu}{\sigma/\sqrt{n}}}{\sqrt{S^2/\sigma^2}}$$

Since $(\bar{X} - \mu)/(\sigma/\sqrt{n}) \sim N(0, 1)$ and $S^2/\sigma^2 \sim \chi^2_{n-1}/(n-1)$, and since $\bar{X}$ and S^2 are independent, we see from Theorem 8-3 that

$$t = \frac{\bar{X} - \mu}{S/\sqrt{n}} \qquad (8\text{-}23)$$

follows a t distribution with $v = n - 1$ degrees of freedom.

A very useful sampling distribution is the F distribution.

Theorem 8-4

Let W and Y be independent chi-square random variables with u and v degrees of freedom, respectively. Then the ratio

$$F = \frac{W/u}{Y/v}$$

has the probability density function

$$h(f) = \frac{\Gamma\left(\dfrac{u + v}{2}\right)\left(\dfrac{u}{v}\right)^{u/2} f^{(u/2)-1}}{\Gamma\left(\dfrac{u}{2}\right)\Gamma\left(\dfrac{v}{2}\right)\left[\left(\dfrac{u}{v}\right)f + 1\right]^{(u+v)/2}} \qquad 0 < f < \infty \qquad (8\text{-}24)$$

and is said to follow the F distribution with u degrees of freedom in the numerator and v degrees of freedom in the denominator. It is usually abbreviated as $F_{u,v}$.

Proof
Since W and Y are independent

$$f(w, y) = \frac{(w)^{(u/2)-1}(y)^{(v/2)-1}}{2^{u/2}\Gamma\left(\dfrac{u}{2}\right)2^{v/2}\Gamma\left(\dfrac{v}{2}\right)} e^{-(w+y)/2} \qquad 0 < w, y < \infty$$

Proceeding as in Section 4-10, define the new random variable $M = Y$. The inverse solutions of $f = (w/u)/(y/v)$ and $m = y$ are

$$w = \frac{umf}{v}$$

and

$$y = m$$

Therefore, the Jacobian

$$J = \begin{vmatrix} \dfrac{um}{v} & \dfrac{uf}{v} \\ 0 & 1 \end{vmatrix} = \frac{u}{v} m$$

Thus,

$$g(f, m) = \frac{\dfrac{u}{v}\left(\dfrac{u}{v} fm\right)^{(u/2)-1} m^{(v/2)-1}}{2^{u/2}\Gamma\left(\dfrac{u}{2}\right) 2^{v/2}\Gamma\left(\dfrac{v}{2}\right)} e^{-(m/2)((u/v)f+1)} \qquad 0 < f, m < \infty$$

and since $h(f) = \int_0^\infty g(f, m)\, dm$, we obtain

$$h(f) = \frac{(u/v)^{u/2} f^{(u/2)-1} \Gamma\left(\dfrac{u+v}{2}\right)}{2^{(u+v)/2}\Gamma\left(\dfrac{u}{2}\right)\Gamma\left(\dfrac{v}{2}\right)\left[\dfrac{\left(\dfrac{u}{v}\right)f+1}{2}\right]^{(u+v)/2}}, \qquad 0 < f < \infty$$

which will simplify to Equation (8-24), completing the proof.

The mean and variance of the F distribution are $\mu = v/(v-2)$ for $v > 2$, and

$$\sigma^2 = \frac{2v^2(u+v-2)}{u(v-2)^2(v-4)} \qquad v > 4$$

Several F distributions are shown in Fig. 8-7. The F random variable is nonnegative and the distribution is skewed to the right.

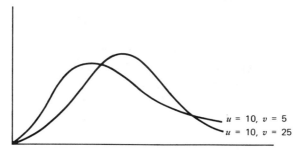

Fig. 8-7. Several F distributions.

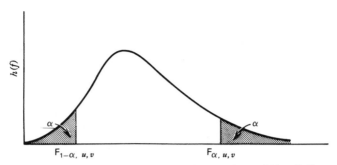

Fig. 8-8. Upper and lower percentage points of the F distribution.

The percentage points of the F distribution are given in Table V of the Appendix. Let $F_{\alpha, u, v}$ be the percentage point of the F distribution, with u and v degrees of freedom such that the probability that the random variable F exceeds this value is

$$P\{F \geq F_{\alpha, u, v}\} = \int_{F_{\alpha, u, v}}^{\infty} h(f)\, df = \alpha$$

This is illustrated in Fig. 8-8. For example, if $u = 5$ and $v = 10$, we find from Table V of the Appendix that

$$P\{F \geq F_{.05, 5, 10}\} = P\{F \geq 3.33\} = .05$$

That is, the upper 5 percentage point of $F_{5, 10}$ is $F_{.05, 5, 10} = 3.33$. Table V contains only upper-tail percentage points (values of $F_{\alpha, u, v}$ for $\alpha \leq .50$). The lower-tail percentage points $F_{1-\alpha, u, v}$ can be found as follows:

$$F_{1-\alpha, u, v} = \frac{1}{F_{\alpha, v, u}} \tag{8-25}$$

For example, to find the lower-tail percentage point $F_{.95, 5, 10}$, note that

$$F_{.95, 5, 10} = \frac{1}{F_{.05, 10, 5}} = \frac{1}{4.74} = .211$$

As an example of a statistic that follows the F distribution, suppose we have two normal populations with variances σ_1^2 and σ_2^2. Let independent random samples of sizes n_1 and n_2 be taken from populations 1 and 2, respectively, and let S_1^2 and S_2^2 be the sample variances. Then the ratio

$$F = \frac{S_1^2/\sigma_1^2}{S_2^2/\sigma_2^2} \tag{8-26}$$

has an F distribution with $n_1 - 1$ numerator degrees of freedom and $n_2 - 1$

denominator degrees of freedom. This follows directly from the fact that $(n_1 - 1)S_1^2/\sigma_1^2 \sim \chi_{n_1-1}^2$ and $(n_2 - 1)S_2^2/\sigma_2^2 \sim \chi_{n_2-1}^2$ and from Theorem 8-4.

8-5 Summary

This chapter has introduced the concepts of population and population sampling. Random samples were defined as samples selected so that the observations in the sample are independent and identically distributed random variables. Most of the statistical techniques to be introduced in subsequent chapters assume that random samples are available.

The frequency distribution and histogram were introduced as graphical data presentation devices. The sample mean $\bar{X}$ and sample variance S^2 are the primary quantitative measures of the central tendency and dispersion of the data. Both the sample mean and sample variance are statistics; that is, functions of the observations in a random sample. In the following chapters we will show how statistics such as these can be used to generalize conclusions about a sample to the population.

The probability distribution of a statistic is called a sampling distribution. For example, the sampling distribution of the sample mean $\bar{X}$ is the normal distribution. Three other important sampling distributions, the chi-square, t, and F distributions, were also introduced. Their use will be illustrated in the next four chapters.

8-6 Exercises

8-1. The weight of bearings produced by a machine is being studied. A random sample of six bearings provided the weights 1.18, 1.21, 1.19, 1.17, 1.20, 1.21 pounds. Find the sample mean, sample variance, sample standard deviation, median, and mode.

Life (days)	Life (days)	Life (days)	Life (days)
125	140	121	141
127	125	127	147
140	124	128	150
135	122	134	132
126	121	140	143
120	127	121	121
121	130	126	124
142	131	124	131
151	141	125	141
160	137	127	127

8-2. The shelf life of a high-speed photographic film is of interest to its producers. From the retail outlets the following data are available.
(*a*) Find the sample mean, sample variance, and sample standard deviation.
(*b*) Find the median and mode.
(*c*) Construct a frequency distribution with class intervals of width 2.
(*d*) Approximate the sample mean, sample variance, mode, and median for the frequency distribution. How do they compare with the exact values?

8-3. The number of homework assignments completed by a random sample of six students in an engineering statistics course at Georgia Tech were as follows: 10, 12, 8, 14, 10, 11. Calculate the sample mean and sample variance.

8-4. The diameter of eight randomly selected piston rings is shown below. Calculate the sample mean, sample variance, and sample standard deviation.

74.001 mm	73.998 mm
74.005	74.000
74.003	74.006
74.001	74.002

8-5. Below are the weekly earnings of all employees in a textile mill.

$120.15	$125.68	$125.49	$122.92
125.34	125.64	129.00	126.16
126.18	125.65	126.00	125.24
120.45	119.30	124.00	128.75
124.53	127.10	128.36	123.08
119.75	125.70	125.25	124.80
126.34	116.70	123.18	122.40
128.00	121.42	124.17	122.65
126.03	117.45	127.67	125.00
130.60	121.55	126.23	126.00
122.48	118.50	120.48	126.50
128.24	124.73	119.35	124.35
124.32	128.83	124.65	124.50
126.67	124.55	124.75	119.70
128.92	126.00	125.16	121.47

(*a*) Compute the sample mean and variance.
(*b*) Construct a frequency distribution.
(*c*) Approximate the mean and variance using the frequency distribution.
(*d*) Approximate the median and mode.

8-6. Given the frequency distribution:

x_i	1	2	3	4	5	6
f_i	10	18	19	20	17	12

    (*a*) Compute the sample mean, sample variance, and sample standard deviation.

    (*b*) Compute the median and mode.

8-7.   Given the frequency distribution:

Class Interval	Frequency
$0 \leq x < 2$	4
$2 \leq x < 4$	6
$4 \leq x < 6$	8
$6 \leq x < 8$	5
$8 \leq x < 10$	2

    compute the approximate sample mean and sample variance.

8-8.   Show that

    (*a*) $\sum\limits_{i=1}^{n} (X_i - \bar{X}) = 0$.

    (*b*) $\sum\limits_{i=1}^{n} (X_i - \bar{X})^2 = \sum\limits_{i=1}^{n} X_i^2 - n\bar{X}^2$.

8-9.   Consider the following frequency distribution.

x_i	115	116	117	118	119	120	121	122	123	124
f_i	4	6	9	13	15	19	20	18	15	10

    (*a*) Calculate the sample mean and variance.

    (*b*) Calculate the median and mode.

8-10. Consider the following frequency distribution.

x_i	−4	−3	−2	−1	0	1	2	3	4
f_i	60	120	180	200	240	190	160	90	30

    (*a*) Calculate the sample mean and variance.

    (*b*) Calculate the median and mode.

8-11. For the two sets of data in Exercises 8-9 and 8-10 compute the sample coefficients of variation.

8-12. Consider the following frequency distribution.

x_i	100	200	300	400	500	600	700
f_i	16	24	12	10	8	5	2

    (*a*) Compute the sample mean and variance.

    (*b*) Compute the median and mode.

8-13. Compute the approximate sample mean, sample variance, median, and mode from the data in the following frequency distribution:

Class Interval	Frequency
$10 \leq x < 20$	121
$20 \leq x < 30$	165
$30 \leq x < 40$	184
$40 \leq x < 50$	173
$50 \leq x < 60$	142
$60 \leq x < 70$	120
$70 \leq x < 80$	118
$80 \leq x < 90$	110
$90 \leq x < 100$	90

8-14. Compute the approximate sample mean, sample variance, median, and mode from the data in the following frequency distribution:

Class Interval	Frequency
$-10 \leq x < 0$	3
$0 \leq x < 10$	8
$10 \leq x < 20$	12
$20 \leq x < 30$	16
$30 \leq x < 40$	9
$40 \leq x < 50$	4
$50 \leq x < 60$	2

8-15. Compute the approximate sample mean, sample standard deviation, sample variance, median, and mode for the data in the following frequency distribution:

Class Interval	Frequency
$600 \leq x < 650$	41
$650 \leq x < 700$	46
$700 \leq x < 750$	50
$750 \leq x < 800$	52
$800 \leq x < 850$	60
$850 \leq x < 900$	64
$900 \leq x < 950$	65
$950 \leq x < 1000$	70
$1000 \leq x < 1050$	72

8-16. Data representing the yearly gross income of 25 corporation presidents and 25 day laborers are given in the following table. Compute the sample variances and coefficients of variation. Which is the best measure for comparing the variability of the two sets of data?

Presidents		Laborers	
$105,200	$100,025	$4250	$4500
98,225	96,350	5140	5175
162,000	87,400	4475	3410
147,500	192,000	3940	4915
150,275	167,000	4010	4000
121,800	174,000	5210	5700
141,350	132,000	5410	4647
93,400	124,000	5600	5115
97,500	204,350	4912	3918
186,550	187,500	3975	4005
173,000	120,000	4745	5157
143,250	143,210	4610	5150
190,100	—	4795	—

8-17. The following table contains the frequency of occurrence of final letters in an article in the *Atlanta Journal*. Construct a histogram from these data. Do any of the numerical descriptors in this chapter have any meaning for these data?

a	2	n	9
b	1	o	3
c	1	p	0
d	10	q	0
e	15	r	5
f	2	s	8
g	2	t	10
h	2	u	0
i	2	v	0
j	0	w	2
k	0	x	0
l	1	y	5
m	2	z	0

8-18. The data below represent the yield of a chemical process. Construct a frequency distribution and histogram for these data. Compute the sample mean and sample variance. Do the data appear to follow the form of any of the distributions we have studied?

94.1	87.3	94.1	92.4	84.6	85.4
93.2	84.1	92.1	90.6	83.6	86.6
90.6	90.1	96.4	89.1	85.4	91.7
91.4	95.2	88.2	88.8	89.7	87.5
88.2	86.1	86.4	86.4	87.6	84.2
86.1	94.3	85.0	85.1	85.1	85.1

95.1	93.2	84.9	84.0	89.6	90.5
90.0	86.7	87.3	93.7	90.0	95.6
92.4	83.0	89.6	87.7	90.1	88.3
87.3	95.3	90.3	90.6	94.3	84.1
86.6	94.1	93.1	89.4	97.3	83.7
91.2	97.8	94.6	88.6	96.8	82.9
86.1	93.1	96.3	84.1	94.4	87.3
90.4	86.4	94.7	82.6	96.1	86.4
89.1	87.6	91.1	83.1	98.0	84.5

8-19. The percentages of cotton in a material used to manufacture men's shirts are given below. Construct a frequency distribution and histogram for the data. Compute the sample mean, sample variance, and median. Do the data appear to follow any of the distributions we have studied?

34.2	33.6	33.8	34.7	37.8	32.6	35.8	34.6
33.1	34.7	34.2	33.6	36.6	33.1	37.6	33.6
34.5	35.0	33.4	32.5	35.4	34.6	37.3	34.1
35.6	35.4	34.7	34.1	34.6	35.9	34.6	34.7
34.3	36.2	34.6	35.1	33.8	34.7	35.5	35.7
35.1	36.8	35.2	36.8	37.1	33.6	32.8	36.8
34.7	35.1	35.0	37.9	34.0	32.9	32.1	34.3
33.6	35.3	34.9	36.4	34.1	33.5	34.5	32.7

8-20. A factory manufactures washing machines. The number of defective machines produced per week has been recorded. Construct a frequency distribution and a frequency histogram from the data. Compute the sample mean, sample variance, range, median, and mode. Do the data seem to follow any of the distributions we have studied?

3	6	4	7	6	7
4	7	8	2	1	4
2	9	4	6	4	8
5	10	10	9	13	7
6	14	14	10	12	3
10	13	8	7	10	6
5	10	12	9	2	7
4	9	4	16	5	8
3	8	5	11	7	4
11	10	14	13	10	12
9	3	2	3	4	6
2	2	8	13	2	17
7	4	6	3	2	5
8	6	10	7	6	10
4	4	8	3	4	8
2	10	6	2	10	9
6	8	4	9	8	11
5	7	6	4	14	7
4	14	15	13	6	2
3	13	4	3	4	8

2	12	7	6	4	10
8	5	5	5	8	7
10	4	3	10	7	4
9	6	2	6	9	3
11	5	6	7	2	6

8-21. Suppose that a random variable is normally distributed with mean μ and variance σ^2. Draw a random sample of five observations. What is the joint density function of the sample?

8-22. Transistors have a life that is exponentially distributed with parameter λ. A random sample of n transistors is taken. What is the joint density function of the sample?

8-23. Suppose that X is uniformly distributed on the interval from 0 to 1. Consider a random sample of size 4 from X. What is the joint density function of the sample?

8-24. A lot consists of N transistors, and of these $M(M \leq N)$ are defective. We randomly select two transistors without replacement from this lot and determine whether they are defective or nondefective. The random variable

$$X_i = \begin{cases} 1, & \text{if the } i\text{th transistor is nondefective} \\ 0, & \text{if the } i\text{th transistor is defective} \end{cases} \quad i = 1, 2$$

Determine the joint probability function for X_1 and X_2. What are the marginal probability functions for X_1 and X_2? Are X_1 and X_2 independent random variables?

8-25. Develop the moment-generating function of the chi-square distribution.

8-26. Derive the mean and variance of the chi-square random variable with u degrees of freedom.

8-27. Derive the mean and variance of the t distribution.

8-28. Derive the mean and variance of the F distribution.

8-29. Order Statistics. Let $X_1, X_2, \ldots, X_n$ be a random sample of size n from X, a random variable having distribution function $F(x)$. Rank the elements in order of increasing numerical magnitude, resulting in $X_{(1)}, X_{(2)}, \ldots, X_{(n)}$, where $X_{(1)}$ is the smallest sample element $(X_{(1)} = \min\{X_1, X_2, \ldots, X_n\})$ and $X_{(n)}$ is the largest sample element $(X_{(n)} = \max\{X_1, X_2, \ldots, X_n\})$. $X_{(i)}$ is called the ith order statistic. Often, the distribution of some of the order statistics is of interest, particularly the minimum and maximum sample values, $X_{(1)}$ and $X_{(n)}$, respectively. Prove that the distribution functions of $X_{(1)}$ and $X_{(n)}$, denoted respectively by $F_{X_{(1)}}(t)$ and $F_{X_{(n)}}(t)$, are

$$F_{X_{(1)}}(t) = 1 - [1 - F(t)]^n$$
$$F_{X_{(n)}}(t) = [F(t)]^n$$

Prove that if X is continuous with probability distribution $f(x)$, then the probability distributions of $X_{(1)}$ and $X_{(n)}$ are

$$f_{x_{(1)}}(t) = n[1 - F(t)]^{n-1}f(t)$$
$$f_{x_{(n)}}(t) = n[F(t)]^{n-1}f(t)$$

8-30. **Continuation of Exercise 8-29.** Let $X_1, X_2, \ldots, X_n$ be a random sample of a Bernoulli random variable with parameter p. Show that

$$P(X_{(n)} = 1) = 1 - (1-p)^n$$
$$P(X_{(1)} = 0) = 1 - p^n$$

Use the results of Exercise 8-29.

8-31. **Continuation of Exercise 8-29.** Let $X_1, X_2, \ldots, X_n$ be a random sample of a normal random variable with mean μ and variance σ^2. Using the results of Exercise 8-29, derive the density functions of $X_{(1)}$ and $X_{(n)}$.

8-32. **Continuation of Exercise 8-29.** Let X_1, X_2, X_3 be a random sample of an exponential random variable with parameter λ. Derive the distribution functions and probability distributions for $X_{(1)}$ and $X_{(n)}$. Use the results of Exercise 8-29.

8-33. Let $X_1, X_2, \ldots, X_n$ be a random sample of a continuous random variable. Find

$$E[F(X_{(n)})]$$

and

$$E[F(X_{(1)})]$$

8-34. Show that if a constant Δ is added to every observation in a sample, the sample variance is unchanged.

8-35. Show that if every observation in a sample is multiplied by a constant Δ, the sample variance is multiplied by Δ^2.

8-36. Find the following values using Table III of the Appendix.
 (a) $\chi^2_{.95, 8}$
 (b) $\chi^2_{.50, 12}$
 (c) $\chi^2_{.025, 20}$
 (d) χ^2_α, such that $P\{\chi^2_{10} \leq \chi^2_{\alpha, 10}\} = .975$

8-37. Find the following values using Table IV of the Appendix.
 (a) $t_{.025, 10}$
 (b) $t_{.25, 20}$
 (c) $t_{\alpha, 10}$, such that $P\{t_{10} \leq t_{\alpha, 10}\} = .95$

8-38. Find the following values using Table V of the Appendix.
 (a) $F_{.25, 4, 9}$
 (b) $F_{.05, 15, 10}$
 (c) $F_{.95, 6, 8}$
 (d) $F_{.90, 24, 24}$

8-39. Let $F_{1-\alpha, u, v}$ denote a lower-tail point ($\alpha \leq .50$) of the $F_{u, v}$ distribution. Prove that $F_{1-\alpha, u, v} = 1/F_{\alpha, v, u}$.

8-40. Suppose that independent random samples of size n_1 and n_2 are taken from two normal populations with means μ_1 and μ_2 and variances σ_1^2 and σ_2^2, respectively. If $\bar{X}_1$ and $\bar{X}_2$ are the sample means, find the sampling distribution of the statistic

$$\frac{\bar{X}_1 - \bar{X}_2 - (\mu_1 - \mu_2)}{\sqrt{(\sigma_1^2/n_1) + (\sigma_2^2/n_2)}}$$

Chapter 9

Parameter Estimation

Statistical inference is the process by which information from sample data is used to draw conclusions about the population from which the sample was selected. The techniques of statistical inference can be divided into two major areas: *parameter estimation* and *hypothesis testing*. This chapter treats parameter estimation, and hypothesis testing is presented in Chapter 10.

As an example of a parameter estimation problem, suppose that civil engineers are analyzing the compressive strength of concrete. There is a natural variability in the strength of each individual concrete specimen. Consequently, the engineers are interested in estimating the average strength for the population consisting of this type of concrete. They may also be interested in estimating the variability of this population. We present methods for obtaining point estimates of parameters such as the population mean and variance and we also discuss methods for obtaining certain kinds of interval estimates of parameters called confidence intervals.

9-1 Point Estimation

A point estimate of a population parameter is a single numerical value of a statistic that corresponds to that parameter. That is, the point estimate is a unique selection for the value of an unknown parameter. More precisely, if X is a random variable with probability distribution $f(x)$, characterized by the unknown parameter θ, and if $X_1, X_2, \ldots, X_n$ is a random sample of size n from X, then the statistic $\hat{\theta} = h(X_1, X_2, \ldots, X_n)$ corresponding to θ is called the *estimator* of θ. Note that the estimator $\hat{\theta}$ is a random variable, because it is a function of sample data. After the sample has been selected, $\hat{\theta}$ takes on a particular numerical value called the point estimate of θ.

As an example, suppose that the random variable X is normally distributed with unknown mean μ and known variance σ^2. The sample mean $\bar{X}$ is a point estimator of the unknown population mean μ. That is, $\hat{\mu} = \bar{X}$. After the sample has been selected, the numerical value $\bar{x}$ is the point estimate of μ. Similarly, if the population variance σ^2 is also unknown, a point estimator for σ^2 is the sample variance S^2, and the numerical value s^2 calculated from sample data is the point estimate of σ^2.

There may be several different potential point estimators for a parameter. For example, if we wish to estimate the mean of a random variable we might consider either the sample mean, the sample median, or perhaps the smallest observation in the sample as point estimators. In order to decide which point estimator of a particular parameter is the best one to use, we need to examine their statistical properties and develop some criteria for comparing estimators.

9-1.1 Properties of Estimators

A desirable property of an estimator is that it should be "close" in some sense to the true value of the unknown parameter. Formally, we say that $\hat{\theta}$ is an *unbiased* estimator of the parameter θ if

$$E(\hat{\theta}) = \theta \tag{9-1}$$

That is, $\hat{\theta}$ is an unbiased estimator of θ if "on the average" its values are equal to θ. Note that this is equivalent to requiring that the mean of the sampling distribution of $\hat{\theta}$ is equal to θ.

● **Example 9-1.** Suppose that X is a random variable with mean μ and variance σ^2. Let $X_1, X_2, \ldots, X_n$ be a random sample of size n from X. Show that the sample mean $\bar{X}$ and sample variance S^2 are unbiased estimators of μ and σ^2, respectively. Consider

$$E(\bar{X}) = E\left(\frac{\sum_{i=1}^{n} X_i}{n}\right)$$

$$= \frac{1}{n} E \sum_{i=1}^{n} X_i$$

$$= \frac{1}{n} \sum_{i=1}^{n} E(X_i)$$

and since $E(X_i) = \mu$, for all $i = 1, 2, \ldots, n$,

$$E(\bar{X}) = \frac{1}{n} \sum_{i=1}^{n} \mu = \mu$$

Therefore, the sample mean $\bar{X}$ is an unbiased estimator of the population mean μ. Now consider

$$E(S^2) = E\left[\frac{\sum\limits_{i=1}^{n}(X_i - \bar{X})^2}{n-1}\right]$$

$$= \frac{1}{n-1}E\sum_{i=1}^{n}(X_i - \bar{X})^2$$

$$= \frac{1}{n-1}E\sum_{i=1}^{n}(X_i^2 + \bar{X}^2 - 2\bar{X}X_i)$$

$$= \frac{1}{n-1}E\left(\sum_{i=1}^{n}X_i^2 - n\bar{X}^2\right)$$

$$= \frac{1}{n-1}\left[\sum_{i=1}^{n}E(X_i^2) - nE(\bar{X}^2)\right]$$

However, since $E(X_i^2) = \mu^2 + \sigma^2$ and $E(\bar{X}^2) = \mu^2 + \sigma^2/n$, we have

$$E(S^2) = \frac{1}{n-1}\left[\sum_{i=1}^{n}(\mu^2 + \sigma^2) - n(\mu^2 + \sigma^2/n)\right]$$

$$= \frac{1}{n-1}(n\mu^2 + n\sigma^2 - n\mu^2 - \sigma^2)$$

$$= \sigma^2$$

Therefore, the sample variance S^2 is an unbiased estimator of the population variance σ^2. However, the sample standard deviation S is a biased estimator of the population standard deviation σ. For large samples this bias is negligible.

The mean square error of an estimator $\hat{\theta}$ is defined as

$$MSE(\hat{\theta}) = E(\hat{\theta} - \theta)^2 \tag{9-2}$$

The mean square error can be rewritten as follows:

$$MSE(\hat{\theta}) = E[\hat{\theta} - E(\hat{\theta})]^2 + [\theta - E(\hat{\theta})]^2$$

$$= V(\hat{\theta}) + (\text{Bias})^2 \tag{9-3}$$

That is, the mean square error of $\hat{\theta}$ is equal to the variance of the estimator plus the squared bias. If $\hat{\theta}$ is an unbiased estimator of $\hat{\theta}$, the mean square error of $\hat{\theta}$ is equal to the variance of $\hat{\theta}$.

The mean square error is an important criterion for comparing two estimators. Let $\hat{\theta}_1$ and $\hat{\theta}_2$ be two estimators of the parameter θ, and let $MSE(\hat{\theta}_1)$ and $MSE(\hat{\theta}_2)$ be the mean squared errors of $\hat{\theta}_1$ and $\hat{\theta}_2$. Then the relative efficiency of $\hat{\theta}_2$ to $\hat{\theta}_1$ is defined as

$$\frac{MSE(\hat{\theta}_1)}{MSE(\hat{\theta}_2)}$$

If this relative efficiency is less than one, we would conclude that $\hat{\theta}_1$ is a more efficient estimator of θ than $\hat{\theta}_2$, in the sense that it has smaller mean square error. For example, suppose that we wish to estimate the mean μ of a population. We have a random sample of n observations $X_1, X_2, \ldots, X_n$, and we wish to compare two possible estimators for μ: the sample mean $\bar{X}$ and a single observation from the sample, say X_i. Note that both $\bar{X}$ and X_i are unbiased estimators of μ; consequently, the mean square error of both estimators is simply the variance. For the sample mean, we have $MSE(\bar{X}) = V(\bar{X}) = \sigma^2/n$, where σ^2 is the population variance; for an individual observation, we have $MSE(X_i) = V(X_i) = \sigma^2$. Therefore, the relative efficiency of X_i to $\bar{X}$ is

$$\frac{MSE(\bar{X})}{MSE(X_i)} = \frac{\sigma^2/n}{\sigma^2} = \frac{1}{n}$$

Since $(1/n) < 1$ for sample sizes $n \geq 2$, we would conclude that the sample mean is a better estimator of μ than a single observation X_i.

Sometimes we find that biased estimators are preferable to unbiased estimators because they have smaller mean square error. That is, we can reduce the variance of the estimator considerably by introducing a relatively small amount of bias. So long as the reduction in variance is greater than the squared bias, an improved estimator in the mean squared error sense will result. An estimator $\hat{\theta}^*$ that has a mean squared error that is less than or equal to the mean squared error of any other estimator $\hat{\theta}$, for all values of the parameter θ, is called an *optimal* estimator of θ.

We have noted that if the estimator $\hat{\theta}$ is unbiased for θ, the mean squared error reduces to the variance of the estimator $\hat{\theta}$. Within the class of unbiased estimators, we would like to find the estimator that has the smallest variance. Such an estimator is called a *minimum variance unbiased estimator*. It is possible to obtain a lower bound on the variance of all unbiased estimators of θ. Let $\hat{\theta}$ be an unbiased estimator of the parameter θ, based on a random sample of n observations, and let $f(x, \theta)$ denote the probability distribution of the random variable X. Then a lower bound on the variance of $\hat{\theta}$ is[1]

$$V(\hat{\theta}) \geq \frac{1}{nE\left[\dfrac{d}{d\theta} \ell n\, f(X, \theta)\right]^2} \tag{9-4}$$

[1]Certain conditions on the function $f(X, \theta)$ are required in obtaining the Cramér-Rao inequality [for example, see Tucker (1962)]. These conditions are satisfied by most of the standard probability distributions.

This inequality is called the Cramér-Rao lower bound. If an unbiased estimator $\hat{\theta}$ satisfies Equation (9-4) as an equality, it is the minimum variance unbiased estimator of θ.

● **Example 9-2.** We will show that the sample mean $\bar{X}$ is the minimum variance unbiased estimator of the mean of a normal distribution with known variance. From Example 9-1, we observe that $\bar{X}$ is an unbiased estimator of μ. Note that

$$\ell n\, f(X, \mu) = \ell n(\sigma\sqrt{2\pi})^{-1} \exp\left[-\frac{1}{2}\left(\frac{X-\mu}{\sigma}\right)^2\right]$$

$$= -\ell n(\sigma\sqrt{2\pi}) - \frac{1}{2}\left(\frac{X-\mu}{\sigma}\right)^2$$

Substituting into Equation (9-4) we obtain

$$V(\bar{X}) \geq \cfrac{1}{nE\left\{\cfrac{d}{d\mu}\left[-\ell n(\sigma\sqrt{2\pi}) - \frac{1}{2}\left(\frac{X-\mu}{\sigma}\right)^2\right]^2\right\}}$$

$$\geq \cfrac{1}{nE\left[\cfrac{(X-\mu)}{\sigma^2}\right]^2}$$

$$\geq \cfrac{1}{n\left[\cfrac{E(X-\mu)^2}{\sigma^4}\right]}$$

$$\geq \cfrac{1}{n\left(\cfrac{\sigma^2}{\sigma^4}\right)}$$

$$\geq \frac{\sigma^2}{n}$$

Since we know that, in general, the variance of the sample mean is $V(\bar{X}) = \sigma^2/n$, we see that $V(\bar{X})$ satisfies the Cramér-Rao lower bound as an equality. Therefore $\bar{X}$ is the minimum variance unbiased estimator of μ for the normal distribution where σ^2 is known.

Another way to define the closeness of an estimator $\hat{\theta}$ to the parameter θ is in terms of *consistency*. If $\hat{\theta}_n$ is an estimator of θ based on a random sample of size n, we say that $\hat{\theta}_n$ is consistent for θ if

$$\lim_{n\to\infty} \dot{P}(|\hat{\theta}_n - \theta| < \epsilon) = 1 \qquad (9\text{-}5)$$

Consistency is a large-sample property, since it describes the limiting behavior of the estimator $\hat{\theta}$ as the sample size tends to infinity. It is usually difficult to prove that an estimator is consistent using the definition of Equation (9-5). However, estimators whose mean square error (or variance, if the estimator is unbiased) tends to zero as the sample size approaches infinity are consistent. For example, $\bar{X}$ is a consistent estimator of the mean of a normal distribution, since $\bar{X}$ is unbiased and $\lim_{n \to \infty} V(\bar{X}) = \lim_{n \to \infty} (\sigma^2/n) = 0$.

9-1.2 The Method of Maximum Likelihood

One of the best methods of obtaining a point estimator is the method of maximum likelihood. Suppose that X is a random variable with probability distribution $f(x, \theta)$, where θ is a single unknown parameter. Let $X_1, X_2, \ldots, X_n$ be the observed values in a random sample of size n. Then the *likelihood function* of the sample is

$$L(\theta) = f(x_1, \theta) \cdot f(x_2, \theta) \cdot \ldots \cdot f(x_n, \theta) \tag{9-6}$$

Note that the likelihood function is now a function of only the unknown parameter θ. The *maximum likelihood estimator* of θ is the value of θ that maximizes the likelihood function $L(\theta)$. Essentially, the maximum likelihood estimator is the value of θ that maximizes the probability of occurrence of the sample results.

● **Example 9-3.** Let X be a Bernoulli random variable. The probability function is

$$p(x) = p^x(1 - p)^{1-x} \quad x = 0, 1$$
$$= 0 \quad\quad\quad\quad \text{otherwise}$$

where p is the parameter to be estimated. The likelihood function of a sample of size n would be

$$L(p) = \prod_{i=1}^{n} p^{x_i}(1 - p)^{1-x_i} = p^{\Sigma x_i}(1 - p)^{n - \Sigma x_i}$$

We observe that if $\hat{p}$ maximizes $L(p)$ then $\hat{p}$ also maximizes $\ell n \, L(p)$. Therefore,

$$\ell n \, L(p) = \sum_{i=1}^{n} x_i \, \ell n \, p + \left(n - \sum_{i=1}^{n} x_i \right) \ell n \, (1 - p)$$

Now

$$\frac{d \, \ell n \, L(p)}{dp} = \frac{\sum_{i=1}^{n} x_i}{p} - \frac{\left(n - \sum_{i=1}^{n} x_i \right)}{(1 - p)}$$

Equating this to zero and solving for p yields the maximum likelihood estimator, $\hat{p}$, as

$$\hat{p} = \frac{1}{n} \sum_{i=1}^{n} X_i$$

● **Example 9-4.** Let X be normally distributed with unknown mean μ and known variance σ^2. The likelihood function of a sample of size n is

$$L(\mu) = \prod_{i=1}^{n} \frac{1}{\sigma\sqrt{2\pi}} e^{-(x_i - \mu)^2/2\sigma^2}$$

$$= \frac{1}{(2\pi\sigma^2)^{n/2}} e^{-(1/2\sigma^2)\sum_{i=1}^{n}(x_i-\mu)^2}$$

Now

$$\ell n \, L(\mu) = (n/2) \, \ell n \, (2\pi\sigma^2) - (2\sigma^2)^{-1} \sum_{i=1}^{n} (x_i - \mu)^2$$

and

$$\frac{d \, \ell n \, L(\mu)}{d\mu} = (\sigma^2)^{-1} \sum_{i=1}^{n} (x_i - \mu)$$

Equating this last result to zero and solving for μ yields

$$\hat{\mu} = \frac{\sum_{i=1}^{n} X_i}{n} = \bar{X}$$

as the maximum likelihood estimator of μ.

It may not always be possible to use calculus methods to determine the maximum of $L(\theta)$. This is illustrated in the following example.

● **Example 9-5.** Let X be uniformly distributed on the interval 0 to a. The likelihood function of a random sample of size n is

$$L(a) = \prod_{i=1}^{n} \frac{1}{a} = \frac{1}{a^n}$$

Note that the slope of this function is not zero anywhere, so we cannot use calculus methods to find the maximum likelihood estimator $\hat{a}$. However, notice that the likelihood function increases as a decreases. Therefore, we would maximize $L(a)$ by setting $\hat{a}$ to the smallest value that it could reasonably assume. Clearly, a can be no smaller than the largest sample value, so we would use the largest observation as $\hat{a}$.

The method of maximum likelihood can be used in situations where there are several unknown parameters, say $\theta_1, \theta_2, \ldots, \theta_k$, to estimate. In such cases, the likelihood function is a function of the k unknown parameters $\theta_1, \theta_2, \ldots, \theta_k$ and the maximum likelihood estimators $\{\hat{\theta}_i\}$ would be found by equating the k first partial derivatives $\partial L(\theta_1, \theta_2, \ldots, \theta_k)/\partial \theta_i$, $i = 1, 2, \ldots, k$, to zero and solving the resulting system of equations.

● **Example 9-6.** Let X be normally distributed with mean μ and variance σ^2, where both μ and σ^2 are unknown. Find the maximum likelihood estimators of μ and σ^2. The likelihood function for a random sample of size n is

$$L(\mu, \sigma^2) = \prod_{i=1}^{n} \frac{1}{\sigma\sqrt{2\pi}} e^{-(x_i-\mu)^2/2\sigma^2}$$

$$= \frac{1}{(2\pi\sigma^2)^{n/2}} e^{-(1/2\sigma^2)\sum_{i=1}^{n}(x_i-\mu)^2}$$

and

$$\ell n\, L(\mu, \sigma^2) = -\frac{n}{2}\,\ell n(2\pi\sigma^2) - \frac{1}{2\sigma^2}\sum_{i=1}^{n}(x_i - \mu)^2$$

Now

$$\frac{\partial\, \ell n\, L(\mu, \sigma^2)}{\partial \mu} = \frac{1}{\sigma^2}\sum_{i=1}^{n}(x_i - \mu) = 0$$

$$\frac{\partial\, \ell n\, L(\mu, \sigma^2)}{\partial(\sigma^2)} = \frac{-n}{2\sigma^2} + \frac{1}{2\sigma^4}\sum_{i=1}^{n}(x_i - \mu)^2 = 0$$

The solutions to the above equations yield the maximum likelihood estimators

$$\hat{\mu} = \frac{1}{n}\sum_{i=1}^{n} X_i = \bar{X}$$

and

$$\hat{\sigma}^2 = \frac{1}{n}\sum_{i=1}^{n}(X_i - \bar{X})^2$$

Maximum likelihood estimators are not necessarily unbiased (see the maximum likelihood estimator of σ^2 in Example 9-6), but they usually may be easily modified to make them unbiased. The bias approaches zero for large samples. In general, maximum likelihood estimators have good large sample or *asymptotic* properties. Specifically, they are asymptotically normally distributed, unbiased, and have a variance that approaches the Cramér-Rao lower bound for large n. More precisely, we say that if $\hat{\theta}$ is the maximum likelihood estimator for θ, then $\sqrt{n}\,(\hat{\theta} - \theta)$ is normally distributed with mean zero and variance

$$V[\sqrt{n}(\hat{\theta} - \theta)] = V(\sqrt{n}\,\hat{\theta}) = \frac{1}{E\left[\dfrac{d}{d\theta}\,\ell n\, f(X, \theta)\right]^2}$$

for large n. Maximum likelihood estimators are also consistent. Furthermore, they have the invariance property; that is, if $\hat{\theta}$ is the maximum likelihood estimator of θ and $u(\theta)$ is a function of θ that has a single-valued inverse, then the maximum likelihood estimator of $u(\theta)$ is $u(\hat{\theta})$.

9-1.3   The Method of Moments

Suppose that X is either a continuous random variable with probability density $f(x; \theta_1, \theta_2, \ldots, \theta_k)$ or a discrete random variable with distribution $p(x; \theta_1, \theta_2, \ldots, \theta_k)$ characterized by k unknown parameters. Let $X_1, X_2, \ldots, X_n$ be a random sample of size n from X, and define the first k sample moments about the origin as

$$m'_t = \frac{\sum_{i=1}^{n} X_i^t}{n} \qquad t = 1, 2, \ldots, k \tag{9-7}$$

The first k population moments about the origin are

$$\mu'_t = E(X^t) = \int_{-\infty}^{\infty} x^t f(x; \theta_1, \theta_2, \ldots, \theta_k)\, dx$$

$$t = 1, 2, \ldots, k \qquad X \text{ continuous}$$

$$= \sum_{x \in R_X} x^t p(x; \theta_1, \theta_2, \ldots, \theta_k)$$

$$t = 1, 2, \ldots, k, \qquad X \text{ discrete} \tag{9-8}$$

The population moments $\{\mu'_t\}$ will, in general, be functions of the k unknown parameters $\{\theta_i\}$. Equating sample moments and population moments will yield k simultaneous equations in k unknowns (the $\{\theta_i\}$); that is,

$$\mu'_t = m'_t \qquad t = 1, 2, \ldots, k \tag{9-9}$$

The solution to Equation (9-9), denoted $\hat{\theta}_1, \hat{\theta}_2, \ldots, \hat{\theta}_k$, yields the moment estimators of $\theta_1, \theta_2, \ldots, \theta_k$.

● **Example 9-7.** Let $X \sim N(\mu, \sigma^2)$ where μ and σ^2 are unknown. To derive estimators for μ and σ^2 by the method of moments, recall that for the normal distribution

$$\mu'_1 = \mu$$

$$\mu'_2 = \sigma^2 + \mu^2$$

The sample moments are $m'_1 = (1/n) \sum_{i=1}^{n} X_i$ and $m'_2 = (1/n) \sum_{i=1}^{n} X_i^2$. From Equation (9-9) we obtain

$$\mu = \frac{1}{n} \sum_{i=1}^{n} X_i$$

$$\sigma^2 + \mu^2 = \frac{1}{n} \sum_{i=1}^{n} X_i^2$$

which have the solution

$$\hat{\mu} = \frac{1}{n} \sum_{i=1}^{n} X_i = \bar{X}$$

$$\hat{\sigma}^2 = \frac{1}{n} \left(\sum_{i=1}^{n} X_i^2 - n\bar{X}^2 \right) = \frac{1}{n} \sum_{i=1}^{n} (X_i - \bar{X})^2$$

● **Example 9-8.** Let X be uniformly distributed on the interval $(0, a)$. To find an estimator of a by the method of moments, we note that the first population moment about zero is

$$\int_0^a x \frac{1}{a} dx = \frac{a}{2}$$

The first sample moment is just $\bar{X}$. Therefore,

$$\hat{a} = 2\bar{X}$$

or the moment estimator of a is just twice the sample mean.

The method of moments often yields estimators that are reasonably good. In Example 9-7, for instance, the moment estimators are identical to the maximum likelihood estimators. In general, moment estimators are asymptotically normally distributed (approximately) and consistent. However, their variance may be larger than the variance of estimators derived by other methods, such as the method of maximum likelihood. Occasionally, the method of moments yields estimators that are very poor, as in Example 9-8. The estimator in that example does not always generate an estimate that is compatible with our knowledge of the situation. For example, if our sample observations were $x_1 = 60$, $x_2 = 10$, and $x_3 = 5$, then $\hat{a} = 50$, which is unreasonable since we know that $a \geq 60$.

9-2 Confidence Interval Estimation

In many situations, a point estimate does not provide enough information about the parameter of interest. For example, if we are interested in estimating the mean compression strength of concrete, a single number may not be very meaningful. An interval estimate of the form $L \leq \mu \leq U$ might be more useful. The end points of this interval will be random variables, since they are functions of sample data.

In general, to construct an interval estimator of the unknown parameter θ, we must find two statistics L and U such that

$$P\{L \leq \theta \leq U\} = 1 - \alpha \tag{9-10}$$

The resulting interval

$$L \leq \theta \leq U \tag{9-11}$$

is called a $100(1 - \alpha)$ percent *confidence interval* for the unknown parameter θ. L and U are called the lower- and upper-*confidence limits*, respectively, and $1 - \alpha$ is called the *confidence coefficient*. The interpretation of a confidence interval is that if many random samples are collected and a $100(1 - \alpha)$ percent confidence interval on θ computed from each sample, then $100(1 - \alpha)$ percent of these intervals will contain the true value of θ. Now in practice, we only obtain one random sample and calculate one confidence interval. Since this interval either will or will not contain the true value of θ, it is not reasonable to attach a probability level to this specific event. The appropriate statement would be that θ lies in the observed interval $[L, U]$ with confidence $100(1 - \alpha)$. This statement has a frequency interpretation; that is, we don't know if the statement is true for this specific sample, but the *method* used to obtain the interval $[L, U]$ yields correct statements $100(1 - \alpha)$ percent of the time.

The confidence interval in Equation (9-11) might be more properly called a *two-sided confidence interval*, as it specifies both a lower and an upper limit on θ. Occasionally, a *one-sided* confidence interval might be more appropriate. A one-sided $100(1 - \alpha)$ percent lower-confidence interval on θ is given by the interval

$$L \leq \theta \tag{9-12}$$

where the lower-confidence limit L is chosen so that

$$P\{L \leq \theta\} = 1 - \alpha \tag{9-13}$$

Similarly, a one-sided $100(1 - \alpha)$ percent upper-confidence interval on θ is given by the interval

$$\theta \leq U \tag{9-14}$$

where the upper-confidence limit U is chosen so that

$$P\{\theta \leq U\} = 1 - \alpha \tag{9-15}$$

The length of the observed confidence interval is an important measure of the quality of the information obtained from the sample. The half-interval width $\theta - L$ or $U - \theta$ is called the *accuracy* of the estimator. The longer the confidence interval, the more confident we are that the interval actually contains the true value of θ. On the other hand, the longer the interval, the less information we have about the true value of θ. In an ideal situation, we obtain a relatively short interval with high confidence.

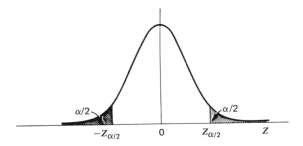

Fig. 9-1.   The distribution of Z.

9-2.1   Confidence Interval on the Mean, Variance Known

Let X be a random variable with unknown mean μ and known variance σ^2, and suppose that a random sample of size n, $X_1, X_2, \ldots, X_n$ is taken. A $100(1 - \alpha)$ percent confidence interval on μ can be obtained by considering the sampling distribution of the sample mean $\bar{X}$. In Section 8-3, we noted that the sampling distribution of $\bar{X}$ is normal if X is normal and approximately normal if the conditions of the central limit theorem are met. The mean of $\bar{X}$ is μ and the variance is σ^2/n. Therefore, the distribution of the statistic

$$Z = \frac{\bar{X} - \mu}{\sigma/\sqrt{n}}$$

is taken to be a standard normal distribution.

The distribution of $Z = (\bar{X} - \mu)/(\sigma/\sqrt{n})$ is shown in Fig. 9-1. From examination of this figure we see that

$$P\{-Z_{\alpha/2} \leq Z \leq Z_{\alpha/2}\} = 1 - \alpha$$

or

$$P\left\{-Z_{\alpha/2} \leq \frac{\bar{X} - \mu}{\sigma/\sqrt{n}} \leq Z_{\alpha/2}\right\} = 1 - \alpha$$

This can be rearranged as

$$P\{\bar{X} - Z_{\alpha/2}\sigma/\sqrt{n} \leq \mu \leq \bar{X} + Z_{\alpha/2}\sigma/\sqrt{n}\} = 1 - \alpha \qquad (9\text{-}16)$$

Comparing Equations (9-16) and (9-10), we see that the $100(1 - \alpha)$ percent two-sided confidence interval on μ is

$$\bar{X} - Z_{\alpha/2}\sigma/\sqrt{n} \leq \mu \leq \bar{X} + Z_{\alpha/2}\sigma/\sqrt{n} \qquad (9\text{-}17)$$

● **Example 9-9.** A quality-control technician is investigating the internal pressure strength of a 1-liter, glass soft drink bottle. Pressure strength is approximately normally distributed with known standard deviation $\sigma = 30$ psi. A random sample of 25 bottles had a mean pressure strength of $\bar{x} = 278$ psi. A 95 percent two-sided confidence interval for μ is

$$\bar{x} - Z_{.025}\sigma/\sqrt{n} \le \mu \le \bar{x} + Z_{.025}\sigma/\sqrt{n}$$

$$278 - 1.96(30)/5 \le \mu \le 278 + 1.96(30)/5$$

or

$$266.24 \le \mu \le 289.76$$

The accuracy of the confidence interval in Equation (9-17) is $Z_{\alpha/2}\,\sigma/\sqrt{n}$. This means that in using $\bar{x}$ to estimate μ, the error $E = |\bar{x} - \mu|$ is less than $Z_{\alpha/2}\,\sigma/\sqrt{n}$ with confidence $100(1 - \alpha)$. In situations where the sample size can be controlled we can choose n to be $100(1 - \alpha)$ percent confident that the error in estimating μ is less than a specified error E. The appropriate sample size is

$$n = \left(\frac{Z_{\alpha/2}\sigma}{E}\right)^2 \tag{9-18}$$

Thus, in Example 9-9, if we wish to be 95 percent confident that the error in estimating the mean pressure strength is less than 10 psi, a sample of size

$$n = \left[\frac{(1.96)30}{10}\right]^2 = 34.57 \simeq 35$$

is required.

It is also possible to obtain one-sided confidence intervals for μ by setting either $L = -\infty$ or $U = \infty$ and replacing $Z_{\alpha/2}$ by Z_α. The $100(1 - \alpha)$ percent upper-confidence interval for μ is

$$\mu \le \bar{X} + Z_\alpha \sigma/\sqrt{n} \tag{9-19}$$

and the $100(1 - \alpha)$ percent lower-confidence interval for μ is

$$\bar{X} - Z_\alpha \sigma/\sqrt{n} \le \mu \tag{9-20}$$

9-2.2 Confidence Interval on the Difference in Two Means, Variances Known

Consider two independent random variables X_1 with unknown mean μ_1 and known variance σ_1^2 and X_2 with unknown mean μ_2 and variance σ_2^2. We wish to find a $100(1 - \alpha)$ percent confidence interval on the difference in means $\mu_1 - \mu_2$. Let $X_{11}, X_{12}, \ldots, X_{1n_1}$ be a random sample of n_1 observations from X_1; and $X_{21}, X_{22}, \ldots, X_{2n_2}$ be a random sample of n_2 observations from X_2. If $\bar{X}_1$ and $\bar{X}_2$ are the sample means, the statistic

$$Z = \frac{\bar{X}_1 - \bar{X}_2 - (\mu_1 - \mu_2)}{\sqrt{\dfrac{\sigma_1^2}{n_1} + \dfrac{\sigma_2^2}{n_2}}}$$

is standard normal if X_1 and X_2 are normal or approximately standard normal if the conditions of the central limit theorem apply respectively. From Fig. 9-1, this implies that

$$P\{-Z_{\alpha/2} \le Z \le Z_{\alpha/2}\} = 1 - \alpha$$

or

$$P\left\{-Z_{\alpha/2} \le \frac{\bar{X}_1 - \bar{X}_2 - (\mu_1 - \mu_2)}{\sqrt{\dfrac{\sigma_1^2}{n_1} + \dfrac{\sigma_2^2}{n_2}}} \le Z_{\alpha/2}\right\} = 1 - \alpha$$

This can be rearranged as

$$P\left\{\bar{X}_1 - \bar{X}_2 - Z_{\alpha/2}\sqrt{\frac{\sigma_1^2}{n_1} + \frac{\sigma_2^2}{n_2}} \le \mu_1 - \mu_2 \le \bar{X}_1 - \bar{X}_2 + Z_{\alpha/2}\sqrt{\frac{\sigma_1^2}{n_1} + \frac{\sigma_2^2}{n_2}}\right\} = 1 - \alpha$$

$$(9\text{-}21)$$

Comparing Equations (9-21) and (9-10), we note that the $100(1 - \alpha)$ percent confidence interval for $\mu_1 - \mu_2$ is

$$\bar{X}_1 - \bar{X}_2 - Z_{\alpha/2}\sqrt{\frac{\sigma_1^2}{n_1} + \frac{\sigma_2^2}{n_2}} \le \mu_1 - \mu_2 \le \bar{X}_1 - \bar{X}_2 + Z_{\alpha/2}\sqrt{\frac{\sigma_1^2}{n_1} + \frac{\sigma_2^2}{n_2}} \qquad (9\text{-}22)$$

One-sided confidence intervals on $\mu_1 - \mu_2$ may also be obtained. A 100 $(1 - \alpha)$ percent upper-confidence interval on $\mu_1 - \mu_2$ is

$$\mu_1 - \mu_2 \le \bar{X}_1 - \bar{X}_2 + Z_\alpha\sqrt{\frac{\sigma_1^2}{n_1}\ \frac{\sigma_2^2}{n_2}} \qquad (9\text{-}23)$$

and a $100(1 - \alpha)$ percent lower-confidence interval is

$$\bar{X}_1 - \bar{X}_2 - Z_\alpha\sqrt{\frac{\sigma_1^2}{n_1} + \frac{\sigma_2^2}{n_2}} \le \mu_1 - \mu_2 \qquad (9\text{-}24)$$

● **Example 9-10.** An ammunition manufacturer is investigating the muzzle velocities of two different types of rifle ammunition. The manufacturer assumes that muzzle velocity is approximately normally distributed and that the standard deviations of the muzzle velocity for ammunition types 1 and 2 are $\sigma_1 = 1.10$ m/s and $\sigma_2 = 1.50$ m/s, respectively. Random samples of 10 type 1 shells and 20 type 2 shells are selected, and these shells are fired in random order from a test rifle. The observed sample means are $\bar{x}_1 = 500$ m/s and $\bar{x}_2 = 494$ m/s, respectively. A 90 percent two-sided confidence interval on the difference in mean muzzle velocities $\mu_1 - \mu_2$ is obtained from Equation (9-22) as

$$\bar{x}_1 - \bar{x}_2 - Z_{\alpha/2}\sqrt{\frac{\sigma_1^2}{n_1} + \frac{\sigma_2^2}{n_2}} \le \mu_1 - \mu_2 \le \bar{x}_1 - \bar{x}_2 + Z_{\alpha/2}\sqrt{\frac{\sigma_1^2}{n_1} + \frac{\sigma_2^2}{n_2}}$$

Then

$$500 - 494 - 1.645 \sqrt{\frac{1.21}{10} + \frac{2.25}{20}} \leq \mu_1 - \mu_2 \leq 500 - 494 + 1.645 \sqrt{\frac{1.21}{10} + \frac{2.25}{20}}$$

or

$$5.21 \leq \mu_1 - \mu_2 \leq 6.79$$

Therefore, we are 90 percent confident that the mean muzzle velocity of ammunition type 1 exceeds that of ammunition type 2 by between 5.21 and 6.79 m/s.

9-2.3 Confidence Interval on the Mean of a Normal Distribution, Variance Unknown

Suppose that X is a normally distributed random variable with unknown mean and variance μ and σ^2, respectively. We wish to find a $100(1 - \alpha)$ percent confidence interval on μ. A random sample of size n, $X_1, X_2, \ldots, X_n$, is available, and $\bar{X}$ and S^2 denote the sample mean and sample variance, respectively. In Section 8-3, we noted that the sampling distribution of the statistic

$$t = \frac{\bar{X} - \mu}{S/\sqrt{n}}$$

is the t distribution with $n - 1$ degrees of freedom. The assumption of normality for X is particularly important for small samples.

The distribution of $t = (\bar{X} - \mu)/(S/\sqrt{n})$ is shown in Fig. 9-2. Letting $t_{\alpha/2, n-1}$ be the upper $\alpha/2$ percentage point of the t distribution with $n - 1$ degrees of freedom, we observe from Fig. 9-2 that

$$P\{-t_{\alpha/2, n-1} \leq t \leq t_{\alpha/2, n-1}\} = 1 - \alpha$$

or

$$P\left\{-t_{\alpha/2, n-1} \leq \frac{\bar{X} - \mu}{S/\sqrt{n}} \leq t_{\alpha/2, n-1}\right\} = 1 - \alpha$$

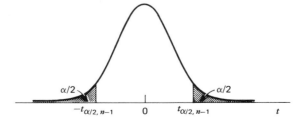

Fig. 9-2. The t distribution.

Rearranging this last equation yields

$$P\{\bar{X} - t_{\alpha/2, n-1}S/\sqrt{n} \leq \mu \leq \bar{X} + t_{\alpha/2, n-1}S/\sqrt{n}\} = 1 - \alpha \qquad (9\text{-}25)$$

Comparing Equations (9-25) and (9-10), we see that a $100(1 - \alpha)$ percent two-sided confidence interval on μ is

$$\bar{X} - t_{\alpha/2, n-1}S/\sqrt{n} \leq \mu \leq \bar{X} + t_{\alpha/2, n-1}S/\sqrt{n} \qquad (9\text{-}26)$$

A $100(1 - \alpha)$ percent lower-confidence interval on μ is given by

$$\bar{X} - t_{\alpha, n-1}S/\sqrt{n} \leq \mu \qquad (9\text{-}27)$$

and a $100(1 - \alpha)$ percent upper-confidence interval on μ is

$$\mu \leq \bar{X} + t_{\alpha, n-1}S/\sqrt{n} \qquad (9\text{-}28)$$

● **Example 9-11.** An engineer is investigating the wear characteristics of a particular type of radial automobile tire used by the company fleet. A random sample of 16 tires is selected, and each tire is run until the wear bars appear. The number of kilometers the tire has been driven at this point is recorded; these data are shown in Table 9-1. We assume that this "useful life" is a normally distributed random variable.
The sample mean and sample variance of these data are

$$\bar{x} = \frac{\sum\limits_{i=1}^{n} x_i}{n} = \frac{657,870}{16} = 41,116.88$$

and

$$\begin{aligned}
s^2 &= \frac{\sum\limits_{i=1}^{n} x_i^2 - \left(\sum\limits_{i=1}^{n} x_i\right)^2}{n-1} \\
&= \frac{27,076,768,350 - \dfrac{(657,870)^2}{16}}{15} \\
&= 1,813,977.99
\end{aligned}$$

TABLE 9-1 **Tire Wear Data, Example 9-11**

Tire Number	Kilometers	Tire Number	Kilometers
1	41,250	9	38,970
2	40,187	10	40,200
3	43,175	11	42,550
4	41,010	12	41,095
5	39,265	13	40,680
6	41,872	14	43,500
7	42,654	15	39,775
8	41,287	16	40,400

The sample standard deviation is $s = 1346.84$ kilometers. A 95 percent lower one-sided confidence interval on the mean wear is computed from Equation (9-27) as

$$\bar{x} - t_{.05, 15} s / \sqrt{n} \leq \mu$$

$$41{,}116.88 - (1.753)\, 1346.84 / \sqrt{16} \leq \mu$$

or

$$40{,}526.63 \leq \mu$$

9-2.4 Confidence Interval on the Difference in Means of Two Normal Distributions, Variances Unknown

We now consider two independent normal random variables, say X_1 with mean μ_1 and variance σ_1^2, and X_2 with mean μ_2 and variance σ_2^2. Both the means μ_1 and μ_2 and the variances σ_1^2 and σ_2^2 are unknown. However, it is reasonable to assume that both variances are equal; that is, $\sigma_1^2 = \sigma_2^2 = \sigma^2$. We wish to find a $100(1 - \alpha)$ percent confidence interval on the difference in means $\mu_1 - \mu_2$.

Random samples of size n_1 and n_2 are taken on X_1 and X_2, respectively; let the sample means be denoted by $\bar{X}_1$ and $\bar{X}_2$ and the sample variances be denoted by S_1^2 and S_2^2. Since both S_1^2 and S_2^2 are estimates of the common variance σ^2, we may obtain a combined (or "pooled") estimator of σ^2 as

$$S_p^2 = \frac{(n_1 - 1)S_1^2 + (n_2 - 1)S_2^2}{n_1 + n_2 - 2} \tag{9-29}$$

To develop the confidence interval for $\mu_1 - \mu_2$, note that the distribution of the statistic

$$t = \frac{\bar{X}_1 - \bar{X}_2 - (\mu_1 - \mu_2)}{S_p \sqrt{\dfrac{1}{n_1} + \dfrac{1}{n_2}}}$$

is the t distribution with $n_1 + n_2 - 2$ degrees of freedom. Therefore,

$$P\{-t_{\alpha/2,\, n_1+n_2-2} \leq t \leq t_{\alpha/2,\, n_1+n_2-2}\} = 1 - \alpha$$

or

$$P\left\{-t_{\alpha/2,\, n_1+n_2-2} \leq \frac{\bar{X}_1 - \bar{X}_2 - (\mu_1 - \mu_2)}{S_p \sqrt{\dfrac{1}{n_1} + \dfrac{1}{n_2}}} \leq t_{\alpha/2,\, n_1+n_2-2}\right\} = 1 - \alpha$$

This may be rearranged as

$$P\left\{\bar{X}_1 - \bar{X}_2 - t_{\alpha/2,\, n_1+n_2-2} S_p \sqrt{\frac{1}{n_1} + \frac{1}{n_2}} \leq \mu_1 - \mu_2 \leq \bar{X}_1 - \bar{X}_2 \right.$$
$$\left. + t_{\alpha/2,\, n_1+n_2-2} S_p \sqrt{\frac{1}{n_1} + \frac{1}{n_2}}\right\} = 1 - \alpha \tag{9-30}$$

Therefore, a $100(1-\alpha)$ percent two-sided confidence interval on the difference in means $\mu_1 - \mu_2$ is

$$\bar{X}_1 - \bar{X}_2 - t_{\alpha/2,\, n_1+n_2-2} S_p \sqrt{\frac{1}{n_1} + \frac{1}{n_2}} \le \mu_1 - \mu_2 \le \bar{X}_1 - \bar{X}_2 + t_{\alpha/2,\, n_1+n_2-2} S_p \sqrt{\frac{1}{n_1} + \frac{1}{n_2}}$$

$$(9\text{-}31)$$

A one-sided $100(1-\alpha)$ percent lower-confidence interval on $\mu_1 - \mu_2$ is

$$\bar{X}_1 - \bar{X}_2 - t_{\alpha,\, n_1+n_2-2} S_p \sqrt{\frac{1}{n_1} + \frac{1}{n_2}} \le \mu_1 - \mu_2 \qquad (9\text{-}32)$$

and a one-sided $100(1-\alpha)$ percent upper-confidence interval on $\mu_1 - \mu_2$ is

$$\mu_1 - \mu_2 \le \bar{X}_1 - \bar{X}_2 + t_{\alpha,\, n_1+n_2-2} S_p \sqrt{\frac{1}{n_1} + \frac{1}{n_2}} \qquad (9\text{-}33)$$

● **Example 9-12.** A chemical process is being studied in an effort to improve the yield. At present catalyst 1 is being used, but a new catalyst 2 is acceptable. An experiment is run in the pilot plant, using $n_1 = 8$ trials for catalyst 1 and $n_2 = 8$ trials for catalyst 2. The observed sample means and variances are $\bar{x}_1 = 91.73$, $s_1^2 = 3.89$, $\bar{x}_2 = 93.75$ and $s_2^2 = 4.02$. To construct a 95 percent confidence interval on $\mu_1 - \mu_2$, we first calculate the pooled estimate of the variance. From Equation (9-29) we have

$$s_p^2 = \frac{(n_1-1)s_1^2 + (n_2-1)s_2^2}{n_1 + n_2 - 2} = \frac{(7)3.89 + (7)4.02}{8+8-2} = 3.96$$

Therefore, the 95 percent confidence interval on $\mu_1 - \mu_2$ is

$$\bar{x}_1 - \bar{x}_2 - t_{.025,\, 14} S_p \sqrt{\frac{1}{n_1} + \frac{1}{n_2}} \le \mu_1 - \mu_2 \le \bar{x}_1 - \bar{x}_2 + t_{.025,\, 14} S_p \sqrt{\frac{1}{n_1} + \frac{1}{n_2}}$$

or

$$91.73 - 93.75 - (2.145)1.99\sqrt{\frac{1}{8} + \frac{1}{8}} \le \mu_1 - \mu_2 \le 91.73 - 93.75 + (2.145)1.99\sqrt{\frac{1}{8} + \frac{1}{8}}$$

which results in

$$-4.15 \le \mu_1 - \mu_2 \le .11$$

Note that this confidence interval includes zero. This indicates that there is no difference in the mean yields when either catalyst 1 or 2 is used. Therefore, there is no reason to switch the catalyst on the basis of yield.

In many situations it is not reasonable to assume that $\sigma_1^2 = \sigma_2^2$. When this assumption is unwarranted, one may still find a $100(1-\alpha)$ percent confidence interval on $\mu_1 - \mu_2$ using the fact that the statistic

$$t^* = \frac{\bar{X}_1 - \bar{X}_2 - (\mu_1 - \mu_2)}{\sqrt{S_1^2/n_1 + S_2^2/n_2}}$$

is distributed approximately as t with degrees of freedom given by

$$\nu = \frac{(S_1^2/n_1 + S_2^2/n_2)^2}{\dfrac{(S_1^2/n_1)^2}{n_1 + 1} + \dfrac{(S_2^2/n_2)^2}{n_2 + 1}} - 2 \tag{9-34}$$

Consequently, an approximate $100(1 - \alpha)$ percent two-sided confidence interval on $\mu_1 - \mu_2$, when $\sigma_1^2 \neq \sigma_2^2$, is

$$\bar{X}_1 - \bar{X}_2 - t_{\alpha/2, \nu}\sqrt{\frac{S_1^2}{n_1} + \frac{S_2^2}{n_2}} \leq \mu_1 - \mu_2 \leq \bar{X}_1 - \bar{X}_2 + t_{\alpha/2, \nu}\sqrt{\frac{S_1^2}{n_1} + \frac{S_2^2}{n_2}} \tag{9-35}$$

Upper (lower) one-sided confidence limits may be found by replacing the lower (upper) confidence limit with $-\infty$ (∞) and changing $\alpha/2$ to α.

9-2.5 Confidence Interval on the Variance of a Normal Distribution

Suppose that X is normally distributed with unknown mean μ and unknown variance σ^2. Let $X_1, X_2, \ldots, X_n$ be a random sample of size n, and let S^2 be the sample variance. It was shown in Section 8-4 that the sampling distribution of

$$\chi^2 = \frac{(n - 1)S^2}{\sigma^2}$$

is chi square with $n - 1$ degrees of freedom. This distribution is shown in Fig. 9-3.
   To develop the confidence interval, we note from Fig. 9-3 that

$$P\{\chi_{1-\alpha/2, n-1}^2 \leq \chi^2 \leq \chi_{\alpha/2, n-1}^2\} = 1 - \alpha$$

or

$$P\left\{\chi_{1-\alpha/2, n-1}^2 \leq \frac{(n - 1)S^2}{\sigma^2} \leq \chi_{\alpha/2, n-1}^2\right\} = 1 - \alpha.$$

This last equation can be rearranged to yield

$$P\left\{\frac{(n - 1)S^2}{\chi_{\alpha/2, n-1}^2} \leq \sigma^2 \leq \frac{(n - 1)S^2}{\chi_{1-\alpha/2, n-1}^2}\right\} = 1 - \alpha \tag{9-36}$$

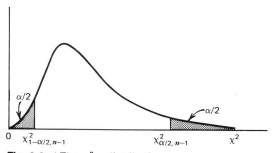

Fig. 9-3. `The χ_{n-1}^2 distribution.

Comparing Equations (9-36) and (9-10), we see that a $100(1 - \alpha)$ percent two-sided confidence interval on σ^2 is

$$\frac{(n - 1)S^2}{\chi^2_{\alpha/2, n-1}} \leq \sigma^2 \leq \frac{(n - 1)S^2}{\chi^2_{1-\alpha/2, n-1}} \tag{9-37}$$

To find a $100(1 - \alpha)$ percent lower-confidence interval on σ^2, set $U = \infty$ and replace $\chi^2_{\alpha/2, n-1}$ with $\chi^2_{\alpha, n-1}$, giving

$$\frac{(n - 1)S^2}{\chi^2_{\alpha, n-1}} \leq \sigma^2 \tag{9-38}$$

The $100(1 - \alpha)$ percent upper-confidence interval is found by setting $L = 0$ and replacing $\chi^2_{1-\alpha/2, n-1}$ with $\chi^2_{1-\alpha, n-1}$, resulting in

$$\sigma^2 \leq \frac{(n - 1)S^2}{\chi^2_{1-\alpha, n-1}} \tag{9-39}$$

● **Example 9-13.** A machine fills cans with a soft drink beverage, and the manufacturer is interested in obtaining a confidence interval estimate of the variance of its fill volume (if the variance is too large, a high percentage of the cans will be filled with a volume of product that is less than the nominal amount). A random sample of 20 cans yields a sample variance of $s^2 = .0225$ (fluid ounces)^2. A 95 percent upper-confidence interval on σ^2 is found from Equation (9-39) as

$$\sigma^2 \leq \frac{(n - 1)s^2}{\chi^2_{95, 19}}$$

or

$$\sigma^2 \leq \frac{(19).0225}{10.117} = .0423$$

9-2.6 Confidence Interval on the Ratio of Variances of Two Normal Distributions

Suppose that X_1 and X_2 are independent normal random variables with unknown means μ_1 and μ_2 and unknown variances σ_1^2 and σ_2^2, respectively. We wish to find a $100(1 - \alpha)$ percent confidence interval on the ratio σ_1^2/σ_2^2. Let two random samples of sizes n_1 and n_2 be taken on X_1 and X_2, and let S_1^2 and S_2^2 denote the sample variances. To find the confidence interval, we note that the sampling distribution of

$$F = \frac{S_2^2/\sigma_2^2}{S_1^2/\sigma_1^2}$$

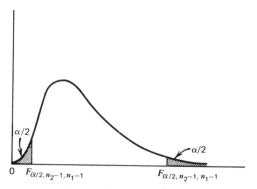

Fig. 9-4. The distribution of F_{n_2-1, n_1-1}.

is F with $n_2 - 1$ and $n_1 - 1$ degrees of freedom. This distribution is shown in Fig. 9-4.

From Fig. 9-4, we note that

$$P\{F_{1-\alpha/2, n_2-1, n_1-1} \leq F \leq F_{\alpha/2, n_2-1, n_1-1}\} = 1 - \alpha$$

or

$$P\left\{F_{1-\alpha/2, n_2-1, n_1-1} \leq \frac{S_2^2/\sigma_2^2}{S_1^2/\sigma_1^2} \leq F_{\alpha/2, n_2-1, n_1-1}\right\} = 1 - \alpha$$

Hence

$$P\left\{\frac{S_1^2}{S_2^2} F_{1-\alpha/2, n_2-1, n_1-1} \leq \frac{\sigma_1^2}{\sigma_2^2} \leq \frac{S_1^2}{S_2^2} F_{\alpha/2, n_2-1, n_1-1}\right\} = 1 - \alpha \qquad (9\text{-}40)$$

Comparing Equations (9-40) and (9-10), we see that a $100(1 - \alpha)$ percent two-sided confidence interval on σ_1^2/σ_2^2 is

$$\frac{S_1^2}{S_2^2} F_{1-\alpha/2, n_2-1, n_1-1} \leq \frac{\sigma_1^2}{\sigma_2^2} \leq \frac{S_1^2}{S_2^2} F_{\alpha/2, n_2-1, n_1-1} \qquad (9\text{-}41)$$

where the lower $1 - \alpha/2$ tail point of the F_{n_2-1, n_1-1} distribution is given by

$$F_{1-\alpha/2, n_2-1, n_1-1} = \frac{1}{F_{\alpha/2, n_1-1, n_2-1}} \qquad (9\text{-}42)$$

We may also construct one-sided confidence intervals. A $100(1 - \alpha)$ percent lower-confidence limit on σ_1^2/σ_2^2 is

$$\frac{S_1^2}{S_2^2} F_{1-\alpha, n_2-1, n_1-1} \leq \frac{\sigma_1^2}{\sigma_2^2} \qquad (9\text{-}43)$$

while a $100(1 - \alpha)$ percent upper-confidence interval on σ_1^2/σ_2^2 is

$$\frac{\sigma_1^2}{\sigma_2^2} \leq \frac{S_1^2}{S_2^2} F_{\alpha, n_2-1, n_1-1}$$

● **Example 9-14.** Consider the process yield data in Example 9-11. We will construct a 90 percent confidence interval on σ_1^2/σ_2^2. Since $n_1 = n_2 = 8$, $s_1^2 = 3.89$, $s_2^2 = 4.02$, $F_{.05,7,7} = 3.79$, and $F_{.95,7,7} = (F_{.05,7,7})^{-1} = (3.79)^{-1} = .264$, we may substitute into Equation (9-41), obtaining

$$\frac{3.89}{4.02}(.264) \leq \frac{\sigma_1^2}{\sigma_2^2} \leq \frac{3.89}{4.02}(3.79)$$

which simplifies to

$$.255 \leq \frac{\sigma_1^2}{\sigma_2^2} \leq 3.67$$

9-2.7 Confidence Interval on a Proportion

It is occasionally necessary to construct a $100(1-\alpha)$ percent confidence interval on a proportion. For example, suppose that a random sample of size n has been taken from an infinite population, and X $(\leq n)$ observations in this sample belongs to a class of interest. Then $\hat{p} = X/n$ is the point estimator of the proportion of the population that belongs to this class. Note that n and p are the parameters of a binomial distribution. Furthermore, in Section 7-5 we saw that the sampling distribution of $\hat{p}$ is approximately normal with mean p and variance $p(1-p)/n$, if p is not too close to either 0 or 1, and if n is relatively large. Thus, the distribution of

$$Z = \frac{\hat{p} - p}{\sqrt{\dfrac{p(1-p)}{n}}}$$

is approximately standard normal.

To construct the confidence interval, note that

$$P\{-Z_{\alpha/2} \leq Z \leq Z_{\alpha/2}\} \simeq 1 - \alpha$$

or

$$P\left\{ -Z_{\alpha/2} \leq \frac{\hat{p} - p}{\sqrt{\dfrac{p(1-p)}{n}}} \leq Z_{\alpha/2} \right\} \simeq 1 - \alpha$$

This may be rearranged as

$$P\left\{ \hat{p} - Z_{\alpha/2}\sqrt{\frac{p(1-p)}{n}} \leq p \leq \hat{p} + Z_{\alpha/2}\sqrt{\frac{p(1-p)}{n}} \right\} \simeq 1 - \alpha \qquad (9\text{-}45)$$

Unfortunately, the upper- and lower-confidence limits derived from Equation (9-45) contain the unknown parameter p. If n is large, a satisfactory solution is obtained by replacing p by $\hat{p}$ in the upper and lower limits in Equation

(9-45). Therefore,

$$P\left\{\hat{p} - Z_{\alpha/2}\sqrt{\frac{\hat{p}(1-\hat{p})}{n}} \le p \le \hat{p} + Z_{\alpha/2}\sqrt{\frac{\hat{p}(1-\hat{p})}{n}}\right\} \approx 1 - \alpha \qquad (9\text{-}46)$$

and the approximate $100(1-\alpha)$ percent two-sided confidence interval on p is

$$\hat{p} - Z_{\alpha/2}\sqrt{\frac{\hat{p}(1-\hat{p})}{n}} \le p \le \hat{p} + Z_{\alpha/2}\sqrt{\frac{\hat{p}(1-\hat{p})}{n}} \qquad (9\text{-}47)$$

An approximate $100(1-\alpha)$ percent lower-confidence interval is

$$\hat{p} - Z_{\alpha}\sqrt{\frac{\hat{p}(1-\hat{p})}{n}} \le p \qquad (9\text{-}48)$$

and an approximate $100(1-\alpha)$ percent upper-confidence interval is

$$p \le \hat{p} + Z_{\alpha}\sqrt{\frac{\hat{p}(1-\hat{p})}{n}} \qquad (9\text{-}49)$$

● **Example 9-15.** In a random sample of 75 axle shafts, 12 have a surface finish that is rougher than the specifications will allow. Therefore, a point estimate of the proportion of shafts in the population that exceed the roughness specifications p is $\hat{p} = x/n = 12/75 = .16$. A 95 percent two-sided confidence interval for p is computed from Equation (9-47) as

$$\hat{p} - Z_{.025}\sqrt{\frac{\hat{p}(1-\hat{p})}{n}} \le p \le \hat{p} + Z_{.025}\sqrt{\frac{\hat{p}(1-\hat{p})}{n}}$$

or

$$.16 - 1.96\sqrt{\frac{.16(.84)}{75}} \le p \le .16 + 1.96\sqrt{\frac{.16(.84)}{75}}$$

which simplifies to

$$.08 \le p \le .24$$

Define the error in estimating p by $\hat{p}$ as $E = |p - \hat{p}|$. Note that we are approximately $100(1-\alpha)$ percent confident that this error is less than $Z_{\alpha/2}\sqrt{p(1-p)/n}$. Therefore, in situations where the sample size can be selected, we may choose n to be $100(1-\alpha)$ percent confident that the error is less than some specified value E. The appropriate sample size is

$$n = \left(\frac{Z_{\alpha/2}}{E}\right)^2 p(1-p) \qquad (9\text{-}50)$$

This function is relatively flat from $p = .3$ to $p = .7$ An estimate of p is required to use Equation (9-50). If an estimate $\hat{p}$ from a previous sample is available, it could be substituted for p in Equation (9-50), or perhaps a subjective estimate could be made. If these alternatives are unsatisfactory, a preliminary sample could be taken, $\hat{p}$ computed, and then Equation (9-50) used to determine how many additional observations are required to estimate p with the desired accuracy. The sample size from Equation (9-50) will always be a maximum for $p = .5$ [that is, $p(1-p) = .25$], and this can be used to obtain an upper bound on n. In other words, we are *at least* $100(1-\alpha)$ percent confident that the error in estimating p by $\hat{p}$ is less than E if the sample size is

$$n = \left(\frac{Z_{\alpha/2}}{E}\right)^2 (.25)$$

● **Example 9-16.** Consider the data in Example 9-15. How large a sample is required if we want to be 95 percent confident that the error in using $\hat{p}$ to estimate p is less than .05? Using $\hat{p} = .16$ as an initial estimate of p, we find from Equation (9-50) that the required sample size is

$$\left(\frac{Z_{.025}}{E}\right)^2 \hat{p}(1-\hat{p}) = \left(\frac{1.96}{.05}\right)^2 .16(.84) = 207$$

We note that the procedures developed in this section depend on the normal approximation to the binomial. In situations where this approximation is inappropriate, particularly cases where n is small, other methods must be used. Tables of the binomial distribution could be used to obtain a confidence interval for p. If n is large but p is small, then the Poisson approximation to the binomial could be used to construct confidence intervals. These procedures are illustrated by Duncan (1974).

9-2.8   Confidence Interval on the Difference in Two Proportions

If there are two proportions of interest, say p_1 and p_2, it is possible to obtain a $100(1-\alpha)$ percent confidence interval on their difference $p_1 - p_2$. If two independent samples of size n_1 and n_2 are taken from infinite populations so that X_1 and X_2 are independent, binomial random variables with parameters (n_1, p_1) and (n_2, p_2), respectively, where X_1 represents the number of sample observations from the first population that belongs to a class of interest and X_2 represents the number of sample observations from the second population that belongs to a class of interest, then $\hat{p}_1 = X_1/n_1$ and $\hat{p}_2 = X_2/n_2$ are independent estimators of p_1 and p_2, respectively. Furthermore, under the assumption that the normal approximation to the binomial applies, the statistic

$$Z = \frac{\hat{p}_1 - \hat{p}_2 - (p_1 - p_2)}{\sqrt{\frac{p_1(1-p_1)}{n_1} + \frac{p_2(1-p_2)}{n_2}}}$$

is distributed approximately as standard normal. Using an approach analogous to that of the previous section, it follows that an approximate $100(1 - \alpha)$ percent two-sided confidence interval for $p_1 - p_2$ is

$$\hat{p}_1 - \hat{p}_2 - Z_{\alpha/2}\sqrt{\frac{\hat{p}_1(1 - \hat{p}_1)}{n_1} + \frac{\hat{p}_2(1 - \hat{p}_2)}{n_2}} \leq p_1 - p_2$$

$$\leq \hat{p}_1 - \hat{p}_2 + Z_{\alpha/2}\sqrt{\frac{\hat{p}_1(1 - \hat{p}_1)}{n_1} + \frac{\hat{p}_2(1 - \hat{p}_2)}{n_2}} \quad (9\text{-}51)$$

An approximate $100(1 - \alpha)$ percent lower-confidence interval for $p_1 - p_2$ is

$$\hat{p}_1 - \hat{p}_2 - Z_{\alpha}\sqrt{\frac{\hat{p}_1(1 - \hat{p}_1)}{n_1} + \frac{\hat{p}_2(1 - \hat{p}_2)}{n_2}} \leq p_1 - p_2 \quad (9\text{-}52)$$

and an approximate $100(1 - \alpha)$ percent upper-confidence interval for $p_1 - p_2$ is

$$p_1 - p_2 \leq \hat{p}_1 - \hat{p}_2 + Z_{\alpha/2}\sqrt{\frac{\hat{p}_1(1 - \hat{p}_1)}{n_1} + \frac{\hat{p}_2(1 - \hat{p}_2)}{n_2}} \quad (9\text{-}53)$$

● **Example 9-17.** Consider the data in Example 9-15. Suppose that a modification is made in the surface finishing process and subsequently a second random sample of 85 axle shafts is obtained. The number of defective shafts in this second sample is 10. Therefore, since $n_1 = 75$, $\hat{p}_1 = .16$, $n_2 = 85$, and $\hat{p}_2 = 10/85 = .12$, we can obtain an approximate 95 percent confidence interval on the difference in the proportion of defectives produced under the two processes from Equation (9-51) as

$$\hat{p}_1 - \hat{p}_2 - Z_{.025}\sqrt{\frac{\hat{p}_1(1 - \hat{p}_1)}{n_1} + \frac{\hat{p}_2(1 - \hat{p}_2)}{n_2}} \leq p_1 - p_2$$

$$\leq \hat{p}_1 - \hat{p}_2 + Z_{.025}\sqrt{\frac{\hat{p}_1(1 - \hat{p}_1)}{n_1} + \frac{\hat{p}_2(1 - \hat{p}_2)}{n_2}}$$

or

$$.16 - .12 - 1.96\sqrt{\frac{.16(.84)}{75} + \frac{.12(.88)}{85}} \leq p_1 - p_2$$

$$\leq .16 - .12 + 1.96\sqrt{\frac{.16(.84)}{75} + \frac{.12(.88)}{85}}$$

This simplifies to

$$-.07 \leq p_1 - p_2 \leq .15$$

This interval includes zero, so it seems unlikely that the changes made in the surface finish process have reduced the proportion of defective axle shafts being produced.

9-2.9 Approximate Confidence Intervals in Maximum Likelihood Estimation

If the method of maximum likelihood is used for parameter estimation, the asymptotic properties of these estimators may be used to obtain approximate confidence intervals. Let $\hat{\theta}$ be the maximum likelihood estimator of θ. In Section 9-1.2, we noted that for large samples $\hat{\theta}$ is approximately normally distributed with mean θ and variance $V(\hat{\theta})$ given by the Cramér-Rao lower bound [Equation (9-4)]. Therefore, an approximate $100(1 - \alpha)$ percent confidence interval for θ is

$$\hat{\theta} - Z_{\alpha/2}[V(\hat{\theta})]^{1/2} \leq \theta \leq \hat{\theta} + Z_{\alpha/2}[V(\hat{\theta})]^{1/2} \tag{9-54}$$

Usually, the $V(\hat{\theta})$ is a function of the unknown parameter θ. In these cases, replace θ by $\hat{\theta}$.

• **Example 9-18.** Recall Example 9-3, where it was shown that the maximum likelihood estimator of the parameter p of a Bernoulli distribution is $\hat{p} = (1/n) \sum_{i=1}^{n} X_i = \bar{X}$. Using the Cramér-Rao lower bound, we may verify that the lower bound for the variance of $\hat{p}$ is

$$V(\hat{p}) \geq \frac{1}{n E\left[\dfrac{d}{dp} \ell n \, p^X(1-p)^{1-X}\right]^2}$$

$$\geq \frac{1}{n E\left[\dfrac{X}{p} - \dfrac{(1-X)}{(1-p)}\right]^2}$$

$$\geq \frac{1}{n E\left[\dfrac{X^2}{p^2} + \dfrac{(1-X)^2}{(1-p)^2} - 2\dfrac{X(1-X)}{p(1-p)}\right]}$$

For the Bernoulli distribution, we observe that $E(X) = p$ and $E(X^2) = p$. Therefore, this last expression simplifies to

$$V(\hat{p}) = \frac{1}{n\left[\dfrac{1}{p} + \dfrac{1}{(1-p)}\right]} = \frac{p(1-p)}{n}$$

Therefore, replacing p in $V(\hat{p})$ by $\hat{p}$, the approximate $100(1 - \alpha)$ percent confidence interval for p is found from Equation (9-54) as

$$\hat{p} - Z_{\alpha/2}\sqrt{\frac{\hat{p}(1-\hat{p})}{n}} \leq p \leq \hat{p} + Z_{\alpha/2}\sqrt{\frac{\hat{p}(1-\hat{p})}{n}}$$

9-2.10 Simultaneous Confidence Intervals

Occasionally it is necessary to construct several confidence intervals on more than one parameter, and we wish the probability to be $(1 - \alpha)$ that *all* such confidence intervals simultaneously produce correct statements. For example, suppose that we are sampling from a normal population with unknown mean and variance, and we wish to construct confidence intervals for μ and σ^2 such that the probability is $(1 - \alpha)$ that both intervals simultaneously yield correct conclusions. Since $\bar{X}$ and S^2 are independent, we could ensure this result by constructing $100(1 - \alpha)^{1/2}$ percent confidence intervals for each parameter separately, and both intervals would simultaneously produce correct conclusions with probability $(1 - \alpha)^{1/2} (1 - \alpha)^{1/2} = (1 - \alpha)$.

If the sample statistics on which the confidence intervals are based are not independent random variables, then the confidence intervals are not independent, and other methods must be used. In general, suppose that m confidence intervals are required. The Bonferroni inequality states that

$$P\{\text{all } m \text{ statements are simultaneously correct}\} \equiv 1 - \alpha \geq 1 - \left(\sum_{i=1}^{m} \alpha_i \right) \qquad (9\text{-}55)$$

where $1 - \alpha_i$ is the confidence level used in the ith confidence interval. In practice, we select a value for the simultaneous confidence level $1 - \alpha$, and then choose the individual α_i such that $\sum_{i=1}^{m} \alpha_i = \alpha$. Usually, we set $\alpha_i = \alpha/m$.

As an illustration, suppose we wished to construct two confidence intervals on the means of two normal distributions such that we are at least 90 percent confident that both statements are simultaneously correct. Therefore, since $1 - \alpha = .90$, we have $\alpha = .10$, and since two confidence intervals are required, each of these should be constructed with $\alpha_i = \alpha/2 = .10/2 = .05$, $i = 1, 2$. That is, two individual 95 percent confidence intervals on μ_1 and μ_2 will *simultaneously* lead to correct statements with probability at least .90.

9-3   Summary

This chapter has introduced the point and interval estimation of unknown parameters. Two methods of obtaining point estimators were discussed: the method of maximum likelihood and the method of moments. The method of maximum likelihood usually leads to estimators that have good statistical properties. Confidence intervals were derived for a variety of parameter estimation problems. These intervals have a frequency interpretation. The two-sided confidence intervals developed in Section 9-2 are summarized in Table 9-2. In some instances, one-sided confidence intervals may be appropriate. These may be obtained by setting one confidence limit in the two-sided confidence interval equal to the lower (or upper) limit of a feasible region for the parameter, and using α instead of $\alpha/2$ as the probability level on the remaining upper (or lower) confidence limit. Approximate confidence intervals in maximum likelihood estimation and simultaneous confidence intervals were also briefly introduced.

TABLE 9-2 **Summary of Confidence Interval Procedures**

Problem Type	Point Estimator	Two-Sided $100(1-\alpha)$ Percent Confidence Interval
Mean μ, variance σ^2 known	$\bar{X}$	$\bar{X} - Z_{\alpha/2}\sigma/\sqrt{n} \leq \mu \leq \bar{X} + Z_{\alpha/2}\sigma/\sqrt{n}$
Difference in two means μ_1 and μ_2, variances σ_1^2 and σ_2^2 known	$\bar{X}_1 - \bar{X}_2$	$\bar{X}_1 - \bar{X}_2 - Z_{\alpha/2}\sqrt{\dfrac{\sigma_1^2}{n_1}+\dfrac{\sigma_2^2}{n_2}} \leq \mu_1 - \mu_2 \leq \bar{X}_1 - \bar{X}_2 + Z_{\alpha/2}\sqrt{\dfrac{\sigma_1^2}{n_1}+\dfrac{\sigma_2^2}{n_2}}$
Mean μ of a normal distribution, variance σ^2 unknown	$\bar{X}$	$\bar{X} - t_{\alpha/2,\,n-1}S/\sqrt{n} \leq \mu \leq \bar{X} + t_{\alpha/2,\,n-1}S/\sqrt{n}$
Difference in means of two normal distributions $\mu_1 - \mu_2$, variance $\sigma_1^2 = \sigma_2^2$ unknown	$\bar{X}_1 - \bar{X}_2$	$\bar{X}_1 - \bar{X}_2 - t_{\alpha/2,\,n_1+n_2-2}S_p\sqrt{\dfrac{1}{n_1}+\dfrac{1}{n_2}} \leq \mu_1 - \mu_2 \leq \bar{X}_1 - \bar{X}_2 + t_{\alpha/2,\,n_1+n_2-2}S_p\sqrt{\dfrac{1}{n_1}+\dfrac{1}{n_2}}$

Variance σ^2 of a normal distribution	S^2	$\dfrac{(n-1)S^2}{\chi^2_{\alpha/2,n-1}} \le \sigma^2 \le \dfrac{(n-1)S^2}{\chi^2_{1-\alpha/2,n-1}}$
Ratio of the variances σ_1^2/σ_2^2 of two normal distributions	$\dfrac{S_1^2}{S_2^2}$	$\dfrac{S_1^2}{S_2^2}F_{1-\alpha/2,n_2-1,n_1-1} \le \dfrac{\sigma_1^2}{\sigma_2^2} \le \dfrac{S_1^2}{S_2^2}F_{\alpha/2,n_2-1,n_1-1}$
Proportion or parameter of a binomial distribution p	$\hat{p}$	$\hat{p} - Z_{\alpha/2}\sqrt{\dfrac{\hat{p}(1-\hat{p})}{n}} \le p \le \hat{p} + Z_{\alpha/2}\sqrt{\dfrac{\hat{p}(1-\hat{p})}{n}}$
Difference in two proportions or two binomial parameters $p_1 - p_2$	$\hat{p}_1 - \hat{p}_2$	$\hat{p}_1 - \hat{p}_2 - Z_{\alpha/2}\sqrt{\dfrac{\hat{p}_1(1-\hat{p}_1)}{n_1} + \dfrac{\hat{p}_2(1-\hat{p}_2)}{n_2}} \le p_1 - p_2 \le \hat{p}_1 - \hat{p}_2 + Z_{\alpha/2}\sqrt{\dfrac{\hat{p}_1(1-\hat{p}_1)}{n_1} + \dfrac{\hat{p}_2(1-\hat{p}_2)}{n_2}}$

9-4 Exercises

9-1. Let X be a random variable with mean μ and variance σ^2. Suppose that $S^{*2} = \Sigma_{i=1}^n (X_i - \bar{X})^2/n$ is used as an estimator of σ^2. Show that this estimator is biased.

9-2. Suppose we have a random sample of size $2n$ from a population denoted by X, and $E(X) = \mu$ and $V(X) = \sigma^2$. Let

$$\bar{X}_1 = \frac{1}{2n} \sum_{i=1}^{2n} X_i \qquad \text{and} \qquad \bar{X}_2 = \frac{1}{n} \sum_{i=1}^n X_i$$

be two estimators of μ. Which is the better estimator of μ? Explain your choice.

9-3. Let $X_1, X_2, \ldots, X_7$ denote a random sample from a population having mean μ and variance σ^2. Consider the following estimators of μ:

$$\hat{\theta}_1 = \frac{X_1 + X_2 + \cdots + X_7}{7}$$

$$\hat{\theta}_2 = \frac{2X_1 - X_6 + X_4}{2}$$

Is either estimator unbiased? Which estimator is "best"? In what sense is it best?

9-4. Suppose that $\hat{\theta}_1$ and $\hat{\theta}_2$ are estimators of the parameter θ. We know that $E(\hat{\theta}_1) = \theta$, $E(\hat{\theta}_2) = \theta/2$, $V(\hat{\theta}_1) = 10$, and $V(\hat{\theta}_2) = 4$. Which estimator is "best"? In what sense is it best?

9-5. Suppose that $\hat{\theta}_1$, $\hat{\theta}_2$, and $\hat{\theta}_3$ are estimators of θ. We know that $E(\hat{\theta}_1) = E(\hat{\theta}_2) = \theta$, $E(\hat{\theta}_3) \neq \theta$, $V(\hat{\theta}_1) = 12$, $V(\hat{\theta}_2) = 10$, and $E(\hat{\theta}_3 - \theta)^2 = 6$. Compare these three estimators. Which do you prefer? Why?

9-6. Let three random samples of sizes $n_1 = 10$, $n_2 = 8$, and $n_3 = 6$ be taken from a population with mean μ and variance σ^2. Let S_1^2, S_2^2, and S_3^2 be the sample variances. Show that

$$S^2 = \frac{10S_1^2 + 8S_2^2 + 6S_3^2}{24}$$

is an unbiased estimator of σ^2.

9-7. **Best Linear Unbiased Estimators.** An estimator $\hat{\theta}$ is called a linear estimator if it is a linear combination of the observations in the sample. $\hat{\theta}$ is called a best linear unbiased estimator, if, of all linear functions of the observations, it is both unbiased and has minimum variance. Show that the sample mean $\bar{X}$ is the best linear unbiased estimator of the population mean μ.

9-8. Find the maximum likelihood estimator of the parameter α of the Poisson distribution, based on a random sample of size n.

9-9. Find the estimator of α in the Poisson distribution by the method of moments, based on a random sample of size n.

9-10. Find the maximum likelihood estimator of the parameter λ in the exponential distribution, based on a random sample of size n.

9-11. Find the estimator of λ in the exponential distribution by the method of moments, based on a random sample of size n.

9-12. Find moment estimators of the parameters r and λ of the gamma distribution, based on a random sample of size n.

9-13. Let X be a geometric random variable with parameter p. Find an estimator of p by the method of moments, based on a random sample of size n.

9-14. Let X be a geometric random variable with parameter p. Find the maximum likelihood estimator of p, based on a random sample of size n.

9-15. Let X be a Bernoulli random variable with parameter p. Find an estimator of p by the method of moments, based on a random sample of size n.

9-16. Let X be a binomial random variable with parameters n (known) and p. Find an estimator of p by the method of moments, based on a random sample of size N.

9-17. Let X be a binomial random variable with parameters n and p, both unknown. Find estimators of n and p by the method of moments, based on a random sample of size N.

9-18. Let X be a binomial random variable with parameters n (unknown) and p. Find the maximum likelihood estimator of p, based on a random sample of size N.

9-19. Set up the likelihood function for a random sample of size n from a Weibull distribution. What difficulties would be encountered in obtaining the maximum likelihood estimators of the three parameters of the Weibull distribution?

9-20. Prove that if $\hat{\theta}$ is an unbiased estimator of θ, and if $\lim_{n \to \infty} V(\hat{\theta}) = 0$, then $\hat{\theta}$ is a consistent estimator of θ.

9-21. Let X be a random variable with mean μ and variance σ^2. Given two random samples of size n_1 and n_2 with sample means $\bar{X}_1$ and $\bar{X}_2$, respectively, show that

$$\bar{X} = a\bar{X}_1 + (1 - a)\bar{X}_2 \qquad 0 < a < 1$$

is an unbiased estimator of μ. Assuming $\bar{X}_1$ and $\bar{X}_2$ to be independent, find the value of a that minimizes the variance of $\bar{X}$.

9-22. Suppose that the random variable X has the probability distribution

$$f(x) = (\gamma + 1)x^{\gamma} \qquad 0 < x < 1$$
$$= 0 \qquad \text{otherwise}$$

Let $X_1, X_2, \ldots, X_n$ be a random sample of size n. Find the maximum likelihood estimator of γ.

9-23. Let X have the truncated (on the left at x_ℓ) exponential distribution

$$f(x) = \lambda e^{-\lambda(x - x_\ell)}, \qquad x > x_\ell > 0$$
$$= 0 \qquad \text{otherwise}$$

Let $X_1, X_2, \ldots, X_n$ be a random sample of size n. Find the maximum likelihood estimator of λ.

9-24. Assume that λ in the previous exercise is known but x_ℓ is unknown. Obtain the maximum likelihood estimator of x_ℓ.

9-25. Let X be a random variable with mean μ and variance σ^2, and let $X_1, X_2, \ldots, X_n$ be a random sample of size n from X. Show that the estimator $G = K \sum_{i=1}^{n-1} (X_{i+1} - X_i)^2$ is unbiased for an appropriate choice for K. Find the appropriate value for K.

9-26. A manufacturer produces piston rings for an automobile engine. It is known that ring diameter is approximately normally distributed and has standard deviation $\sigma = .002$ mm. A random sample of 15 rings has a mean diameter of $\bar{x} = 74.036$ mm.

(a) Construct a 99 percent two-sided confidence interval on the mean piston ring diameter.

(b) Construct a 95 percent lower-confidence limit on the mean piston ring diameter.

9-27. The life in hours of a 75-watt light bulb is known to be approximately normally distributed, with standard deviation $\sigma = 25$ hours. A random sample of 20 bulbs has a mean life of $\bar{x} = 1014$ hours.

(a) Construct a 95 percent two-sided confidence interval on the mean life.

(b) Construct a 95 percent lower-confidence interval on the mean life.

9-28. A civil engineer is analyzing the compressive strength of concrete. Compressive strength is approximately normally distributed with variance $\sigma^2 = 1000$ (psi)^2. A random sample of 12 specimens has a mean compressive strength of $\bar{x} = 3250$ psi.

(a) Construct a 95 percent two-sided confidence interval on mean compressive strength.

(b) Construct a 99 percent two-sided confidence interval on mean compressive strength. Compare the width of this confidence interval with the width of the one found in part (a).

9-29. Suppose that in Exercise 9-27 we wanted to be 95 percent confident that the error in estimating the mean life is less than five hours. What sample size should be used?

9-30. Suppose that in Exercise 9-27 we wanted the total width of the confidence interval on mean life to be eight hours. What sample size should be used?

9-31. Suppose that in Exercise 9-28, it is desired to estimate the compressive strength with an error that is less than 15 psi. What sample size is required?

9-32. Two machines are used to fill plastic bottles with dishwashing detergent. The standard deviations of fill volume are known to be $\sigma_1 = .15$ fluid ounces and $\sigma_2 = .12$ fluid ounces for the two machines, respectively. Two random samples of $n_1 = 12$ bottles from machine 1 and $n_2 = 10$ bottles from machine 2 are selected, and the sample mean fill volumes are $\bar{x}_1 = 30.87$ fluid ounces and 30.68 fluid ounces.

(a) Construct a 90 percent two-sided confidence interval on the mean difference in fill volume.

(b) Construct a 95 percent two-sided confidence interval on the mean difference in fill volume. Compare the width of this interval to the width of the interval in part (a).

(c) Construct a 95 percent upper-confidence interval on the mean difference in fill volume.

9-33. The burning rates of two different solid-fuel rocket propellants are being studied. It is known that both propellants have approximately the same standard deviation of burning rate; that is, $\sigma_1 = \sigma_2 = .5$ cm/s. Two random samples of $n_1 = 20$ and $n_2 = 20$ specimens are tested, and the sample mean burning rates are $\bar{x}_1 = 18$ cm/s and $\bar{x}_2 = 24$ cm/s. Construct a 99 percent confidence interval on the mean difference in burning rate.

9-34. Two different formulations of a lead-free gasoline are being tested to study their road octane numbers. The variance of road octane number for formulation 1 is $\sigma_1^2 = 1.8$ and for formulation 2 it is $\sigma_2^2 = 1.2$. Two random samples of size $n_1 = 15$ and $n_2 = 20$ are tested, and the mean road octane numbers observed are $\bar{x}_1 = 89.6$ and $\bar{x}_2 = 92.5$. Construct a 95 percent two-sided confidence interval on the difference in mean road octane number.

9-35. The range of a new type of mortar shell is being investigated. The observed ranges, in meters, of 16 randomly selected shells are shown below. It is assumed that range is normally distributed.

2216	2237	2249	2204
2225	2301	2281	2263
2318	2255	2275	2295
2250	2238	2300	2217

(a) Construct a 95 percent two-sided confidence interval on the mean range.

(b) Construct a 95-percent lower-confidence interval on the mean range.

(c) Construct a 95 percent two-sided confidence interval on the mean range assuming that $\sigma = 36$. Compare this interval with the one from part (a).

9-36. A machine produces metal rods used in an automobile suspension system. A random sample of 15 rods is selected and the diameter is measured. The resulting data are shown below. Assume that rod diameter is normally distributed. Construct a 95 percent two-sided confidence interval on the mean rod diameter.

8.24 mm	8.23 mm	8.20 mm
8.21	8.20	8.28
8.23	8.26	8.24
8.25	8.19	8.25
8.26	8.23	8.24

9-37. The wall thickness of 25 glass 2-liter bottles was measured by a quality-control engineer. The sample mean was $\bar{x} = 4.02$ mm and the sample standard deviation was $s = .09$ mm. Find a 90 percent lower-confidence interval on the mean wall thickness.

9-38. An industrial engineer is interested in estimating the mean time required to assemble an electronic component. How large a sample is required if the engineer wishes to be 95 percent confident that the error in estimating the mean is less than .25 minutes? The standard deviation of assembly time is .45 minutes.

9-39. A random sample of size 15 from a normal population has mean $\bar{x} = 250$ and variance $s^2 = 49$. Find the following:
(a) A 95 percent two-sided confidence interval on μ.
(b) A 95 percent lower-confidence interval on μ.
(c) A 95 percent upper-confidence interval on μ.

9-40. A post-mix beverage machine is adjusted to release a certain amount of syrup into a chamber where it is mixed with carbonated water. A random sample of 20 beverages were found to have a mean syrup content of $\bar{x} = 1.15$ fluid ounces and a standard deviation of $s = .025$ fluid ounces. Find a 90 percent two-sided confidence interval on the mean quantity of syrup mixed with each drink.

9-41. Two independent random samples of sizes $n_1 = 15$ and $n_2 = 20$ are taken from two normal populations. The sample means are $\bar{x}_1 = 200$ and $\bar{x}_2 = 195$. We know that the variances are $\sigma_1^2 = 10$ and $\sigma_2^2 = 12$. Find the following:
(a) A 95 percent two-sided confidence interval on $\mu_1 - \mu_2$.
(b) A 95 percent lower-confidence interval on $\mu_1 - \mu_2$.
(c) A 95 percent upper-confidence interval on $\mu_1 - \mu_2$.

9-42. The output voltage from two different types of batteries is being investigated. Ten batteries of each type are selected at random and the voltage measured. The sample means are $\bar{x}_1 = 12.13$ volts and $\bar{x}_2 = 12.05$ volts. We know that the variances of output voltage for the two types of batteries are $\sigma_1^2 = .5$ and $\sigma_2^2 = .8$, respectively. Construct a 99 percent two-sided confidence interval on the difference in mean voltage.

9-43. Random samples of size 10 were drawn from two independent normal populations. The sample means and standard deviations were $\bar{x}_1 = 20.0$, $s_1 = 1.5$, $\bar{x}_2 = 21.5$, and $s_2 = 1.3$. Assuming that $\sigma_1^2 = \sigma_2^2$, find the following:
(a) A 95 percent two-sided confidence interval on $\mu_1 - \mu_2$.
(b) A 95 percent upper-confidence interval on $\mu_1 - \mu_2$.
(c) A 95 percent lower-confidence interval on $\mu_1 - \mu_2$.

9-44. The diameter of steel rods manufactured on two different extrusion machines is being investigated. Two random samples of sizes $n_1 = 12$ and $n_2 = 18$ are selected, and the sample means and sample variances are $\bar{x}_1 = 8.75$, $s_1^2 = .29$, $\bar{x}_2 = 8.63$, and $s_2^2 = .34$, respectively. Assuming that $\sigma_1^2 = \sigma_2^2$, construct a 95 percent two-sided confidence interval on the difference in mean rod diameter.

9-45. Random samples of sizes $n_1 = 15$ and $n_2 = 20$ are drawn from two independent normal populations. The sample means and variances are $\bar{x}_1 = 300$, $s_1^2 = 16$, $\bar{x}_2 = 315$, $s_2^2 = 49$. Assuming that $\sigma_1^2 \neq \sigma_2^2$, construct a 95 percent two-sided confidence interval on $\mu_1 - \mu_2$.

9-46. Consider the data in Exercise 9-35. Construct the following:
(a) A 95 percent two-sided confidence interval on σ^2.

(b) A 95 percent lower-confidence interval on σ^2.

(c) A 95 percent upper-confidence interval on σ^2.

9-47. Consider the data in Exercise 9-36. Construct the following:

(a) A 99 percent two-sided confidence interval on σ^2.

(b). A 99 percent lower-confidence interval on σ^2.

(c) A 99 percent upper-confidence interval on σ^2.

9-48. Construct a 95 percent two-sided confidence interval on the variance of the wall thickness data in Exercise 9-37.

9-49. In a random sample of 100 light bulbs, the sample standard deviation of bulb life was found to be 12.6 hours. Compute a 90 percent upper-confidence interval on the variance of bulb life.

9-50. Consider the data in Exercise 9-43. Construct a 95 percent two-sided confidence interval on the ratio of the population variances σ_1^2/σ_2^2.

9-51. Consider the data in Exercise 9-44. Construct the following:

(a) A 90 percent two-sided confidence interval on σ_1^2/σ_2^2.

(b) A 95 percent two-sided confidence interval on σ_1^2/σ_2^2. Compare the width of this interval with the width of the interval in part (a).

(c) A 90 percent lower-confidence interval on σ_1^2/σ_2^2.

(d) A 90 percent upper-confidence interval on σ_1^2/σ_2^2.

9-52. Construct a 95 percent two-sided confidence interval on the ratio of the variances σ_1^2/σ_2^2 using the data in Exercise 9-45.

9-53. Of 1000 randomly selected cases of lung cancer, 699 resulted in death. Construct a 95 percent two-sided confidence interval on the death rate from lung cancer.

9-54. How large a sample would be required in Exercise 9-53 to be 95 percent confident that the error in estimating the death rate from lung cancer is less than .03?

9-55. A manufacturer of electronic calculators is interested in estimating the fraction of defective units produced. A random sample of 8000 calculators contains 24 defectives. Compute a 99 percent upper-confidence interval on the fraction defective.

9-56. A study is to be conducted of the percentage of homeowners who own at least two television sets. How large a sample is required if we wish to be 99 percent confident that the error in estimating this quantity is less than .01?

9-57. A study is conducted to determine the effectiveness of Swine flu vaccine. A random sample of 2000 subjects were given the vaccine, and of this group 140 contracted the flu. A control group of 2500 randomly selected subjects were not vaccinated, and of this group 170 contracted the flu. Construct a 95 percent confidence interval on the difference in proportions $p_1 - p_2$.

9-58. The fraction of defective product produced by two production lines is being analyzed. A random sample of 1000 units from line 1 has 12 defectives, while a random sample of 1500 units from line 2 has 24 defectives. Find a 99 percent confidence interval on the difference in fraction defective produced by the two lines.

9-59. Consider the data in Exercise 9-37. Find confidence intervals on μ and σ^2 such that we are at least 90 percent confident that both intervals simultaneously lead to correct conclusions.

9-60. Consider the data in Exercise 9-43. Suppose that a random sample of size $n_3 = 15$ is obtained from a third normal population, with $\bar{x}_3 = 20.5$ and $s_3 = 1.2$. Find two-sided confidence intervals on $\mu_1 - \mu_2$, $\mu_1 - \mu_3$, and $\mu_2 - \mu_3$ such that the probability is at least .95 that all three intervals simultaneously lead to correct conclusions.

Chapter 10

Tests of Hypotheses

Many problems require that we decide whether or not a statement about some parameter is true or false. The statement is usually called a *hypothesis*, and the decision-making procedure about the truth or falsity of the hypothesis is called *hypothesis testing*. This is one of the most useful aspects of statistical inference, since many types of decision problems can be formulated as hypothesis-testing problems. This chapter will develop hypothesis-testing procedures for several important situations.

10-1 Introduction

10-1.1 Statistical Hypotheses

A statistical hypothesis is a statement about the probability distribution of a random variable. Statistical hypotheses often involve one or more parameters of this distribution. For example, suppose that we are interested in the mean compressive strength of a particular type of concrete. Specifically, we are interested in deciding whether or not the mean compressive strength (say μ) is 2500 psi. We may express this formally as

$$H_0: \mu = 2500 \text{ psi}$$
$$H_1: \mu \neq 2500 \text{ psi} \tag{10-1}$$

The statement $H_0: \mu = 2500$ psi in Equation (10-1) is called the *null hypothesis*, and the statement $H_1: \mu \neq 2500$ psi is called the *alternative hypothesis*. Since the alternative hypothesis specifies values of μ that could be either greater than 2500 psi or less than 2500 psi, it is called a *two-sided alternative hypothesis*. In some situations, we may wish to formulate a *one-sided alternative hypothesis*, as in

$$H_0: \mu = 2500 \text{ psi}$$
$$H_1: \mu > 2500 \text{ psi} \tag{10-2}$$

It is important to remember that hypotheses are always statements about the population or distribution under study, not statements about the sample. The value of the population parameter specified in the null hypothesis (2500 psi in the above example) is usually determined in one of three ways. First, it may result from past experience or knowledge of the process, or even from prior experimentation. The objective of hypothesis testing then is usually to determine whether the experimental situation has changed. Second, this value may be determined from some theory or model regarding the process under study. Here the objective of hypothesis testing is to verify the theory or model. A third situation arises when the value of the population parameter results from external considerations, such as design or engineering specifications, or from contractual obligations. In this situation, the objective of hypothesis testing is conformance testing.

We are interested in making a decision about the truth or falsity of a hypothesis. A procedure leading to such a decision is called a *test of a hypothesis*. Hypothesis-testing procedures rely on using the information in a random sample from the population of interest. If this information is consistent with the hypothesis, then we would conclude that the hypothesis is true; however, if this information is inconsistent with the hypothesis, we would conclude that the hypothesis is false.

To test a hypothesis, we must take a random sample, compute an appropriate test statistic from the sample data, and then use the information contained in this test statistic to make a decision. For example, in testing the null hypothesis concerning the mean compressive strength of concrete in Equation (10-1), suppose that a random sample of 10 concrete specimens is tested and the sample mean $\bar{x}$ is used as a test statistic. If $\bar{x} > 2550$ psi or if $\bar{x} < 2450$ psi, we will consider the mean compressive strength of this particular type of concrete to be different from 2500 psi. That is, we would *reject* the null hypothesis $H_0: \mu = 2500$. Rejecting H_0 implies that the alternative hypothesis H_1 is true. The set of all possible values of $\bar{x}$ that are either greater than 2550 psi or less than 2450 psi is called the *critical region* or *rejection region* for the test. Alternatively, if 2450 psi $\leq \bar{x} \leq 2550$ psi, then we would *accept* the null hypothesis $H_0: \mu = 2500$. Thus, the interval [2450 psi, 2550 psi] is called the *acceptance region* for the test. Note that the boundaries of the critical region, 2450 psi and 2550 psi (often called the *critical values* of the test statistic), have been determined somewhat arbitrarily. In subsequent sections we will show how to construct an appropriate test statistic to determine the critical region for several hypothesis-testing situations.

10-1.2 Type I and Type II Errors

The decision to accept or reject the null hypothesis is based on a test statistic computed from the data in a random sample. When a decision is made using

TABLE 10-1   **Decisions in Hypothesis Testing**

	H_0 is True	H_0 is False
Accept H_0	No error	Type II error
Reject H_0	Type I error	No error

the information in a random sample, this decision is subject to error. Two kinds of errors may be made when testing hypotheses. If the null hypothesis is rejected when it is true, then a type I error has been made. If the null hypothesis is accepted when it is false, then a type II error has been made. The situation is described in Table 10-1.

The probabilities of occurrence of type I and type II errors are given special symbols:

$$\alpha = P \text{ \{type I error\}} = P \text{ \{reject } H_0|H_0 \text{ is true\}} \qquad (10\text{-}3)$$

$$\beta = P \text{ \{type II error\}} = P \text{ \{accept } H_0|H_0 \text{ is false\}} \qquad (10\text{-}4)$$

Sometimes it is more convenient to work with the *power* of the test, where

$$\text{Power} = 1 - \beta = P \text{ \{reject } H_0|H_0 \text{ is false\}} \qquad (10\text{-}5)$$

Note that the power of the test is the probability that a false null hypothesis is correctly rejected. Because the results of a test of a hypothesis are subject to error, we cannot "prove" or "disprove" a statistical hypothesis. However, it is possible to design test procedures that control the error probabilities α and β to suitably small values.

The probability of type I error α is often called the *significance level* or *size* of the test. In the concrete-testing example, a type I error would occur if the sample mean $\bar{x} > 2550$ psi or if $\bar{x} < 2450$ psi when in fact the true mean compressive strength $\mu = 2500$ psi. Generally, the type I error probability is controlled by the location of the critical region. Thus, it is usually easy in practice for the analyst to set the type I error probability at (or near) any desired value. Since the probability of wrongly rejecting H_0 is directly controlled by the decision maker, rejection of H_0 is always a *strong conclusion*. Now suppose that the null hypothesis $H_0: \mu = 2500$ psi is false. That is, the true mean compressive strength μ is some value other than 2500 psi. The probability of type II error is not a constant but depends on the true mean compressive strength of the concrete. If μ denotes the true mean compressive strength, then $\beta(\mu)$ denotes the type II error probability corresponding to μ. The function $\beta(\mu)$ is evaluated by finding the probability that the test statistic (in this case $\bar{x}$) falls in the acceptance region given a

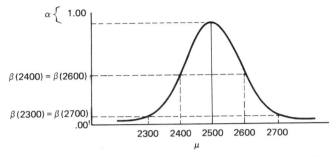

Fig. 10-1. Operating characteristic curve for the concrete testing example.

particular value of μ. We define the *operating characteristic curve* (or OC curve) of a test as the plot of $\beta(\mu)$ against μ. An example of an operating characteristic curve for the concrete-testing problem is shown in Fig. 10-1. From this curve, we see that the type II error probability depends on the extent to which $H_0: \mu = 2500$ psi is false. For example, note that $\beta(2700) < \beta(2600)$. Thus we can think of the type II error probability as a measure of the ability of the test procedure to detect a particular deviation from the null hypothesis H_0. Small deviations are harder to detect than large ones. We also observe that, since this is a two-sided alternative hypothesis, the operating characteristic curve is symmetric; that is, $\beta(2400) = \beta(2600)$. Furthermore, when $\mu = 2500$ the probability of type II error $\beta = 1 - \alpha$.

The probability of type II error is also a function of sample size, as illustrated in Fig. 10-2. From this figure, we see that for a given value of the type I error probability α and a given value of mean compressive strength the type II error probability decreases as the sample size n increases. That is, a specified deviation of the true mean from the value specified in the null

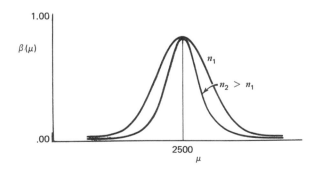

Fig. 10-2. The effect of sample size on the operating characteristic curve.

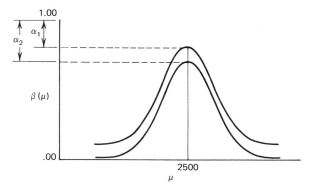

Fig. 10-3. The effect of type I error on the operating characteristic curve.

hypothesis is easier to detect for larger sample sizes than for small ones. The effect of the type I error probability α on the type II error probability β for a given sample size n is illustrated in Fig. 10-3. Decreasing α causes β to increase, and increasing α causes β to decrease.

Because the type II error probability β is a function of both the sample size and the extent to which the null hypothesis H_0 is false, it is customary to think of the decision to accept H_0 as a *weak conclusion*, unless we know that β is acceptably small. Therefore, rather than saying we "accept H_0," we prefer the terminology "*fail to reject H_0*." Failing to reject H_0 implies that we have not found sufficient evidence to reject H_0, that is, to make a strong statement. Thus failing to reject H_0 does not necessarily mean that there is a high probability that H_0 is true. It may imply that more data are required to reach a strong conclusion. This can have important implications for the formulation of hypotheses.

10-1.3 One-Sided and Two-Sided Hypotheses

Because rejecting H_0 is always a strong conclusion while failing to reject H_0 can be a weak conclusion unless β is known to be small, we usually prefer to construct hypotheses such that the statement about which a strong conclusion is desired is in the alternative hypothesis H_1. Problems for which a two-sided alternative hypothesis is appropriate do not really present the analyst with a choice of formulation. That is, if we wish to test the hypothesis that the mean of a distribution μ equals some arbitrary value, say μ_0, and it is important to detect values of the true mean μ that could be either greater than μ_0 or less than μ_0, then one must use the two-sided alternative in

$$H_0 : \mu = \mu_0$$
$$H_1 : \mu \neq \mu_0$$

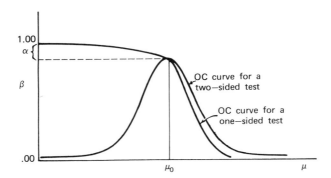

Fig. 10-4. Operating characteristic curves for two-sided and one-sided tests.

Many hypothesis-testing problems naturally involve a one-sided alternative hypothesis. For example, suppose that we want to reject H_0 only when the true value of the mean exceeds μ_0. The hypotheses would be

$$H_0: \mu = \mu_0$$
$$H_1: \mu > \mu_0 \tag{10-6}$$

This would imply that the critical region is located in the upper tail of the distribution of the test statistic. That is, if the decision is to be based on the value of the sample mean $\bar{x}$, then we would reject H_0 in Equation (10-6) if $\bar{x}$ is too large. The operating characteristic curve for the test for this hypothesis is shown in Fig. 10-4, along with the operating characteristic curve for a two-sided test. We observe that when the true mean μ exceeds μ_0 (that is, the alternative hypothesis $H_1: \mu > \mu_0$) is true, the one-sided test is superior to the two-sided test in the sense that it has a steeper operating characteristic curve. When the true mean $\mu = \mu_0$, both the one-sided and two-sided tests are equivalent. However, when the true mean μ is less than μ_0, the two operating characteristic curves differ. If $\mu < \mu_0$, the two-sided test has a higher probability of detecting this departure from μ_0 than the one-sided test. This is intuitively appealing, as the one-sided test is designed assuming that either μ cannot be less than μ_0, or if μ is less than μ_0 then it is desirable to accept the null hypothesis.

In effect, there are two different models that can be used for the one-sided alternative hypothesis. For the case where the alternative hypothesis is $H_1: \mu > \mu_0$, these two models are:

$$H_0: \mu = \mu_0$$
$$H_1: \mu > \mu_0 \tag{10-7}$$

and

$$H_0: \mu \leq \mu_0$$
$$H_1: \mu > \mu_0 \tag{10-8}$$

In Equation (10-7), we are assuming that μ cannot be less than μ_0, and the operating characteristic curve is undefined for values of $\mu < \mu_0$. In Equation (10-8), we are assuming that μ can be less than μ_0 and that in such a situation it would be desirable to accept H_0. Thus for Equation (10-8) the operating characteristic curve is defined for all values of $\mu \leq \mu_0$. Specifically, if $\mu \leq \mu_0$, we have $\beta(\mu) = 1 - \alpha(\mu)$, where $\alpha(\mu)$ is the significance level as a function of μ. For situations in which the model of Equation (10-8) is appropriate, we define the significance level of the test as the maximum value of the type I error probability α; that is, the value of α at $\mu = \mu_0$. In situations where one-sided alternative hypotheses are appropriate, we will usually write the null hypothesis with an equality; for example, $H_0: \mu = \mu_0$. This will be interpreted as including the cases $H_0: \mu \leq \mu_0$ or $H_0: \mu \geq \mu_0$, as appropriate.

In problems where one-sided test procedures are indicated, analysts occasionally experience difficulty in choosing an appropriate formulation of the alternative hypothesis. For example, suppose that a soft drink beverage bottler purchases 10-ounce nonreturnable bottles from a glass company. The bottler wants to be sure that the bottles exceed the specification on mean internal pressure or bursting strength, which for 10-ounce bottles is 200 psi. The bottler has decided to formulate the decision procedure for a specific lot of bottles as a hypothesis-testing problem. There are two possible formulations for this problem, either

$$H_0: \mu \leq 200 \text{ psi}$$
$$H_1: \mu > 200 \text{ psi} \tag{10-9}$$

or

$$H_0: \mu \geq 200 \text{ psi}$$
$$H_1: \mu < 200 \text{ psi} \tag{10-10}$$

Consider the formulation in Equation (10-9). If the null hypothesis is rejected, the bottles will be judged satisfactory; while if H_0 is not rejected, the implication is that the bottles do not conform to specifications and should not be used. Because rejecting H_0 is a strong conclusion, this formulation forces the bottle manufacturer to "demonstrate" that the mean bursting strength of the bottles exceeds the specification. Now consider the formulation in Equation (10-10). In this situation, the bottles will be judged satisfactory *unless* H_0 is rejected. That is, we would conclude that the bottles are satisfactory unless there is strong evidence to the contrary.

Which formulation is correct, the one of Equation (10-9) or Equation (10-10)? The answer is "it depends." For Equation (10-9), there is some probability that H_0 will be accepted (i.e., we would decide that the bottles are not satisfactory) even though the true mean is slightly greater than 200 psi. This formulation implies that we want the bottle manufacturer to *demonstrate* that the product meets or exceeds our specifications. Such a formulation could be appropriate if the manufacturer has experienced difficulty in meeting specifications in the past, or if product safety considerations force us to hold tightly to the 200 psi specification. On the other hand, for the formulation of Equation (10-10) there is some probability that H_0 will be accepted and the bottles judged satisfactory even though the true mean is slightly less than 200 psi. We would only conclude that the bottles are unsatisfactory when there is strong evidence that the mean does not exceed 200 psi; that is, when $H_0: \mu \geq 200$ psi is rejected. This formulation assumes that we are relatively happy with the bottle manufacturer's past performance and that small deviations from the specification of $\mu \geq 200$ psi are not harmful.

In formulating one-sided alternative hypotheses, we should remember that rejecting H_0 is always a strong conclusion, and, consequently, we should put the statement about which it is important to make a strong conclusion in the alternative hypothesis. Often this will depend on our point of view and experience with the situation.

10-2 Tests of Hypotheses on the Mean, Variance Known

10-2.1 Statistical Analysis

Suppose that the random variable X represents some process or population of interest. We assume that the distribution of X is either normal or that, if it is nonnormal, the conditions of the central limit theorem hold. In addition, we assume that the mean μ of X is unknown but that the variance σ^2 is known. We are interested in testing the hypothesis

$$H_0: \mu = \mu_0$$
$$H_1: \mu \neq \mu_0 \tag{10-11}$$

where μ_0 is a specified constant.

A random sample of size n, $X_1, X_2, \ldots, X_n$, is available. Each observation in this sample has unknown mean μ and known variance σ^2. The test procedure for $H_0: \mu = \mu_0$ uses the test statistic

$$Z_0 = \frac{\bar{X} - \mu_0}{\sigma/\sqrt{n}} \tag{10-12}$$

If the null hypothesis $H_0: \mu = \mu_0$ is true, then $E(\bar{X}) = \mu_0$, and it follows that

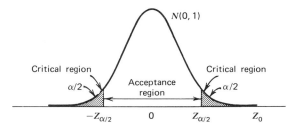

Fig. 10-5. The distribution of Z_0, when $H_0 : \mu = \mu_0$ is true.

the distribution of Z_0 is $N(0, 1)$. Consequently, if $H_0 : \mu = \mu_0$ is true, the probability is $1 - \alpha$ that a value of the test statistic Z_0 falls between $-Z_{\alpha/2}$ and $Z_{\alpha/2}$, where $Z_{\alpha/2}$ is the percentage point of the standard normal distribution such that $P\{Z \geq Z_{\alpha/2}\} = \alpha/2$ (i.e., $Z_{\alpha/2}$ is the $100\,\alpha/2$ percentage point of the standard normal distribution). The situation is illustrated in Fig. 10-5. Note that the probability is α that a value of the test statistic Z_0 would fall in the region $Z_0 > Z_{\alpha/2}$ or $Z_0 < -Z_{\alpha/2}$ when $H_0 : \mu = \mu_0$ is true. Clearly, a sample producing a value of the test statistic that falls in the tails of the distribution of Z_0 would be unusual if $H_0 : \mu = \mu_0$ is true; it is also an indication that H_0 is false. Thus, we should reject H_0 if either

$$Z_0 > Z_{\alpha/2} \tag{10-13a}$$

or

$$Z_0 < -Z_{\alpha/2} \tag{10-13b}$$

and fail to reject H_0 if

$$-Z_{\alpha/2} \leq Z_0 \leq Z_{\alpha/2} \tag{10-14}$$

Equation (10-14) defines the *acceptance region* for H_0 and Equation (10-13) defines the *critical region* or *rejection region*. The type I error probability for this test procedure is α.

● **Example 10-1.** The burning rate of a rocket propellant is being studied. Specifications require that the mean burning rate must be 40 cm/s. Further-more, suppose that we know that the variance of the burning rate is 4.0. The experimenter decides to specify a type I error probability of $\alpha = .05$, and he will base the test on a random sample of size $n = 25$. The hypotheses we wish to test are

$$H_0 : \mu = 40 \text{ cm/s}$$
$$H_1 : \mu \neq 40 \text{ cm/s}$$

Twenty-five specimens are tested, and the sample mean burning rate obtained is $\bar{x} = 41.25$ cm/s. The value of the test statistic in Equation (10-12) is

$$Z_0 = \frac{\bar{x} - \mu_0}{\sigma/\sqrt{n}}$$

$$= \frac{41.25 - 40}{2/\sqrt{25}} = 3.125$$

Since $\alpha = .05$, the boundaries of the critical region are $Z_{.025} = 1.96$ and $-Z_{.025} = -1.96$, and we note that Z_0 falls in the critical region. Therefore, H_0 is rejected, and we conclude that the mean burning rate is not equal to 40 cm/s.

Now suppose that we wish to test the one-sided alternative, say

$$H_0: \mu = \mu_0$$
$$H_1: \mu > \mu_0 \tag{10-15}$$

(Note that we could also write $H_0: \mu \le \mu_0$.) In defining the critical region for this test, we observe that a negative value of the test statistic Z_0 would never lead us to conclude that $H_0: \mu = \mu_0$ is false. Therefore, we would place the critical region in the upper tail of the $N(0, 1)$ distribution and reject H_0 on values of Z_0 that are too large. That is, we should reject H_0 if

$$Z_0 > Z_\alpha \tag{10-16}$$

Similarly, to test

$$H_0: \mu = \mu_0$$
$$H_1: \mu < \mu_0 \tag{10-17}$$

we would calculate the test statistic Z_0 and reject H_0 on values of Z_0 that are too small. That is, the critical region is in the lower tail of the $N(0, 1)$ distribution, and we reject H_0 if

$$Z_0 < -Z_\alpha \tag{10-18}$$

10-2.2 Choice of Sample Size

In testing the hypotheses of Equations (10-11), (10-15), and (10-17), the type I error probability α is directly selected by the analyst. However, the probability of type II error β depends on the choice of sample size. In this section, we will show how to select the sample size in order to arrive at a specified value of β.

Consider the two-sided hypothesis

$$H_0: \mu = \mu_0$$
$$H_1: \mu \ne \mu_0$$

Suppose that the null hypothesis is false and that the true value of the mean is

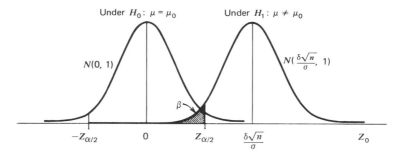

Fig. 10-6.   The distribution of Z_0 under H_0 and H_1.

$\mu = \mu_0 + \delta$, say, where $\delta > 0$. Now since H_1 is true, the distribution of the test statistic Z_0 is

$$Z_0 \sim N \left(\frac{\delta \sqrt{n}}{\sigma}, 1\right) \tag{10-19}$$

The distribution of the test statistic Z_0 under both the null hypothesis H_0 and the alternative hypothesis H_1 is shown in Fig. 10-6. From examining this figure, we note that if H_1 is true, a type II error will be made only if $-Z_{\alpha/2} \leq Z_0 \leq Z_{\alpha/2}$ where $Z_0 \sim N(\delta\sqrt{n}/\sigma, 1)$. That is, the probability of the type II error β is the probability that Z_0 falls between $-Z_{\alpha/2}$ and $Z_{\alpha/2}$ *given that H_1 is true.* This probability is shown as the shaded portion of Fig. 10-6. Expressed mathematically, this probability is

$$\beta = \Phi \left(Z_{\alpha/2} - \frac{\delta \sqrt{n}}{\sigma}\right) - \Phi \left(-Z_{\alpha/2} - \frac{\delta \sqrt{n}}{\sigma}\right) \tag{10-20}$$

where $\Phi(z)$ denotes the probability to the left of z on the standard normal distribution. Note that Equation (10-20) was obtained by evaluating the probability that Z_0 falls in the interval $[-Z_{\alpha/2}, Z_{\alpha/2}]$ on the distribution of Z_0 when H_1 is true. These two points were standardized to produce Equation (10-20). Furthermore, note that Equation (10-20) also holds if $\delta < 0$, due to the symmetry of the normal distribution.

While Equation (10-20) could be used to evaluate the type II error, it is more convenient to use the operating characteristic curves in Appendix Charts VI*a* and VI*b*. These curves plot β as calculated from Equation (10-20) against a parameter d for various sample sizes n. Curves are provided for both $\alpha = .05$ and $\alpha = .01$. The parameter d is defined as

$$d = \frac{|\mu - \mu_0|}{\sigma} = \frac{|\delta|}{\sigma} \tag{10-21}$$

We have chosen d so that one set of operating characteristic curves can be used for all problems regardless of the value of μ_0 and σ. From examining the operating characteristic curves or Equation (10-20) and Fig. 10-6 we note that:

1. The further the true value of the mean μ is from μ_0, the smaller the probability of type II error β for a given n and α. That is, we see that for a specified sample size and α, large differences in the mean are easier to detect than small ones.

2. For a given δ and α, the probability of type II error β decreases as n increases. That is, to detect a specified difference in the mean δ, we may make the test more powerful by increasing the sample size.

● **Example 10-2.** Consider the rocket propellant problem in Example 10-1. Suppose that the analyst is concerned about the probability of type II error if the true mean burning rate is $\mu = 41$ cm/s. We may use the operating characteristic curves to find β. Note that $\delta = 41 - 40 = 1$, $n = 25$, $\sigma = 2$, and $\alpha = .05$. Then

$$d = \frac{|\mu - \mu_0|}{\sigma} = \frac{|\delta|}{\sigma} = \frac{1}{2}$$

and from Appendix Chart VIa, with $n = 25$, we find that $\beta = .30$. That is, if the true mean burning rate is $\mu = 41$ cm/s, then there is approximately a 30 percent chance that this will not be detected by the test with $n = 25$.

● **Example 10-3.** Once again, consider the rocket propellant problem in Example 10-1. Suppose that the analyst would like to design the test so that if the true mean burning rate differs from 40 cm/s by as much as 1 cm/s, the test will detect this (i.e., reject $H_0: \mu = 40$) with a high probability, say .90. The operating characteristic curves can be used to find the sample size that will give such a test. Since $d = |\mu - \mu_0|/\sigma = 1/2$, $\alpha = .05$, and $\beta = .10$, we find from Appendix Chart VIa that the required sample size is $n = 40$, approximately.

In general, the operating characteristic curves involve three parameters: β, δ, and n. Given any two of these parameters, the value of the third can be determined. There are two typical applications of these curves:

1. For a given n and δ, find β. This was illustrated in Example 10-2. This kind of problem is often encountered when the analyst is concerned about the sensitivity of an experiment already performed, or when sample size is restricted by economic or other factors.

2. For a given β and δ, find n. This was illustrated in Example 10-3. This kind of problem is usually encountered when the analyst has the opportunity to select the sample size at the outset of the experiment.

Operating characteristic curves are given in Appendix Charts VIc and VId for the one-sided alternatives. If the alternative hypothesis is $H_1: \mu > \mu_0$, then the abscissa scale on these charts is

$$d = \frac{\mu - \mu_0}{\sigma} \tag{10-22}$$

When the alternative hypothesis is $H_1: \mu < \mu_0$, the corresponding abscissa scale is

$$d = \frac{\mu_0 - \mu}{\sigma} \tag{10-23}$$

It is also possible to derive formulas to determine the appropriate sample size to use to obtain a particular value of β for a given δ and α. These formulas are alternatives to using the operating characteristic curves. For the two-sided alternative, we know from Equation (10-20) that

$$\beta = \Phi\left(Z_{\alpha/2} - \frac{\delta\sqrt{n}}{\sigma}\right) - \Phi\left(-Z_{\alpha/2} - \frac{\delta\sqrt{n}}{\sigma}\right),$$

or if $\delta > 0$,

$$\beta \simeq \Phi\left(Z_{\alpha/2} - \frac{\delta\sqrt{n}}{\sigma}\right) \tag{10-24}$$

since $\Phi(-Z_{\alpha/2} - \delta\sqrt{n}/\sigma) \simeq 0$ when δ is positive. From Equation (10-24), we have

$$-Z_\beta \simeq Z_{\alpha/2} - \frac{\delta\sqrt{n}}{\sigma}$$

or

$$n \simeq \frac{(Z_{\alpha/2} + Z_\beta)^2 \sigma^2}{\delta^2} \tag{10-25}$$

This approximation is good when $\Phi(-Z_{\alpha/2} - \delta\sqrt{n}/\sigma)$ is small compared to β. For either of the one-sided alternative hypotheses in Equation (10-15) or Equation (10-17), the sample size required to produce a specified type II error with probability β given δ and α is

$$n = \frac{(Z_\alpha + Z_\beta)^2 \sigma^2}{\delta^2} \tag{10-26}$$

● **Example 10-4.** Returning to the rocket propellant problem of Example 10-3, we note that $\sigma = 2$, $\delta = 41 - 40 = 1$, $\alpha = .05$, and $\beta = .10$. Since $Z_{\alpha/2} = Z_{.025} = 1.96$ and $Z_\beta = Z_{.10} = 1.28$, the sample size required to detect this departure from $H_0 : \mu = 40$ is found from Equation (10-25) as

$$n \simeq \frac{(Z_{\alpha/2} + Z_\beta)^2 \sigma^2}{\delta^2} = \frac{(1.96 + 1.28)^2 2^2}{(1)^2} = 42$$

which is in close agreement with the value determined from the operating characteristic curve. Note that the approximation is good, since $\Phi(-Z_{\alpha/2} - \delta\sqrt{n}/\sigma) = \Phi(-1.96 - (1)\sqrt{42}/2) = \Phi(-5.20) \simeq 0$, which is small relative to β.

10-2.3   The Relationship Between Tests of Hypotheses and Confidence Intervals

There is a close relationship between the test of a hypothesis about a parameter θ and the confidence interval for θ. If $[L, U]$ is a $100(1 - \alpha)$ percent confidence interval for the parameter θ, then the test of size α of the hypothesis

$$H_0 : \theta = \theta_0$$
$$H_1 : \theta \neq \theta_0$$

will lead to rejection of H_0 if and only if θ_0 is not in the interval $[L, U]$. As an illustration, consider the rocket propellant problem in Example 10-1. The null hypothesis $H_0 : \mu = 40$ was rejected, using $\alpha = .05$. The 95 percent two-sided confidence interval on μ for these data may be computed from Equation (9-17) as $40.47 \leq \mu \leq 42.03$. That is, the interval $[L, U]$ is $[40.47, 42.03]$, and since $\mu_0 = 40$ is not included in this interval, the null hypothesis $H_0 : \mu = 40$ is rejected.

10-3   Tests of Hypotheses on the Equality of Two Means, Variances Known

10-3.1   Statistical Analysis

Suppose that there are two populations of interest, say X_1 and X_2. We assume that X_1 has unknown mean μ_1 and known variance σ_1^2, and that X_2 has unknown mean μ_2 and known variance σ_2^2. We will be concerned with testing the hypothesis that the means μ_1 and μ_2 are equal. It is assumed that the random variables X_1 and X_2 are either normally distributed or if they are nonnormal then the conditions of the central limit theorem apply.

Consider first the two-sided alternative hypothesis

$$H_0 : \mu_1 = \mu_2$$
$$H_1 : \mu_1 \neq \mu_2 \tag{10-27}$$

Suppose that a random sample of size n_1 is drawn from X_1, say $X_{11}, X_{12}, \ldots, X_{1n_1}$, and that a second random sample of size n_2 is drawn from

X_2, say $X_{21}, X_{22}, \ldots, X_{2n_2}$. It is assumed that the $\{X_{1j}\}$ are independently distributed with mean μ_1 and variance σ_1^2, that the $\{X_{2j}\}$ are independently distributed with mean μ_2 and variance σ_2^2, and that all of the $\{X_{1j}\}$ and $\{X_{2j}\}$ are independent. The test procedure is based on the distribution of the difference in sample means, say $\bar{X}_1 - \bar{X}_2$. In general, we know that

$$\bar{X}_1 - \bar{X}_2 \sim N\left(\mu_1 - \mu_2, \frac{\sigma_1^2}{n_1} + \frac{\sigma_2^2}{n_2}\right)$$

Thus, if the null hypothesis $H_0: \mu_1 = \mu_2$ is true, the test statistic

$$Z_0 = \frac{\bar{X}_1 - \bar{X}_2}{\sqrt{\dfrac{\sigma_1^2}{n_1} + \dfrac{\sigma_2^2}{n_2}}} \tag{10-28}$$

follows the $N(0, 1)$ distribution. Therefore, the procedure for testing $H_0: \mu_1 = \mu_2$ is to calculate the test statistic Z_0 in Equation (10-28) and reject the null hypothesis if

$$Z_0 > Z_{\alpha/2} \tag{10-29a}$$

or

$$Z_0 < -Z_{\alpha/2} \tag{10-29b}$$

The one-sided alternative hypotheses are analyzed similarly. To test

$$H_0: \mu_1 = \mu_2$$
$$H_1: \mu_1 > \mu_2 \tag{10-30}$$

the test statistic Z_0 in Equation (10-28) is calculated, and $H_0: \mu_1 = \mu_2$ is rejected if

$$Z_0 > Z_\alpha \tag{10-31}$$

To test the other one-sided alternative hypothesis

$$H_0: \mu_1 = \mu_2$$
$$H_1: \mu_1 < \mu_2 \tag{10-32}$$

use the test statistic Z_0 in Equation (10-28) and reject $H_0: \mu_1 = \mu_2$ if

$$Z_0 < -Z_\alpha \tag{10-33}$$

● **Example 10-5.** The plant manager of an orange juice canning facility is interested in comparing the performance of two different production lines in her plant. As line number 1 is relatively new, she suspects that its output in number of cases per day is greater than the number of cases produced by the older line 2. Ten days of data are selected at random for each line, for which it is found that $\bar{x}_1 = 824.9$ cases per day and $\bar{x}_2 = 818.6$ cases per day. From experience with operating this type of equipment it is known that $\sigma_1^2 = 40$ and $\sigma_2^2 = 50$. We wish to test

$$H_0: \mu_1 = \mu_2$$
$$H_1: \mu_1 > \mu_2$$

The value of the test statistic is

$$Z_0 = \frac{\bar{x}_1 - \bar{x}_2}{\sqrt{\dfrac{\sigma_1^2}{n_1} + \dfrac{\sigma_2^2}{n_2}}} = \frac{824.9 - 818.6}{\sqrt{\dfrac{40}{10} + \dfrac{50}{10}}} = 2.10$$

Using $\alpha = .05$ we find that $Z_{.05} = 1.645$, and since $Z_0 > Z_{.05}$, we would reject H_0 and conclude that the mean number of cases per day produced by the new production line is greater than the mean number of cases per day produced by the old line.

10-3.2 Choice of Sample Size

The operating characteristic curves in Appendix Charts VIa, VIb, VIc, and VId may be used to evaluate the type II error probability for the hypothesis in Equations (10-27), (10-30), and (10-32). These curves are also useful in sample size determination. Curves are provided for $\alpha = .05$ and $\alpha = .01$. For the two-sided alternative hypothesis in Equation (10-27), the abscissa scale of the operating characteristic curve in Charts VIa and VIb is d, where

$$d = \frac{|\mu_1 - \mu_2|}{\sqrt{\sigma_1^2 + \sigma_2^2}} = \frac{|\delta|}{\sqrt{\sigma_1^2 + \sigma_2^2}} \qquad (10\text{-}34)$$

and one must choose equal sample sizes, say $n = n_1 = n_2$. The one-sided alternative hypotheses require the use of Charts VIc and VId. For the one-sided alternative hypothesis $H_1 : \mu_1 > \mu_2$ in Equation (10-30), the abscissa scale is

$$d = \frac{\mu_1 - \mu_2}{\sqrt{\sigma_1^2 + \sigma_2^2}} = \frac{\delta}{\sqrt{\sigma_1^2 + \sigma_2^2}} \qquad (10\text{-}35)$$

with $n = n_1 = n_2$. The other one-sided alternative hypothesis, $H_1: \mu_1 < \mu_2$, requires that d be defined as

$$d = \frac{\mu_2 - \mu_1}{\sqrt{\sigma_1^2 + \sigma_2^2}} = \frac{\delta}{\sqrt{\sigma_1^2 + \sigma_2^2}} \qquad (10\text{-}36)$$

and $n = n_1 = n_2$.

It is not unusual to encounter problems where the costs of collecting data differ substantially between the two populations, or where one population variance is much greater than the other. In those cases, one often uses unequal sample sizes. If $n_1 \neq n_2$, the operating characteristic curves may be entered with an *equivalent* value of n computed from

$$n = \frac{\sigma_1^2 + \sigma_2^2}{\sigma_1^2/n_1 + \sigma_2^2/n_2} \tag{10-37}$$

If $n_1 \neq n_2$, and their values are fixed in advance, then Equation (10-37) is used directly to calculate n, and the operating characteristic curves are entered with a specified d to obtain β. If we are given d and it is necessary to determine n_1 and n_2 to obtain a specified β, say β^*, then one guesses at trial values of n_1 and n_2, calculates n in Equation (10-37), and enters the curves with the specified value of d and finds β. If $\beta = \beta^*$, then the trial values of n_1 and n_2 are satisfactory. If $\beta \neq \beta^*$, then adjustments to n_1 and n_2 are made and the process repeated.

- **Example 10-6.** Consider the orange juice production line problem in Example 10-5. If the true difference in mean production rates were 10 cases per day, find the sample sizes required to detect this difference with probability .90. The appropriate value of the abscissa parameter is

$$d = \frac{\mu_1 - \mu_2}{\sqrt{\sigma_1^2 + \sigma_2^2}} = \frac{10}{\sqrt{40 + 50}} = 1.05$$

and since $\alpha = .05$, we find from Chart VIc that $n = n_1 = n_2 = 8$.

It is also possible to derive formulas to obtain the sample size required to obtain a specified β for a given δ and α. These formulas occasionally are useful supplements to the operating characteristic curves. For the two-sided alternative, the sample size $n_1 = n_2 = n$ is

$$n \approx \frac{(Z_{\alpha/2} + Z_\beta)^2(\sigma_1^2 + \sigma_2^2)}{\delta^2} \tag{10-38}$$

This approximation is valid when $\Phi(-Z_{\alpha/2} - \delta\sqrt{n}/\sqrt{\sigma_1^2 + \sigma_2^2})$ is small compared to β. For a one-sided alternative, we have $n_1 = n_2 = n$, where

$$n = \frac{(Z_\alpha + Z_\beta)^2(\sigma_1^2 + \sigma_2^2)}{\delta^2} \tag{10-39}$$

The derivations of Equations (10-38) and (10-39) closely follow the single-sample case in Section 10-2.2. To illustrate the use of these equations, consider the situation in Example 10-6. We have a one-sided alternative with $\alpha = .05$, $\delta = 10$, $\sigma_1^2 = 40$, $\sigma_2^2 = 50$, and $\beta = .10$. Thus $Z_{.05} = 1.645$, $Z_\beta = 1.28$, and the required sample size is found from Equation (10-39) as

$$n = \frac{(Z_\alpha + Z_\beta)^2(\sigma_1^2 + \sigma_2^2)}{\delta^2} = \frac{(1.645 + 1.28)^2(40 + 50)}{(10)^2} = 8$$

which agrees with the results obtained in Example 10-6.

10-4 Tests of Hypotheses on the Mean of a Normal Distribution, Variance Unknown

10-4.1 Statistical Analysis

Suppose that X is a normally distributed random variable with unknown mean μ and variance σ^2. We wish to test the hypothesis that μ equals a constant μ_0. Note that this situation is similar to that treated in Section 10-2, except that now *both μ and σ^2* are unknown. Because σ^2 is now unknown, we must require that X be normally distributed in order to develop a test procedure. While the normality assumption is required in theory, in practice, slight departures from normality are not serious, particularly if the sample size is not small. Assume that a random sample of size n, say $X_1, X_2, \ldots, X_n$ is available, and let $\bar{X}$ and S^2 be the sample mean and variance, respectively.

Suppose that we wish to test the two-sided alternative

$$H_0 : \mu = \mu_0$$
$$H_1 : \mu \neq \mu_0 \tag{10-40}$$

The test procedure is based on the statistic

$$t_0 = \frac{\bar{X} - \mu_0}{S/\sqrt{n}} \tag{10-41}$$

which follows the t distribution with $n - 1$ degrees of freedom if the null hypothesis $H_0 : \mu = \mu_0$ is true. To test $H_0 : \mu = \mu_0$ in Equation (10-40), the test statistic t_0 in Equation (10-41) is calculated, and H_0 is rejected if either

$$t_0 > t_{\alpha/2, n-1} \tag{10-42a}$$

or if

$$t_0 < -t_{\alpha/2, n-1} \tag{10-42b}$$

where $t_{\alpha/2, n-1}$ and $-t_{\alpha/2, n-1}$ are the upper and lower $\alpha/2$ percentage points of the t distribution with $n - 1$ degrees of freedom.

For the one-sided alternative hypothesis

$$H_0 : \mu = \mu_0$$
$$H_1 : \mu > \mu_0 \tag{10-43}$$

we calculate the test statistic t_0 from Equation (10-41) and reject H_0 if

$$t_0 > t_{\alpha, n-1} \tag{10-44}$$

For the other one-sided alternative

$$H_0 : \mu = \mu_0$$
$$H_1 : \mu < \mu_0 \tag{10-45}$$

we would reject H_0 if

$$t_0 < -t_{\alpha, n-1} \tag{10-46}$$

● **Example 10-7.** The breaking strength of a textile fiber is a normally distributed random variable. Specifications require that the mean breaking strength should equal 150 psi. The manufacturer would like to detect any significant departure from this value. Thus, he wishes to test

$$H_0: \mu = 150 \text{ psi}$$
$$H_1: \mu \neq 150 \text{ psi}$$

A random sample of 15 fiber specimens is selected and their breaking strengths determined. The sample mean and variance are computed from the sample data as $\bar{x} = 152.18$ and $s^2 = 16.63$. Therefore, the test statistic is

$$t_0 = \frac{\bar{x} - \mu_0}{s/\sqrt{n}} = \frac{152.18 - 150}{\sqrt{16.63/15}} = 2.07$$

The type I error is specified as $\alpha = .05$. Therefore $t_{.025, 14} = 2.145$ and $-t_{.025, 14} = -2.145$, and we would conclude that there is not sufficient evidence to reject the hypothesis that $\mu = 150$ psi.

10-4.2 Choice of Sample Size

The type II error probability for tests on the mean of a normal distribution with unknown variance depends on the distribution of the test statistic in Equation (10-41) when the null hypothesis $H_0: \mu = \mu_0$ is false. When the true value of the mean is $\mu = \mu_0 + \delta$, note that the test statistic can be written as

$$t_0 = \frac{\bar{X} - \mu_0}{S/\sqrt{n}}$$

$$= \frac{\dfrac{[\bar{X} - (\mu_0 + \delta)]\sqrt{n}}{\sigma} + \dfrac{\delta\sqrt{n}}{\sigma}}{S/\sigma}$$

$$= \frac{Z + \dfrac{\delta\sqrt{n}}{\sigma}}{W} \tag{10-47}$$

The distributions of Z and W in Equation (10-47) are $N(0, 1)$ and $\sqrt{\chi_{n-1}^2/n - 1}$, respectively, and Z and W are independent random variables. However, $\delta\sqrt{n}/\sigma$ is a nonzero constant, so that the numerator of Equation (10-47) is a $N(\delta\sqrt{n}/\sigma, 1)$ random variable. The resulting distribution is called the *noncentral t* distribution with $n - 1$ degrees of freedom and noncentrality parameter $\delta\sqrt{n}/\sigma$. Note that if $\delta = 0$, then the noncentral *t* distribution reduces to the usual or central t distribution. Therefore, the type II error of the two-sided alternative (for example) would be

$$\beta = P\{-t_{\alpha/2, n-1} \leq t_0 \leq t_{\alpha/2, n-1} | \delta \neq 0\}$$
$$= P\{-t_{\alpha/2, n-1} \leq t_0' \leq t_{\alpha/2, n-1}\}$$

where t_0' denotes the noncentral t random variable. Finding the type II error for the t-test involves finding the probability contained between two points on the noncentral t distribution.

The operating characteristic curves in Appendix Charts VIe, VIf, VIg, and VIh plot β against a parameter d for various sample sizes n. Curves are provided for both the two-sided and one-sided alternatives and for $\alpha = .05$ or $\alpha = .01$. For the two-sided alternative in Equation (10-40), the abscissa scale factor d on Charts VIe and VIf is defined as

$$d = \frac{|\mu - \mu_0|}{\sigma} = \frac{|\delta|}{\sigma} \tag{10-48}$$

For the one-sided alternatives, if rejection is desired, $\mu > \mu_0$ as in Equation (10-43), we use Charts VIg and VIh with

$$d = \frac{\mu - \mu_0}{\sigma} = \frac{\delta}{\sigma} \tag{10-49}$$

while if rejection is desired when $\mu < \mu_0$, as in Equation (10-45),

$$d = \frac{\mu_0 - \mu}{\sigma} = \frac{\delta}{\sigma} \tag{10-50}$$

We note that d depends on the unknown parameter σ^2. There are several ways to avoid this difficulty. In some cases, we may use the results of a previous experiment or prior information to make a rough initial estimate of σ^2. If we are interested in examining the operating characteristic after the data has been collected, we could use the sample variance S^2 to estimate σ^2. If analysts do not have any previous experience on which to draw in estimating σ^2, they can define the difference in the mean δ that they wish to detect relative to σ. For example, if one wishes to detect a small difference in the mean, one might use a value of $d = |\delta|/\sigma \leq 1$ (for example), while if one is interested in detecting only moderately large differences in the mean, one might select $d = |\delta|/\sigma = 2$ (for example). That is, it is the value of the ratio $|\delta|/\sigma$ that is important in determining sample size, and if it is possible to specify the relative size of the difference in means that we are interested in detecting, then a proper value of d can usually be selected.

● **Example 10-8.** Consider the fiber-testing problem in Example 10-7. If the breaking strength of this fiber differs from 150 psi by as much as 2.5 psi, the analyst would like to reject the null hypothesis $H_0: \mu = 150$ psi with probability at least .90. Is the sample size $n = 15$ adequate to ensure that the test is this sensitive? If we use the sample standard deviation $s = \sqrt{16.68} = 4.08$ to estimate σ, then $d = |\delta|/\sigma = 2.5/4.08 = .61$. By referring to the operating characteristic curves in Chart VIe, with $d = .61$ and $n = 15$, we find $\beta \simeq .45$. Thus, the probability of rejecting $H_0: \mu = 150$ psi if the true mean differs from this value by ± 2.5 psi is $1 - \beta = 1 - .45 = .55$, approximately, and we would conclude that a sample size of $n = 15$ is not adequate. To find the sample size required to give the desired degree of protection, enter the operating characteristic curves in Chart VIe with $d = .61$ and $\beta = .10$, and read the corresponding sample size as $n = 35$, approximately.

10-5 Tests of Hypotheses on the Means of Two Normal Distributions, Variances Unknown

We now consider tests of hypotheses on the equality of the means μ_1 and μ_2 of two normal distributions where the variances σ_1^2 and σ_2^2 are unknown. A t statistic will be used to test these hypotheses. As noted in Section 10-4.1, the normality assumption is required to develop the test procedure, but moderate departures from normality do not adversely affect the procedure. There are two different situations that must be treated. In the first case, we assume that the variances of the two normal distributions are unknown but equal; that is, $\sigma_1^2 = \sigma_2^2 = \sigma^2$. In the second, we assume that σ_1^2 and σ_2^2 are unknown and not necessarily equal.

10-5.1 Case 1: $\sigma_1^2 = \sigma_2^2 = \sigma^2$

Let X_1 and X_2 be two independent normal populations with unknown means μ_1 and μ_2, and unknown but equal variances, $\sigma_1^2 = \sigma_2^2 = \sigma^2$. We wish to test

$$H_0: \mu_1 = \mu_2$$
$$H_1: \mu_1 \neq \mu_2 \tag{10-51}$$

Suppose that $X_{11}, X_{12}, \ldots, X_{1n_1}$ is a random sample of n_1 observations from X_1, and $X_{21}, X_{22}, \ldots, X_{2n_2}$ is a random sample of n_2 observations from X_2. Let $\bar{X}_1, \bar{X}_2, S_1^2$, and S_2^2 be the sample means and sample variances, respectively. Since both S_1^2 and S_2^2 estimate the common variance σ^2, we may combine them to yield a single estimate, say

$$S_p^2 = \frac{(n_1 - 1)S_1^2 + (n_2 - 1)S_2^2}{n_1 + n_2 - 2} \tag{10-52}$$

This combined or "pooled" estimator was introduced in Section 9-2.4. To test $H_0: \mu_1 = \mu_2$ in Equation (10-51), compute the test statistic

$$t_0 = \frac{\bar{X}_1 - \bar{X}_2}{S_p \sqrt{\dfrac{1}{n_1} + \dfrac{1}{n_2}}} \tag{10-53}$$

If $H_0: \mu_1 = \mu_2$ is true, t_0 is distributed as $t_{n_1+n_2-2}$. Therefore, if

$$t_0 > t_{\alpha/2, n_1+n_2-2} \tag{10-54a}$$

or if

$$t_0 < -t_{\alpha/2, n_1+n_2-2} \tag{10-54b}$$

we reject $H_0: \mu_1 = \mu_2$.

The one-sided alternatives are treated similarly. To test

$$H_0: \mu_1 = \mu_2$$
$$H_1: \mu_1 > \mu_2 \tag{10-55}$$

compute the test statistic t_0 in Equation (10-53) and reject $H_0: \mu_1 = \mu_2$ if

$$t_0 > t_{\alpha, n_1+n_2-2} \tag{10-56}$$

For the other one-sided alternative,

$$H_0: \mu_1 = \mu_2$$
$$H_1: \mu_1 < \mu_2 \tag{10-57}$$

calculate the test statistic t_0 and reject $H_0: \mu_1 = \mu_2$ if

$$t_0 < -t_{\alpha, n_1+n_2-2} \tag{10-58}$$

The two-sample t-test given in this section is often called the *pooled* t-test, because the sample variances are combined or pooled to estimate the common variance. It is also known as the *independent* t-test, because the two normal populations X_1 and X_2 are assumed to be independent.

● **Example 10-9.** In Example 9-11, two different catalysts are being analyzed to determine how they affect the mean yield of a chemical process. Specifically, catalyst 1 is currently in use, but catalyst 2 is acceptable. Since catalyst 2 is cheaper, if it does not change the process yield, it should be adopted. Suppose we wish to test the hypotheses

$$H_0: \mu_1 = \mu_2$$
$$H_1: \mu_1 \neq \mu_2$$

Pilot plant data yields $n_1 = 8$, $\bar{x}_1 = 91.73$, $s_1^2 = 3.89$, $n_2 = 8$, $\bar{x}_2 = 93.75$, and $s_2^2 = 4.02$. From Equation (10-52), we find

$$s_p^2 = \frac{(n_1 - 1)s_1^2 + (n_2 - 1)s_2^2}{n_1 + n_2 - 2} = \frac{(7)3.89 + 7(4.02)}{8 + 8 - 2} = 3.96$$

The test statistic is

$$t_0 = \frac{\bar{x}_1 - \bar{x}_2}{s_p \sqrt{\dfrac{1}{n_1} + \dfrac{1}{n_2}}} = \frac{91.73 - 93.75}{1.99\sqrt{\dfrac{1}{8} + \dfrac{1}{8}}} = -2.03$$

Using $\alpha = .05$, we find that $t_{.025, 14} = 2.145$ and $-t_{.025, 14} = -2.145$, and, consequently, $H_0: \mu_1 = \mu_2$ cannot be rejected. That is, we do not have strong evidence to conclude that catalyst 2 results in a mean yield that differs from the mean yield when catalyst 1 is used. Referring to Example 9-11, we note that the 95 percent confidence interval on $\mu_1 - \mu_2$ includes zero; therefore, considering the equivalence between tests of hypotheses and confidence intervals noted in Section 10-2.3, failing to reject $H_0: \mu_1 = \mu_2$ is anticipated.

10-5.2 Case 2: $\sigma_1^2 \neq \sigma_2^2$

In some situations, we cannot reasonably assume that the unknown variances σ_1^2 and σ_2^2 are equal. There is not an exact t statistic available for testing $H_0: \mu_1 = \mu_2$ in this case. However, the statistic

$$t_0^* = \frac{\bar{X}_1 - \bar{X}_2}{\sqrt{\dfrac{S_1^2}{n_1} + \dfrac{S_2^2}{n_2}}} \tag{10-59}$$

is distributed approximately as t with degrees of freedom given by

$$\nu = \frac{\left(\dfrac{S_1^2}{n_1} + \dfrac{S_2^2}{n_2}\right)^2}{\dfrac{(S_1^2/n_1)^2}{n_1 + 1} + \dfrac{(S_2^2/n_2)^2}{n_2 + 1}} - 2 \tag{10-60}$$

if the null hypothesis $H_0: \mu_1 = \mu_2$ is true. Therefore, if $\sigma_1^2 \neq \sigma_2^2$, the hypotheses of Equations (10-51), (10-55), and (10-57) are tested as in Section 10-5.1, except that t_0^* is used as the test statistic and $n_1 + n_2 - 2$ is replaced by ν in determining the degrees of freedom for the test. This general problem is often called the Behrens–Fisher problem.

● **Example 10-10.** Suppose that there are two normal populations, say $X_1 \sim N(\mu_1, \sigma_1^2)$, and $X_2 \sim N(\mu_2, \sigma_2^2)$ where $\sigma_1^2 \neq \sigma_2^2$ and both variances are unknown. We wish to test

$$H_0: \mu_1 = \mu_2$$
$$H_1: \mu_1 < \mu_2$$

Two random samples yield $n_1 = 15$, $\bar{x}_1 = 2.0$, $s_1^2 = 10$, and $n_2 = 10$, $\bar{x}_2 = 1.0$, and $s_2^2 = 20$. The test statistic t_0^* is

$$t_0^* = \frac{\bar{x}_1 - \bar{x}_2}{\sqrt{\dfrac{s_1^2}{n_1} + \dfrac{s_2^2}{n_2}}} = \frac{2.0 - 1.0}{\sqrt{\dfrac{10}{15} + \dfrac{20}{10}}} = .61$$

The degrees of freedom on t_0^* are found from Equation (10-60) as

$$t_0^* = \frac{\left(\dfrac{s_1^2}{n_1}+\dfrac{s_2^2}{n_2}\right)^2}{\dfrac{(s_1^2/n_1)^2}{n_1+1}+\dfrac{(s_2^2/n_2)^2}{n_2+1}} - 2 = \frac{\left(\dfrac{10}{15}+\dfrac{20}{10}\right)^2}{\dfrac{(10/15)^2}{16}+\dfrac{(20/10)^2}{11}} - 2 = 16$$

Since $t_0^* < -t_{.05, 16}$, we cannot reject $H_0: \mu_1 = \mu_2$.

10-5.3   Choice of Sample Size

The operating characteristic curves in Appendix Charts VIe, VIf, VIg, and VIh are used to evaluate the type II error for the case where $\sigma_1^2 = \sigma_2^2 = \sigma^2$. Unfortunately, when $\sigma_1^2 \neq \sigma_2^2$, the distribution of t_0^* is unknown if the null hypothesis is false, and no operating characteristic curves are available for this case.

For the two-sided alternative in Equation (10-51), when $\sigma_1^2 = \sigma_2^2 = \sigma^2$ and $n_1 = n_2 = n$, Charts VIe and VIf are used with

$$d = \frac{|\mu_1 - \mu_2|}{2\sigma} = \frac{|\delta|}{2\sigma} \tag{10-61}$$

To use these curves, they must be entered with the sample size $n^* = 2n - 1$. For the one-sided alternative hypothesis of Equation (10-55), we use Charts VIg and VIh and define

$$d = \frac{\mu_1 - \mu_2}{2\sigma} = \frac{\delta}{2\sigma} \tag{10-62}$$

while for the other one-sided alternative hypothesis of Equation (10-57), we use

$$d = \frac{\mu_2 - \mu_1}{2\sigma} = \frac{\delta}{2\sigma} \tag{10-63}$$

It is noted that the parameter d is a function of σ, which is unknown. As in the single-sample t-test (Section 10-4), we may have to rely on a prior estimate of σ, or use a subjective estimate. Alternatively, we could define the differences in the mean that we wish to detect relative to σ.

● **Example 10-11.** Consider the catalyst experiment in Example 10-9. Suppose that if catalyst 2 produces a yield that differs from the yield of catalyst 1 by 3.0 percent we would like to reject the null hypothesis with probability at least .85. What sample size is required? Using $s_p = 1.99$ as a rough estimate of the common standard deviation σ, we have $d = |\delta|/2\sigma = |3.00|/(2)(1.99) = .75$. From Appendix Chart VI$e$ with $d = .75$ and $\beta = .15$, we find $n^* = 20$, approximately. Therefore, since $n^* = 2n - 1$,

$$n = \frac{n^* + 1}{2} = \frac{20 + 1}{2} = 10.5 \simeq 11 \text{ (say)}$$

and we would use sample sizes of $n_1 = n_2 = n = 11$.

10-6 The Paired t-Test

A special case of the two-sample t-tests of Section 10-5 occurs when the observations on the two populations of interest are collected in pairs. Each pair of observations, say (X_{1j}, X_{2j}), are taken under homogeneous conditions, but these conditions may change from one pair to another. For example, suppose that we are interested in comparing two different types of tips for a hardness-testing machine. This machine presses the tip into a metal specimen with a known force. By measuring the depth of the depression caused by the tip, the hardness of the specimen can be determined. If several specimens were selected at random, half tested with tip 1, half tested with tip 2, and the pooled or independent t-test in Section 10-5 applied, the results of the test could be invalid. That is, the metal specimens could have been cut from bar stock that was produced in different heats, or they may not be homogeneous in some other way that might affect hardness; then the observed difference between mean hardness readings for the two tip types also includes hardness differences between specimens.

The correct experimental procedure is to collect the data in *pairs*; that is, to make two hardness readings on each specimen, one with each tip. The test procedure would then consist of analyzing the *differences* between hardness readings on each specimen. If there is no difference between tips, then the mean of the differences should be zero. This test procedure is called the *paired t-test*.

Let (X_{11}, X_{21}), (X_{12}, X_{22}), ..., (X_{1n}, X_{2n}) be a set of n paired observations, where we assume that $X_1 \sim N(\mu_1, \sigma_1^2)$ and $X_2 \sim N(\mu_2, \sigma_2^2)$. Define the differences between each pair of observations as $D_j = X_{1j} - X_{2j}, j = 1, 2, \ldots, n$. Assume that the differences are normally and independently distributed random variables with mean μ_D and variance σ_D^2. Testing the hypothesis $H_0: \mu_1 = \mu_2$ against $H_1: \mu_1 \neq \mu_2$ is equivalent to testing

$$H_0: \mu_D = 0$$
$$H_1: \mu_D \neq 0 \tag{10-64}$$

The appropriate test statistic for Equation (10-64) is

$$t_0 = \frac{\bar{D}}{S_D/\sqrt{n}} \tag{10-65}$$

where

$$\bar{D} = \frac{\sum\limits_{i=1}^{n} D_i}{n} \qquad (10\text{-}66)$$

and

$$S_D^2 = \frac{\sum\limits_{i=1}^{n} D_i^2 - \left[\left(\sum\limits_{j=1}^{n} D_j\right)^2 / n\right]}{n-1} \qquad (10\text{-}67)$$

are the sample mean and variance of the differences. We would reject $H_0: \mu_D = 0$ (implying that $\mu_1 \neq \mu_2$) if $t_0 > t_{\alpha/2, n-1}$ or if $t_0 < -t_{\alpha/2, n-1}$. One-sided alternatives would be treated similarly.

● **Example 10-12.** Consider the hardness-testing experiment discussed above. Suppose that eight metal specimens are selected, and each specimen is tested with both tips. The resulting data (after coding) are shown in the following table.

Specimen	Tip 1	Tip 2	Difference, d_i
1	4	3	1
2	3	3	0
3	3	5	-2
4	4	3	1
5	4	4	0
6	3	2	1
7	2	4	-2
8	2	2	0

The sample mean and variance of the differences are

$$\bar{d} = \frac{\sum\limits_{i=1}^{8} d_i}{8} = \frac{-1}{8} = -.125$$

and

$$s_d^2 = \frac{\sum\limits_{j=1}^{8} d_j^2 - \left[\left(\sum\limits_{j=1}^{8} d_j\right)^2 / 8\right]}{7} = \frac{11 - \frac{(-1)^2}{8}}{7} = 1.55$$

Therefore the test statistic is

$$t_0 = \frac{\bar{d}}{s_d/\sqrt{n}} = \frac{-.125}{\sqrt{1.55/8}} = -.28$$

and since $t_{.025, 7} = 2.365$, we cannot reject H_0. Therefore, we conclude that the two tips produce the same mean hardness readings.

10-7 Tests of Hypotheses on the Variance of a Normal Distribution

10-7.1 Statistical Analysis

Suppose that we wish to test the hypothesis that the variance of a normal distribution σ^2 equals a specified value, say σ_0^2. Unlike the t-tests of Section 10-4, 10-5, and 10-6, tests on variances are rather sensitive to the normality assumption. Let $X \sim N(\mu, \sigma^2)$, where μ and σ^2 are unknown, and let $X_1, X_2, \ldots, X_n$ be a random sample of n observations from X. To test

$$H_0: \sigma^2 = \sigma_0^2$$
$$H_1: \sigma^2 \neq \sigma_0^2 \tag{10-68}$$

we use the test statistic

$$\chi_0^2 = \frac{(n-1)S^2}{\sigma_0^2} \tag{10-69}$$

where S^2 is the sample variance. Now if $H_0: \sigma^2 = \sigma_0^2$ is true, then the test statistic χ_0^2 follows the chi-square distribution with $n-1$ degrees of freedom. Therefore, $H_0: \sigma^2 = \sigma_0^2$ would be rejected if

$$\chi_0^2 > \chi_{\alpha/2, n-1}^2 \tag{10-70a}$$

or if

$$\chi_0^2 < \chi_{1-\alpha/2, n-1}^2 \tag{10-70b}$$

where $\chi_{\alpha/2, n-1}^2$ and $\chi_{1-\alpha/2, n-1}^2$ are the upper and lower $\alpha/2$ percentage points of the chi-square distribution with $n-1$ degrees of freedom.

The same test statistic is used for the one-sided alternatives. For the one-sided hypothesis

$$H_0: \sigma^2 = \sigma_0^2$$
$$H_1: \sigma^2 > \sigma_0^2$$

we would reject H_0 if

$$\chi_0^2 > \chi_{\alpha, n-1}^2 \tag{10-72}$$

For the other one-sided hypothesis

$$H_0: \sigma^2 = \sigma_0^2$$
$$H_1: \sigma^2 < \sigma_0^2 \tag{10-73}$$

we would reject H_0 if

$$\chi_0^2 < \chi_{1-\alpha, n-1}^2 \tag{10-74}$$

● **Example 10-13.** Consider the machine described in Example 9-12, which is used to fill cans with a soft drink beverage. If the variance of the fill volume exceeds .02 (fluid ounces)^2, then an unacceptably large percentage of the cans will be underfilled. The bottler is interested in testing the hypothesis

$$H_0: \sigma^2 = .02$$
$$H_1: \sigma^2 > .02$$

A random sample of $n = 20$ cans yields a sample variance of $s^2 = .0225$. Thus, the test statistic is

$$\chi_0^2 = \frac{(n-1)s^2}{\sigma_0^2} = \frac{(19).0225}{.02} = 21.38$$

If we choose $\alpha = .05$, we find that $\chi_{.05, 19}^2 = 30.14$, and we would conclude that there is no strong evidence that the variance of fill volume exceeds .02 (fluid ounces)^2.

10-7.2   Choice of Sample Size

Operating characteristic curves for the χ^2 tests in Section 10-7.1 are provided in Appendix Charts VIi through VIn for $\alpha = .05$ and $\alpha = .01$. For the two-sided alternative hypothesis of Equation (10-68), Charts VIi and VIj plot β against an abscissa parameter

$$\lambda = \frac{\sigma}{\sigma_0} \qquad (10\text{-}75)$$

for various sample sizes n, where σ denotes the true value of the standard deviation. Charts VIk and VIl are for the one-sided alternative $H_1: \sigma^2 > \sigma_0^2$, while Charts VI$m$ and VIn are for the other one-sided alternative $H_1: \sigma^2 < \sigma_0^2$. In using these charts, we think of σ as the value of the standard deviation that we want to detect.

● **Example 10-14.** In Example 10-13, find the probability of rejecting $H_0: \sigma^2 = .02$ if the true variance is as large as $\sigma^2 = .03$. Since $\sigma = \sqrt{.03} = .1732$ and $\sigma_0 = \sqrt{.02} = .1414$, the abscissa parameter is

$$\lambda = \frac{\sigma}{\sigma_0} = \frac{.1732}{.1414} = 1.23$$

From Chart VIk, with $\lambda = 1.23$ and $n = 20$, we find that $\beta \approx .60$. That is, there is only about a 40 percent chance that $H_0: \sigma^2 = .02$ will be rejected if the variance is really as large as $\sigma^2 = .03$. To reduce β, a large sample size must be used. From the operating characteristic curve, we note that to reduce β to .20 a sample size of 75 is necessary.

10-8   Tests of Hypotheses on the Variances of Two Normal Distributions

10-8.1   Statistical Analysis

Suppose that two independent normal populations are of interest, say $X_1 \sim N(\mu_1, \sigma_1^2)$ and $X_2 \sim N(\mu_2, \sigma_2^2)$, where μ_1, σ_1^2, μ_2, and σ_2^2 are unknown. We wish to test hypotheses about the equality of the two variances, say $H_0: \sigma_1^2 = \sigma_2^2$. Assume that two random samples of size n_1 from population 1 and of size n_2 from population 2 are available, and let S_1^2 and S_2^2 be the sample variances. To test the two-sided alternative

$$H_0: \sigma_1^2 = \sigma_2^2$$
$$H_1: \sigma_1^2 \neq \sigma_2^2 \tag{10-76}$$

we use the fact that the statistic

$$F_0 = \frac{S_1^2}{S_2^2} \tag{10-77}$$

is distributed as F, with $n_1 - 1$ and $n_2 - 1$ degrees of freedom, if the null hypothesis $H_0: \sigma_1^2 = \sigma_2^2$ is true. Therefore, we would reject H_0 if

$$F_0 > F_{\alpha/2, n_1-1, n_2-1} \tag{10-78a}$$

or if

$$F_0 < F_{1-\alpha/2, n_1-1, n_2-1} \tag{10-78b}$$

where $F_{\alpha/2, n_1-1, n_2-1}$ and $F_{1-\alpha/2, n_1-1, n_2-1}$ are the upper and lower $\alpha/2$ percentage points of the F distribution with $n_1 - 1$ and $n_2 - 1$ degrees of freedom. Appendix Table V gives only the upper tail points of F, so to find $F_{1-\alpha/2, n_1-1, n_2-1}$ we must use

$$F_{1-\alpha/2, n_1-1, n_2-1} = \frac{1}{F_{\alpha/2, n_2-1, n_1-1}} \tag{10-79}$$

The same test statistic can be used to test one-sided alternative hypotheses. Since the notation X_1 and X_2 is arbitrary, let X_1 denote the population that may have the largest variance. Therefore, the one-sided alternative hypothesis is

$$H_0: \sigma_1^2 = \sigma_2^2$$
$$H_1: \sigma_1^2 > \sigma_2^2 \tag{10-80}$$

If

$$F_0 > F_{\alpha, n_1-1, n_2-1} \tag{10-81}$$

we would reject $H_0: \sigma_1^2 = \sigma_2^2$.

● **Example 10-15.** Consider the chemical process yield data in Example 9-11, where X_1 and X_2 represent process yields when two different catalysts are used. Suppose that we wish to test

$$H_0: \sigma_1^2 = \sigma_2^2$$
$$H_1: \sigma_1^2 \neq \sigma_2^2$$

Two samples of sizes $n_1 = n_2 = 8$ yield $s_1^2 = 3.89$ and $s_2^2 = 4.02$.

$$F_0 = \frac{s_1^2}{s_2^2} = \frac{3.89}{4.02} = .97$$

If $\alpha = .05$, we find that $F_{.025, 7, 7} = 4.99$ and $F_{.975, 7, 7} = (F_{.025, 7, 7})^{-1} = (4.99)^{-1} = .20$. Therefore, we cannot reject $H_0: \sigma_1^2 = \sigma_2^2$, and we can conclude that there is no strong evidence that the variance of the yield is affected by the catalyst.

10-8.2 Choice of Sample Size

Appendix Charts VIo, VIp, VIq, and VIr provide operating characteristic curves for the F-test given in Section 10-8.1, for $\alpha = .05$ and $\alpha = .01$, assuming that $n_1 = n_2 = n$. Charts VIo and VIp are used with the two-sided alternative of Equation (10-76). They plot β against the abscissa parameter

$$\lambda = \frac{\sigma_1}{\sigma_2} \tag{10-82}$$

for various $n_1 = n_2 = n$. Charts VIq and VIr are used for the one-sided alternative of Equation (10-80).

● **Example 10-16.** For the chemical process yield analyses problem in Example 10-15, suppose that if one of the catalysts affected the variance of the yield so that one of the variances was four times the other, we wished to detect this with probability at least .80. What sample size should be used? Note that if one variance is four times the other, then

$$\lambda = \frac{\sigma_1}{\sigma_2} = 2$$

By referring to Chart VIo, with $\beta = .20$ and $\lambda = 2$, we find that a sample size of $n_1 = n_2 = 20$, approximately, is necessary.

10-9 Tests of Hypotheses on a Proportion

In many engineering and management science problems, we are concerned with a random variable that follows the binomial distribution. For example, consider a production process that manufactures items that are classified as either acceptable or defective. It is usually reasonable to model the occurrence of defectives with the binomial distribution, where the binomial parameter p represents the proportion of defective items produced.

We will consider testing

$$H_0: p = p_0$$

$$H_1: p \neq p_0 \tag{10-83}$$

An approximate test based on the normal approximation to the binomial will be given. This approximate procedure will be valid as long as p is not extremely close to zero or 1, and if the sample size is relatively large. Let X be the number of observations in a random sample of size n that belongs to the class associated with p. Then, if the null hypothesis $H_0: p = p_0$ is true, we have $X \sim N(np_0, np_0(1-p_0))$, approximately. To test $H_0: p = p_0$ calculate the test statistic

$$Z_0 = \frac{X - np_0}{\sqrt{np_0(1-p_0)}} \tag{10-84}$$

and reject $H_0: p = p_0$ if

$$Z_0 > Z_{\alpha/2} \quad \text{or} \quad \text{if } Z_0 < -Z_{\alpha/2} \tag{10-85}$$

Critical regions for the one-sided alternative hypotheses would be located in the usual manner.

● **Example 10-17.** An electronics firm produces transistors. The contract with their customer calls for a fraction defective of no more than .05. They wish to test

$$H_0: p = .05$$

$$H_1: p > .05$$

A random sample of 200 transistors yields six defectives. The test statistic presented in Equation (10-84) is

$$Z_0 = \frac{x - np_0}{\sqrt{np_0(1-p_0)}} = \frac{6 - 200(.05)}{\sqrt{200(.05)(.95)}} = -1.30$$

Using $\alpha = .05$, we find that $Z_{.05} = 1.645$, and so we cannot reject the null hypothesis that $p = .02$.

10-10 Tests of Hypotheses on Two Proportions

The tests of Section 10-9 can be extended to the case where there are two binomial parameters of interest, say p_1 and p_2, and we wish to test that they are equal. That is, we wish to test

$$H_0: p_1 = p_2$$

$$H_1: p_1 \neq p_2 \tag{10-86}$$

Once again, we assume that the normal approximation to the binomial applies.

Let two random samples of sizes n_1 and n_2 be taken from two independent binomial populations, and let $\hat{p}_1 = X_1/n_1$ and $\hat{p}_2 = X_2/n_2$ be the estimates of the corresponding binomial parameters. Now, if the null hypothesis is true, then using the fact that $p_1 = p_2 = p$, the random variable

$$Z = \frac{\hat{p}_1 - \hat{p}_2}{\sqrt{p(1-p)\left[\dfrac{1}{n_1} + \dfrac{1}{n_2}\right]}}$$

is distributed approximately $N(0, 1)$. An estimate of the common parameter p is

$$\hat{p} = \frac{X_1 + X_2}{n_1 + n_2}$$

The test statistic for $H_0: p_1 = p_2$ is then

$$Z_0 = \frac{\hat{p}_1 - \hat{p}_2}{\sqrt{\hat{p}(1-\hat{p})\left[\dfrac{1}{n_1} + \dfrac{1}{n_2}\right]}} \tag{10-87}$$

If

$$Z_0 > Z_{\alpha/2} \quad \text{or} \quad \text{if } Z_0 < -Z_{\alpha/2} \tag{10-88}$$

the null hypothesis is rejected.

● **Example 10-18.** Two different types of fire control computers are being considered for use by the U.S. Army in 6-gun 105 mm batteries. The two computer systems are subjected to an operational test in which the total number of hits of the target are counted. Computer system 1 gave 250 hits out of 300 rounds, while computer system 2 gave 178 hits out of 260 rounds. Is there reason to believe that the two computer systems differ? To answer this question, we test

$$H_0: p_1 = p_2$$
$$H_1: p_1 \neq p_2$$

Note that $\hat{p}_1 = 250/300 = .8333$, $\hat{p}_2 = 178/260 = .6846$, and

$$\hat{p} = \frac{x_1 + x_2}{n_1 + n_2} = \frac{250 + 178}{300 + 260} = .7643$$

The value of the test statistic is

$$Z = \frac{\hat{p}_1 - \hat{p}_2}{\sqrt{\hat{p}(1-\hat{p})\left[\dfrac{1}{n_1} + \dfrac{1}{n_2}\right]}} = \frac{.8333 - .6846}{\sqrt{.7643(.2357)\left[\dfrac{1}{300} + \dfrac{1}{260}\right]}} = 4.13$$

If we use $\alpha = .05$, then $Z_{.025} = 1.96$ and $-Z_{.025} = -1.96$, and we would reject H_0, concluding that there is a significant difference in the two computer systems.

10-11   Testing for Goodness of Fit

The hypothesis-testing procedures that we have discussed in previous sections are for problems in which the form of the density function of the random variable is known, and the hypotheses involve the parameters of the distribution. Another kind of hypothesis is often encountered: we do not know the probability distribution of the random variable under study, say X, and we wish to test the hypothesis that X follows a particular probability distribution. For example, we might wish to test the hypothesis that X follows the normal distribution.

One procedure for testing this type of hypothesis is the chi-square "goodness of fit" test. The test procedure consists of obtaining a random sample of size n of the random variable X, whose probability density function is unknown. These n observations are arrayed in a frequency histogram, having k class intervals. Let O_i be the observed frequency in the ith class interval. From the hypothesized probability distribution we compute the expected frequency in the ith class interval, denoted E_i. The test statistic is

$$\chi_0^2 = \sum_{i=1}^{k} \frac{(O_i - E_i)^2}{E_i} \qquad (10\text{-}89)$$

It can be shown that χ_0^2 approximately follows the chi-square distribution with $k - p - 1$ degrees of freedom, where p represents the number of parameters of the hypothesized distribution estimated by sample statistics. This approximation improves as n increases. We would reject the hypothesis that X conforms to the hypothesized distribution if $\chi_0^2 > \chi_{\alpha, k-p-1}^2$.

One point to be noted in the application of this test procedure concerns the magnitude of the expected frequencies. If these expected frequencies are too small, then χ_0^2 will not reflect the departure of observed from expected, but only the smallness of the expected frequencies. There is no general agreement regarding the minimum value of expected frequencies, but values of 3, 4, and 5 are widely used as minimal. Should an expected frequency be too small, it can be combined with the expected frequency in an adjacent class interval. The corresponding observed frequencies would then be combined also, and k would be reduced by 1. Class intervals are not required to be of equal width.

● **Example 10-19.** The number of failures per shift of looms in a weaving mill is hypothesized to follow a Poisson distribution. The following data have been collected:

Number of Failures	Observed Frequency
0	32
1	15
2	9
3	4

The estimate of the mean is the average of the observations, that is, $(32 \cdot 0 + 15 \cdot 1 + 9 \cdot 2 + 4 \cdot 3)/60 = .75$. From the cumulative Poisson distribution with parameter .75 we may compute the expected frequencies as $E_i = np_i$, where p_i is the theoretical, hypothesized probability associated with the ith class interval, and n is the total number of observations. The appropriate hypotheses are

$$H_0 : p(x) = \frac{e^{-.75}(.75)^x}{x!} \qquad x = 0, 1, 2, \ldots$$

$H_1 : p(x)$ is not of the form stated in H_0

We may compute the expected frequencies as follows:

Number of Failures	Probability	Expected Frequency
0	.472	28.32
1	.354	21.24
2	.133	7.98
3	.033	1.98

The expected frequencies are obtained by multiplying the sample size times the respective probabilities. Since the expected frequency in the last cell is less than, say 3, we combine the last two cells:

Number of Failures	Observed Frequency	Expected Frequency
0	32	28.32
1	15	21.24
≥ 2	13	9.96

The test statistic becomes

$$\chi_0^2 = \frac{(32-28.32)^2}{28.32} + \frac{(15-21.24)^2}{21.24} + \frac{(13-9.96)^2}{9.96} = 3.24$$

and since $\chi_{.05,1}^2 = 3.84$, we cannot reject the hypothesis that the number of loom failures follows a Poisson distribution with mean .75.

Graphical methods are also useful when selecting a probability distribution to describe data. Probability plotting is a graphical method for determining whether the data conform to a hypothesized distribution based on a subjective visual examination of the data. The general procedure is very simple and can be performed quickly. Probability plotting requires special graph paper, known as *probability* paper, that has been designed for the hypothesized distribution. Probability paper is widely available for the normal, lognormal, Weibull, and various chi-square and gamma distributions. To construct a probability plot, the observations in the sample are first ranked from smallest to largest. That is, the sample $X_1, X_2, \ldots, X_n$ is arranged as $X_{(1)}, X_{(2)}, \ldots, X_{(n)}$, where $X_{(j)} \le X_{(j+1)}$. The ordered observations $X_{(j)}$ are then plotted against their observed cumulative frequency $(j-.5)/n$ on the appropriate probability paper. If the hypothesized distribution adequately describes the data, the plotted points will fall approximately along a straight line; if the plotted points deviate significantly from a straight line, then the hypothesized model is not appropriate. Usually, the determination of whether or not the data plot is a straight line is subjective.

● **Example 10-20.** To illustrate probability plotting, consider the following data:

$$-.314, 1.080, .863, -.179, -1.390, -.563, 1.436, 1.153, .504, -.801$$

We hypothesize that these data are adequately modeled by a normal distribution. The observations are arranged in ascending order and their cumulative frequencies $(j-.5)/n$ calculated as follows:

j	$X_{(j)}$	$(j-.5)/n$
1	−1.390	.05
2	−.801	.15
3	−.563	.25
4	−.314	.35
5	−.179	.45
6	.504	.55
7	.863	.65
8	1.080	.75
9	1.153	.85
10	1.436	.95

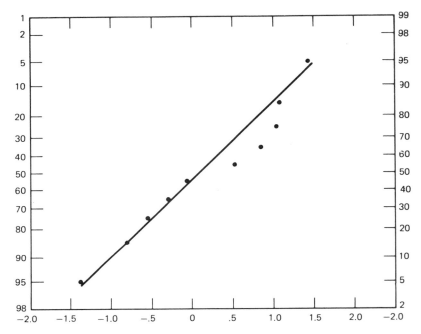

Fig. 10-7. Normal probability plot for the data.

The pairs of values $X_{(j)}$ and $(j - .5)/n$ are plotted on normal probability paper, as in Fig. 10-7. A straight line, chosen subjectively, has been drawn through these points. Since the points fall generally near the line, we conclude that a normal distribution describes the data.

We can obtain an estimate of the mean and standard deviation directly from the normal probability plot. The mean is estimated as the 50th percentile of the sample, or $\hat{\mu} = .10$ approximately, and the standard deviation is estimated as the difference between the 84th and 50th percentiles, or $\hat{\sigma} = .95 - .10 = .85$, approximately.

The choice of the distribution hypothesized to fit the data is important. Sometimes analysts can use their knowledge of the physical phenomena to choose a distribution to model the data. For example, in studying the loom failure data in Example 10-19, a Poisson distribution was hypothesized to describe the data, because failures are an "event per unit time" phenomena, and such phenomena are often well-modeled by a Poisson distribution. Sometimes previous experience can suggest the choice of distribution.

In situations where there is no previous experience or theory to suggest a distribution that describes the data, analysts must rely on other methods.

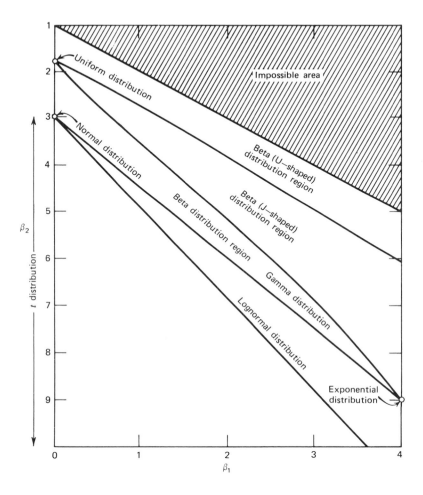

Fig. 10-8. Regions in the β_1, β_2 plane for various standard distributions. (Adapted from G. J. Hahn and S. S. Shapiro, *Statistical Models in Engineering,* John Wiley & Sons, New York, 1967; used with permission of the publisher and Professor E. S. Pearson, University of London.)

Inspection of a frequency histogram can often suggest an appropriate distribution. One may also use the display in Fig. 10-8 to assist in selecting a distribution that describes the data. This figure shows the regions in the β_1, β_2 plane for several standard probability distributions where

$$\sqrt{\beta_1} = \frac{E(X - \mu)^3}{(\sigma^2)^{3/2}}$$

is a standardized measure of skewness and

$$\beta_2 = \frac{E(X - \mu)^4}{\sigma^4}$$

is a standardized measure of kurtosis (or peakedness). To use Fig. 10-8, calculate the sample estimates of β_1 and β_2, say

$$\sqrt{\hat{\beta}_1} = \frac{M_3}{(M_2)^{3/2}}$$

and

$$\hat{\beta}_2 = \frac{M_4}{M_2^2}$$

where

$$M_j = \frac{\sum_{i=1}^{n} (X_i - \bar{X})^j}{n} \qquad j = 1, 2, 3, 4$$

and plot the point $\hat{\beta}_1, \hat{\beta}_2$ on Fig. 10-8. If this plotted point falls reasonably close to a point, line, or area that corresponds to one of the distributions given in the figure, then this distribution is a logical candidate to model the data.

From inspecting Fig. 10-8 we note that all normal distributions are represented by the point $\beta_1 = 0$ and $\beta_2 = 3$. This is reasonable, since all normal distributions have the same shape. Similarly, the exponential and uniform distributions are represented by a single point in the β_1, β_2 plane. The gamma and lognormal distributions are represented by lines, because their shapes depend on their parameter values. Note that these lines are close together, which may explain why some data sets are modeled equally well by either distribution. We also observe that there are regions of the β_1, β_2 plane for which none of the distributions in Fig. 10-8 are appropriate. Other, more general distributions, such as the Johnson and Pearson families of distributions, may be required in these cases. Procedures for fitting these families of distributions and figures similar to Fig. 10-8 are given in Hahn and Shapiro (1967).

10-12 Tests for Independence

Many times, the n elements of a sample may be classified according to two different criteria. It is then of interest to know whether the two methods of classification are statistically independent. Assume that the first method of classification has r levels and the second method of classification has c levels. We shall let O_{ij} be the observed frequency for level i of the first classification method and level j of the second classification method. The data would, in general, appear as in Table 10-2. Such a table is commonly called an $r \times c$ *contingency table*.

TABLE 10-2   **An *r* × *c* Contingency Table**

		Columns			
		1	2	$\cdots$	*c*
Rows	1	O_{11}	O_{12}	$\cdots$	O_{1c}
	2	O_{21}	O_{22}	$\cdots$	O_{2c}
	$\vdots$	$\vdots$	$\vdots$	$\cdots$	$\vdots$
	r	O_{r1}	O_{r2}	$\cdots$	O_{rc}

We are interested in testing the hypothesis that the row and column methods of classification are independent. If we reject this hypothesis, we conclude there is some *interaction* between the two criteria of classification. The exact test procedures are difficult to obtain, but an approximate test statistic is valid for large *n*. Assume the O_{ij} to be multinomial random variables and p_{ij} to be the probability that a randomly selected element falls in the *ij*th cell, given that the two classifications are independent. Then $p_{ij} = u_i v_j$, where u_i is the probability that a randomly selected element falls in row class *i* and v_j is the probability that a randomly selected element falls in column class *j*. Now, assuming independence, the maximum likelihood estimators of u_i and v_j are

$$\hat{u}_i = \frac{1}{n} \sum_{j=1}^{c} O_{ij}$$

$$\hat{v}_j = \frac{1}{n} \sum_{i=1}^{r} O_{ij} \tag{10-90}$$

Therefore, the expected number in each cell is

$$E_{ij} = n\hat{u}_i\hat{v}_j = \frac{1}{n} \sum_{j=1}^{c} O_{ij} \sum_{i=1}^{r} O_{ij} \tag{10-91}$$

Then, for large *n*, the statistic

$$\chi_0^2 = \sum_{i=1}^{r} \sum_{j=1}^{c} \frac{(O_{ij} - E_{ij})^2}{E_{ij}} \sim \chi_{(r-1)(c-1)}^2 \tag{10-92}$$

and we would reject the hypothesis of independence if $\chi_0^2 > \chi_{\alpha,\,(r-1)(c-1)}^2$.

● **Example 10-21.** A company has to choose among three pension plans. Management wishes to know whether the preference for plans is independent

TABLE 10-3 **Observed Data for Example 10-21**

	Pension Plan			
	1	2	3	Total
Salaried workers	160	140	40	340
Hourly workers	40	60	60	160
Totals	200	200	100	500

of job classification. The opinions of a random sample of 500 workers are shown in Table 10-3.

We may compute $\hat{u}_1 = (340/500) = .68$, $\hat{u}_2 = (160/500) = .32$, $\hat{v}_1 = (200/500) = .40$, $\hat{v}_2 = (200/500) = .40$, and $\hat{v}_3 = (100/500) = .20$. The expected frequencies may be computed from Equation (10-91). For example, the expected number of salaried workers favoring pension plan 1 is

$$E_{11} = n\hat{u}_1\hat{v}_1 = 500(.68)(.40) = 136$$

The expected frequencies are shown in Table 10-4.

The test statistic is computed as follows:

$$\chi_0^2 = \sum_{i=1}^{2} \sum_{j=1}^{3} \frac{(O_{ij} - E_{ij})^2}{E_{ij}}$$

$$= \frac{(160 - 136)^2}{136} + \frac{(140 - 136)^2}{136} + \frac{(40 - 68)^2}{68} + \frac{(40 - 64)^2}{64}$$

$$+ \frac{(60 - 64)^2}{64} + \frac{(60 - 32)^2}{32} = 49.63$$

TABLE 10-4 **Expected Frequencies for Example 10-21**

	Pension Plan			
	1	2	3	Total
Salaried workers	136	136	68	340
Hourly workers	64	64	32	160
Totals	200	200	100	500

Since $\chi^2_{.05,2} = 5.99$, we reject the hypothesis of independence and conclude that the preference for pension plans is not independent of job classification.

The chi-square test statistic in Equation (10-92) is only an approximation. If the contingency table has only two rows and two columns, a correction known as Yates' continuity correction should be employed. The test statistic becomes

$$\chi^2_0 = \sum_{i=1}^{2} \sum_{j=1}^{2} \frac{(|O_{ij} - E_{ij}| - .5)^2}{E_{ij}} \qquad (10\text{-}93)$$

Contingency tables may also be used to test the equality of k binomial parameters, say $p_1, p_2, \ldots, p_k$. The test procedure consists of taking a random sample of n_i observations from the ith population, $i = 1, 2, \ldots, k$, classifying the sample observations into two categories (one associated with p, and one with $1 - p$), and arranging the results as a $2 \times k$ contingency table. Expected frequencies are calculated from Equation (10-91). Then the test statistic of Equation (10-92) is used to test the equality of the binomial parameters $p_1, p_2, \ldots, p_k$.

10-13   Nonparametric Tests

Most of the statistical tests that we have presented so far assume that the observations are taken from a normal population. While many of these procedures still work reasonably well when the underlying population is nonnormal, analysts are becoming more interested in procedures that do not require any distributional assumptions regarding the underlying population. Such procedures are usually called nonparametric tests. Most nonparametric tests require only that the underlying distribution is continuous.

Nonparametric tests have the advantage of computational simplicity. Furthermore, they may be applied to qualitative data, or data expressed as ranks. However, a nonparametric test will have a larger type II error than the corresponding parametric test when both procedures can be used. In this section we present three useful nonparametric tests: the Wilcoxon two-sample test, the sign test, and the signed-rank test.

10-13.1   The Wilcoxon Two-Sample Test for Independent Samples

Suppose that we have two independent nonnormal populations that may differ only in their means μ_1 and μ_2. We wish to test the hypothesis

$$H_0 : \mu_1 = \mu_2$$
$$H_1 : \mu_1 \neq \mu_2 \qquad (10\text{-}94)$$

A random sample of size n_1 is taken from population 1 and a random sample of size n_2 is taken from population 2. We assume that $n_1 \leq n_2$. Arrange all $n_1 + n_2$ observations in ascending order of magnitude and assign ranks to them. If two or more observations are tied (identical), then use the mean of the ranks that would have been assigned if the observations differed. Let R_1 be the sum of the ranks in the smaller sample (1), and define

$$R_2 = n_1(n_1 + n_2 + 1) - R_1 \qquad (10\text{-}95)$$

Now if the sample means do not differ, we would expect the sum of the ranks to be nearly equal for both. Consequently, if the sums of the ranks differ greatly, we would conclude that the means are not equal.

Appendix Table IX contains the critical value of the rank sums for $\alpha = .05$ and $\alpha = .01$. Refer to Table IX with the appropriate sample sizes n_1 and n_2 and the critical value R_α^* can be obtained. The null hypothesis in Equation (10-94) is rejected if either R_1 or R_2 is less than or equal to the tabulated critical value R_α^*.

The procedure can also be used for one-sided alternatives. If the alternative is $H_1: \mu_1 < \mu_2$, then reject H_0 if $R_1 \leq R_\alpha^*$; while for $H_0: \mu_1 > \mu_2$, reject H_0 if $R_2 \leq R_\alpha^*$. For these one-sided tests the tabulated critical values R_α^* correspond to levels of significance of $\alpha = .025$ and $\alpha = .005$.

When the sample sizes n_1 and n_2 are large, the sampling distribution of R_1 approaches the normal with mean

$$\mu_R = \frac{n_1(n_1 + n_2 + 1)}{2}$$

and variance

$$\sigma_R^2 = \frac{n_1 n_2(n_1 + n_2 + 1)}{12}$$

Therefore, for large sample sizes we could use

$$Z_0 = \frac{R_1 - \mu_R}{\sigma_R} \qquad (10\text{-}96)$$

as a test statistic, and the appropriate critical region as $|Z_0| > Z_{\alpha/2}$, $Z_0 > Z_\alpha$, or $Z < -Z_\alpha$, depending on whether the test is a two-tailed, upper-tail, or lower-tail test.

In many practical situations, we find that the two populations being studied are not even approximately normally distributed, so the parametric t-test is not valid. Furthermore, if the two populations *are* normal, the Wilcoxon two-sample test is almost as efficient as the t-test.

● **Example 10-22.** The mean axial stress in tensile members formed of two different types of aluminum alloy is being studied. Ten specimens of each alloy type are tested, and the axial stress measured. The sample data are assembled in the following table:

Alloy 1		Alloy 2	
3238 psi	3254 psi	3261 psi	3248 psi
3195	3229	3187	3215
3246	3225	3209	3226
3190	3217	3212	3240
3204	3241	3258	3234

The data are arranged in ascending order and ranked as follows.

Alloy Number	Axial Stress	Rank
2	3187 psi	1
1	3190	2
1	3195	3
1	3204	4
2	3209	5
2	3212	6
2	3215	7
1	3217	8
1	3225	9
2	3226	10
1	3229	11
2	3234	12
1	3238	13
2	3240	14
1	3241	15
1	3246	16
2	3248	17
1	3254	18
2	3258	19
2	3261	20

The sum of the ranks for alloy 1 are

$$R_1 = 2+3+4+8+9+11+13+15+16+18 = 99$$

and for alloy 2,

$$R_2 = n_1(n_1 + n_2 + 1) - R_1 = 10(10+10+1) - 99 = 111$$

From Appendix Table IX, with $n_1 = n_2 = 10$ and $\alpha = .05$, we find that $R_{.05}^* = 78$. Since neither R_1 nor R_2 is less than $R_{.05}^*$, we cannot reject the hypothesis that both aluminum alloys exhibit the same mean axial stress.

10-13.2 The Sign Test

The sign test is applied to paired observations drawn from continuous populations. Let (X_{1j}, X_{2j}), $j = 1, 2, \ldots, n$ be a collection of paired observations, and let $D_j = X_{1j} - X_{2j}$, $j = 1, 2, \ldots, n$ be the differences for each pair. The hypothesis to be tested is that the differences D_j have a probability distribution with a zero median. If the underlying distributions are assumed to be symmetric, then the hypothesis tested is that the mean of the distribution of differences μ_D is zero. Note that this is equivalent to testing that the means of the underlying populations μ_1 and μ_2 do not differ.

The test procedure is as follows. For each difference D_j record the *sign* of the difference. Let R^+ be the number of positive signs and R^- be the number of negative signs. Denote $R = \min(R^+, R^-)$. Ties in the data ($D_j = 0$) should not appear, because the underlying populations are assumed to be continuous, although they sometimes appear due to the way the observations are measured. When ties occur, they should be disregarded and the test applied to the remaining data. Let R_α^* be the critical value corresponding to a level of significance α for the number of times the less frequent sign appears. A table of critical values is given in Appendix Table X. If the test statistic $R \le R_\alpha^*$, then the null hypothesis that the two distributions have the same means (or medians) is rejected.

Appendix Table X can also be applied to the one-sided alternative hypothesis. If the alternative is $H_1 : \mu_1 > \mu_2$ (i.e., $H_1 : \mu_D > 0$), then reject H_0 if $R^- < R_\alpha^*$; if the alternative is $H_1 : \mu_1 < \mu_2$ (i.e., $H_1 : \mu_D < 0$), then reject H_0 if $R^+ < R_\alpha^*$. The level of significance on one-sided tests is one-half the value appearing in Table X.

● **Example 10-23.** An automotive engineer is investigating two different types of metering devices for an electronic fuel injection system to determine if they differ in their fuel mileage performance. The system is installed on 12 different cars, and a test is run with each metering system on each car. The observed fuel mileage performance data, corresponding differences, and their signs are shown in the table at the top of page 311. Note that $R^+ = 8$ and $R^- = 4$. Therefore $R = \min(R^+, R^-) = \min(8, 4) = 4$. From Appendix Table X, with $n = 12$, we find the critical value for $\alpha = .05$ is $R_{.05}^* = 2$. Since R is not less than the critical value $R_{.05}^*$, we cannot reject the null hypothesis that the two metering devices produce different fuel mileage performance.

For large sample sizes, say $n > 40$, the distribution of R is approximately normal with mean $n/2$ and variance $n/4$. Therefore, in these cases the null

	Metering Device			
Car	1	2	Difference, D_i	Sign
1	17.6	16.8	.8	+
2	19.4	20.0	−.6	−
3	19.5	18.2	1.3	+
4	17.1	16.4	.7	+
5	15.3	16.0	−.7	−
6	15.9	15.4	.5	+
7	16.3	16.5	−.2	−
8	18.4	18.0	.4	+
9	17.3	16.4	.9	+
10	19.1	20.1	−.1	−
11	17.8	16.7	1.1	+
12	18.2	17.9	.3	+

hypothesis can be tested with the statistic

$$Z_0 = \frac{2R - n}{\sqrt{n}} \tag{10-97}$$

and a critical region selected to reflect the sense of the alternate hypothesis and desired type I error.

A disadvantage of the sign test is that it considers only the signs of the differences and not their magnitudes. It is possible to devise a better test that treats the signed ranks of the differences. This test is described in the next section.

10-13.3 The Wilcoxon Signed-Rank Test

We consider the same hypothesis-testing problem as in Section 10-13.2. In the Wilcoxon signed-rank test, the differences are first ranked in ascending order of their absolute values, and then the ranks are given the signs of the differences. Ties are assigned average ranks. Let R^+ be the sum of the positive ranks and R^- be the absolute value of the sum of the negative ranks, and $R = \min(R^+, R^-)$. Appendix Table XI contains critical values of R, say R_α^*. If $R \leq R_\alpha^*$, the hypothesis is rejected.

For one-sided tests, if the alternative is $H_1: \mu_1 > \mu_2$ (or $H_1: \mu_D > 0$), reject H_0 if $R^- < R_\alpha^*$; and if $H_1: \mu_1 < \mu_2$ (or $H_1: \mu_D < 0$), reject H_0 if $R^+ < R_\alpha^*$. Note that the significance level for one-sided tests is one-half the value given in Table XI.

Like the Wilcoxon two-sample test for independent samples, the Wilcoxon signed-rank test can be used in situations where the usual t-tests are not valid.

Furthermore, it compares favorably with the t-test if the populations are normally distributed. Therefore, it would be useful in many situations.

● **Example 10-24.** Consider the automobile data in Example 10-23. The signed ranks are shown in the following table:

Car	Difference	Signed Rank
7	−.2	−1
12	.3	2
8	.4	3
6	.5	4
2	−.6	−5
4	.7	6.5
5	−.7	−6.5
1	.8	8
9	.9	9
10	−1.0	−10
11	1.1	11
3	1.3	12

Note that $R^+ = 55.5$ and $R^- = 22.5$; therefore, $R = \min(R^+, R^-) = \min(55.5, 22.5) = 22.5$. From Appendix Table XI, with $n = 12$ and $\alpha = .05$, we find the critical value $R^*_{.05} = 13$. Since R exceeds $R^*_{.05}$, we cannot reject the null hypothesis that the two metering devices produce the same mileage performance.

For large sample sizes, say $n > 50$, we may use the fact that the distribution of R is approximately normal with mean

$$\mu_R = \frac{n(n+1)}{4}$$

and variance

$$\sigma_R^2 = \frac{n(n+1)(2n+1)}{24}$$

Therefore, a test can be based on the statistic

$$Z_0 = \frac{R - \mu_R}{\sigma_R} \tag{10-98}$$

and an appropriate critical region can be obtained from the table of the standard normal distribution.

TABLE 10-5 **Summary of Hypothesis Testing Procedures on Means and Variances**

Null Hypothesis	Test Statistic	Alternative Hypothesis	Criteria for Rejection	OC Curve Parameter
$H_0 : \mu = \mu_0$ σ^2 known	$Z_0 = \dfrac{\bar{X} - \mu_0}{\sigma/\sqrt{n}}$	$H_1 : \mu \neq \mu_0$ $H_1 : \mu > \mu_0$ $H_1 : \mu < \mu_0$	$\lvert Z_0 \rvert > Z_{\alpha/2}$ $Z_0 > Z_\alpha$ $Z_0 < -Z_\alpha$	$d = \lvert \mu - \mu_0 \rvert / \sigma$ $d = (\mu - \mu_0)/\sigma$ $d = (\mu_0 - \mu)/\sigma$
$H_0 : \mu = \mu_0$ σ^2 unknown	$t_0 = \dfrac{\bar{X} - \mu_0}{S/\sqrt{n}}$	$H_1 : \mu \neq \mu_0$ $H_1 : \mu > \mu_0$ $H_1 : \mu < \mu_0$	$\lvert t_0 \rvert > t_{\alpha/2, n-1}$ $t_0 > t_{\alpha, n-1}$ $t_0 < -t_{\alpha, n-1}$	$d = \lvert \mu - \mu_0 \rvert / \sigma$ $d = (\mu - \mu_0)/\sigma$ $d = (\mu_0 - \mu)/\sigma$
$H_0 : \mu_1 = \mu_2$ σ_1^2 and σ_2^2 known	$Z_0 = \dfrac{\bar{X}_1 - \bar{X}_2}{\sqrt{\dfrac{\sigma_1^2}{n_1} + \dfrac{\sigma_2^2}{n_2}}}$	$H_1 : \mu_1 \neq \mu_2$ $H_1 : \mu_1 > \mu_2$ $H_1 : \mu_1 < \mu_2$	$\lvert Z_0 \rvert > Z_{\alpha/2}$ $Z_0 > Z_\alpha$ $Z_0 < -Z_\alpha$	$d = \lvert \mu_1 - \mu_2 \rvert / \sqrt{\sigma_1^2 + \sigma_2^2}$ $d = (\mu_1 - \mu_2)/\sqrt{\sigma_1^2 + \sigma_2^2}$ $d = (\mu_2 - \mu_1)/\sqrt{\sigma_1^2 + \sigma_2^2}$
$H_0 : \mu_1 = \mu_2$ $\sigma_1^2 = \sigma_2^2 = \sigma^2$ unknown	$t_0 = \dfrac{\bar{X}_1 - \bar{X}_2}{S_p \sqrt{\dfrac{1}{n_1} + \dfrac{1}{n_2}}}$	$H_1 : \mu_1 \neq \mu_2$ $H_1 : \mu_1 > \mu_2$ $H_1 : \mu_1 < \mu_2$	$\lvert t_0 \rvert > t_{\alpha/2, n_1+n_2-2}$ $t_0 > t_{\alpha, n_1+n_2-2}$ $t_0 < -t_{\alpha, n_1+n_2-2}$	$d = \lvert \mu_1 - \mu_2 \rvert / 2\sigma$ $d = (\mu_1 - \mu_2)/2\sigma$ $d = (\mu_2 - \mu_1)/2\sigma$
$H_0 : \mu_1 = \mu_2$ $\sigma_1^2 \neq \sigma_2^2$ unknown	$t_0 = \dfrac{\bar{X}_1 - \bar{X}_2}{\sqrt{\dfrac{S_1^2}{n_1} + \dfrac{S_2^2}{n_2}}}$ $\nu = \dfrac{\left(\dfrac{S_1^2}{n_1} + \dfrac{S_2^2}{n_2}\right)^2}{\dfrac{(S_1^2/n_1)^2}{n_1 + 1} + \dfrac{(S_2^2/n_2)^2}{n_2 + 1}} - 2$	$H_1 : \mu_1 \neq \mu_2$ $H_1 : \mu_1 > \mu_2$ $H_1 : \mu_1 < \mu_2$	$\lvert t_0 \rvert > t_{\alpha/2, \nu}$ $t_0 > t_{\alpha, \nu}$ $t_0 < -t_{\alpha, \nu}$	— — —
$H_0 : \sigma^2 = \sigma_0^2$	$\chi_0^2 = \dfrac{(n-1)S^2}{\sigma_0^2}$	$H_1 : \sigma^2 \neq \sigma_0^2$ $H_1 : \sigma^2 > \sigma_0^2$ $H_1 : \sigma^2 < \sigma_0^2$	$\chi_0^2 > \chi_{\alpha/2, n-1}^2$ or $\chi_0^2 < \chi_{1-\alpha/2, n-1}^2$ $\chi_0^2 > \chi_{\alpha, n-1}^2$ $\chi_0^2 < \chi_{1-\alpha, n-1}^2$	$\lambda = \sigma/\sigma_0$ $\lambda = \sigma/\sigma_0$ $\lambda = \sigma/\sigma_0$
$H_0 : \sigma_1^2 = \sigma_2^2$	$F_0 = S_1^2/S_2^2$	$H_1 : \sigma_1^2 \neq \sigma_2^2$ $H_1 : \sigma_1^2 > \sigma_2^2$	$F_0 > F_{\alpha/2, n_1-1, n_2-1}$ or $\bar{F}_0 < F_{1-\alpha/2, n_1-1, n_2-1}$ $F_0 > F_{\alpha, n_1-1, n_2-1}$	$\lambda = \sigma_1/\sigma_2$ $\lambda = \sigma_1/\sigma_2$

10-14 Summary

This chapter has introduced hypothesis testing. Procedures for testing hypotheses on means and variances are summarized in Table 10-5. The chi-square goodness of fit test was introduced to test the hypothesis that an empirical distribution follows a particular probability law. Graphical methods are also useful in goodness of fit testing, particularly when sample sizes are small. Contingency tables for testing the hypothesis that two methods of classification of a sample are independent were also introduced. Nonparametric statistical tests that do not rely on the assumption of any particular probability distribution for the underlying populations were also presented. Two of these tests, the Wilcoxon test for two independent samples and the Wilcoxon rank sum test for paired samples, are almost as efficient as the parametric t-test when the underlying populations are normal and superior when they are nonnormal. Consequently, these tests are good alternatives to the t-test in many practical situations.

10-15 Exercises

10-1. The breaking strength of a certain type of fiber used in manufacturing cloth is required to be not less than 160 psi. Past experience has indicated that the standard deviation of breaking strength is 3 psi. A random sample of four specimens is tested and the average breaking strength is found to be 158 psi.
(a) Should the fiber be judged acceptable with $\alpha = .05$?
(b) What is the probability of accepting $H_0: \mu \leq 160$ if the fiber has a true breaking strength of 165 psi?

10-2. The yield of a chemical process is being studied. The variance of yield is known from previous experience with this process to be 5 (units of $\sigma^2 =$ percentage^2). The past five days of plant operation have resulted in the following yields (in percentages): 91.6, 88.75, 90.8, 89.95, 91.3.
(a) Is there reason to believe the yield is less than 90 percent?
(b) What sample size would be required to detect a true mean yield of 85 percent with probability .95?

10-3. The diameters of bolts produced by a certain manufacturing process are known to have a standard deviation of .0001 inch. A random sample of 10 bolts yields an average diameter of .2546 inch.
(a) Test the hypothesis that the true mean diameter of bolts equals .255 inch, using $\alpha = .05$.
(b) What size sample would be necessary to detect a true mean bolt diameter of .2552 inch with probability at least .90?

10-4. Consider the data in Exercise 9-26.
(a) Test the hypothesis that the mean piston ring diameter is 74.035 mm. Use $\alpha = .01$.
(b) What sample size is required to detect a true meaning diameter of 74.030 with probability at least .95?

10-5. Consider the data in Exercise 9-27. Test the hypothesis that the mean life of the light bulbs is 1000 hours. Use $\alpha = .05$.

10-6. Consider the data in Exercise 9-28. Test the hypothesis that mean compressive strength equals 3500 psi. Use $\alpha = .01$.

10-7. Two machines are used for filling plastic bottles with a net volume of 16.0 ounces. The filling processes can be assumed normal, with standard deviations $\sigma_1 = .015$ and $\sigma_2 = .018$. The quality control department suspects that both machines fill to the same net volume, whether or not this volume is 16.0 ounces. A random sample is taken from the output of each machine.

Machine 1		Machine 2	
16.03	16.01	16.02	16.03
16.04	15.96	15.97	16.04
16.05	15.98	15.96	16.02
16.05	16.02	16.01	16.01
16.02	15.99	15.99	16.00

(a) Do you think the quality control department is correct? Use $\alpha = .05$.

(b) Assuming equal sample sizes, what sample size should be used to assure that $\beta = .05$ if the true difference in means is .075? Assume that $\alpha = .05$.

(c) What is the power of the test in (a) for a true difference in means of .075?

10-8. Two types of plastic are suitable for use by an electronics component manufacturer. The breaking strength of this plastic is important. It is known that $\sigma_1 = \sigma_2 = 1.0$ psi. From a random sample of size $n_1 = 10$ and $n_2 = 12$ we obtain $\bar{x}_1 = 162.5$ and $\bar{x}_2 = 155.0$. The company will not adopt plastic 1 unless its breaking strength exceeds that of plastic 2 by at least 10 psi. Based on the sample information, should they use plastic 1?

10-9. Consider the data in Exercise 9-32. Test the hypothesis that both machines fill to the same volume. Use $\alpha = .10$.

10-10. Consider the data in Exercise 9-33. Test $H_0: \mu_1 = \mu_2$ against $H_1: \mu_1 > \mu_2$, using $\alpha = .05$.

10-11. Consider the gasoline road octane number data in Exercise 9-34. If formulation 2 produces a higher road octane number than formulation 1, the manufacturer would like to detect this. Formulate and test an appropriate hypothesis, using $\alpha = .05$.

10-12. The lateral deviation in yards of a certain type of mortar shell is being investigated by the propellant manufacturer. The following data have been observed.

Round	Deviation	Round	Deviation
1	11.28	6	−9.48
2	−10.42	7	6.25
3	−8.51	8	10.11
4	1.95	9	−8.65
5	6.47	10	−.68

Test the hypothesis that the mean lateral deviation of these mortar shells is zero. Assume that lateral deviation is normally distributed.

10-13. The shelf life of a battery is of interest to retail sales outlets. The manufacturer observes the following shelf life for eight batteries chosen at random from the current production. Assume that shelf life is normally distributed.

108 days	138 days
124	163
124	159
106	134

(a) Is there any evidence that the mean shelf life is greater than or equal to 125 days?

(b) If it is important to detect a ratio of δ/σ of 1.0 with probability .90, is the sample size sufficient?

10-14. The copper content of an alloy is being studied in the hope of ultimately reducing the manufacturing cost. An analysis of six recent melts chosen at random produces the following copper contents.

8.031%	7.745%
9.994	11.652
9.920	14.640

Is there any evidence that the copper content is greater than 9.5 percent?

10-15. The time to repair an electronic instrument is a normally distributed random variable measured in hours. The repair times for 16 such instruments, chosen at random, are as follows.

Hours			
159	280	101	212
224	379	179	264
222	362	168	250
149	260	485	170

Does it seem reasonable that the true mean repair time is greater than 225 hours?

10-16. The percentage of scrap produced in a metal finishing operation is hypothesized to be less than 7 percent. Several days were chosen at random and the percentages of scrap were calculated.

5.5151%	7.3257%
6.4970	8.8199
6.4601	8.5618
5.3723	

(a) In your opinion, is the true scrap rate less than 7 percent?

(b) If it is important to detect a ratio of δ/σ of 1.5 with probability at least .90, what is the minimum sample size that can be used?

(c) For δ/σ of 2.0, what is the power of the above test?

10-17. Suppose that we must test the hypotheses

$$H_0 : \mu \geq 15$$
$$H_1 : \mu < 15$$

where it is known that $\sigma^2 = 2.5$. If $\alpha = .05$ and the true mean is 13, what sample size is necessary to assure a type II error of 5 percent?

10-18. An engineer desires to test the hypothesis that the melting point of an alloy is 1000°C. If the true melting point differs from this by more than 25° he must change the alloy's composition. If we assume that the melting point is a normally distributed random variable, $\alpha = .05$, and $\sigma = 10°C$, how many observations should be taken?

10-19. Two methods for producing gasoline from crude oil are being investigated. The yields of both processes are assumed to be normally distributed. The following yield data have been obtained from the pilot plant.

Process	Yields (%)					
1	24.2	26.6	25.7	24.8	25.9	26.5
2	21.0	22.1	21.8	20.9	22.4	22.0

(a) Is there reason to believe that process 1 has a greater mean yield? Use $\alpha = .01$. Assume that both variances are equal.

(b) Assuming that in order to adopt process 1 it must produce a mean yield that is at least 5 percent greater than that of process 2, what are your recommendations?

(c) Find the power of the test in part (a) if the mean yield of process 1 is 5 percent greater than that of process 2.

(d) What sample size is required for the test in part (a) to ensure that the null hypothesis will be rejected with probability .90 if the mean yield of process 1 exceeds the mean yield of process 2 by 5 percent.

10-20. The resistance of two wires is being investigated, with the following sample information.

Wire	Resistance (ohms)					
1	.140	.141	.139	.140	.138	.144
2	.135	.138	.140	.139	—	—

Assuming that the two variances are equal, what conclusions can be drawn regarding the mean resistance of the wires?

10-21. The following are the burning times of flares of two different types.

Type 1		Type 2	
65	82	64	56
81	67	71	69
57	59	83	74
66	75	59	82
82	70	65	79

(a) Test the hypothesis that the two variances are equal. Use $\alpha = .05$.

(b) Using the results of (a), test the hypothesis that the mean burning times are equal.

10-22. A new filtering device is installed in a chemical unit. Before its installation, a random sample yielded the following information about the percentage of impurity: $\bar{x}_1 = 12.5$, $s_1^2 = 101.17$, and $n_1 = 8$. After installation, a random sample yielded $\bar{x}_2 = 10.2$, $s_2^2 = 94.73$, $n_2 = 9$.

(a) Can you conclude that the two variances are equal?

(b) Has the filtering device reduced the percentage of impurity significantly?

10-23. Suppose that two random samples were drawn from normal populations with equal variances. The sample data yields $\bar{x}_1 = 20.0$, $n_1 = 10$, $\Sigma(x_{1i} - \bar{x}_1)^2 = 1480$, $\bar{x}_2 = 15.8$, $n_2 = 10$, and $\Sigma(x_{2i} - \bar{x}_2)^2 = 1401$.

(a) Test the hypothesis that the two means are equal. Use $\alpha = .01$.

(b) Find the probability that the null hypothesis in (a) will be rejected if the true difference in means is 10.

(c) What sample size is required to detect a true difference in means of 5 with probability at least .80 if it is known at the start of the experiment that a rough estimate of the common variance is 150?

10-24. Consider the data in Exercise 9-43.

(a) Test the hypothesis that the means of the two normal distributions are equal. Use $\alpha = .05$ and assume that $\sigma_1^2 = \sigma_2^2$.

(b) What sample size is required to detect a difference in means of 2.0 with probability at least .85?

(c) Test the hypothesis that the variances of the two distributions are equal. Use $\alpha = .05$.

(d) Find the power of the test in (c) if the variance of a population is four times the other.

10-25. Consider the data in Exercise 9-44. Assuming that $\sigma_1^2 = \sigma_2^2$, test the hypothesis that the mean diameters of rods produced on the two different types of machines do not differ. Use $\alpha = .05$.

10-26. A chemical company produces a certain drug whose weight has a standard deviation of 4 milligrams. A new method of producing this drug has been proposed, although some additional cost is involved. Management will authorize a change in production technique only if the standard deviation of the weight in the new process is less than 4 milligrams. If the standard deviation of weight in the new process is as small as 3 milligrams, the company would like to switch production methods with probability at least .90. Assuming weight to be normally distributed and $\alpha = .05$, how many observations should be taken? Suppose the researchers choose $n = 10$ and obtain the data below. Is this a good choice for n? What should be their decision?

16.628 grams	16.630 grams
16.622	16.631
16.627	16.624
16.623	16.622
16.618	16.626

10-27. A manufacturer of precision measuring instruments claims that the standard deviation in the use of the instrument is .00002 inch. An analyst, who is unaware of the claim, uses the instrument 8 times and obtains a sample standard deviation of .00005 inch.

(a) Using $\alpha = .01$, is the claim justified?

(b) Compute a 99 percent confidence interval for the true variance.

(c) What is the power of the test if the true standard deviation equals .00004?

(d) What is the smallest sample size that can be used to detect a true standard deviation of .00004 with probability at least .95? Use $\alpha = .01$.

10-28. The standard deviation of measurements made by a special thermocouple is supposed to be .005 degree. If the standard deviation is as great as .010 we wish to detect it with probability at least .90. Use $\alpha = .01$. What sample size should be used? If this sample size is used and the sample standard deviation $s = .007$, what is your conclusion, using $\alpha = .01$? Construct a 95 percent upper-confidence interval for the true variance.

10-29. Suppose that $X \sim N(\mu, \sigma^2)$, and the following random sample is available.

5.34	6.65	4.76
6.00	7.55	5.54
5.97	7.35	5.44
5.25	6.35	4.61

(a) Test the hypothesis that $\sigma^2 = 1.0$. Use $\alpha = .05$.

(b) If the true value of $\sigma^2 = 1.5$, what is the probability that the hypothesis in (a) will be rejected?

10-30. For the data in Exercise 10-7, test the hypothesis that the two variances are equal, using $\alpha = .01$. Does the result of this test influence the manner in which a test on means would be conducted? What sample size is necessary to detect $\sigma_1^2/\sigma_2^2 = 2.5$, with probability at least .90?

10-31. Consider the following two samples, drawn from two normal populations.

Sample 1	Sample 2
4.34	1.87
5.00	2.00
4.97	2.00
4.25	1.85
5.55	2.11
6.55	2.31
6.37	2.28
5.55	2.07
3.76	1.76
—	1.91
—	2.00

Is there evidence to conclude that the variance of population 1 is greater than the variance of population 2? Use $\alpha = .01$. Find the probability of detecting $\sigma_1^2/\sigma_2^2 = 4.0$.

10-32. Two machines produce metal parts. The variance of the weight of these parts is of interest. The following data have been collected.

Machine 1	Machine 2
$n_1 = 75$	$n_2 = 60$
$\bar{x}_1 = .984$	$\bar{x}_2 = .907$
$s_1^2 = 13.46$	$s_2^2 = 9.65$

(a) Test the hypothesis that the variances of the two machines are equal. Use $\alpha = .05$.

(b) Test the hypothesis that the two machines produce parts having the same mean weight. Use $\alpha = .05$.

10-33. In a particular type of hardness test a steel ball is pressed into the material being tested at a standard load. The diameter of the indentation is measured, which is related to the strength. Two types of steel balls are available, and their performance is compared on 10 specimens. Each specimen is tested twice, once with each ball. The results are given here.

Ball x	75	46	57	43	58	32	61	56	34	65
Ball y	52	41	43	47	32	49	52	44	57	60

Test the hypothesis that the two steel balls give the same hardness measurement. Use $\alpha = .05$.

10-34. The diameter of a ball bearing was measured by twelve individuals, using two different kinds of calipers. The results are shown here.

Caliper 1	Caliper 2
.265 .267	.264 .264
.265 .267	.265 .265
.266 .265	.264 .265
.267 .268	.266 .267
.267 .268	.267 .268
.265 .265	.268 .269

Is there a significant difference between the means of the population of measurements represented by the two samples? Use $\alpha = .05$.

10-35. An automobile designer has theoretical evidence that painting a racing car reduces its top speed. He selects six cars from the factory and tests them with and without paint. The results are shown here.

	Top Speed (mph)	
Car	Painted	Not Painted
1	186	189
2	185	186
3	179	183
4	184	188
5	183	185
6	186	188

Do the data support the designer's theory? Use $\alpha = .05$.

10-36. Two different kinds of shoe leather are being investigated for possible use in shoes by the U.S. Army. Since one shoe leather is substantially cheaper than the other, the Army is interested in the wear characteristics. Sixteen men are selected at random and each man wears one shoe of each type. After three months of simulated use, the wear is noted.

Man	Shoe Leather 1	Shoe Leather 2
1	.01	.02
2	.02	.04
3	.03	.03
4	.01	.01
5	.01	.04
6	.01	.06
7	.02	.01
8	.04	.01
9	.05	.05
10	.03	.01
11	.02	.04
12	.01	.03
13	.06	.04
14	.01	.06
15	.01	.03
16	.04	.01

What are your conclusions regarding the wear? Use $\alpha = .01$.

10-37. Consider the data in Exercise 9-53. Test the hypothesis that the death rate from lung cancer is 70 percent. Use $\alpha = .05$.

10-38. Consider the data in Exercise 9-55. Test the hypothesis that the fraction of defective calculators produced is $2\frac{1}{2}$ percent.

10-39. Suppose that we wish to test the hypothesis $H_0: \mu_1 = \mu_2$ against the alternative $H_1: \mu_1 \neq \mu_2$, where both variances σ_1^2 and σ_2^2 are known. A total of $n_1 + n_2 = N$ observations can be taken. How should these observations be allocated to the two populations to maximize the probability that H_0 will be rejected if H_1 is true?

10-40. Consider the swine flu vaccine study described in Exercise 9-57. Test the hypothesis that the proportion of subjects contracting swine flu in the vaccinated group does not differ from the proportion of individuals contracting swine flu in the control group. Use $\alpha = .05$.

10-41. Using the data in Exercise 9-58, is it reasonable to conclude that production line 2 produced a higher fraction of defective product than line 1? Use $\alpha = .01$.

10-42. Two different types of injection-molding machines are used to form plastic bottles. A bottle is considered defective if the neck ring is improperly formed. Two random samples, each of size 500, are selected, and 21 defective bottles are found in the sample from machine 1, while 32 defective bottles are found in the sample from machine 2. Is it reasonable to conclude that both machines produce the same fraction of defective bottles?

10-43. Suppose that we wish to test $H_0: \mu_1 = \mu_2$ against $H_1: \mu_1 \neq \mu_2$, where σ_1^2 and σ_2^2 are known. The total sample size N is fixed, but the allocation of observations to the two populations such that $n_1 + n_2 = N$ is to be made on the basis of cost. If the cost of sampling for populations 1 and 2 are C_1 and C_2, respectively, find the minimum cost sample sizes that provide a specified variance for the difference in sample means.

10-44. A manufacturer of a new pain relief tablet would like to demonstrate that her product works twice as fast as her competitors' product. Specifically, she would like to test

$$H_0 : \mu_1 = 2\mu_2$$
$$H_1 : \mu_1 > 2\mu_2$$

where μ_1 is the mean absorption time of the competitive product and μ_2 is the mean absorption time of the new product. Assuming that the variances σ_1^2 and σ_2^2 are known, suggest a procedure for testing this hypothesis.

10-45. Derive an expression similar to Equation (10-20) for the β error for the test on the variance of a normal distribution. Assume that the two-sided alternative is specified.

10-46. Derive an expression similar to Equation (10-20) for the β error for the test of the equality of the variances of two normal distributions. Assume that the two-sided alternative is specified.

10-47. The number of defective items found each day by the inspectors in a manufacturing plant is recorded.

Number of Defectives per Day	Times Observed
0–10	6
11–15	11
16–20	16
21–25	28
26–30	22
31–35	19
36–40	11
41–45	4

(a) Is it reasonable to conclude that these data come from a normal distribution? Use the chi-square goodness of fit test.

(b) Plot the data on normal probability paper. Does an assumption of normality seem justified?

10-48. The number of cars passing southbound through the intersection of North Avenue and Peachtree Street has been tabulated by the students of the Civil Engineering School at Georgia Tech. They have obtained the following data.

Vehicles per Minute	Times Observed	Vehicles per Minute	Times Observed
40	14	53	102
41	24	54	96
42	57	55	90
43	111	56	81
44	194	57	73
45	256	58	64
46	296	59	61
47	378	60	59
48	250	61	50
49	185	62	42
50	171	63	29
51	150	64	18
52	110	65	15

Does the assumption of a Poisson distribution seem appropriate as a probability model for this process?

10-49. A pseudorandom number generator is designed so that the integers 0 through 9 have equal probability of occurrence. The first 10,000 numbers are:

0	1	2	3	4	5	6	7	8	9
967	1008	975	1022	1003	989	1001	981	1043	1011

Does this generator seem to be working properly?

10-50. The cycle time of an automatic machine has been observed and recorded.

Seconds	2.10	2.11	2.12	2.13	2.14	2.15	2.16	2.17	2.18	2.19	2.20
Frequency	16	28	41	74	149	256	137	82	40	19	11

(a) Does the normal distribution seem to be a reasonable probability model for the cycle time? Use the chi-square goodness of fit test.

(b) Plot the data on normal probability paper. Does the assumption of normality seem reasonable?

10-51. A soft drink bottler is studying the internal pressure strength of 1-liter glass nonreturnable bottles. A random sample of 16 bottles is tested and the

226.16 psi	211.14 psi
202.20	203.62
219.54	188.12
193.73	224.39
208.15	221.31
195.45	204.55
193.71	202.21
200.81	201.63

pressure strengths obtained. The data are shown below. Plot these data on normal probability paper. Does it seem reasonable to conclude that pressure strength is normally distributed?

10-52. A company operates four machines three shifts each day. From production records, the following data on the number of breakdowns are collected.

	Machines			
Shift	A	B	C	D
1	41	20	12	16
2	31	11	9	14
3	15	17	16	10

Test the hypothesis that breakdowns are independent of the shift.

10-53. Patients in a hospital are classified as surgical or medical. A record is kept of the number of times patients require nursing service during the night and whether these patients are on Medicare or not. The data are presented here.

	Patient Category	
Medicare	Surgical	Medical
Yes	46	52
No	36	43

Test the hypothesis that calls by surgical–medical patients are independent of whether the patients are receiving Medicare.

10-54. Grades in a statistics course and an operations research course taken simultaneously were as follows for a group of students.

Statistics Grade	Operations Research Grade			
	A	B	C	Other
A	25	6	17	13
B	17	16	15	6
C	18	4	18	10
Other	10	8	11	20

Are the grades in statistics and operations research related?

10-55. An experiment with artillery shells yields the following data on the characteristics of lateral deflections and ranges.

Would you conclude that deflection and range are independent?

	Lateral Deflection		
Range (yards)	Left	Normal	Right
0–1,999	6	14	8
2,000–5,999	9	11	4
6,000–11,999	8	17	6

10-56. A study is being made of the failures of an electronic component. There are four types of failures possible and two mounting positions for the device. The following data have been taken.

	Failure Type			
Mounting Position	A	B	C	D
1	22	46	18	9
2	4	17	6	12

Would you conclude that the type of failure is independent of the mounting position?

10-57. A random sample of students are asked their opinions on a proposed core curriculum change. The results are presented here.

	Opinion	
Class	Favoring	Opposing
Freshman	120	80
Sophomore	70	130
Junior	60	70
Senior	40	60

Test the hypothesis that the opinions are independent of the class grouping.

10-58. Fabric is graded into three classifications: A, B, and C. The results below were obtained from five looms. Is fabric classification independent of the loom?

10-59. Consider the data in Exercise 10-21. Using the Wilcoxon test for two independent samples, is it reasonable to conclude that the mean burning times of the two types of flares are the same? Use $\alpha = .05$.

	Number of Pieces of Fabric in Fabric Classification		
Loom	A	B	C
1	185	16	12
2	190	24	21
3	170	35	16
4	158	22	7
5	185	22	15

10-60. Consider the data in Exercise 10-31. Use the Wilcoxon test for two independent samples to test the hypothesis that the means of the two populations are equal. Use $\alpha = .05$.

10-61. Consider the data in Exercise 10-35. Test the hypothesis that the mean speeds of painted and unpainted cars are equal, using the sign test.

10-62. Consider the data in Exercise 10-36. Using the sign test, test the hypothesis that the mean wear of both types of shoe leather is the same.

10-63. Consider the data in Exercise 10-35. Using the Wilcoxon signed-rank test, test the hypothesis that the mean speeds of painted and unpainted cars are equal.

10-64. Consider the data in Exercise 10-36. Using the Wilcoxon signed-rank test, test the hypothesis that the mean wear of both types of shoe leather is the same.

Chapter 11

Analysis of Variance

In Chapter 10 we presented several methods for testing hypotheses about either one or two parameters. Many decision problems require that more than two parameters be considered. For example, suppose that a civil engineer is investigating the effect of five different curing methods on the mean compressive strength of concrete. That is, he is interested in testing the equality of five means. Initially, we might think that this problem could be solved by performing a t-test on all possible pairs of means. Since there are five means, there are $\binom{5}{2} = 10$ possible pairs of means to test. However, testing each pair of means using the t-test would lead to considerable distortion in the type I error probability for the entire experiment. That is, if all five means are really equal, and the probability of correctly accepting the null hypothesis for each individual test is $1 - \alpha = .95$, then the probability of correctly accepting the null hypothesis for all 10 individual tests is $(.95)^{10} = .60$, assuming that the tests are independent. Thus the type I error probability for the complete experiment is .40. There is a very good chance that one or more individual t-tests will indicate that at least one pair of means differ when they really do not.

The appropriate procedure for testing the equality of several population means is the analysis of variance. This chapter will present an introduction to the analysis of variance for the situation where only one factor (such as curing methods in the above example) is of interest. However, the analysis of variance has many other important applications. Some of these applications will be presented in subsequent chapters.

11-1 The One-Way Classification Analysis of Variance

Suppose we have a different levels of a single factor that we wish to compare. The different levels of the factor are often called *treatments*. The observed response from each of the a treatments is a random variable. The data would appear as in Table 11-1. An entry in Table 11-1, say y_{ij}, represents the jth

TABLE 11-1   **Typical Data for One-Way Classification Analysis of Variance**

	Observation			
	1	y_{11}	y_{12}	$\cdots$	y_{1n}
	2	y_{21}	y_{22}	$\cdots$	y_{2n}
Treatment	·	·	·	$\cdots$	·
	·	·	·	$\cdots$	·
	·	·	·	$\cdots$	·
	a	y_{a1}	y_{a2}	$\cdots$	y_{an}

observation taken under treatment *i*. We initially consider the case where there is an equal number of observations, *n*, on each treatment.

We may describe the observations in Table 11-1 by the linear statistical model

$$y_{ij} = \mu + \tau_i + \epsilon_{ij} \begin{cases} i = 1, 2, \ldots, a \\ j = 1, 2, \ldots, n \end{cases} \tag{11-1}$$

where y_{ij} is the (*ij*)th observation, μ is a parameter common to all treatments called the *overall mean*, τ_i is a parameter associated with the *i*th treatment called the *i*th *treatment effect*, and ϵ_{ij} is a random error component. Note that y_{ij} represents both the random variable and its realization. We would like to test certain hypotheses about the treatment effects and to estimate them. For hypothesis testing, the model errors are assumed to be normally and independently distributed random variables with mean zero and variance σ^2 [abbreviated as NID(0, σ^2)]. The variance σ^2 is assumed constant for all levels of the factor.

The model of Equation (11-1) is called the *one-way classification* analysis of variance, because only one factor is investigated. Furthermore, we will require that the observations be taken in random order so that the environment in which the treatments are used (often called the experimental units) is as uniform as possible. This arrangement is called a completely randomized experimental design. There are two different ways that the *a* factor levels in the experiment could have been chosen. First, the *a* treatments could have been specifically chosen by the experimenter. In this situation we wish to test hypotheses about the τ_i, and conclusions will apply only to the factor levels considered in the analysis. The conclusions cannot be extended to similar treatments that were not considered. Also, we may wish to estimate the τ_i.

This is called the *fixed effects* model. Alternatively, the a treatments could be a random sample from a larger population of treatments. In this situation we would like to be able to extend the conclusions (which are based on the sample of treatments) to all treatments in the population, whether they were explicitly considered in the analysis or not. Here the τ_i are random variables, and knowledge about the particular ones investigated is relatively useless. Instead, we test hypotheses about the variability of the τ_i and try to estimate this variability. This is called the *random effects*, or *components of variance*, model.

11-2 The Fixed Effects Model

11-2.1 Statistical Analysis

In this section we will develop the analysis of variance for the fixed effects model, one-way classification. In the fixed effects model, the treatment effects τ_i are usually defined as deviations from the overall mean, so that

$$\sum_{i=1}^{a} \tau_i = 0 \tag{11-2}$$

Let $y_{i.}$ represent the total of the observations under the ith treatment and $\bar{y}_{i.}$ represent the average of the observations under the ith treatment. Similarly, let $y_{..}$ represent the grand total of all observations and $\bar{y}_{..}$ represent the grand mean of all observations. Expressed mathematically,

$$y_{i.} = \sum_{j=1}^{n} y_{ij} \qquad \bar{y}_{i.} = y_{i.}/n \qquad i = 1, 2, \ldots, a$$

$$y_{..} = \sum_{i=1}^{a} \sum_{j=1}^{n} y_{ij} \qquad \bar{y}_{..} = y_{..}/N \tag{11-3}$$

where $N = an$ is the total number of observations. Thus the "dot" subscript notation implies summation over the subscript that it replaces.

We are interested in testing the equality of the a treatment effects. Using Equation (11-2), the appropriate hypotheses are

$$H_0: \quad \tau_1 = \tau_2 = \cdots = \tau_a = 0$$

$$H_1: \quad \tau_i \neq 0 \text{ for at least one } i \tag{11-4}$$

That is, if the null hypothesis is true, then each observation is made up of the overall mean μ plus a realization of the random error ϵ_{ij}.

The test procedure for the hypotheses in Equation (11-4) is called the analysis of variance. The name "analysis of variance" results from partitioning total variability in the data into its component parts. The total corrected sum of squares, which is a measure of total variability in the data, may be written as

$$\sum_{i=1}^{a} \sum_{j=1}^{n} (y_{ij} - \bar{y}_{..})^2 = \sum_{i=1}^{a} \sum_{j=1}^{n} [(\bar{y}_{i.} - \bar{y}_{..}) + (y_{ij} - \bar{y}_{i.})]^2 \tag{11-5}$$

or

$$\sum_{i=1}^{a} \sum_{j=1}^{n} (y_{ij} - \bar{y}_{..})^2 = n \sum_{i=1}^{a} (\bar{y}_{i.} - \bar{y}_{..})^2 + \sum_{i=1}^{a} \sum_{j=1}^{n} (y_{ij} - \bar{y}_{i.})^2$$

$$+ \sum_{i=1}^{a} \sum_{j=1}^{n} (\bar{y}_{i.} - \bar{y}_{..})(y_{ij} - \bar{y}_{i.}) \tag{11-6}$$

Note that the cross-product term in Equation (11-6) is zero, since

$$\sum_{j=1}^{n} (y_{ij} - \bar{y}_{i.}) = y_{i.} - n\bar{y}_{i.} = y_{i.} - n(y_{i.}/n) = 0$$

Therefore, we have

$$\sum_{i=1}^{a} \sum_{j=1}^{n} (y_{ij} - \bar{y}_{..})^2 = n \sum_{i=1}^{a} (\bar{y}_{i.} - \bar{y}_{..})^2 + \sum_{i=1}^{a} \sum_{j=1}^{n} (y_{ij} - \bar{y}_{i.})^2 \tag{11-7}$$

Equation (11-7) shows that the total variability in the data, measured by the total corrected sum of squares, can be partitioned into a sum of squares of differences between treatment means and the grand mean and a sum of squares of differences of observations within treatments and the treatment mean. Differences between observed treatment means and the grand mean measure the differences between treatments, while differences of observations within a treatment from the treatment mean can be due only to random error. Therefore, we write Equation (11-7) symbolically as

$$SS_T = SS_{\text{treatments}} + SS_E$$

where SS_T is the total sum of squares, $SS_{\text{treatments}}$ is called the sum of squares due to treatments (i.e., *between* treatments), and SS_E is called the sum of squares due to error (i.e., *within* treatments). There are $an = N$ total observations; thus SS_T has $N - 1$ degrees of freedom. There are a levels of the factor, so $SS_{\text{treatments}}$ has $a - 1$ degrees of freedom. Finally, within any treatment there are n replicates providing $n - 1$ degrees of freedom with which to estimate the experimental error. Since there are a treatments, we have $a(n - 1) = an - a = N - a$ degrees of freedom for error.

Now consider the distributional properties of these sums of squares. Since we have assumed that the errors ϵ_{ij} are NID$(0, \sigma^2)$, the observations y_{ij} are NID $(\mu + \tau_i, \sigma^2)$. Thus SS_T/σ^2 is distributed as chi-square with $N - 1$ degrees of freedom, since SS_T is a sum of squares in normal random variables. We may also show that $SS_{\text{treatments}}/\sigma^2$ is chi-square with $a - 1$ degrees of freedom, if H_0 is true, and SS_E/σ^2 is chi-square with $N - a$ degrees of freedom.

However, all three sums of squares are not independent, since $SS_{\text{treatments}}$ and SS_E add up to SS_T. The following theorem, which is a special form of one due to Cochran, is useful in developing the test procedure.

Theorem 11-1 (Cochran)

Let Z_i be $NID(0, 1)$ for $i = 1, 2, \ldots, \nu$ and

$$\sum_{i=1}^{\nu} Z_i^2 = Q_1 + Q_2 + \cdots + Q_s$$

where $s < \nu$, and Q_i has ν_i degrees of freedom $(i = 1, 2, \ldots, s)$. Then Q_1, $Q_2, \ldots, Q_s$ are independent chi-square random variables with $\nu_1, \nu_2, \ldots, \nu_s$ degrees of freedom, respectively, if and only if

$$\nu = \nu_1 + \nu_2 + \cdots + \nu_s$$

Using this theorem, we note that the degrees of freedom for $SS_{\text{treatments}}$ and SS_E add up to $N - 1$, so that $SS_{\text{treatments}}/\sigma^2$ and SS_E/σ^2 are independently distributed chi-square random variables. Therefore, under the null hypothesis, the statistic

$$F_0 = \frac{SS_{\text{treatments}}/(a - 1)}{SS_E/(N - a)} = \frac{MS_{\text{treatments}}}{MS_E} \tag{11-8}$$

follows the $F_{a-1, N-a}$ distribution. The quantities $MS_{\text{treatments}}$ and MS_E are called *mean squares*.

The expected values of the mean squares are used to show that F_0 in Equation (11-8) is an appropriate test statistic for $H : \tau_i = 0$ and to determine the criterion for rejecting this null hypothesis. Consider

$$E(MS_E) = E\left(\frac{SS_E}{N - a}\right) = \frac{1}{N - a} E\left[\sum_{i=1}^{a} \sum_{j=1}^{n} (y_{ij} - \bar{y}_{i.})^2\right]$$

$$= \frac{1}{N - a} E\left[\sum_{i=1}^{a} \sum_{j=1}^{n} (y_{ij}^2 - 2y_{ij}\bar{y}_{i.} + \bar{y}_{i.}^2)\right]$$

$$= \frac{1}{N - a} E\left[\sum_{i=1}^{a} \sum_{j=1}^{n} y_{ij}^2 - 2n\sum_{i=1}^{a} \bar{y}_{i.}^2 + n\sum_{i=1}^{a} \bar{y}_{i.}^2\right]$$

$$= \frac{1}{N - a} E\left[\sum_{i=1}^{a} \sum_{j=1}^{n} y_{ij}^2 - \frac{1}{n}\sum_{i=1}^{a} y_{i.}^2\right]$$

Substituting the model, Equation (11-1), into this equation we obtain

$$E(MS_E) = \frac{1}{N - a} E\left[\sum_{i=1}^{a} \sum_{j=1}^{n} (\mu + \tau_i + \epsilon_{ij})^2 - \frac{1}{n}\sum_{i=1}^{a}\left(\sum_{j=1}^{n} \mu + \tau_i + \epsilon_{ij}\right)^2\right]$$

Now on squaring and taking expectation of the quantities within brackets, we see that terms involving ϵ_{ij}^2 and $\sum_{j=1}^{n} \epsilon_{ij}^2$ are replaced by σ^2 and $n\sigma^2$, respectively, because $E(\epsilon_{ij}) = 0$. Furthermore, all cross products involving ϵ_{ij} have zero expectation. Therefore, after squaring and taking expectation, we have

$$E(MS_E) = \frac{1}{N-a}\left[N\mu^2 + n\sum_{i=1}^{a}\tau_i^2 + N\sigma^2 - N\mu^2 - n\sum_{i=1}^{a}\tau_i^2 - a\sigma^2\right]$$

or

$$E(MS_E) = \sigma^2$$

Using a similar approach, we may show that

$$E(MS_{\text{treatments}}) = \sigma^2 + \frac{n\sum_{i=1}^{a}\tau_i^2}{a-1}$$

From the expected mean square we see that MS_E is an unbiased estimator of σ^2. Also, under the null hypothesis, $MS_{\text{treatments}}$ is an unbiased estimator of σ^2. However, if the null hypothesis is false, then the expected value of $MS_{\text{treatments}}$ is greater than σ^2. Therefore, under the alternative hypothesis the expected value of the numerator of the test statistic [Equation (11-8)] is greater than the expected value of the denominator. Consequently, we should reject H_0 if the test statistic is large. This implies an upper-tail, one-tail critical region. Therefore, we would reject H_0 if

$$F_0 > F_{\alpha, a-1, N-a}$$

where F_0 is computed from Equation (11-8).

Efficient computational formulas for the sums of squares may be obtained by expanding and simplifying the definitions of $SS_{\text{treatments}}$ and SS_T in Equation (11-7). This yields

$$SS_T = \sum_{i=1}^{a}\sum_{j=1}^{n} y_{ij}^2 - \frac{y_{..}^2}{N} \tag{11-9}$$

and

$$SS_{\text{treatments}} = \sum_{i=1}^{a}\frac{y_{i.}^2}{n} - \frac{y_{..}^2}{N} \tag{11-10}$$

The error sum of squares is obtained by subtraction as

$$SS_E = SS_T - SS_{\text{treatments}} \tag{11-11}$$

The test procedure is summarized in Table 11-2. This is called an analysis of variance table.

TABLE 11-2 **The Analysis of Variance for the One-Way Classification Fixed-Effects Model**

Source of Variation	Sum of Squares	Degrees of Freedom	Mean Square	F_0
Between treatments	$SS_{treatments}$	$a - 1$	$MS_{treatments}$	$F_0 = \dfrac{MS_{treatments}}{MS_E}$
Error (within treatments)	SS_E	$N - a$	MS_E	
Total	SS_T	$N - 1$		

● **Example 11-1.** A manufacturer of paper used for grocery bags is interested in improving the tensile strength of her product. She suspects that the tensile strength is a function of the hardwood concentration in the pulp. She decides to investigate five different hardwood concentrations: 5%, 10%, 15%, 20%, and 25%. Five observations are to be taken under each hardwood concentration, using a pilot plant. The 25 required observations are run in random order, and the data obtained are shown in Table 11-3.

The sums of squares are computed using Equations (11-9), (11-10), and (11-11) as follows:

$$SS_T = \sum_{i=1}^{5} \sum_{j=1}^{5} y_{ij}^2 - \frac{y_{..}^2}{N}$$
$$= (7)^2 + (7)^2 + (15)^2 + \cdots + (15)^2 + (11)^2 - \frac{(376)^2}{25} = 636.96$$

$$SS_{treatments} = \sum_{i=1}^{5} \frac{y_{i.}^2}{n} - \frac{y_{..}^2}{N}$$
$$= \frac{(49)^2 + \cdots + (54)^2}{5} - \frac{(376)^2}{25} = 475.76$$

$$SS_E = SS_T - SS_{treatments}$$
$$= 636.96 - 475.76 = 161.20$$

TABLE 11-3 **Tensile Strength of Paper (psi)**

Hardwood Concentration (%)	Observations					Totals $y_{i.}$
	1	2	3	4	5	
5	7	7	15	11	9	49
10	12	17	12	18	18	77
15	14	18	18	19	19	88
20	19	25	22	19	23	108
25	7	10	11	15	11	54
						$376 = y_{..}$

TABLE 11-4 **Analysis of Variance for the Tensile Strength Data**

Source of Variation	Sum of Squares	Degrees of Freedom	Mean Square	F_0
Treatments	475.76	4	118.94	$F_0 = 14.76$
Error	161.20	20	8.06	
Total	636.96	24		

The analysis of variance is summarized in Table 11-4. Since $F_{.01, 4, 20} = 4.43$, we reject H_0 and conclude that hardwood concentration in the pulp significantly affects the strength of the paper.

11-2.2 Estimation of the Model Parameters

It is possible to derive estimators for the parameters in the one-way classification model

$$y_{ij} = \mu + \tau_i + \epsilon_{ij}$$

An appropriate estimation criterion is to estimate μ and τ_i such that the sum of the squares of the errors or deviations ϵ_{ij} is a minimum. This method of parameter estimation is called the method of *least squares*. In estimating μ and τ_i by least squares, the normality assumption on the errors ϵ_{ij} is not needed. To find the least squares estimators of μ and τ_i, we form the sum of squares of the errors

$$L = \sum_{i=1}^{a} \sum_{j=1}^{n} \epsilon_{ij}^2 = \sum_{i=1}^{a} \sum_{j=1}^{n} (y_{ij} - \mu - \tau_i)^2 \tag{11-12}$$

and find values of μ and τ_i, say $\hat{\mu}$ and $\hat{\tau}_j$, that minimize L. The values $\hat{\mu}$ and $\hat{\tau}_i$ are the solutions to the $a + 1$ simultaneous equations

$$\left. \frac{\partial L}{\partial \mu} \right|_{\hat{\mu}, \hat{\tau}_i} = 0$$

$$\left. \frac{\partial L}{\partial \tau_i} \right|_{\hat{\mu}, \hat{\tau}_i} = 0 \quad i = 1, 2, \ldots, a$$

Differentiating Equation (11-12) with respect to μ and τ_i and equating to zero, we obtain

$$-2 \sum_{i=1}^{a} \sum_{j=1}^{n} (y_{ij} - \hat{\mu} - \hat{\tau}_i) = 0$$

and

$$-2 \sum_{j=1}^{n} (y_{ij} - \hat{\mu} - \hat{\tau}_i) = 0 \qquad i = 1, 2, \ldots, a$$

After simplification these equations become

$$N\hat{\mu} + n\hat{\tau}_1 + n\hat{\tau}_2 + \cdots + n\hat{\tau}_a = y_{..}$$
$$n\hat{\mu} + n\hat{\tau}_1 \qquad\qquad = y_{1.}$$
$$n\hat{\mu} \qquad + n\hat{\tau}_2 \qquad = y_{2.}$$
$$\cdot \qquad\qquad \cdot \qquad\qquad \cdot$$
$$\cdot \qquad\qquad \cdot \qquad\qquad \cdot$$
$$\cdot \qquad\qquad \cdot \qquad\qquad \cdot$$
$$n\hat{\mu} \qquad\qquad + n\hat{\tau}_a = y_{a.} \qquad\qquad (11\text{-}13)$$

Equations (11-13) are called the *least squares normal equations*. Notice that if we add the last a normal equations we obtain the first normal equation. Therefore, the normal equations are not linearly independent, and there are no unique estimates for $\mu, \tau_1, \ldots, \tau_a$. One way to overcome this difficulty is to impose a constraint on the solution to the normal equations. There are many ways to choose this constraint. Since we have defined the treatment effects as deviations from the overall mean, it seems reasonable to apply the constraint

$$\sum_{i=1}^{a} \hat{\tau}_i = 0 \qquad\qquad (11\text{-}14)$$

Using this constraint, we obtain as the solution to the normal equations

$$\hat{\mu} = \bar{y}_{..}$$
$$\hat{\tau}_i = \bar{y}_{i.} - \bar{y}_{..} \qquad i = 1, 2, \ldots, a \qquad\qquad (11\text{-}15)$$

This solution has considerable intuitive appeal, since the overall mean is estimated by the grand average of the observations and the estimate of any treatment effect is just the difference between the treatment average and the grand average.

This solution is obviously not unique because it depends on the constraint [Equation (11-14)] that we have chosen. At first this may seem unfortunate, because two different experimenters could analyze the same data and obtain different results if they apply different constraints. However, certain *functions* of the model parameter are uniquely estimated, regardless of the constraint. Some examples are $\tau_i - \tau_j$, which would be estimated by $\hat{\tau}_i - \hat{\tau}_j = \bar{y}_{i.} - \bar{y}_{j.}$, and $\mu + \tau_i$, which would be estimated by $\hat{\mu} + \hat{\tau}_i = \bar{y}_{i.}$. Since we are usually interested in differences among the treatment effects rather than their actual values, it causes no concern that the τ_i cannot be uniquely estimated. In general, any function of the model parameters that is a linear combination of the left-hand side of the normal equations can be uniquely estimated. Functions that are uniquely estimated, regardless of which constraint is used, are called *estimable* functions.

Frequently, we would like to construct a confidence interval for the ith treatment mean. The mean of the ith treatment is

$$\mu_i = \mu + \tau_i \qquad i = 1, 2, \ldots, a$$

A point estimator of μ_i would be $\hat{\mu}_i = \hat{\mu} + \hat{\tau}_i = \bar{y}_{i.}$. Now, if we assume that the errors are normally distributed, each $\bar{y}_{i.}$ is $\text{NID}(\mu_i, \sigma^2/n)$. Thus, if σ^2 were known, we could use the normal distribution to construct a confidence interval. Using MS_E as an estimator of σ^2, we would base the confidence interval on the t distribution. Therefore, a $100(1 - \alpha)$ percent confidence interval on the ith treatment mean μ_i is

$$[\bar{y}_{i.} \pm t_{\alpha/2, N-a} \sqrt{MS_E/n}] \qquad (11\text{-}16)$$

A $100(1 - \alpha)$ percent confidence interval on the difference in any two treatment means, say $\mu_i - \mu_j$, would be

$$[\bar{y}_{i.} - \bar{y}_{j.} \pm t_{\alpha/2, N-a} \sqrt{2MS_E/n}] \qquad (11\text{-}17)$$

● **Example 11-2.** Using the data in Example 11-1, we may find the estimates of the overall mean and the treatment effects as $\hat{\mu} = 376/25 = 15.04$ and

$$\hat{\tau}_i = \bar{y}_{1.} - \bar{y}_{..} = 9.80 - 15.04 = -5.24$$
$$\hat{\tau}_2 = \bar{y}_{2.} - \bar{y}_{..} = 15.40 - 15.04 = +.36$$
$$\hat{\tau}_3 = \bar{y}_{3.} - \bar{y}_{..} = 17.60 - 15.04 = +2.56$$
$$\hat{\tau}_4 = \bar{y}_{4.} - \bar{y}_{..} = 21.60 - 15.04 = +6.56$$
$$\hat{\tau}_5 = \bar{y}_{5.} - \bar{y}_{..} = 10.80 - 15.04 = -4.24$$

A 95 percent confidence interval on the mean of treatment 4 is computed from Equation (11-16) as

$$[21.60 \pm (2.086) \sqrt{8.06/5}]$$

or

$$[21.60 \pm 2.65]$$

Thus, the desired confidence interval is $18.95 \le \mu_4 \le 24.25$.

11-2.3   The Unbalanced Case

In some single-factor experiments the number of observations taken under each treatment may be different. We then say that the design is *unbalanced*. The analysis of variance described above is still valid, but slight modifications must be made in the sums of squares formulas. Let n_i observations be taken under treatment $i(i = 1, 2, \ldots, a)$, and let the total number of observations $N = \sum_{i=1}^{a} n_i$. The computational formulas for SS_T and $SS_{\text{treatments}}$ become

$$SS_T = \sum_{i=1}^{a} \sum_{j=1}^{n_i} y_{ij}^2 - \frac{y_{..}^2}{N}$$

and

$$SS_{\text{treatments}} = \sum_{i=1}^{a} \frac{y_{i.}^2}{n_i} - \frac{y_{..}^2}{N}$$

In solving the normal equations, the constraint $\sum_{i=1}^{a} n_i \hat{\tau}_i = 0$ is used. No other changes are required in the analysis of variance.

There are two important advantages in choosing a balanced design. First, the test statistic is relatively insensitive to small departures from the assumption of equality of variances if the sample sizes are equal. This is not the case for unequal sample sizes. Second, the power of the test is maximized if the samples are of equal size.

11-3 Tests on Individual Treatment Means

11-3.1 Orthogonal Contrasts

Rejecting the null hypothesis in the fixed effects model analysis of variance implies that there are differences between the a treatment means, but the exact nature of the differences is not specified. In this situation, further comparisons between groups of treatment means may be useful. The ith treatment mean is defined as $\mu_i = \mu + \tau_i$, and μ_i is estimated by $\bar{y}_{i.}$. Comparisons between treatment means are usually made in terms of the treatment totals $\{y_{i.}\}$.

Consider the paper tensile strength experiment presented in Example 11-1. Since the hypothesis $H_0: \tau_i = 0$ was rejected, we know that some hardwood concentrations produce different tensile strengths than others, but which ones actually cause this difference? We might suspect at the outset of the experiment that hardwood concentrations 4 and 5 produce the same tensile strength, implying that we would like to test the hypothesis

$$H_0: \mu_4 = \mu_5$$
$$H_1: \mu_4 \neq \mu_5$$

This hypothesis could be tested by using a linear combination of treatment totals, say

$$y_{4.} - y_{5.} = 0$$

If we had suspected that the *average* of hardwood concentrations 1 and 3 did not differ from the average of hardwood concentrations 4 and 5, then the hypothesis would have been

$$H_0: \mu_1 + \mu_3 = \mu_4 + \mu_5$$
$$H_1: \mu_1 + \mu_3 \neq \mu_4 + \mu_5$$

which implies the linear combination of treatment totals

$$y_{1.} + y_{3.} - y_{4.} - y_{5.} = 0$$

In general, the comparison of treatment means of interest will imply a linear combination of treatment totals such as

$$C = \sum_{i=1}^{a} c_i y_{i.}$$

with the restriction that $\sum_{i=1}^{a} c_i = 0$. These linear combinations are called *contrasts*. The sum of squares for any contrast is

$$SS_C = \frac{\left(\sum\limits_{i=1}^{a} c_i y_{i.}\right)^2}{n \sum\limits_{i=1}^{a} c_i^2} \tag{11-18}$$

and has a single degree of freedom. If the design is unbalanced, then the comparison of treatment means requires that $\sum_{i=1}^{a} n_i c_i = 0$, and Equation (11-18) becomes

$$SS_C = \frac{\left(\sum\limits_{i=1}^{a} c_i y_{i.}\right)^2}{\sum\limits_{i=1}^{a} n_i c_i^2} \tag{11-19}$$

A contrast is tested by comparing its sum of squares to the mean square error. The resulting statistic would be distributed as F, with 1 and $N - a$ degrees of freedom.

A very important special case of the above procedure is that of *orthogonal contrasts*. Two contrasts with coefficients $\{c_i\}$ and $\{d_i\}$ are orthogonal if

$$\sum_{i=1}^{a} c_i d_i = 0$$

or, for an unbalanced design, if

$$\sum_{i=1}^{a} n_i c_i d_i = 0$$

For a treatments a set of $a - 1$ orthogonal contrasts will partition the sum of squares due to treatments into $a - 1$ independent single-degree-of-freedom components. Thus, tests performed on orthogonal contrasts are independent.

There are many ways to choose the orthogonal contrast coefficients for a set of treatments. Usually, something in the nature of the experiment should suggest which comparison will be of interest. For example, if there are $a = 3$

treatments with treatment 1 a control and treatments 2 and 3 actual levels of the factor of interest to the experimenter, then appropriate orthogonal contrasts might be as follows.

Treatment	Orthogonal Contrasts	
1 (control)	-2	0
2 (level 1)	1	-1
3 (level 2)	1	1

Note that contrast 1 with $c_i = -2, 1, 1$ compares the average effect of the factor with the control, while contrast 2 with $d_i = 0, -1, 1$ compares the two levels of the factor of interest.

Contrast coefficients must be chosen prior to running the experiment, for if these comparisons are selected after examining the data, most experimenters would construct tests that compare large observed differences in means. These large differences could be due to the presence of real effects or they could be due to random error. If experimenters always pick the largest differences to compare, they will inflate the type I error of the test, since it is likely that in an unusually high percentage of the comparisons selected the observed differences will be due to error.

● **Example 11-3.** Consider the data in Example 11-1. There are five treatment means and four degrees of freedom between these treatments. One set of comparisons between these means and the associated orthogonal contrasts are

$$H_0: \mu_4 = \mu_5 \qquad\qquad C_1 = \qquad\qquad - y_{4.} + y_{5.}$$
$$H_0: \mu_1 + \mu_3 = \mu_4 + \mu_5 \qquad C_2 = y_{1.} \qquad + y_{3.} - y_{4.} - y_{5.}$$
$$H_0: \mu_1 = \mu_3 \qquad\qquad C_3 = y_{1.} \qquad - y_{3.}$$
$$H_0: 4\mu_2 = \mu_1 + \mu_3 + \mu_4 + \mu_5 \qquad C_4 = -y_{1.} + 4y_{2.} - y_{3.} - y_{4.} - y_{5.}$$

Notice that the contrast coefficients are orthogonal. Using the data in Table 11-3, we find the numerical values of the contrasts and the sums of squares as follows:

$$C_1 = \qquad\qquad -1(108) + 1(54) = -54 \qquad SS_{C_1} = \frac{(-54)^2}{5(2)} = 291.60$$

$$C_2 = +1(49) \qquad + 1(88) - 1(108) - 1(54) = -25 \qquad SS_{C_2} = \frac{(-25)^2}{5(4)} = 31.25$$

$$C_3 = +1(49) \qquad - 1(88) \qquad\qquad = -39 \qquad SS_{C_3} = \frac{(-39)^2}{5(2)} = 152.10$$

$$C_4 = -1(49) + 4(77) - 1(88) - 1(108) - 1(54) = \quad 9 \qquad SS_{C_4} = \frac{(9)^2}{5(20)} = \qquad 81$$

TABLE 11-5 **Analysis of Variance for the Tensile Strength Data**

Source of Variation	Sum of Squares	Degrees of Freedom	Mean Square	F_0
Hardwood concentration percentage	475.76	4	118.94	14.76[a]
$C_1: \mu_4 = \mu_5$	(291.60)	1	191.60	36.18[a]
$C_2: \mu_1 + \mu_3 = \mu_4 + \mu_5$	(31.25)	1	31.25	3.88
$C_3: \mu_1 = \mu_3$	(152.10)	1	152.10	18.87[a]
$C_4: 4\mu_2 = \mu_1 + \mu_3 + \mu_4 + \mu_5$	(.81)	1	.81	.10
Error	161.20	20	8.06	
Total	636.96	24		

[a] Significant at 1 percent.

These contrast sums of squares completely partition the treatment sum of squares; that is, $SS_{\text{treatments}} = SS_{C_1} + SS_{C_2} + SS_{C_3} + SS_{C_4}$. The tests on the contrasts are usually incorporated in the analysis of variance, such as shown in Table 11-5. From this analysis, we conclude that there are significant differences between hardwood concentration 4 and 5, and 1 and 3; but that the *average* of 1 and 3 does not differ from the average of 4 and 5, nor does 2 differ from the average of the other four hardwood concentrations.

11-3.2 Duncan's Multiple Range Test

Frequently analysts do not know in advance how to construct appropriate orthogonal contrasts, or they may wish to test more than $a - 1$ comparisons using the same data. For example, analysts may want to test all possible pairs of means. The null hypotheses would then be $H_0: \mu_i = \mu_j$, for all $i \neq j$. If we test all possible pairs of means using t-tests, the probability of type I error for the entire set of comparisons can be greatly increased. There are several procedures available that avoid this problem. Among the more popular of these procedures are the Newman-Keuls test [Newman (1939); Keuls (1952)], Tukey's test [Tukey (1953)], and Duncan's multiple range test [Duncan (1955)]. Here we describe Duncan's multiple range test.

To apply Duncan's multiple range test for equal sample sizes, the a treatment means are arranged in ascending order, and the standard error of each mean is determined as

$$S_{\bar{y}_{i.}} = \sqrt{\frac{MS_E}{n}} \tag{11-20}$$

From Duncan's table of significant ranges (Appendix Table XI), obtain the values $r_\alpha(p, f)$, for $p = 2, 3, \ldots, a$, where α is the significance level and f is the number of degrees of freedom for error. Convert these ranges into a set of

$a - 1$ least significant ranges (e.g., R_p), for $p = 2, 3, \ldots, a$, by calculating

$$R_p = r_\alpha(p, f)S_{\bar{y}_i.} \qquad \text{for } p = 2, 3, \ldots, a$$

Then, the observed ranges between means are tested, beginning with largest versus smallest, which would be compared with the least significant range R_a. Next, the range of the largest and the second smallest is computed and compared with the least significant range R_{a-1}. These comparisons are continued until all means have been compared with the largest mean. The range of the second largest mean and the smallest is then computed and compared against the least significant range R_{a-1}. This process is continued until the ranges of all possible $a(a - 1)/2$ pairs of means have been considered. If an observed range is greater than the corresponding least significant range, then we conclude that the pair of means in question are significantly different. To prevent contradictions, no differences between a pair of means is considered significant if the two means involved fall between two other means that do not differ significantly.

- **Example 11-4.** We will apply Duncan's multiple range test to the data of Example 11-1. Recall that $MS_E = 8.06$, $N = 25$, $n = 5$, and there are 20 error degrees of freedom. The treatment means displayed in ascending order are

$$\bar{y}_{1.} = 9.8$$
$$\bar{y}_{5.} = 10.8$$
$$\bar{y}_{2.} = 15.4$$
$$\bar{y}_{3.} = 17.6$$
$$\bar{y}_{4.} = 21.6$$

The standard error of each mean is $S_{\bar{y}_i.} = \sqrt{8.06/5} = 1.27$. From the table of significant ranges in Appendix Table XI, for 20 degrees of freedom and $\alpha = .05$, we obtain $r_{.05}(2, 20) = 2.95$, $r_{.05}(3, 20) = 3.10$, $r_{.05}(4, 20) = 3.18$, and $r_{.05}(5, 20) = 3.25$. Therefore, the least significant ranges are

$$R_2 = r_{.05}(2, 20)S_{\bar{y}_i.} = (2.95)(1.27) = 3.75$$
$$R_3 = r_{.05}(3, 20)S_{\bar{y}_i.} = (3.10)(1.27) = 3.94$$
$$R_4 = r_{.05}(4, 20)S_{\bar{y}_i.} = (3.18)(1.27) = 4.04$$
$$R_5 = r_{.05}(5, 20)S_{\bar{y}_i.} = (3.25)(1.27) = 4.13$$

The comparisons between the treatment means are as follows:

$$4 \text{ vs. } 1 = 21.6 - 9.8 = 11.8 > 4.13 \ (R_5)$$
$$4 \text{ vs. } 5 = 21.6 - 10.8 = 10.8 > 4.04 \ (R_4)$$
$$4 \text{ vs. } 2 = 21.6 - 15.4 = 6.2 > 3.94 \ (R_3)$$

$$4 \text{ vs. } 3 = 21.6 - 17.6 = 4.0 > 3.75 \ (R_2)$$
$$3 \text{ vs. } 1 = 17.6 - 9.8 = 7.8 > 4.04 \ (R_4)$$
$$3 \text{ vs. } 5 = 17.6 - 10.8 = 6.8 > 3.95 \ (R_3)$$
$$3 \text{ vs. } 2 = 17.6 - 15.4 = 2.2 < 3.75 \ (R_2)$$
$$2 \text{ vs. } 1 = 15.4 - 9.8 = 5.6 > 3.94 \ (R_3)$$
$$2 \text{ vs. } 5 = 15.4 - 10.8 = 4.6 > 3.75 \ (R_2)$$
$$5 \text{ vs. } 1 = 10.8 - 9.8 = 1.0 < 3.75 \ (R_2)$$

From the analysis we see that there are significant differences between all pairs of means except 3 and 2, and 5 and 1. It is helpful to underline those means that are not significantly different as in Fig. 11-1. From this figure we see that treatment 4 produces a significantly greater tensile strength than the other treatments.

$\bar{y}_1.$	$\bar{y}_5.$	$\bar{y}_2.$	$y_3.$	$\bar{y}_4.$
9.8	10.8	15.4	17.6	21.6

Fig. 11-1.   Results of Duncan's multiple range test.

The difference between two means is a contrast. In some experiments after examining the data, we may wish to test more elaborate contrasts involving several means. Scheffé (1953) has proposed a method for testing such comparisons. His test gives the experimenter freedom to examine the data and test any estimable function that seems interesting. In general, Scheffé's test does not perform very well with respect to detecting true differences between pairs of means.

11-4   The Random Effects Model

In many situations, the factor of interest has a large number of possible levels. The analyst is interested in drawing conclusions about the entire *population* of factor levels. If the experimenter randomly selects a of these levels from the population of factor levels, then we say that the factor is a *random* factor. Because the levels of the factor actually used in the experiment were chosen randomly, the conclusions reached will be valid about the entire population of factor levels. We will assume that the population of factor levels is either of infinite size, or is large enough to be considered infinite.

The linear statistical model is

$$y_{ij} = \mu + \tau_i + \epsilon_{ij} \begin{cases} i = 1, 2, \ldots, a \\ j = 1, 2, \ldots, n \end{cases} \tag{11-21}$$

where τ_i and ϵ_{ij} are independent random variables. Note that the model is identical in structure to the fixed effects case, but the parameters have a different interpretation. If the variance of τ_i is σ_τ^2, then the variance of any observation is

$$V(y_{ij}) = \sigma_\tau^2 + \sigma^2$$

The variances σ_τ^2 and σ^2 are called *variance components*, and the model, Equation (11-21), is called the *components of variance* or the *random effects model*. To test hypotheses in this model, we require that the $\{\epsilon_{ij}\}$ are NID$(0, \sigma^2)$, that the $\{\tau_i\}$ are NID$(0, \sigma_\tau^2)$, and that τ_i and ϵ_{ij} are independent.[1]

The sum of squares identity

$$SS_T = SS_{\text{treatments}} + SS_E \qquad (11\text{-}22)$$

still holds. That is, we partition the total variability in the observations into a component that measures variation between treatments ($SS_{\text{treatments}}$) and a component that measures variation within treatments (SS_E). However, instead of testing hypotheses about individual treatment effects, we test the hypotheses

$$H_0: \sigma_\tau^2 = 0$$

$$H_1: \sigma_\tau^2 > 0$$

If $\sigma_\tau^2 = 0$, all treatments are identical; but if $\sigma_\tau^2 > 0$, then there is variability between treatments. SS_E/σ^2 is distributed as chi-square with $N - a$ degrees of freedom, and under the null hypothesis $SS_{\text{treatments}}/\sigma^2$ is distributed as a chi-square with $a - 1$ degrees of freedom. Both random variables are independent. Thus, under the null hypothesis, the ratio

$$F_0 = \frac{\dfrac{SS_{\text{treatments}}}{a-1}}{\dfrac{SS_E}{N-a}} = \frac{MS_{\text{treatments}}}{MS_E} \qquad (11\text{-}23)$$

is distributed as F with $a - 1$ and $N - a$ degrees of freedom. By examining the expected mean squares we can determine the critical region for this statistic.

Consider

$$E(MS_{\text{treatments}}) = \frac{1}{a-1} E(SS_{\text{treatments}}) = \frac{1}{a-1} E\left[\sum_{i=1}^{a} \frac{y_{i.}^2}{n} - \frac{y_{..}^2}{N} \right]$$

$$= \frac{1}{a-1} E\left[\frac{1}{n} \sum_{i=1}^{a} \left(\sum_{j=1}^{n} \mu + \tau_i + \epsilon_{ij} \right)^2 - \frac{1}{N} \left(\sum_{i=1}^{a} \sum_{j=1}^{n} \mu + \tau_i + \epsilon_{ij} \right)^2 \right]$$

[1]The assumption that the $\{\tau_i\}$ are independent random variables implies that the usual assumption of $\sum_{i=1}^{a} \tau_i = 0$ from the fixed effects model does not apply to the random effects model.

If we square and take expectation of the quantities in brackets, we see that terms involving τ_i^2 are replaced by σ_τ^2, as $E(\tau_i) = 0$. Also, terms involving $\Sigma_{j=1}^n \epsilon_{ij}^2$, $\Sigma_{i=1}^a \Sigma_{j=1}^n \epsilon_{ij}^2$, and $\Sigma_{i=1}^a \Sigma_{j=1}^n \tau_i^2$ are replaced by $n\sigma^2$, $an\sigma^2$, and $an^2\sigma_\tau^2$, respectively. Finally, all cross-product terms involving τ_i and ϵ_{ij} have zero expectation. This leads to

$$E(MS_{\text{treatments}}) = \frac{1}{a-1}[N\mu^2 + N\sigma_\tau^2 + a\sigma^2 - N\mu^2 - n\sigma_\tau^2 - \sigma^2]$$

or

$$E(MS_{\text{treatments}}) = \sigma^2 + n\sigma_\tau^2 \tag{11-24}$$

A similar approach will show that

$$E(MS_E) = \sigma^2 \tag{11-25}$$

From the expected mean squares, we see that if H_0 is true both numerator and denominator of the test statistic, Equation (11-23), are unbiased estimators of σ^2, while if H_1 is true the expected value of the numerator is greater than the expected value of the denominator. Therefore, we should reject H_0 for values of F_0 that are too large. This implies an upper-tail, one-tail critical region, so we reject H_0 if $F_0 > F_{\alpha, a-1, N-a}$.

The computational procedure and analysis of the variance table for the random effects model are identical to the fixed effects case. The conclusions, however, are quite different because they apply to the entire population of treatments.

We usually need to estimate the variance components (σ^2 and σ_τ^2) in the model. The procedure used to estimate σ^2 and σ_τ^2 is called the "analysis of variance method," because it uses the lines in the analysis of variance table. It does not require the normality assumption on the observations. The procedure consists of equating the expected mean squares to their observed value in the analysis of the variance table and solving for the variance components. When equating observed and expected mean squares in the one-way classification random effects model, we obtain

$$MS_{\text{treatments}} = \sigma^2 + n\sigma_\tau^2$$

and

$$MS_E = \sigma^2$$

Therefore, the estimators of the variance components are

$$\hat{\sigma}^2 = MS_E \tag{11-26}$$

and

$$\hat{\sigma}_\tau^2 = \frac{MS_{\text{treatments}} - MS_E}{n} \tag{11-27}$$

For unequal sample sizes, replace n in Equation (11-27) by

$$n_0 = \frac{1}{a-1} \left[\sum_{i=1}^{a} n_i - \frac{\sum_{i=1}^{a} n_i^2}{\sum_{i=1}^{a} n_i} \right]$$

Sometimes the analysis of variance method produces a negative estimate of a variance component. Since variance components are by definition non-negative, a negative estimate of a variance component is disturbing. One course of action is to accept the estimate and use it as evidence that the true value of the variance component is zero, assuming that sampling variation led to the negative estimate. While this has intuitive appeal, it will disturb the statistical properties of other estimates. Another alternative is to reestimate the negative variance component with a method that always yields non-negative estimates. Still another possibility is to consider the negative estimate as evidence that the assumed linear model is incorrect, requiring that a study of the model and its assumptions be made to find a more appropriate model.

● **Example 11-5.** A manufacturer suspects that the nitrogen content of a product varies from batch to batch. He selects a random sample of four batches, and makes five determinations of nitrogen content on each. The resulting data is presented in Table 11-6.

Since the batches of product were selected at random, this is a random effects model. The analysis of variance is shown in Table 11-7.

Since $F_0 = 35.00 > F_{.01, 3, 16} = 5.29$, we would reject H_0 and conclude that there is variability in nitrogen content from batch to batch. The variance components σ^2 and σ_τ^2 may be estimated using Equations (11-26) and (11-27) as follows:

$$\hat{\sigma}^2 = MS_E = .09$$

$$\hat{\sigma}_\tau^2 = \frac{MS_{\text{treatments}} - MS_E}{n} = \frac{3.15 - .09}{5} = .61$$

TABLE 11-6   **Nitrogen Content Data for Example 11-6**

Batch	Observations					$y_{i.}$
1	26.15	26.25	26.39	26.18	26.20	131.17
2	24.95	25.01	24.89	24.85	25.13	124.83
3	25.00	25.36	25.20	25.09	25.12	125.77
4	26.81	26.75	26.15	26.50	26.70	132.89
						$y_{..} = 514.66$

TABLE 11-7 **Analysis of Variance for the Data in Table 11-6**

Source of Variation	Sum of Squares	Degrees of Freedom	Mean Square	F_0
Batches	9.44	3	3.15	35.00
Error	1.51	16	.09	
Total	10.95	19		

Therefore, the variance of any observation on nitrogen content is estimated by $\hat{\sigma}^2 + \hat{\sigma}_\tau^2 = .09 + .61 = .70$. Most of the variability in nitrogen content, in fact approximately $(.61/.70)100 = 87.14$ percent, is due to the batch-to-batch variability.

11-5 Departures from Assumptions in Analysis of Variance

The assumptions required in the development of the analysis of variance are that treatment effects are additive, and that experimental errors are normally and independently distributed, with constant error variance. We can never be sure that all of these assumptions are satisfied, and often we suspect that some are false. A number of studies have been made concerning the effect of departures from the assumptions in the analysis of variance. These studies indicate that there is little effect when the assumptions underlying the analysis of variance are not exactly satisfied. That is, slight departures from the assumptions are of little concern. Departures from the underlying assumptions effect both the significance level and power of statistical tests. For example, when experimenters think they are testing at the 5 percent level, the test may actually be at the 7 percent or 8 percent level. Usually, the true type I error is larger than the specified one, and, as a result, too many significant differences in treatments are reported. The power is affected in that a more powerful test could be obtained if the correct statistical model were known. Therefore, we should consider error probabilities and confidence intervals only as approximate.

The assumption of homogeneity of variance may cause serious problems. The treatments may have different variances if they produce erratic effects, or if the data follows a nonnormal, skewed distribution, because the variance in a skewed distribution is usually related to the mean. There are a number of statistical procedures that may be used to test for inequality of variance. Here we present a procedure known as *Bartlett's test*.

Suppose that there are a treatments, and we wish to test the hypothesis

$$H_0: \sigma_1^2 = \sigma_2^2 = \cdots = \sigma_a^2$$

H_1: above not true for at least one σ_i^2

The test procedure uses a statistic whose sampling distribution is approximated by the chi-square distribution with $a - 1$ degrees of freedom when the a random samples are from independent normal populations. The test statistic is

$$\chi_0^2 = 2.3026 \frac{q}{c} \tag{11-28}$$

where

$$q = (N - a) \log S_p^2 - \sum_{i=1}^{a} (n_i - 1) \log S_i^2$$

$$c = 1 + \frac{1}{3(a - 1)} \sum_{i=1}^{a} (n_i - 1)^{-1} - (N - a)^{-1}$$

$$S_p^2 = \frac{\sum_{i=1}^{a} (n_i - 1)S_i^2}{N - a}$$

and S_i^2 is the sample variance in the ith treatment. The quantity q is large if the sample variances S_i^2 differ greatly; q is equal to zero if all S_i^2 are equal. Therefore, we should reject H_0 if

$$\chi_0^2 > \chi_{\alpha, a-1}^2$$

Bartlett's test is very sensitive to the normality assumption, and it should not be applied when the normality assumption is doubtful. Other tests for equality of variance are reviewed by Anderson and McLean (1974).

● **Example 11-6.** Here we apply Bartlett's test to the data of Example 11-1. The sample variances in each treatment are $s_1^2 = 11.2$, $s_2^2 = 9.8$, $s_3^2 = 4.3$, $s_4^2 = 6.8$, and $s_5^2 = 8.2$. Then

$$s_p^2 = \frac{4(11.2) + 4(9.8) + 4(4.3) + 4(6.8) + 4(8.2)}{20} = 8.06$$

$$q = 20 \log (8.06) - 4[\log 11.2 + \log 11.2 + \log 9.8 + \log 4.3 + \log 6.8 + \log 8.2]$$
$$= 0.45$$

$$c = 1 + \frac{1}{3(4)}\left(\frac{5}{4} - \frac{1}{20}\right) = 1.10$$

and the test statistic is

$$\chi_0^2 = 2.3026 \frac{(.45)}{(1.10)} = .93$$

Since $\chi_{.05, 4}^2 = 9.49$, we cannot reject the null hypothesis and conclude that the variances are homogeneous.

If inequality of variances is a problem, the usual approach recommended is to transform the observations and apply the analysis of variance to the transformed data. Note that the conclusions obtained apply to the transformed data, not the *original* data. It is possible to give some general guidelines for selecting an appropriate transformation. If we know the true distribution of the data, we may utilize this information in choosing a transformation. For example, if the observations follow the Poisson distribution then the square root transformation $y_{ij}^* = \sqrt{y_{ij}}$ or $y_{ij}^* = \sqrt{1 + y_{ij}}$ would be used. If the data follow the lognormal distribution then the logarithmic transformation $y_{ij}^* = \log y_{ij}$ is appropriate. For binomial data expressed as fractions the arcsine transformation $y_{ij}^* = \arcsin y_{ij}$ is useful. When there is no obvious transformation, the experimenter usually seeks a transformation that equalizes the variance regardless of the value of the mean. For additional discussion of transformations, refer to Box and Cox (1964) and Draper and Hunter (1969).

If the treatment effects are not additive, the transformations discussed above may also be useful. There are several causes of nonadditivity. For example, the true treatment effects may be multiplicative; that is, $y_{ij} = \mu \tau_i \epsilon_{ij}$, or interaction effects may exist that have not been included in the model. Transformations made for nonadditivity will also affect the distribution of the experimental errors. In most cases, this transformation will make the error distribution more nearly normal.

The assumption of independence is generally critical. Proper randomization of the experiment usually introduces independence in the experimental errors; one way to obtain proper randomization is to randomly assign the order in which the observations are run.

11-6 Power of the Analysis of Variance

In this section we will show how to use operating characteristic curves to determine the power of the analysis of variance. An important use of the curves is to provide guidance in selecting the number of replicates, so that the test will be sensitive to important potential differences in the treatments.

We first consider the power of the fixed effects model for the case of equal sample sizes per treatment. The power of the test is

$$1 - \beta = P\{\text{Reject } H_0 | H_0 \text{ is false}\}$$
$$= P\{F_0 > F_{\alpha, a-1, N-a} | H_0 \text{ is false}\} \qquad (11\text{-}29)$$

To evaluate this probability statement, we need to know the distribution of the test statistic F_0 if the null hypothesis is false. It can be shown that, if H_0 is false, the statistic $F_0 = MS_{\text{treatments}}/MS_E$ is distributed as a noncentral F random variable, with $a - 1$ and $N - a$ degrees of freedom and a noncentrality parameter δ. If $\delta = 0$ then the noncentral F distribution becomes the usual *central F* distribution.

Operating characteristic curves in Chart VII of the Appendix are used to calculate the power of the test for the fixed effects model. These curves plot the probability of type II error (β) against Φ, where

$$\Phi^2 = \frac{n \sum_{i=1}^{a} \tau_i^2}{a\sigma^2} \tag{11-30}$$

The parameter Φ^2 is related to the noncentrality parameter δ. Curves are available for $\alpha = .05$ and $\alpha = .01$, and for several values of degrees of freedom for numerator and denominator.

In using the operating characteristic curves, we must define the difference in means that we wish to detect in terms of $\sum_{i=1}^{a} \tau_i^2$. Also, the error variance σ^2 is usually unknown. In such cases, we must choose ratios of $\sum_{i=1}^{a} \tau_i^2/\sigma^2$ that we wish to detect. Alternatively, if an estimate of σ^2 is available, one may replace σ^2 with this estimate. For example, if we were interested in the sensitivity of an experiment that has already been performed, we might use MS_E as the estimate of σ^2.

● **Example 11-7.** Suppose that five means are being compared in an analysis of variance with $\alpha = .01$. The experimenter would like to know how many replicates to run if it is important to reject H_0 with probability at least .90 if $\sum_{i=1}^{5} \tau_i^2/\sigma^2 = 5.0$. The parameter Φ^2 is, in this case,

$$\Phi^2 = \frac{n \sum_{i=1}^{a} \tau_i^2}{a\sigma^2} = \frac{n}{5}(5) = n$$

and for the operating characteristic curve for $a - 1 = 5 - 1 = 4$, and $N - a = a(n - 1) = 5(n - 1)$ error degrees of freedom refer to in Appendix Chart VII. As a first guess, try $n = 4$ replicates. This yields $\Phi^2 = 4$, $\Phi = 2$, and $5(3) = 15$ error degrees of freedom. Consequently, from Chart VII, we find that $\beta \approx .38$. Therefore, the power of the test is approximately $1 - \beta = 1 - .38 = .62$, which is less than the required .90, and so we conclude that $n = 4$ replicates are not sufficient. Proceeding in a similar manner, we can construct the following display.

n	Φ^2	Φ	$a(n-1)$	β	Power($1 - \beta$)
4	4	2.00	15	.38	.62
5	5	2.24	20	.18	.82
6	6	2.45	25	.06	.94

Thus, at least $n = 6$ replicates must be run in order to obtain a test with the required power.

The power of the test for the random effects model is

$$1 - \beta = P\{\text{Reject } H_0 | H_0 \text{ is false}\}$$
$$= P\{F_0 > F_{\alpha, a-1, N-a} | \sigma_\tau^2 > 0\} \qquad (11\text{-}31)$$

Once again the distribution of the test statistic F_0 under the alternative hypothesis is needed. It can be shown that if H_1 is true ($\sigma_\tau^2 > 0$) the distribution of F_0 is central F, with $a - 1$ and $N - a$ degrees of freedom.

Since the power of the random effects model is based on the central F distribution, we could use the tables of the F distribution in the Appendix to evaluate Equation (11-31). However, it is much easier to evaluate the power of the test by using the operating characteristic curves in Chart VIII of the Appendix. These curves plot the probability of the type II error against λ, where

$$\lambda = \sqrt{1 + \frac{n\sigma_\tau^2}{\sigma^2}} \qquad (11\text{-}32)$$

Since σ^2 is usually unknown, we may either use a prior estimate or define the value of σ_τ^2 that we are interested in detecting in terms of the ratio σ_τ^2/σ^2.

- **Example 11-8.** Suppose we have five treatments selected at random, with six observations per treatment and $\alpha = .05$ and we wish to determine the power of the test if σ_τ^2 is equal to σ^2. Since $a = 5$, $n = 6$, and $\sigma_\tau^2 = \sigma^2$, we may compute

$$\lambda = \sqrt{1 + 6(1)} = 2.646$$

From the operating characteristic curve with $a - 1 = 4$, $N - a = 25$ degrees of freedom, and $\alpha = .05$, we find that

$$\beta \simeq 20$$

Therefore, the power is approximately .80.

11-7 Summary

This chapter has introduced the analysis of variance for an experiment with a single factor. Both the fixed effects and random effects models were presented. The primary difference between the two models is the inference space. In the fixed effects model inferences are valid only about the factor levels specifically considered in the analysis, while in the random effects model the conclusions may be extended to the population of factor levels. Orthogonal contrasts and Duncan's multiple range test were suggested for making comparisons between factor level means in the fixed effects model. A procedure was also given for estimating the variance components in the random effects model.

11-8   Exercises

11-1.   The compressive strength of concrete is being studied. Four different mixing techniques are being investigated. The following data have been collected.

Mixing Technique	Compressive Strength (psi)			
1	3129	3000	2865	2890
2	3200	3300	2975	3150
3	2800	2900	2985	3050
4	2600	2700	2600	2765

(a) Test the hypothesis that mixing techniques affect the strength of the concrete. Use $\alpha = .05$.

(b) Use Duncan's multiple range test to make comparisons between pairs of means. Estimate the treatment effects.

11-2.   A textile mill has a large number of looms. Each loom is supposed to provide the same output of cloth per minute. To investigate this assumption, five looms are chosen at random and their output measured at different times. The following data are obtained.

Loom	Output (lb/min)				
1	4.0	4.1	4.2	4.0	4.1
2	3.9	3.8	3.9	4.0	4.0
3	4.1	4.2	4.1	4.0	3.9
4	3.6	3.8	4.0	3.9	3.7
5	3.8	3.6	3.9	3.8	4.0

(a) Is this a fixed or random effects experiment? Are the looms equal in output?

(b) Estimate the variability between looms.

(c) Estimate the experimental error variance.

(d) What is the probability of accepting H_0 if σ_τ^2 is four times the experimental error variance?

11-3. An experiment was run to determine whether four specific firing temperatures affect the density of a certain type of brick. The experiment led to the following data.

Temperature (°F)	Density						
100	21.8	21.9	21.7	21.6	21.7	21.5	21.8
125	21.7	21.4	21.5	21.5	—	—	—
150	21.9	21.8	21.8	21.6	21.5	—	—
175	21.9	21.7	21.8	21.7	21.6	21.8	—

(a) Does the firing temperature affect the density of the bricks?

(b) Estimate the components in the model.

11-4. An electronics engineer is interested in the effect on tube conductivity of five different types of coating for cathode ray tubes used in a telecommunications system display device. The following conductivity data are obtained.

Coating Type	Conductivity			
1	143	141	150	146
2	152	149	137	143
3	134	133	132	127
4	129	127	132	129
5	147	148	144	142

(a) Is there any difference in conductivity due to coating type? Use $\alpha = .05$.

(b) Estimate the overall mean and the treatment effects.

(c) Compute a 95 percent interval estimate of the mean coating type 1. Compute a 99 percent interval estimate of the mean difference between coating types 1 and 4.

(d) Test all pairs of means using Duncan's multiple range test, with $\alpha = .05$.

(e) Assuming that coating type 4 is currently in use, what are your recommendations to the manufacturer? We wish to minimize conductivity.

11-5. The response time in milliseconds was determined for three different types of circuits used in an electronic calculator. The results are recorded here.

Circuit Type	Response Time				
1	19	22	20	18	25
2	20	21	33	27	40
3	16	15	18	26	17

(a) Test the hypothesis that the three circuit types have the same response time.

(b) Use Duncan's multiple range test to compare pairs of treatment means.

(c) Construct a set of orthogonal contrasts, assuming that at the outset of the experiment you suspected the response time of circuit type 2 to be different from the other two.

(d) What is the power of this test for detecting $\sum_{i=1}^{3} \tau_i^2/\sigma^2 = 3.0$?

11-6. Five different digital computer circuit designs are being studied in order to compare the amount of noise present. The following data have been obtained.

Circuit Type	Noise Observed				
1	19	20	19	30	8
2	80	61	73	56	80
3	47	26	25	35	50
4	95	46	83	78	97
5	19	17	21	5	32

(a) Is the amount of noise present the same for all five circuits?

(b) Estimate the components of the appropriate model for this problem. Estimate all possible differences between pairs of treatments. Are these treatment differences uniquely estimated? If so, why?

(c) Which circuit would you select for use?

11-7. A manufacturer suspects that the batches of raw material furnished by her supplier differ significantly in potassium content. There are a large number of batches currently in the warehouse. Five of these are randomly selected for study. A chemist makes five determinations on each batch and obtains the following data.

Batch 1	Batch 2	Batch 3	Batch 4	Batch 5
23.46	23.59	23.51	23.28	23.29
23.48	23.46	23.64	23.40	23.46
23.56	23.42	23.46	23.37	23.37
23.39	23.49	23.52	23.46	23.32
23.40	23.50	23.49	23.39	23.38

(a) Is there a significant variation in potassium content from batch to batch?

(b) Estimate the components of variance.

(c) If it is important to detect a ratio of σ_τ^2/σ^2 of 2.5 with a high probability, is the number of replicates adequate?

11-8. A large number of ovens in a metal working shop are used to heat metal specimens. All ovens are supposed to operate at the same temperature, although it is suspected that this may not be true. Three ovens are selected at random and their temperatures on successive heats is noted. These data are shown here.

Oven	Temperature					
1	491.50	498.30	498.12	490.50	493.60	492.48
2	488.50	484.65	479.92	477.33	—	—
3	490.10	484.80	488.26	473.01	471.83	—

(a) Is there a significant temperature variation between ovens?

(b) How much of the variability in the data is due to differences between ovens?

11-9. A brewer uses three different production lines to can beer. He suspects that the mean net contents of the cans differ from line to line. He randomly selects five cans from each line and measures the net contents. The data found is presented here.

Line	Contents (fluid ounces)				
1	12.11	12.17	12.20	12.15	12.18
2	12.05	12.09	12.12	12.10	12.08
3	12.15	12.10	12.09	12.10	12.11

(a) Is there a significant variation in mean net contents from line to line?

(b) Construct a 95 percent confidence interval on the mean net contents from line 2.

11-10. Three chemists are asked to determine the percentage of calcium in a certain chemical compound. Each chemist makes four determinations, and the results are shown in the following tables.

Chemist	Percentage of Calcium			
1	84.99	84.04	84.38	84.24
2	85.15	85.13	84.88	84.69
3	84.72	84.48	85.16	84.72

(a) Do chemists differ significantly? Use $\alpha = .05$.

(b) What is the power of this test for detecting $\sum_{i=1}^{3} \tau_i^2/\sigma^2 = 3.5$?

(c) If chemist 2 is a new employee, construct a meaningful set of orthogonal contrasts that we might have thought useful at the start of the experiment.

(d) If it is desired to detect $\sum_{i=1}^{3} \tau_i^2/\sigma^2 = 1.5$, with probability at least .95, what sample size should be used?

11-11. Three brands of batteries are under study. It is suspected that the life (in weeks) of the three brands is different. Five batteries of each brand are tested with the following results.

Weeks of Life		
Brand 1	Brand 2	Brand 3
125	119	127
124	120	125
123	119	123
124	121	124
123	122	125

(a) Are the lives of these brands of batteries different?

(b) Estimate the components of the appropriate statistical model.

(c) Construct a 95 percent interval estimate for the mean life of battery brand 2. Construct a 99 percent interval estimate for the mean difference between the life of battery brands 2 and 3.

(d) Which brand would you select for use? If the manufacturer will replace without charge any battery that fails in less than 120 weeks, what percentage would he expect to replace?

11-12. Four catalysts that may affect the concentration of one component in a three-component liquid mixture are being investigated. The following concentrations are obtained.

Catalyst			
1	2	3	4
8.2	6.3	.1	2.9
7.2	4.5	4.2	9.9
8.4	7.0	5.4	.0
5.8	5.3	—	1.7
4.9	—	—	—

(a) Do the four catalysts have the same effect on the concentration?

(b) Estimate the components of the appropriate statistical model.

(c) Construct a 99 percent confidence interval estimate of the mean response for catalyst 1.

11-13. The purity of a chemical product is suspected to vary from one batch to another. A random sample of five batches is selected and several determinations made on each batch.

Batch 1	Batch 2	Batch 3	Batch 4	Batch 5
93.88	94.53	95.40	93.16	93.75
93.33	94.39	95.88	93.71	93.38
93.16	94.16	95.89	93.16	94.01
—	93.99	—	93.67	93.91
—	—	—	93.08	—

(a) Does the purity vary significantly from batch to batch?

(b) Estimate the between- and within-batch variation.

11-14. Three types of tubes are used in a particular industrial application. Ten specimens of each tube type are selected and their life in weeks noted.

	Weeks of Life				
Type 1		Type 2		Type 3	
25.33	40.31	28.96	29.18	37.79	31.44
29.78	39.14	40.51	27.15	27.65	29.50
29.61	32.30	21.96	32.71	24.65	32.51
24.68	21.38	25.47	30.22	29.07	30.68
33.54	26.69	24.55	23.54	37.74	23.41

(a) Are the mean lives of the three tube types the same?

(b) Estimate the components of the appropriate statistical model.

(c) If it is important to detect $\sum_{i=1}^{3} \tau_i^2/\sigma^2 = 4.0$, with probability .95, are ten replicates sufficient?

11-15. Suppose that four normal populations have the common variance $\sigma^2 = 25$ and means $\mu_1 = 50$, $\mu_2 = 60$, $\mu_3 = 50$, and $\mu_4 = 60$. How many observations should be taken on each population so that the probability of rejecting the hypothesis of equality of means is at least .90? Use $\alpha = .05$.

11-16. Suppose that five normal populations have common variance $\sigma^2 = 100$ and means $\mu_1 = 175$, $\mu_2 = 190$, $\mu_3 = 160$, $\mu_4 = 200$, and $\mu_5 = 215$. How many observations per population must be taken so that the probability of rejecting the hypothesis of equality of means is at least .95? Use $\alpha = .01$.

11-17. Use Bartlett's test to determine if the assumption of homogeneous variances is reasonable in Exercise 11-11.

11-18. Use Bartlett's test to determine if the assumption of homogeneous variances is reasonable in Exercise 11-14.

11-19. Consider testing the equality of the means of two normal populations where the variances are unknown, but assumed equal. The appropriate test procedure is the two-sample t-test. Show that the two-sample t-test is equivalent to the one-way classification analysis of variance.

11-20. Show that the variance of the linear combination $\sum_{i=1}^{a} c_i y_{i.}$ is $\sigma^2 \sum_{i=1}^{a} n_i c_i^2$.

11-21. In a fixed effects model, suppose that there are n observations for each of four treatments. Let Q_1^2, Q_2^2 and Q_3^2 be single degree-of-freedom components for the orthogonal contrasts. Prove that $SS_{\text{treatments}} = Q_1^2 + Q_2^2 + Q_3^2$.

11-22. Consider the data shown in Exercise 11-5.

(a) Write out the least squares normal equations for this problem, and solve them for $\hat{\mu}$ and $\hat{\tau}_i$, making the usual constraint ($\sum_{i=1}^{3} \hat{\tau}_i = 0$). Estimate $\tau_1 - \tau_2$.

(b) Solve the equations in (a) using the constraint $\hat{\tau}_3 = 0$. Are the estimators $\hat{\tau}_i$ and $\hat{\mu}$ the same as you found in (a)? Why? Now estimate $\tau_1 - \tau_2$ and compare your answer with (a). What statement can you make about estimating contrasts in the τ_i?

(c) Estimate $\mu + \tau_1$, $2\tau_1 - \tau_2 - \tau_3$, and $\mu + \tau_1 + \tau_2$, using the two solutions to the normal equations. Compare the results obtained in each case.

Chapter 12

Simple Linear Regression and Correlation

In many problems there are two or more variables that are inherently related, and it is necessary to explore the nature of this relationship. Regression analysis is a statistical technique for modeling and investigating the relationship between two or more variables. For example, in a chemical process, suppose that the yield of product is related to the process operating temperature. Regression analysis can be used to build a model that expresses yield as a function of temperature. This model can then be used to predict yield at a given temperature level. It could also be used for process optimization or process control purposes.

In general, suppose that there is a single *dependent variable* or response y that is related to k *independent variables*, say $x_1, x_2, \ldots, x_k$. The dependent variable y is a random variable, while the independent variables $x_1, x_2, \ldots, x_k$ are measured with negligible error. The $\{x_i\}$ are called *mathematical* variables and are frequently controlled by the experimenter. Regression analysis can also be used in situations where $y, x_1, x_2, \ldots, x_k$ are jointly distributed random variables, such as when the data are collected as different measurements on a common experimental unit. The relationship between these variables is characterized by a mathematical model called a *regression equation*. More precisely, we speak of the regression of y on $x_1, x_2, \ldots, x_k$. This regression model is fitted to a set of data. In some instances, the experimenter will know the exact form of the true functional relationship between y and $x_1, x_2, \ldots, x_k$, say, $y = \phi(x_1, x_2, \ldots, x_k)$. However, in most cases, the true functional relationship is unknown, and the experimenter will choose an appropriate function to approximate ϕ. A polynomial model is usually employed as the approximating function.

In this chapter, we discuss the case where only a single independent variable, x, is of interest. Chapter 13 will present the case involving more than one independent variable.

12-1 Simple Linear Regression

We wish to determine the relationship between a single independent variable x and a dependent variable y. The independent variable x is assumed to be a continuous mathematical variable, controllable by the experimenter. Suppose that the true relationship between y and x is a straight line, and that the observation y at each level of x is a random variable. Now, the expected value of y for each value of x is

$$E(y|x) = \beta_0 + \beta_1 x \tag{12-1}$$

where the intercept β_0 and the slope β_1 are unknown constants. We assume that each observation, y, can be described by the model

$$y = \beta_0 + \beta_1 x + \epsilon \tag{12-2}$$

where ϵ is a random error with mean zero and variance σ^2. The $\{\epsilon\}$ are also assumed to be uncorrelated random variables. The regression model of Equation (12-2) involving only a single independent variable x is often called the *simple* linear regression model.

Suppose that we have n pairs of observations, say (y_1, x_1), (y_2, x_2), $\ldots$, (y_n, x_n). These data may be used to estimate the unknown parameters β_0 and β_1 in Equation (12-2). Our estimation procedure will be the method of least squares. That is, we will estimate β_0 and β_1 so that the sum of squares of the deviations between the observations and the regression line is a minimum. Now using Equation (12-2), we may write

$$y_i = \beta_0 + \beta_1 x_i + \epsilon_i, \qquad i = 1, 2, \ldots, n$$

and the sum of squares of the deviations of the observations from the true regression line is

$$L = \sum_{i=1}^{n} \epsilon_i^2 = \sum_{i=1}^{n} (y_i - \beta_0 - \beta_1 x_i)^2 \tag{12-3}$$

Minimizing the least squares function L is simplified if we rewrite the model, Equation (12-2), as

$$y = \beta_0' + \beta_1(x - \bar{x}) + \epsilon \tag{12-4}$$

where $\bar{x} = (1/n)\sum_{i=1}^{n} x_i$ and $\beta_0' = \beta_0 + \beta_1\bar{x}$. In Equation (12-4) we have corrected the independent variable for its mean, resulting in a transformation on the intercept. Equation (12-4) is frequently called the transformed simple linear regression model.

Employing the transformed model, the least squares function is

$$L = \sum_{i=1}^{n} [y_i - \beta_0' - \beta_1(x_i - \bar{x})]^2 \tag{12-5}$$

The least squares estimators of β_0 and β_1, say $\hat{\beta}_0$ and $\hat{\beta}_1$, must satisfy

$$\left.\frac{\partial L}{\partial \beta_0'}\right|_{\hat{\beta}_0,\hat{\beta}_1} = -2\sum_{i=1}^{n}[y_i - \hat{\beta}_0' - \hat{\beta}_1(x_i - \bar{x})] = 0$$

$$\left.\frac{\partial L}{\partial \beta_1}\right|_{\hat{\beta}_0,\hat{\beta}_1} = -2\sum_{i=1}^{n}[y_i - \hat{\beta}_0' - \hat{\beta}_1(x_i - \bar{x})](x_i - \bar{x}) = 0$$

Simplifying these two equations yields

$$n\hat{\beta}_0' = \sum_{i=1}^{n} y_i$$

$$\hat{\beta}_1 \sum_{i=1}^{n} (x_i - \bar{x})^2 = \sum_{i=1}^{n} y_i(x_i - \bar{x}) \tag{12-6}$$

Equations (12-6) are called the *least squares normal equations*. The solution to the normal equations is

$$\hat{\beta}_0' = \frac{1}{n}\sum_{i=1}^{n} y_i = \bar{y} \tag{12-7}$$

$$\hat{\beta}_1 = \frac{\sum_{i=1}^{n} y_i(x_i - \bar{x})}{\sum_{i=1}^{n}(x_i - \bar{x})^2} \tag{12-8}$$

Therefore, $\hat{\beta}_0'$ and $\hat{\beta}_1$ are the least squares estimators of the transformed intercept and slope, respectively. The estimated simple linear regression model is then

$$\hat{y} = \hat{\beta}_0' + \hat{\beta}_1(x - \bar{x}) \tag{12-9}$$

To present the results in terms of the original intercept β_0, note that

$$\hat{\beta}_0 = \hat{\beta}_0' - \hat{\beta}_1\bar{x}$$

and the corresponding estimated simple linear regression model is

$$\hat{y} = \hat{\beta}_0 + \hat{\beta}_1 x \tag{12-10}$$

Equations (12-9) and (12-10) are equivalent; that is, they both produce the same value of $\hat{y}$ for a given value of x.

Notationally, it is convenient to give special symbols to the numerator and denominator of Equation (12-8). That is, let

$$S_{xx} = \sum_{i=1}^{n} (x_i - \bar{x})^2 = \sum_{i=1}^{n} x_i^2 - \frac{\left(\sum_{i=1}^{n} x_i\right)^2}{n} \tag{12-11}$$

and

$$S_{xy} = \sum_{i=1}^{n} y_i(x_i - \bar{x}) = \sum_{i=1}^{n} x_i y_i - \frac{\left(\sum_{i=1}^{n} x_i\right)\left(\sum_{i=1}^{n} y_i\right)}{n} \qquad (12\text{-}12)$$

We call S_{xx} the corrected sum of squares of x and S_{xy} the corrected sum of cross products of x and y. The extreme right-hand sides of Equations (12-11) and (12-12) are the usual computational formulas. Using this new notation, the least squares estimator of the slope is

$$\hat{\beta}_1 = \frac{S_{xy}}{S_{xx}} \qquad (12\text{-}13)$$

● **Example 12-1.** A chemical engineer is investigating the effect of process operating temperature on product yield. The study results in the following data:

Temperature, °C (x)	100	110	120	130	140	150	160	170	180	190
Yield, % (y)	45	51	54	61	66	70	74	78	85	89

These pairs of points are plotted in Fig. 12-1. Such a display is called a *scatter*

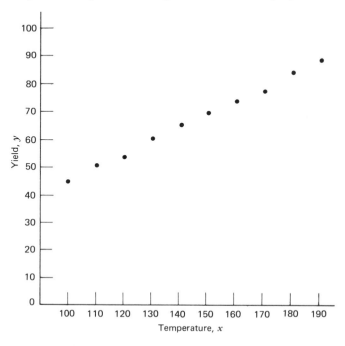

Fig. 12-1. Scatter diagram of yield versus temperature.

diagram. Examination of this scatter diagram indicates that there is a strong relationship between yield and temperature, and the tentative assumption of the straight-line model $y = \beta_0 + \beta_1 x + \epsilon$ appears to be reasonable. The following quantities may be computed:

$$n = 10 \qquad \sum_{i=1}^{10} x_i = 1450 \qquad \sum_{i=1}^{10} y_i = 673 \qquad \bar{x} = 145 \qquad \bar{y} = 67.3$$

$$\sum_{i=1}^{10} x_i^2 = 218{,}500 \qquad \sum_{i=1}^{10} y_i^2 = 47{,}225 \qquad \sum_{i=1}^{10} x_i y_i = 101{,}570$$

From Equations (12-11) and (12-12), we find

$$S_{xx} = \sum_{i=1}^{10} x_i^2 - \frac{\left(\sum_{i=1}^{10} x_i\right)^2}{10} = 218{,}500 - \frac{(1450)^2}{10} = 8250$$

and

$$S_{xy} = \sum_{i=1}^{10} x_i y_i - \frac{\left(\sum_{i=1}^{10} x_i\right)\left(\sum_{i=1}^{10} y_i\right)}{10} = 101{,}570 - \frac{(1450)(673)}{10} = 3985$$

Therefore, the least squares estimates of the slope and intercept are

$$\hat{\beta}_1 = \frac{S_{xy}}{S_{xx}} = \frac{3985}{8250} = .48303$$

and

$$\hat{\beta}_0' = \bar{y} = 67.3$$

The estimated simple linear regression model is

$$\hat{y} = \hat{\beta}_0' + \hat{\beta}_1(x - \bar{x})$$

or

$$\hat{y} = 67.3 + .48303(x - 145)$$

To express the model in terms of the original intercept, note that

$$\hat{\beta}_0 = \hat{\beta}_0' - \hat{\beta}_1 \bar{x}$$

$$= 67.3 - .48303(145) = -2.73939$$

and, consequently, we have

$$\hat{y} = -2.73939 + .48303x$$

Since we have only tentatively assumed the straight-line model to be appropriate, we will want to investigate the adequacy of the model. The statistical properties of the least squares estimators $\hat{\beta}_0$ (or $\hat{\beta}_0'$) and $\hat{\beta}_1$ are useful in assessing model adequacy. The estimators $\hat{\beta}_0$ (or $\hat{\beta}_0'$) and $\hat{\beta}_1$ are random variables, since they are just linear combinations of the y_i, and the y_i are random variables. We will investigate the bias and variance properties of these estimators. Consider first $\hat{\beta}_1$. The expected value of $\hat{\beta}_1$ is

$$E(\hat{\beta}_1) = E\left(\frac{S_{xy}}{S_{xx}}\right)$$

$$= \frac{1}{S_{xx}}E\left[\sum_{i=1}^{n} y_i(x_i - \bar{x})\right]$$

$$= \frac{1}{S_{xx}}E\left[\sum_{i=1}^{n}(\beta_0' + \beta_1(x_i - \bar{x}) + \epsilon_i)(x_i - \bar{x})\right]$$

$$= \frac{1}{S_{xx}}\left\{E\left[\beta_0'\sum_{i=1}^{n}(x_i - \bar{x})\right] + E\left[\beta_1\sum_{i=1}^{n}(x_i - \bar{x})^2\right] + E\left[\sum_{i=1}^{n}\epsilon_i(x_i - \bar{x})\right]\right\}$$

$$= \frac{1}{S_{xx}}\beta_1 S_{xx}$$

$$= \beta_1$$

since $\sum_{i=1}^{n}(x_i - \bar{x}) = 0$, and by assumption $E(\epsilon_i) = 0$. Thus, $\hat{\beta}_1$ is an *unbiased* estimator of the true slope β_1. Now consider the variance of $\hat{\beta}_1$. Since we have assumed that $V(\epsilon_i) = \sigma^2$, it follows that $V(y_i) = \sigma^2$, and

$$V(\hat{\beta}_1) = V\left(\frac{S_{xy}}{S_{xx}}\right)$$

$$= \frac{1}{S_{xx}^2}V\left[\sum_{i=1}^{n} y_i(x_i - \bar{x})\right] \tag{12-14}$$

The random variables $\{y_i\}$ are uncorrelated because the $\{\epsilon_i\}$ are uncorrelated. Therefore, the variance of the sum in Equation (12-14) is just the sum of the variances, and the variance of each term in the sum, say $V[y_i(x_i - \bar{x})]$, is $\sigma^2(x_i - \bar{x})^2$. Thus,

$$V(\hat{\beta}_1) = \frac{1}{S_{xx}^2}\sigma^2\sum_{i=1}^{n}(x_i - \bar{x})^2$$

$$= \frac{\sigma^2}{S_{xx}} \tag{12-15}$$

By using a similar approach, we can show that

$$E(\hat{\beta}_0') = \beta_0' \qquad V(\hat{\beta}_0') = \frac{\sigma^2}{n} \tag{12-16}$$

and

$$E(\beta_0) = \beta_0 \qquad V(\hat{\beta}_0) = \sigma^2\left[\frac{1}{n} + \frac{\bar{x}^2}{S_{xx}}\right] \tag{12-17}$$

To find $V(\hat{\beta}_0')$, we must make use of the result $\text{Cov}(\hat{\beta}_0', \hat{\beta}_1) = 0$. However, the covariance of $\hat{\beta}_0$ and $\hat{\beta}_1$ is not zero; in fact, $\text{Cov}(\hat{\beta}_0, \hat{\beta}_1) = -\sigma^2 \bar{x}/S_{xx}$. (Refer to Exercises 12-14 and 12-15.) Note that $\hat{\beta}_0'$ and $\hat{\beta}_0$ are unbiased estimators of β_0' and β_0, respectively.

It is usually necessary to obtain an estimate of σ^2. The difference between the observation y_i and the corresponding predicted value $\hat{y}_i$, say $e_i = y_i - \hat{y}_i$, is called a *residual*. The sum of the squares of the residuals, or the error sum of squares, would be

$$SS_E = \sum_{i=1}^{n} e_i^2$$

$$= \sum_{i=1}^{n} (y_i - \hat{y}_i)^2 \tag{12-18}$$

A more convenient computing formula for SS_E may be found by substituting the estimated model $\hat{y}_i = \bar{y} + \hat{\beta}_1(x_i - \bar{x})$ into Equation (12-18) and simplifying as follows:

$$SS_E = \sum_{i=1}^{n} [y_i - \bar{y} - \hat{\beta}_1(x_i - \bar{x})]^2$$

$$= \sum_{i=1}^{n} [y_i^2 + \bar{y}^2 + \hat{\beta}_1^2(x_i - \bar{x})^2 - 2\bar{y}y_i - 2\hat{\beta}_1 y_i(x_i - \bar{x}) - 2\hat{\beta}_1\bar{y}(x_i - \bar{x})]$$

$$= \sum_{i=1}^{n} y_i^2 + n\bar{y}^2 + \hat{\beta}_1^2 S_{xx} - 2\bar{y}\sum_{i=1}^{n} y_i - 2\hat{\beta}_1 S_{xy} - 2\hat{\beta}_1\bar{y}\sum_{i=1}^{n}(x_i - \bar{x}) \tag{12-19}$$

The last term in Equation (12-19) is zero, $2\bar{y}\sum_{i=1}^{n} y_i = 2n\bar{y}^2$, and $\hat{\beta}_1^2 S_{xx} = \hat{\beta}_1(S_{xy}/S_{xx})S_{xx} = \hat{\beta}_1 S_{xy}$. Therefore, Equation (12-19) becomes

$$SS_E = \sum_{i=1}^{n} y_i^2 - n\bar{y}^2 - \hat{\beta}_1 S_{xy}$$

But $\sum_{i=1}^{n} y_i^2 - n\bar{y}^2 = \sum_{i=1}^{n} (y_i - \bar{y})^2 \equiv S_{yy}$, say, so we may write SS_E as

$$SS_E = S_{yy} - \hat{\beta}_1 S_{xy} \tag{12-20}$$

The expected value of SS_E is $E(SS_E) = (n-2)\sigma^2$. Therefore,

$$\hat{\sigma}^2 = \frac{SS_E}{n-2} \equiv MS_E \tag{12-21}$$

is an unbiased estimator of σ^2.

Regression analysis is widely used, and frequently *misused*. There are several common abuses of regression that should be briefly mentioned. Care should be taken in selecting variables with which to construct regression models and in determining the form of the approximating function. It is quite possible to develop statistical relationships among variables that are completely unrelated in a practical sense. For example, one might attempt to relate the shear strength of spot welds with the number of boxes of punched cards used by the data processing department. A straight line may even appear to provide a good fit to the data, but the relationship is an unreasonable one on which to rely. A strong observed association between variables does not necessarily imply that a *causal* relationship exists between those variables. One should not necessarily infer causality from a regression model without further analysis and investigation.

Regression relationships are valid only for values of the independent variable within the range of the original data. The linear relationship that we have tentatively assumed may be valid over the original range of x, but it may be unlikely to remain so as we encounter x values beyond that range. In other words, as we move beyond the range of values of x for which data were collected, we become less certain about the validity of the assumed model. Regression models are not necessarily valid for extrapolation purposes.

Finally, one occasionally feels that the model $y = \beta x + \epsilon$ is appropriate. The omission of the intercept from this model implies, of course, that $y = 0$ when $x = 0$. This is a very strong assumption that often is unjustified. Even when two variables, such as the height and weight of men, would seem to qualify for the use of this model, we would usually obtain a better fit by including the intercept, because of the limited range of data on the independent variable.

12-2   Hypothesis Testing in Simple Linear Regression

An important part of assessing the adequacy of the simple linear regression model is testing statistical hypotheses about the model parameters and constructing certain confidence intervals. Hypothesis testing is discussed in this section, and Section 12-3 presents methods for constructing confidence intervals. To test hypotheses about the slope and intercept of the regression model, we must make the additional assumption that the error component ϵ_i is normally distributed. Thus, the complete assumptions are that the errors are $NID(0, \sigma^2)$. Later we will discuss how these assumptions can be checked through *residual analysis*.

Suppose the experimenter wishes to test the hypothesis that the slope equals a constant, say β_{10}. The appropriate hypotheses are

$$H_0: \beta_1 = \beta_{10}$$

$$H_1: \beta_1 \neq \beta_{10} \tag{12-22}$$

where we have assumed a two-sided alternative. Now since the ϵ_i are $NID(0, \sigma^2)$ it follows directly that the observations y_i are $NID(\beta_0 + \beta_1 x_i, \sigma^2)$. From Equation (12-18) we observe that $\hat{\beta}_1$ is a linear combination of the observations y_i. Thus, $\hat{\beta}_1$ is a linear combination of independent normal random variables and, consequently, $\hat{\beta}_1$ is $N(\beta_1, \sigma^2/S_{xx})$, using the bias and variance properties of $\hat{\beta}_1$ from Section 12-1. Furthermore, $\hat{\beta}_1$ is independent of MS_E. Then, as a result of the normality assumption, the statistic

$$t_0 = \frac{\hat{\beta}_1 - \beta_{10}}{\sqrt{MS_E/S_{xx}}} \tag{12-23}$$

follows the t distribution with $n - 2$ degrees of freedom under $H_0: \beta_1 = \beta_{10}$. We would reject $H_0: \beta_1 = \beta_{10}$ if

$$|t_o| > t_{\alpha/2, n-2} \tag{12-24}$$

where t_0 is computed from Equation (12-23).

A similar procedure can be used to test hypotheses about the intercept. To test

$$H_0: \beta_0 = \beta_{00}$$
$$H_1: \beta_0 \neq \beta_{00} \tag{12-25}$$

we would use the statistic

$$t_0 = \frac{\hat{\beta}_0 - \beta_{00}}{\sqrt{MS_E\left[\dfrac{1}{n} + \dfrac{\bar{x}^2}{S_{xx}}\right]}} \tag{12-26}$$

and reject the null hypothesis if $|t_0| > t_{\alpha/2, n-2}$.

A very important special case of the hypothesis of Equation (12-22) is

$$H_0: \beta_1 = 0$$
$$H_1: \beta_1 \neq 0 \tag{12-27}$$

This hypothesis relates to the significance of regression. Failing to reject $H_0: \beta_1 = 0$ is equivalent to concluding that there is no linear relationship between x and y. This situation is illustrated in Fig. 12-2. Note that this may imply either that x is of little value in explaining the variation in y and that the best estimator of y for any x is $\hat{y} = \bar{y}$ (Fig. 12-2a), or that the true relationship between x and y is not linear (Fig. 12-2b). Alternatively, if $H_0: \beta_1 = 0$ is rejected, this implies that x is of value in explaining the variability in y. This is illustrated in Fig. 12-3. However, rejecting $H_0: \beta_1 = 0$ could mean either that the straight-line model is adequate (Fig. 12-3a), or that even though there is a linear effect of x, better results could be obtained with the addition of higher-order polynomial terms in x (Fig. 12-3b).

The test procedure for $H_0: \beta_1 = 0$ may be developed from two approaches. The first approach starts with the following partitioning of the total corrected

368   Simple Linear Regression and Correlation

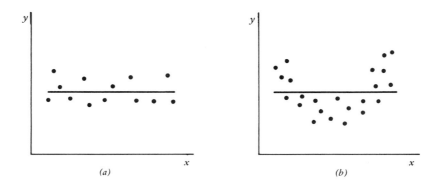

Fig. 12-2. The hypothesis $H_0 : \beta_1 = 0$ is not rejected.

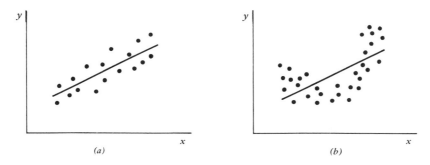

Fig. 12-3. The hypothesis $H_0 : \beta_1 = 0$ is rejected.

sum of squares for y:

$$S_{yy} \equiv \sum_{i=1}^{n} (y_i - \bar{y})^2 = \sum_{i=1}^{n} (\hat{y}_i - \bar{y})^2 + \sum_{i=1}^{n} (y_i - \hat{y}_i)^2 \qquad (12\text{-}28)$$

The two components of S_{yy} measure, respectively, the amount of variability in the y_i, accounted for by the regression line, and the residual variation left unexplained by the regression line. We usually define $SS_E = \sum_{i=1}^{n}(y_i - \hat{y}_i)^2$ as the *error* sum of squares and $SS_R = \sum_{i=1}^{n}(\hat{y}_i - \bar{y})^2$ as the *regression* sum of squares. Thus, Equation (12-28) may be written as

$$S_{yy} = SS_R + SS_E \qquad (12\text{-}29)$$

Comparing Equation (12-29) with Equation (12-20), we note that the regression sum of squares SS_R is

$$SS_R = \hat{\beta}_1 S_{xy} \qquad (12\text{-}30)$$

TABLE 12-1 Analysis of Variance for Testing Significance of Regression

Source of Variation	Sum of Squares	Degrees of Freedom	Mean Square	F_0
Regression	$SS_R = \hat{\beta}_1 S_{xy}$	1	MS_R	MS_R/MS_E
Error or residual	$SS_E = S_{yy} - \hat{\beta}_1 S_{xy}$	$n-2$	MS_E	
Total	S_{yy}	$n-1$		

S_{yy} has $n-1$ degrees of freedom, and SS_R and SS_E have 1 and $n-2$ degrees of freedom, respectively.

We may show that $E[SS_E/(n-2)] = \sigma^2$ and $E(SS_R) = \sigma^2 + \beta_1^2 S_{xx}$, and that SS_E and SS_R are independent. Thus, if $H_0: \beta_1 = 0$ is true, the statistic

$$F_0 = \frac{SS_R/1}{SS_E/(n-2)} = \frac{MS_R}{MS_E} \tag{12-31}$$

follows the $F_{1,n-2}$ distribution, and we would reject H_0 if $F_0 > F_{\alpha,1,n-2}$. The test procedure is usually arranged in an analysis of variance table, such as Table 12-1.

The test for significance of regression may also be developed from Equation (12-23) with $\beta_{10} = 0$, say

$$t_0 = \frac{\hat{\beta}_1}{\sqrt{MS_E/S_{xx}}} \tag{12-32}$$

Squaring both sides of Equation (12-32), we obtain

$$t_0^2 = \frac{\hat{\beta}_1^2 S_{xx}}{MS_E} = \frac{\hat{\beta}_1 S_{xy}}{MS_E} = \frac{MS_R}{MS_E} \tag{12-33}$$

Note that t_0^2 in Equation (12-33) is identical to F_0 in Equation (12-31). It is true, in general, that the square of a t random variable with f degrees of freedom is an F random variable, with one and f degrees of freedom in the numerator and denominator, respectively. Thus, the test using t_0 is equivalent to the test based on F_0.

● **Example 12-2.** We will test the model developed in Example 12-1 for significance of regression. The estimated model is $\hat{y} = -2.73939 + .48303x$, and S_{yy} is computed as

$$S_{yy} = \sum_{i=1}^{n} y_i^2 - \frac{\left(\sum_{i=1}^{n} y_i\right)^2}{n}$$

$$= 47,225 - \frac{(673)^2}{10}$$

$$= 1932.10$$

TABLE 12-2 **Testing for Significance of Regression, Example 12-2**

Source of Variation	Sum of Squares	Degree of Freedom	Mean Square	F_0
Regression	1924.87	1	1924.87	2138.74
Error	7.23	8	.90	
Total	1932.10	9		

The regression sum of squares is

$$SS_R = \hat{\beta}_1 S_{xy} = (.48303)(3985) = 1924.87$$

and the error sum of squares is

$$SS_E = S_{yy} - SS_R$$

$$= 1932.10 - 1924.87$$

$$= 7.23$$

The analysis of variance for testing $H_0: \beta_1 = 0$ is summarized in Table 12-2. Noting that $F_0 = 2138.74 > F_{.01,1,8} = 11.26$, we reject H_0 and conclude that $\beta_1 \neq 0$.

12-3 Interval Estimation in Simple Linear Regression

In addition to point estimates of the slope and intercept, it is possible to obtain confidence interval estimates of these parameters. The width of these confidence intervals is a measure of the overall quality of the regression line. If the ϵ_i are normally and independently distributed, then

$$(\hat{\beta}_1 - \beta_1)/\sqrt{MS_E/S_{xx}} \quad \text{and} \quad (\hat{\beta}_0 - \beta_0)\bigg/\sqrt{MS_E\left[\frac{1}{n} + \frac{\bar{x}^2}{S_{xx}}\right]}$$

are both distributed as t with $n - 2$ degrees of freedom. Therefore, a $100(1 - \alpha)$ percent confidence interval on the slope β_1 is given by

$$\hat{\beta}_1 - t_{\alpha/2,n-2}\sqrt{\frac{MS_E}{S_{xx}}} \leq \beta_1 \leq \hat{\beta}_1 + t_{\alpha/2,n-2}\sqrt{\frac{MS_E}{S_{xx}}} \tag{12-34}$$

Similarly, a $100(1 - \alpha)$ percent confidence interval on the intercept β_0 is

$$\hat{\beta}_0 - t_{\alpha/2,n-2}\sqrt{MS_E\left[\frac{1}{n} + \frac{\bar{x}^2}{S_{xx}}\right]} \leq \beta_0 \leq \hat{\beta}_0 + t_{\alpha/2,n-2}\sqrt{MS_E\left[\frac{1}{n} + \frac{\bar{x}^2}{S_{xx}}\right]} \tag{12-35}$$

● **Example 12-3.** We will find a 95 percent confidence interval on the slope of the regression line using the data in Example 12-1. Recall that $\hat{\beta}_1 = .48303$, $S_{xx} = 8250$, and $MS_E = .90$ (see Table 12-2). Then, from Equation (12-34) we find

$$\hat{\beta}_1 - t_{.025,8}\sqrt{\frac{MS_E}{S_{xx}}} \le \beta_1 \le \hat{\beta}_1 + t_{.025,8}\sqrt{\frac{MS_E}{S_{xx}}}$$

or

$$.48303 - 2.306\sqrt{\frac{.90}{8250}} \le \beta_1 \le .48303 + 2.306\sqrt{\frac{.90}{8250}}$$

This simplifies to

$$.45894 \le \beta_1 \le .50712$$

A confidence interval may be constructed for the mean response at a specified x, say x_0. This is a confidence interval about $E(y|x_0)$ and is often called a confidence interval about the regression line. Since $E(y|x_0) = \beta_0' + \beta_1(x_0 - \bar{x})$, we may obtain a point estimate of $E(y|x_0)$ from the estimated model as

$$\widehat{E(y|x_0)} \equiv \hat{y}_0 = \hat{\beta}_0' + \hat{\beta}_1(x_0 - \bar{x})$$

Now $\hat{y}_0$ is an unbiased point estimator of $E(y|x_0)$. That is, $E(\hat{y}_0) = \beta_0' + \beta_1(x_0 - \bar{x})$, since $\hat{\beta}_0'$ and $\hat{\beta}_1$ are unbiased estimators of β_0' and β_1. The variance of $\hat{y}_0$ is

$$V(\hat{y}_0) = \sigma^2\left[\frac{1}{n} + \frac{(x_0 - \bar{x})^2}{S_{xx}}\right]$$

since $\text{Cov}(\hat{\beta}_0'\hat{\beta}_1) = 0$. Also, $\hat{y}_0$ is normally distributed, as $\hat{\beta}_0'$ and $\hat{\beta}_1$ are normally distributed. Therefore, a $100(1-\alpha)$ percent confidence interval about the true regression line at $x = x_0$ may be computed from

$$\hat{y}_0 - t_{\alpha/2,n-2}\sqrt{MS_E\left(\frac{1}{n} + \frac{(x_0 - \bar{x})^2}{S_{xx}}\right)} \le E(y|x_0)$$

$$\le \hat{y}_0 + t_{\alpha/2,n-2}\sqrt{MS_E\left(\frac{1}{n} + \frac{(x_0 - \bar{x})^2}{S_{xx}}\right)} \qquad (12\text{-}36)$$

The width of the confidence interval for $E(y|x_0)$ is a function of x_0. The interval width is a minimum for $x_0 = \bar{x}$ and widens as $|x_0 - \bar{x}|$ increases.

● **Example 12-4.** We will construct a 95 percent confidence interval about the regression line for the data in Example 12-1. The estimated model is $\hat{y}_0 = -2.73939 + .48303x_0$, and the 95 percent confidence interval on $E(y|x_0)$ is

TABLE 12-3 **Confidence Interval about the Regression Line, Example 12-4**

x_0	100	110	120	130	140	150	160	170	180	190
$\hat{y}_0$	45.56	50.39	55.22	60.05	64.88	69.72	74.55	79.38	84.21	89.04
95% confidence limits	±1.30	±1.10	±.93	±.79	±.71	±.71	±.79	±.93	±1.10	±1.30

found from Equation (12-36) as

$$\left[\hat{y}_0 \pm 2.306 \sqrt{.90\left(\frac{1}{10} + \frac{(x_0 - 145)^2}{8250}\right)} \right]$$

The fitted values $\hat{y}_0$ and the corresponding 95 percent confidence limits for the points $x_0 = x_i$, $i = 1, 2, \ldots, 10$, are displayed in Table 12-3. To illustrate the use of this table, we may find the 95 percent confidence interval on the true mean process yield at $x_0 = 140\,°C$ (say) as

$$64.88 - .71 \le E(y|x_0 = 140) \le 64.88 + .71$$

or

$$64.17 \le E(y|x_0 = 140) \le 65.49$$

The estimated model and the 95 percent confidence interval about the regression line are shown in Fig. 12-4.

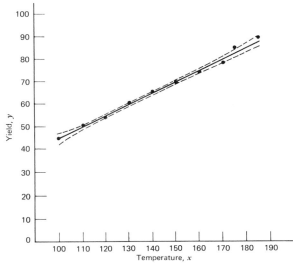

Fig. 12-4. A 95 percent confidence interval about the regression line for Example 12-4.

12-4   Prediction of New Observations

An important application of the regression model is the prediction of new or future observations y corresponding to a specified level of the independent variable x. If x_0 is the value of the independent variable of interest, then

$$\hat{y}_0 = \hat{\beta}_0 + \hat{\beta}_1 x_0 \qquad (12\text{-}37)$$

is the point estimate of the new or future value of the response y_0.

Now consider obtaining an interval estimate of this future observation y_0. This new observation is independent of the observations used to develop the regression model. Therefore, the confidence interval about the regression line, Equation (12-36), is inappropriate, since it is based only on the data used to fit the regression model. The confidence interval about the regression line refers to the true mean response at $x = x_0$ (that is, a *population parameter*), not to *future* observations.

Let y_0 be the future observations at $x = x_0$, and let $\hat{y}_0$ given by Equation (12-37) be the estimator of y_0. Note that the random variable

$$\psi = y_0 - \hat{y}_0$$

is normally distributed with mean zero and variance

$$V(\psi) = V(y_0 - \hat{y}_0)$$

$$= \sigma^2 \left[1 + \frac{1}{n} + \frac{(x_0 - \bar{x})^2}{S_{xx}} \right]$$

because y_0 is independent of $\hat{y}_0$. Thus, the $100(1 - \alpha)$ percent prediction interval on a future observations at x_0 is

$$\hat{y}_0 - t_{\alpha/2, n-2} \sqrt{MS_E \left[1 + \frac{1}{n} + \frac{(x_0 - \bar{x})^2}{S_{xx}} \right]} \leq y_0$$

$$\leq \hat{y}_0 + t_{\alpha/2, n-2} \sqrt{MS_E \left[1 + \frac{1}{n} + \frac{(x_0 - \bar{x})^2}{S_{xx}} \right]} \qquad (12\text{-}38)$$

Notice that the prediction interval is of minimum width at $x_0 = \bar{x}$ and widens as $|x_0 - \bar{x}|$ increases. By comparing Equation (12-38) with Equation (12-36), we observe that the prediction interval at x_0 is always wider than the confidence interval at x_0. This results because the prediction interval depends on both the error from the estimated model and the error associated with future observations.

We may also find a $100(1 - \alpha)$ percent prediction interval on the *mean* of k future observations on the response at $x = x_0$. Let $\bar{y}_0$ be the mean of k future observations at $x = x_0$. The $100(1 - \alpha)$ percent prediction interval on $\bar{y}_0$ is

$$\hat{y}_0 - t_{\alpha/2, n-2} \sqrt{MS_E \left[\frac{1}{k} + \frac{1}{n} + \frac{(x_0 - \bar{x})^2}{S_{xx}} \right]} \leq \bar{y}_0$$

$$\leq \hat{y}_0 + t_{\alpha/2, n-2} \sqrt{MS_E \left[\frac{1}{k} + \frac{1}{n} + \frac{(x_0 - \bar{x})^2}{S_{xx}} \right]} \qquad (12\text{-}39)$$

To illustrate the construction of a prediction interval, suppose we use the data in Example 12-1 and find a 95 percent prediction interval on the next observation on the process yield at $x_0 = 160°C$. Using Equation (12-38), we find that the prediction interval is

$$74.55 - 2.306\sqrt{.90\left[1 + \frac{1}{10} + \frac{(160-145)^2}{8250}\right]} \leq y_0$$
$$\leq 74.55 + 2.306\sqrt{.90\left[1 + \frac{1}{10} + \frac{(160-145)^2}{8250}\right]}$$

which simplifies to

$$72.21 \leq y_0 \leq 76.89$$

12-5   Measuring the Adequacy of the Regression Model

Fitting a regression model requires several assumptions. Estimation of the model parameters requires the assumption that the errors are uncorrelated random variables with mean zero and constant variance. Tests of hypotheses and interval estimation require that the errors are normally distributed. In addition, we assume that the order of the model is correct; that is, if we fit a first-order polynomial, then we are assuming that the phenomena actually behaves in a first-order manner.

The analyst should always consider the validity of these assumptions to be doubtful and conduct analyses to examine the adequacy of the model that has been tentatively entertained. In this section we discuss methods useful in this respect.

12-5.1   Residual Analysis

We define the residuals as $e_i = y_i - \hat{y}_i$, $i = 1, 2, \ldots, n$, where y_i is an observation and $\hat{y}_i$ is the corresponding estimated value from the regression model. Analysis of the residuals is frequently helpful in checking the assumption that the errors are $NID(0, \sigma^2)$ and in determining if additional terms in the model would be useful.

As an approximate check of normality, the experimenter can construct a frequency histogram of the residuals or plot them on normal probability paper. It requires judgment to assess the abnormality of such plots. One may also standardize the residuals by computing $d_i = e_i/\sqrt{MS_E}$, $i = 1, 2, \ldots, n$. If the errors are $NID(0, \sigma^2)$, then approximately 95 percent of the standardized residuals should fall in the interval $(-2, +2)$. Residuals far outside this interval may indicate the presence of an *outlier*; that is, an observation that is not typical of the rest of the data. Various rules have been proposed for discarding outliers. However, sometimes outliers provide important information about unusual circumstances of interest to the experimenter and

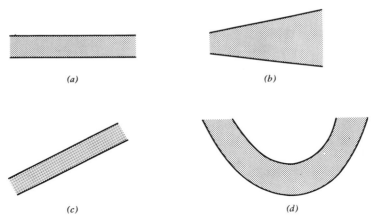

Fig. 12-5. Patterns for residual plots. (Adapted with permission from *Applied Regression Analysis*, by N. R. Draper and H. Smith, John Wiley & Sons, New York, 1966.)

should not be discarded. For further discussion of outliers, see Draper and Smith (1966).

It is frequently helpful to plot the residuals (1) in time sequence (if known), (2), against the $\hat{y}_i$, and (3) against the independent variable x. These graphs will usually look like one of the four general patterns in Fig. 12-5. Pattern (a) in Fig. 12-5 represents the normal situation, while patterns (b), (c), and (d) represent anomalies. If the residuals appear as in (b), then the variance of the observations may be increasing with time or with the magnitude of the y_i or x_i. If a plot of the residuals against time has the appearance of (c), then a linear term in time should be added to the model. Plots against $\hat{y}_i$ and x_i that look like (c) indicate an error in calculation or analysis. Residual plots that look like (d) indicate model inadequacy; that is, higher-order terms should be added to the model.

● **Example 12-5.** The residuals for the regression model in Example 12-1 are computed as follows:

$$e_1 = 45.00 - 45.56 = \quad -.56 \qquad e_6 = 70.00 - 69.72 = \quad .28$$
$$e_2 = 51.00 - 50.39 = \quad .61 \qquad e_7 = 74.00 - 74.55 = \quad -.55$$
$$e_3 = 54.00 - 55.22 = -1.22 \qquad e_8 = 78.00 - 79.38 = -1.38$$
$$e_4 = 61.00 - 60.05 = \quad .95 \qquad e_9 = 85.00 - 84.21 = \quad .79$$
$$e_5 = 66.00 - 64.88 = \quad 1.12 \qquad e_{10} = 89.00 - 89.04 = \quad -.04$$

These residuals are plotted on normal probability paper in Fig. 12-6. Since the

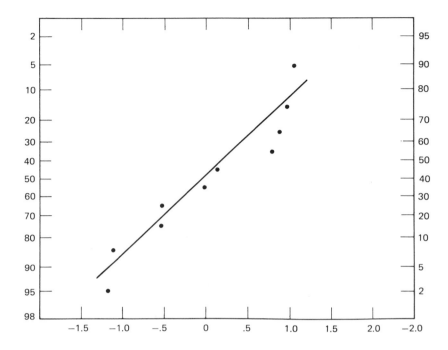

Fig. 12-6.   Residuals plotted on normal probability paper.

residuals fall approximately along a straight line in Fig. 12-6, we conclude that there is no severe departure from normality. The residuals are also plotted against $\hat{y}_i$ in Fig. 12-7a and against x_i in Fig. 12-7b. These plots do not indicate any serious model inadequacies.

12-5.2   The Lack-of-Fit Test

Regression models are often fit to data when the true functional relationship is unknown. Naturally, we would like to know whether the order of the model tentatively assumed is correct. This section will describe a test for the validity of this assumption.

The danger of using a regression model that is a poor approximation of the true functional relationship is illustrated in Fig. 12-8. Obviously, a polynomial of degree two or greater should have been used in this situation.

We present a test for the "goodness of fit" of a regression model. Specifically, the hypotheses we wish to test are

H_0: The model adequately fits the data
H_1: The model does not fit the data

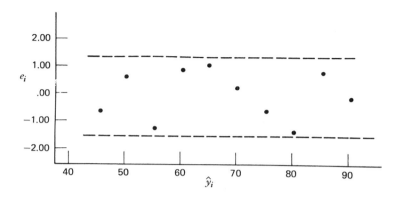

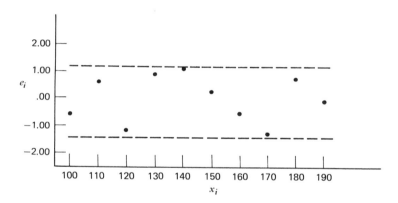

Fig. 12-7. Residual plots for Example 12-5. (*a*) Plot against $\hat{y}_i$. (*b*) Plot against x_i.

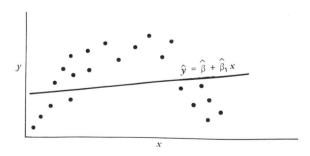

Fig. 12-8. A regression model displaying lack of fit.

The test involves partitioning the error or residual sum of squares into the following two components:

$$SS_E = SS_{PE} + SS_{LOF}$$

where SS_{PE} is the sum of squares attributable to "pure" error, and SS_{LOF} is the sum of squares attributable to the lack of fit of the model. To compute SS_{PE} we must have repeated observations on y for at least one level of x. Suppose that we have n total observations such that

$$y_{11}, y_{12}, \ldots, y_{1n_1} \qquad \text{repeated observations at } x_1$$

$$y_{21}, y_{22}, \ldots, y_{2n_2} \qquad \text{repeated observations at } x_2$$

$$\vdots \qquad\qquad\qquad \vdots$$

$$y_{m1}, y_{m2}, \ldots, y_{mn_m} \qquad \text{repeated observations at } x_m$$

Note that there are m distinct levels of x. The contribution to the pure-error sum of squares at x_1 (say) would be

$$\sum_{u=1}^{n_1} (y_{1u} - \bar{y}_1)^2 \tag{12-40}$$

The total sum of squares for pure error would be obtained by summing Equation (12-40) over all levels of x as

$$SS_{PE} = \sum_{i=1}^{m} \sum_{u=1}^{n_i} (y_{iu} - \bar{y}_i)^2 \tag{12-41}$$

There are $n_e = \sum_{i=1}^{m}(n_i - 1) = n - m$ degrees of freedom associated with the pure-error sum of squares. The sum of squares for lack of fit is simply

$$SS_{LOF} = SS_E - SS_{PE} \tag{12-42}$$

with $n - 2 - n_e = m - 2$ degrees of freedom. The test statistic for lack of fit would then be

$$F_0 = \frac{SS_{LOF}/(m-2)}{SS_{PE}/(n-m)} = \frac{MS_{LOF}}{MS_{PE}} \tag{12-43}$$

and we would reject it if $F_0 > F_{\alpha, m-2, n-m}$.

This test procedure may be easily introduced into the analysis of variance conducted for the significance of regression. If the null hypothesis of model adequacy is rejected, then the model must be abandoned and attempts made to find a more appropriate model. If H_0 is not rejected, then there is no apparent reason to doubt the adequacy of the model, and MS_{PE} and MS_{LOF} are often combined to estimate σ^2.

● **Example 12-6.** Suppose we have the following data:

x	1.0	1.0	2.0	3.3	3.3	4.0	4.0	4.0	4.7	5.0
y	2.3	1.8	2.8	1.8	3.7	2.6	2.6	2.2	3.2	2.0

x	5.6	5.6	5.6	6.0	6.0	6.5	6.9
y	3.5	2.8	2.1	3.4	3.2	3.4	5.0

We may compute $S_{yy} = 10.97$, $S_{xy} = 13.62$, $S_{xx} = 52.53$, $\bar{y} = 2.847$, and $\bar{x} = $ ' 4.382. The regression model is $\hat{y} = 1.708 + .260x$, and the regression sum of squares is $SS_R = \hat{\beta}_1 S_{xy} = (.260)(13.62) = 3.541$. The pure-error sum of squares is computed as follows:

Level of x	$\Sigma(y_i - \bar{y})^2$	Degrees of Freedom
1.0	.1250	1
3.3	1.8050	1
4.0	.1066	2
5.6	.9800	2
6.0	.0200	1
Totals:	3.0366	7

The analysis of variance is summarized in Table 12-4. Since $F_{.25,8,7} = 1.70$, we cannot reject the hypothesis that the tentative model adequately describes the data. We will pool lack-of-fit and pure-error mean squares to form the denominator mean square in the test for significance of regression. Also, since $F_{.05,1,15} = 4.54$, we must conclude that $\beta_1 \neq 0$.

In fitting a regression model to experimental data, a good practice is to use the

TABLE 12-4 **Analysis of Variance for Example 12-6**

Source of Variation	Sum of Squares	Degrees of Freedom	Mean Square	F_0
Regression	3.541	1	3.541	7.15
Residual	7.429	15	.4952	
(Lack of fit)	4.3924	8	.5491	1.27
(Pure error)	3.0366	7		
Total	10.970	16		

lowest degree model that adequately describes the data. The lack-of-fit test may be useful in this respect. However, it is always possible to fit a polynomial of degree $n - 1$ to n data points, and the experimenter should not consider using a model that is "saturated," that is, that has very nearly as many independent variables as observations on y.

12-5.3 The Coefficient of Determination

The quantity

$$R^2 = \frac{SS_R}{S_{yy}} = 1 - \frac{SS_E}{S_{yy}} \qquad (12\text{-}44)$$

is called the coefficient of determination, and it is often used to judge the adequacy of a regression model. (We will see subsequently that in the case where x and y are jointly distributed random variables R^2 is the square of the correlation coefficient between x and y.) Clearly $0 \le R^2 \le 1$. We often refer loosely to R^2 as the amount of variability in the data explained or accounted for by the regression model. For the data in Example 12-1, we have $R^2 = SS_R/S_{yy} = 1924.87/1932.10 = .9963$; that is, 99.63 percent of the variability in the data is accounted for by the model.

The statistic R^2 should be used with caution, since it is always possible to make R^2 unity by simply adding enough terms to the model. For example, we can obtain a "perfect" fit to n data points with a polynomial of degree $n - 1$. Also, R^2 will always increase if we add a variable to the model, but this does not necessarily mean the new model is superior to the old one. Unless the error sum of squares in the new model is reduced by an amount equal to the original error mean square, the new model will have a larger error mean square than the old one, because of the loss of one degree of freedom. Thus the new model will actually be worse than the old one.

There are several important misconceptions about R^2. In general, R^2 does not measure the magnitude of the slope of the regression line. A large value of R^2 does not imply a steep slope. Furthermore, R^2 does not measure the appropriateness of the model, since it can be artificially inflated by adding higher-order polynomial terms. Even if y and x are related in a nonlinear fashion, R^2 will often be large. For example, R^2 for the regression equation in Fig. 12-3b will be relatively large, even though the linear approximation is poor. Finally, even though R^2 is large, this does not necessarily imply that the regression model will provide accurate predictions of future observations.

12-6 Transformations to a Straight Line

We occasionally find that the straight-line regression model $y = \beta_0 + \beta_1 x + \epsilon$ is inappropriate because the true regression function is nonlinear. Sometimes this is visually determined from the scatter diagram, and sometimes we know in advance that the model is nonlinear because of prior experience or underlying theory. In some situations a nonlinear function can be expressed as a straight line by using a suitable transformation. Such nonlinear models are called *intrinsically linear.*

As an example of a nonlinear model that is intrinsically linear, consider the exponential function

$$y = \beta_0 e^{\beta_1 x} \epsilon$$

This function is intrinsically linear, since it can be transformed to a straight line by a logarithmic transformation

$$\ln y = \ln \beta_0 + \beta_1 x + \ln \epsilon$$

This transformation requires that the transformed error terms $\ln \epsilon$ are normally and independently distributed with mean 0 and variance σ^2.

Another intrinsically linear function is

$$y = \beta_0 + \beta_1 \left(\frac{1}{x}\right) + \epsilon$$

By using the reciprocal transformation $z = 1/x$, the model is linearized to

$$y = \beta_0 + \beta_1 z + \epsilon$$

Sometimes the logarithmic and reciprocal transformations can be employed jointly to linearize a function. For example, consider the function

$$y = \frac{1}{\exp(\beta_0 + \beta_1 x + \epsilon)}$$

Letting $y^* = 1/y$, we have the linearized form

$$\ln y^* = \beta_0 + \beta_1 x + \epsilon$$

Several other examples of nonlinear models that are intrinsically linear are given by Daniel and Wood (1971).

12-7 Correlation

Our development of regression analysis thus far has assumed that x is a mathematical variable, measured with negligible error, and that y is a random variable. Many applications of regression analysis involve situations where *both* x and y are random variables. In these situations, it is usually assumed that the observations (y_i, x_i), $i = 1, 2, \ldots, n$ are jointly distributed random variables obtained from the distribution $f(y, x)$. For example, suppose we wish to develop a regression model relating the shear strength of spot welds to the weld diameter. In this example, weld diameter cannot be controlled. We would randomly select n spot welds and observe a diameter (x_i) and a shear strength (y_i) for each. Therefore, (y_i, x_i) are jointly distributed random variables.

We usually assume that the joint distribution of y_i and x_i is the bivariate normal distribution. That is,

$$f(y, x) = \frac{1}{2\pi\sigma_1\sigma_2\sqrt{1-\rho^2}} \exp\left\{ -\frac{1}{2(1-\rho^2)}\left[\left(\frac{y-\mu_1}{\sigma_1}\right)^2 + \left(\frac{x-\mu_2}{\sigma_2}\right)^2 - 2\rho\left(\frac{y-\mu_1}{\sigma_1}\right) \times \left(\frac{x-\mu_2}{\sigma_2}\right) \right]\right\}$$ (12-45)

where μ_1 and σ_1^2 are the mean and variance of y, μ_2 and σ_2^2 are the mean and variance of x, and ρ is the correlation coefficient between y and x. Recall from Chapter 4 that the correlation coefficient is defined as

$$\rho = \frac{\sigma_{12}}{\sigma_1\sigma_2}$$

where σ_{12} is the covariance between y and x.

The conditional distribution of y for a given value of x is (see Chapter 7)

$$f(y|x) = \frac{1}{\sqrt{2\pi}\sigma_{1.2}} \exp\left[-\frac{1}{2}\left(\frac{y-\beta_0-\beta_1 x}{\sigma_{1.2}}\right)^2 \right]$$ (12-46)

where

$$\beta_0 = \mu_1 - \mu_2\rho\frac{\sigma_1}{\sigma_2}$$ (12-47a)

$$\beta_1 = \frac{\sigma_1}{\sigma_2}\rho$$ (12-47b)

and

$$\sigma_{1.2}^2 = \sigma_1^2(1-\rho^2)$$ (12-47c)

That is, the conditional distribution of y given x is normal with mean

$$E(y|x) = \beta_0 + \beta_1 x$$ (12-48)

and variance $\sigma_{1.2}^2$. Note that the mean of the conditional distribution of y given x is a straight line regression model. Furthermore, there is a relationship between the correlation coefficient ρ and the slope β_1. From Equation (12-47c) we see that if $\rho = 0$ then $\beta_1 = 0$, which implies that there is no regression of y on x. That is, knowledge of x does not assist us in predicting y.

The method of maximum likelihood may be used to estimate the parameters β_0 and β_1. It may be shown that the maximum likelihood estimators of these parameters are

$$\hat{\beta}_0 = \bar{y} - \hat{\beta}_1\bar{x}$$ (12-49a)

and

$$\hat{\beta}_1 = \frac{\sum\limits_{i=1}^{n} y_i(x_i - \bar{x})}{\sum\limits_{i=1}^{n}(x_i - \bar{x})^2} = \frac{S_{xy}}{S_{xx}} \tag{12-49b}$$

We note that the estimators of the intercept and slope in Equation (12-49) are identical to those given by the method of least squares in the case where x was assumed to be a mathematical variable. That is, the regression model with y and x jointly normally distributed is equivalent to the model with x considered as a mathematical variable. This follows, because the random variables y given x are independently and normally distributed with mean $\beta_0 + \beta_1 x$ and constant variance $\sigma_{1.2}^2$. These results will also hold for *any* joint distribution of y and x such that the conditional distribution of y given x is normal.

It is possible to draw inferences about the correlation coefficient ρ in this model. The estimator of ρ is the *sample correlation coefficient*

$$r = \frac{\sum\limits_{i=1}^{n} y_i(x_i - \bar{x})}{\left[\sum\limits_{i=1}^{n} (x_i - \bar{x})^2 \sum\limits_{i=1}^{n} (y_i - \bar{y})^2\right]^{1/2}}$$

$$= \frac{S_{xy}}{[S_{xx}S_{yy}]^{1/2}} \tag{12-50}$$

Note that

$$\hat{\beta}_1 = \left(\frac{S_{yy}}{S_{xx}}\right)^{1/2} r \tag{12-51}$$

so that the slope $\hat{\beta}_1$ is just the sample correlation coefficient r multiplied by a scale factor that is the square root of the "spread" of the y values divided by the "spread" of the x values. Thus $\hat{\beta}_1$ and r are closely related, although they provide somewhat different information. The sample correlation coefficient r measures the *linear* association between y and x, while $\hat{\beta}_1$ measures the predicted change in y for a unit change in x. In the case of a mathematical variable x, r has no meaning because the magnitude of r depends on the choice of spacing for x. We may also write, from Equation (12-51),

$$R^2 \equiv r^2 = \hat{\beta}_1^2 \frac{S_{xx}}{S_{yy}}$$

$$= \frac{\hat{\beta}_1 S_{xy}}{S_{yy}}$$

$$= \frac{SS_R}{S_{yy}}$$

which we recognize from Equation (12-44) as the coefficient of determination. That is, the coefficient of determination R^2 is just the square of the correlation coefficient between y and x.

It is often useful to test the hypothesis

$$H_0: \rho = 0$$

$$H_1: \rho \neq 0 \tag{12-52}$$

The appropriate test statistic for this hypothesis is

$$t_0 = \frac{r\sqrt{n-2}}{\sqrt{1-r^2}} \tag{12-53}$$

which follows the t distribution with $n-2$ degrees of freedom if $H_0: \rho = 0$ is true. Therefore, we would reject the null hypothesis if $|t_0| > t_{\alpha/2, n-2}$. This test is equivalent to the test of the hypothesis $H_0: \beta_1 = 0$ given in Section 12-2. This equivalence follows directly from Equation (12-51).

The test procedure for the hypothesis

$$H_0: \rho = \rho_0$$

$$H_1: \rho \neq \rho_0 \tag{12-54}$$

where $\rho_0 \neq 0$ is somewhat more complicated. For moderately large samples (say $n \geq 25$) the statistic

$$Z = \text{arctanh } r = \frac{1}{2} \ln \frac{1+r}{1-r} \tag{12-55}$$

is approximately normally distributed with mean

$$\mu_Z = \text{arctanh } \rho = \frac{1}{2} \ln \frac{1+\rho}{1-\rho}$$

and variance

$$\sigma_Z^2 = (n-3)^{-1}$$

Therefore, to test the hypothesis $H_0: \rho = \rho_0$, we may compute the statistic

$$Z_0 = (\text{arctanh } r - \text{arctanh } \rho_0)(n-3)^{1/2} \tag{12-56}$$

and reject $H_0: \rho = \rho_0$ if $|Z_0| > Z_{\alpha/2}$.

It is also possible to construct a $100(1-\alpha)$ percent confidence interval for ρ, using the transformation in Equation (12-55). The $100(1-\alpha)$ percent confidence interval is

$$\tanh\left(\text{arctanh } r - \frac{Z_{\alpha/2}}{\sqrt{n-3}}\right) \leq \rho \leq \tanh\left(\text{arctanh } r + \frac{Z_{\alpha/2}}{\sqrt{n-3}}\right) \tag{12-57}$$

where tanh $u = (e^u - e^{-u})/(e^u + e^{-u})$.

● **Example 12-7.** An industrial engineer employed by a soft drink bottler is studying the product delivery and service operations for vending machines. She suspects that the time required to load and service a machine is related to the number of cases of product delivered. A random sample of 25 retail outlets having vending machines is selected, and the in–outlet delivery time (in minutes) and volume of product delivered (in cases) is observed for each outlet. The data are shown in Table 12-5. We assume that delivery time and volume of product delivered are jointly normally distributed.

Using the data in Table 12-5, we may calculate

$$S_{yy} = 6105.9447 \quad S_{xx} = 698.5600 \quad \text{and} \quad S_{xy} = 2027.7132$$

The regression model is

$$\hat{y} = 5.1145 + 2.9027x$$

The sample correlation coefficient between x and y is computed from Equation (12-50) as

$$r = \frac{S_{xy}}{[S_{xx}S_{yy}]^{1/2}} = \frac{2027.7132}{[(698.5600)(6105.9447)]^{1/2}} = .9818$$

Note that $R^2 = (.9818)^2 = .9640$, or that approximately 96.40 percent of the variability in delivery time is explained by the linear relationship with delivery

TABLE 12-5 **Data for Example 12-7**

Observation	Delivery Time, y	Number of Cases, x	Observation	Delivery Time, y	Number of Cases, x
1	9.95	2	14	11.66	2
2	24.45	8	15	21.65	4
3	31.75	11	16	17.89	4
4	35.00	10	17	69.00	20
5	25.02	8	18	10.30	1
6	16.86	4	19	34.93	10
7	14.38	2	20	46.59	15
8	9.60	2	21	44.88	15
9	24.35	9	22	54.12	16
10	27.50	8	23	56.63	17
11	17.08	4	24	22.13	6
12	37.00	11	25	21.15	5
13	41.95	12			

volume. To test the hypothesis

$$H_0: \rho = 0$$

$$H_1: \rho \neq 0$$

we can compute the test statistic of Equation (12-53) as follows:

$$t_0 = \frac{r\sqrt{n-2}}{\sqrt{1-r^2}} = \frac{.9818\sqrt{23}}{\sqrt{1-.9640}} = 24.80$$

Since $t_{.025,23} = 2.069$, we reject H_0 and conclude that the correlation coefficient $\rho \neq 0$. Finally, we may construct an approximate 95 percent confidence interval on ρ from Equation (12-57). Since arctanh $r =$ arctanh $.9818 = 2.3452$, Equation (12-57) becomes

$$\tanh\left(2.3452 - \frac{1.96}{\sqrt{22}}\right) \leq \rho \leq \tanh\left(2.3452 + \frac{1.96}{\sqrt{22}}\right)$$

which reduces to

$$.9585 \leq \rho \leq .9921$$

12-8 Summary

This chapter has introduced the simple linear regression model and shown how least squares estimates of the model parameters may be obtained. Hypothesis-testing procedures and confidence interval estimates of the model parameters have also been developed. Tests of hypotheses and confidence intervals require the assumption that the observations y are normally and independently distributed random variables. Procedures for testing model adequacy, including a lack-of-fit test and residual analysis, were presented. The correlation model was also introduced to deal with the case where x and y are jointly normally distributed. The equivalence of the regression model parameter estimation problem for the case where x and y are jointly normal to the case where x is a mathematical variable was also discussed. Procedures for obtaining point and interval estimates of the correlation coefficient and for testing hypotheses about the correlation coefficient were developed.

12-9 Exercises

12-1. The yield of a chemical process is thought to be a function of the amount of catalyst added to the reaction. An experiment is run and the following data are obtained.

Yield (%)	60.54	63.86	63.76	60.15	66.66	71.66	70.81	65.72
Catalyst (lb)	.9	1.4	1.6	1.7	1.8	2.0	2.1	2.3

(a) Fit a simple linear regression model to the data.

(b) Test for significance of regression.

(c) Construct a 95 percent confidence interval on the slope β_1.

(d) Calculate R^2 for this model.

12-2. The strength of paper used in the manufacture of cardboard boxes (y) is related to the percentage of hardwood concentration in the original pulp (x). Under controlled conditions, a pilot plant manufactures 16 samples, each from a different batch of pulp, and measures the tensile strength. The data are shown here.

y	101.4	117.4	117.1	106.2	131.9	146.9	146.8	133.9
x	1.0	1.5	1.5	1.5	2.0	2.0	2.2	2.4

y	111.3	123.0	125.1	145.2	134.3	144.5	143.7	146.9
x	2.5	2.5	2.8	2.8	3.0	3.0	3.2	3.3

(a) Fit a simple linear regression model to the data.

(b) Test for lack of fit and significance of regression.

(c) Construct a 90 percent confidence interval on the slope β_1.

(d) Construct a 90 percent confidence interval on the intercept β_0.

(e) Construct a 95 percent confidence interval on the true regression line at $x = 2.5$.

12-3. An electronics engineer suspects that the life of an electronic component (y) is linearly related to the temperature in the operating environment (x). She performs an experiment with 10 components and records the following data.

y	673.58	699.95	704.33	696.21	738.76
x	175	185	190	200	210
y	768.67	769.36	753.83	719.04	743.59
x	215	220	230	235	240

(a) Fit a simple linear regression model to the data.

(b) Test for significance of regression.

(c) Calculate R^2 for this model.

12-4. The surface finish of a metal part is thought to be linearly related to the cutting speed of the machine which produces it. Surface finish is measured on an arbitrary scale from 0 to 20, with 0 being the roughest finish. The following data have been observed.

Surface finish	4.893	5.948	6.323	5.998	7.701	9.496	9.725	9.503
Speed	12	13	14	15	16	18	19	20

(a) Fit a simple linear regression model to the data.

(b) Test for significance of regression.

(c) Test the hypothesis that the slope $\beta_1 = .40$.

(d) Construct a 90 percent prediction interval on the next observation of the surface finish at $x = 15$.

(e) Construct a 90 percent prediction interval on the mean surface finish of the next four parts produced at a speed of $x = 15$.

12-5. The number of pounds of steam used per month by a chemical plant is thought to be related to the average ambient temperature for that month. The past year's usage and temperature are shown in the following table.

Month	Temp.	Usage/1000	Month	Temp.	Usage/1000
Jan.	21	185.79	July	68	621.55
Feb.	24	214.47	Aug.	74	675.06
Mar.	32	288.03	Sept.	62	562.03
Apr.	47	424.84	Oct.	50	452.93
May	50	454.58	Nov.	41	369.95
June	59	539.03	Dec.	30	273.98

(a) Fit a simple linear regression model to the data.

(b) Test for significance of regression.

(c) Test the hypothesis that the slope $\beta_1 = 10$.

(d) Construct a 99 percent confidence interval about the true regression line at $x = 58$.

(e) Construct a 99 percent prediction interval on the steam usage in the next month having a mean ambient temperature of 58°.

12-6. The percentage of impurity in oxygen gas produced by a distilling process is thought to be related to the percentage of hydrocarbon in the main condenser of the processor. One month's operating data are available, as shown here.

Purity (%)	86.91	89.85	90.28	86.34	92.58	87.33	86.29	91.86	95.61	89.86
Hydrocarbon (%)	1.02	1.11	1.43	1.11	1.01	.95	1.11	.87	1.43	1.02

Purity (%)	96.73	99.42	98.66	96.07	93.65	87.31	95.00	96.85	85.20	90.56
Hydrocarbon (%)	1.46	1.55	1.55	1.55	1.40	1.15	1.01	.99	.95	.98

(a) Fit a simple linear regression model to the data.

(b) Test for lack of fit and significance of regression.

(c) Calculate R^2 for this model.

(d) Calculate a 95 percent confidence interval for the slope β_1.

12-7. Compute the residuals for the data in Exercise 12-4.

(a) Plot the residuals on normal probability paper and draw appropriate conclusions.

(b) Plot the residuals against $\hat{y}$ and x. Interpret these displays.

12-8. Compute the residuals for the data in Exercise 12-5.

(a) Plot the residuals on normal probability paper and draw appropriate conclusions.

(b) Plot the residuals against $\hat{y}$ and x. Interpret these displays.

12-9 A public utility manager thinks that a relationship exists between the number of residential customers added per year and the total construction expenditure made by the company. He has seven years' data available.

Costs/10^4	7.4361	9.3661	8.4884	8.0220	10.9280	13.4605	11.9366
Δ Customers	435	501	418	512	550	610	491

(a) Fit a simple linear regression model to the data.
(b) Construct 90 percent confidence interval estimates of the slope and intercept.

12-10. Total inventory costs per year in a small manufacturing industry are thought to be linearly related to the average inventory level for that year. Five years' data are available.

Costs ($)	159.18	175.99	139.36	157.06	134.68
Average inventory ($/100)	105.0	113.0	94.7	104.0	92.0

(a) Fit a simple linear regression model to the data.
(b) Test for significance of regression.

12-11. An ice cream company suspects that storing ice cream at low temperatures for extended periods of time has a linear effect on the weight loss of the product. The following data were collected from the company's storage plant.

Weight loss (oz)	1.01	1.32	1.30	1.06	1.00	1.30	1.26
Time (weeks)	26	32	35	27	25	31	30

(a) Fit a simple linear regression model to the data.
(b) Test for significance of regression.
(c) Calculate R^2 for this model.
(d) Construct 90 percent confidence interval estimates of the slope and intercept.

12-12. The number of pounds of a certain chemical product produced per day is thought to be related to the reaction time (in hours). A pilot plant obtains the following data.

Pounds	Time of Reaction	Pounds	Time of Reaction
140.90	2.0	185.71	2.7
147.61	2.1	199.68	2.8
152.60	2.5	196.70	2.9
175.74	2.5	210.00	3.0
162.39	2.4	199.41	3.0
180.51	2.4	187.21	3.0
188.59	2.6	167.96	3.0
170.36	2.6	200.32	3.1

(a) Fit a simple linear regression model and test for significance of regression and lack of fit.

(b) Construct a 99 percent interval estimate of the slope.

(c) Construct a 95 percent prediction interval for the mean of six new observations if $x = 2.3$.

12-13. The final averages for 20 randomly selected students taking a course in engineering statistics and a course in operations research at Georgia Tech are shown here. Assume that the final averages are jointly normally distributed.

Statistics	86	75	69	75	90	94	83	86	71	65
OR	80	81	75	81	92	95	80	81	76	72
Statistics	84	71	62	90	83	75	71	76	84	97
OR	85	72	65	93	81	70	73	72	80	98

(a) Find the regression line relating the statistics final average to the OR final average.

(b) Estimate the correlation coefficient.

(c) Test the hypothesis that $\rho = 0$.

(d) Test the hypothesis that $\rho = .5$.

(e) Construct a 95 percent confidence interval estimate of the correlation coefficient.

12-14. The weight and systolic blood pressure of 26 randomly selected males in the age group 25 to 30 are shown in the following table. Assume that weight and blood pressure are jointly normally distributed.

Subject	Weight	Systolic BP	Subject	Weight	Systolic BP
1	165	130	14	172	153
2	167	133	15	159	128
3	180	150	16	168	132
4	155	128	17	174	149
5	212	151	18	183	158
6	175	146	19	215	150
7	190	150	20	195	163
8	210	140	21	180	156
9	200	148	22	143	124
10	149	125	23	240	170
11	158	133	24	235	165
12	169	135	25	192	160
13	170	150	26	187	159

(a) Find a regression line relating systolic blood pressure to weight.

(b) Estimate the correlation coefficient.

(c) Test the hypothesis that $\rho = 0$.

(d) Test the hypothesis that $\rho = .6$.

(e) Construct a 95 percent confidence interval estimate of the correlation coefficient.

12-15. Consider the simple linear regression model $y = \beta_0 + \beta_1 x + \epsilon$. Show that $E(MS_R) = \sigma^2 + \beta_1^2 S_{xx}$.

12-16. Suppose that we have assumed the straight-line regression model

$$y = \beta_0 + \beta_1 x_1 + \epsilon$$

but the response is affected by a second variable x_2 such that the true regression function is

$$E(y) = \beta_0 + \beta_1 x_1 + \beta_2 x_2$$

Is the estimator of the slope in the simple linear regression model unbiased?

12-17. Suppose that we are fitting a straight line and we wish to make the variance of the slope $\hat{\beta}_1$ as small as possible. Where should the observations x_i, $i = 1, 2, \ldots, n$, be taken so as to minimize $V(\hat{\beta}_1)$? Discuss the practical implications of this allocation of the x_i.

12-18. **Weighted Least Squares.** Suppose that we are fitting the straight line $y = \beta_0 + \beta_1 x + \epsilon$, but the variance of the y values now depends on the level of x; that is,

$$V(y_i \mid x_i) = \sigma_i^2 = \frac{\sigma^2}{w_i} \qquad i = 1, 2, \ldots, n$$

where the w_i are unknown constants, often called *weights*. Show that the resulting least squares normal equations are

$$\hat{\beta}_0 \sum_{i=1}^{n} w_i + \hat{\beta}_1 \sum_{i=1}^{n} w_i x_i = \sum_{i=1}^{n} w_i y_i$$

$$\hat{\beta}_0 \sum_{i=1}^{n} w_i x_i + \hat{\beta}_1 \sum_{i=1}^{n} w_i x_i^2 = \sum_{i=1}^{n} w_i x_i y_i$$

12-19. Consider the data shown below. Suppose that the relationship between y and x is hypothesized to be $y = (\beta_0 + \beta_1 x + \epsilon)^{-1}$. Fit an appropriate model to the data. Does the assumed model form seem appropriate?

x	10	15	18	12	9	8	11	6
y	.17	.13	.09	.15	.2	.21	.18	.24

Chapter 13

Multiple Regression

Many regression problems involve more than one independent variable. Regression models that employ more than one independent variable are called *multiple regression models*. Multiple regression is one of the most widely used statistical techniques. This chapter presents the basic techniques of parameter estimation and hypothesis testing for the multiple regression model, including several methods for measuring model adequacy and methods for the selection of variables and model building. A number of the special problems that one frequently encounters in using multiple regression are also discussed.

13-1 Multiple Regression Models

A regression model that involves more than one independent variable is called a *multiple regression model*. As an example, suppose that the effective life of a cutting tool depends on the cutting speed and the tool angle. A multiple regression model that might describe this relationship is

$$y = \beta_0 + \beta_1 x_1 + \beta_2 x_2 + \epsilon \tag{13-1}$$

where y denotes the effective tool life, x_1 denotes the cutting speed, and x_2 denotes the tool angle. This is a multiple linear regression model with *two* independent variables. The term "linear" is used because Equation (13-1) is a linear function of the unknown parameters β_0, β_1, and β_2. The independent variables x_1 and x_2 are often called *regressor variables* or *carriers*. Note that the model describes a plane in the two-dimensional space of the independent variables x_1 and x_2. The parameter β_0 is the intercept of the regression plane. The parameter β_1 indicates the expected change in response (y) per unit change in x_1 when x_2 is held constant. Similarly, β_2 measures the expected change in y per unit change in x_2 when x_1 is held constant.

In general, the dependent variable or response y may be related to k independent variables. The model

$$y = \beta_0 + \beta_1 x_1 + \beta_2 x_2 + \cdots + \beta_k x_k + \epsilon \qquad (13\text{-}2)$$

is called a multiple linear regression model with k independent variables. The parameters β_j, $j = 0, 1, \ldots, k$, are called the regression coefficients. This model describes a hyperplane in the k-dimensional space of the independent variables $\{x_j\}$. The parameter β_j represents the expected change in response y per unit change in x_j when all the remaining independent variables x_i ($i \neq j$) are held constant. The parameters β_j, $j = 1, 2, \ldots, k$, are often called *partial* regression coefficients, because they describe the partial effect of one independent variable when the other independent variables in the model are held constant.

Multiple linear regression models are often used as approximating functions. That is, the true functional relationship between y and $x_1, x_2, \ldots, x_k$ is unknown, but over certain ranges of the independent variables the linear regression model is an adequate approximation.

Models that are more complex in appearance than Equation (13-2) may often still be analyzed by multiple linear regression techniques. For example, consider the cubic polynomial model in one independent variable

$$y = \beta_0 + \beta_1 x + \beta_2 x^2 + \beta_3 x^3 + \epsilon \qquad (13\text{-}3)$$

If we let $x_1 = x$, $x_2 = x^2$, and $x_3 = x^3$, then Equation (13-3) can be written as

$$y = \beta_0 + \beta_1 x_1 + \beta_2 x_2 + \beta_3 x_3 + \epsilon \qquad (13\text{-}4)$$

which is a multiple linear regression model with three regressor variables. Models that include interaction effects may also be analyzed by multiple linear regression methods. For example, suppose that the model is

$$y = \beta_0 + \beta_1 x_1 + \beta_2 x_2 + \beta_{12} x_1 x_2 + \epsilon \qquad (13\text{-}5)$$

If we let $x_3 = x_1 x_2$ and $\beta_3 = \beta_{12}$, then Equation (13-5) can be written as

$$y = \beta_0 + \beta_1 x_1 + \beta_2 x_2 + \beta_3 x_3 + \epsilon \qquad (13\text{-}6)$$

which is a linear regression model. In general, any regression model that is linear in the *parameters* (the β values) is a linear regression model, regardless of the shape of the surface that it generates.

13-2 Estimation of the Parameters

The method of least squares is used to estimate the regression coefficients in Equation (13-2). Suppose that $n > k$ observations are available, and let x_{ij} denote the ith observation or level of variable x_j. The data will appear as in

TABLE 13-1 **Data for Multiple Linear Regression**

y	x_1	x_2	$\cdots$	x_k
y_1	x_{11}	x_{12}	$\cdots$	x_{1k}
y_2	x_{21}	x_{22}	$\cdots$	x_{2k}
$\vdots$	$\vdots$	$\vdots$		$\vdots$
y_n	x_{n1}	x_{n2}	$\cdots$	x_{nk}

Table 13-1. We assume that the error term ϵ in the model has $E(\epsilon) = 0$, $V(\epsilon) = \sigma^2$, and that the $\{\epsilon_i\}$ are uncorrelated random variables.

We may write the model, Equation (13-2), in terms of the observations as

$$
\begin{aligned}
y_i &= \beta_0 + \beta_1 x_{i1} + \beta_2 x_{i2} + \cdots + \beta_k x_{ik} + \epsilon_i \\
&= \beta_0 + \sum_{j=1}^{k} \beta_j x_{ij} + \epsilon_i \qquad i = 1, 2, \ldots, n
\end{aligned}
\tag{13-7}
$$

The least squares function is

$$
\begin{aligned}
L &= \sum_{i=1}^{n} \epsilon_i^2 \\
&= \sum_{i=1}^{n} \left(y_i - \beta_0 - \sum_{j=1}^{n} \beta_j x_{ij} \right)^2
\end{aligned}
\tag{13-8}
$$

The function L is to be minimized with respect to $\beta_0, \beta_1, \ldots, \beta_k$. The least squares estimators of $\beta_0, \beta_1, \ldots, \beta_k$ must satisfy

$$
\left. \frac{\partial L}{\partial \beta_0} \right|_{\hat{\beta}_0, \hat{\beta}_1, \ldots, \hat{\beta}_k} = -2 \sum_{i=1}^{n} \left(y_i - \hat{\beta}_0 - \sum_{j=1}^{k} \hat{\beta}_j x_{ij} \right) = 0
\tag{13-9a}
$$

and

$$
\left. \frac{\partial L}{\partial \beta_j} \right|_{\hat{\beta}_0, \hat{\beta}_1, \ldots, \hat{\beta}_k} = -2 \sum_{i=1}^{n} \left(y_i - \hat{\beta}_0 - \sum_{j=1}^{n} \hat{\beta}_j x_{ij} \right) x_{ij} = 0 \qquad j = 1, 2, \ldots, k
\tag{13-9b}
$$

Simplifying Equation (13-9), we obtain the least squares normal equations

$$
n\hat{\beta}_0 + \hat{\beta}_1 \sum_{i=1}^{n} x_{i1} + \hat{\beta}_2 \sum_{i=1}^{n} x_{i2} + \cdots + \hat{\beta}_k \sum_{i=1}^{n} x_{ik} = = \sum_{i=1}^{n} y_i
$$

$$
\hat{\beta}_0 \sum_{i=1}^{n} x_{i1} + \hat{\beta}_1 \sum_{i=1}^{n} x_{i1}^2 + \hat{\beta}_2 \sum_{i=1}^{n} x_{i1} x_{i2} + \cdots + \hat{\beta}_k \sum_{i=1}^{n} x_{i1} x_{ik} = \sum_{i=1}^{n} x_{i1} y_i
\tag{13-10}
$$

$$
\vdots \qquad \vdots \qquad \vdots \qquad \vdots \qquad \vdots
$$

$$
\hat{\beta}_0 \sum_{i=1}^{n} x_{ik} + \hat{\beta}_1 \sum_{i=1}^{n} x_{ik} x_{i1} + \hat{\beta}_2 \sum_{i=1}^{n} x_{ik} x_{i2} + \cdots + \hat{\beta}_k \sum_{i=1}^{n} x_{ik}^2 = \sum_{i=1}^{n} x_{ik} y_i
$$

Note that there are $p = k + 1$ normal equations, one for each of the unknown regression coefficients. The solution to the normal equations will be the least squares estimators of the regression coefficients, $\hat{\beta}_0, \hat{\beta}_1, \ldots, \hat{\beta}_k$.

It is simpler to solve the normal equations if they are expressed in matrix notation. We now give a matrix development of the normal equations that parallels the development of Equation (13-10). The model in terms of the observations, Equation (13-7), may be written in matrix notation as

$$\mathbf{y} = \mathbf{X}\boldsymbol{\beta} + \boldsymbol{\epsilon}$$

where

$$\mathbf{y} = \begin{bmatrix} y_1 \\ y_2 \\ \vdots \\ y_n \end{bmatrix} \qquad \mathbf{X} = \begin{bmatrix} 1 & x_{11} & x_{12} \cdots x_{1k} \\ 1 & x_{21} & x_{22} \cdots x_{2k} \\ \vdots & \vdots & \vdots & \vdots \\ 1 & x_{n1} & x_{n2} \cdots x_{nk} \end{bmatrix}$$

$$\boldsymbol{\beta} = \begin{bmatrix} \beta_0 \\ \beta_1 \\ \vdots \\ \beta_k \end{bmatrix} \quad \text{and} \quad \boldsymbol{\epsilon} = \begin{bmatrix} \epsilon_1 \\ \epsilon_2 \\ \vdots \\ \epsilon_n \end{bmatrix}$$

In general, $\mathbf{y}$ is an $(n \times 1)$ vector of the observations, $\mathbf{X}$ is an $(n \times p)$ matrix of the levels of the independent variables, $\boldsymbol{\beta}$ is a $(p \times 1)$ vector of the regression coefficients, and $\boldsymbol{\epsilon}$ is an $(n \times 1)$ vector of random errors.

We wish to find the vector of least squares estimators, $\hat{\boldsymbol{\beta}}$, that minimizes

$$L = \sum_{i=1}^{n'} \epsilon_i^2 = \boldsymbol{\epsilon}'\boldsymbol{\epsilon} = (\mathbf{y} - \mathbf{X}\boldsymbol{\beta})'(\mathbf{y} - \mathbf{X}\boldsymbol{\beta})$$

Note that L may be expressed as

$$L = \mathbf{y}'\mathbf{y} - \boldsymbol{\beta}'\mathbf{X}'\mathbf{y} - \mathbf{y}'\mathbf{X}\boldsymbol{\beta} + \boldsymbol{\beta}'\mathbf{X}'\mathbf{X}\boldsymbol{\beta}$$
$$= \mathbf{y}'\mathbf{y} - 2\boldsymbol{\beta}'\mathbf{X}'\mathbf{y} + \boldsymbol{\beta}'\mathbf{X}'\mathbf{X}\boldsymbol{\beta} \tag{13-11}$$

since $\boldsymbol{\beta}'\mathbf{X}'\mathbf{y}$ is a (1×1) matrix, or a scalar, and its transpose $(\boldsymbol{\beta}'\mathbf{X}'\mathbf{y})' = \mathbf{y}'\mathbf{X}\boldsymbol{\beta}$ is the same scalar. The least squares estimators must satisfy

$$\frac{\partial L}{\partial \boldsymbol{\beta}}\bigg|_{\hat{\boldsymbol{\beta}}} = -2\mathbf{X}'\mathbf{y} + 2\mathbf{X}'\mathbf{X}\hat{\boldsymbol{\beta}} = 0$$

which simplifies to

$$\mathbf{X}'\mathbf{X}\hat{\boldsymbol{\beta}} = \mathbf{X}'\mathbf{y} \tag{13-12}$$

Equations (13-12) are the least squares normal equations. They are identical to Equations (13-10). To solve the normal equations, multiply both sides of Equation (13-12) by the inverse of $\mathbf{X}'\mathbf{X}$. Thus, the least squares estimator of $\boldsymbol{\beta}$ is

$$\hat{\boldsymbol{\beta}} = (\mathbf{X}'\mathbf{X})^{-1}\mathbf{X}'\mathbf{y} \tag{13-13}$$

It is easy to see that the matrix form of the normal equations is identical to the scalar form. Writing out Equation (13-12) in detail we obtain

$$
\begin{bmatrix}
n & \sum\limits_{i=1}^{n} x_{i1} & \sum\limits_{i=1}^{n} x_{i2} & \cdots & \sum\limits_{i=1}^{n} x_{ik} \\
\sum\limits_{i=1}^{n} x_{i1} & \sum\limits_{i=1}^{n} x_{i1}^{2} & \sum\limits_{i=1}^{n} x_{i1}x_{i2} & \cdots & \sum\limits_{i=1}^{n} x_{i1}x_{ik} \\
\vdots & \vdots & \vdots & & \vdots \\
\sum\limits_{i=1}^{n} x_{ik} & \sum\limits_{i=1}^{n} x_{ik}x_{i1} & \sum\limits_{i=1}^{n} x_{ik}x_{i2} & \cdots & \sum\limits_{i=1}^{n} x_{ik}^{2}
\end{bmatrix}
\begin{bmatrix}
\hat{\beta}_0 \\
\hat{\beta}_1 \\
\vdots \\
\hat{\beta}_k
\end{bmatrix}
=
\begin{bmatrix}
\sum\limits_{i=1}^{n} y_i \\
\sum\limits_{i=1}^{n} x_{i1}y_i \\
\vdots \\
\sum\limits_{i=1}^{n} x_{ik}y_i
\end{bmatrix}
$$

If the indicated matrix multiplication is performed, the scalar form of the normal equations [that is, Equation (13-10)] will result. In this form it is easy to see that $\mathbf{X'X}$ is a $(p \times p)$ symmetric matrix and $\mathbf{X'y}$ is a $(p \times 1)$ column vector. Note the special structure of the $\mathbf{X'X}$ matrix. The diagonal elements of $\mathbf{X'X}$ are the sums of squares of the elements in the columns of $\mathbf{X}$, and the off-diagonal elements are the sums of cross products of the elements in the columns of $\mathbf{X}$. Furthermore, note that the elements of $\mathbf{X'y}$ are the sums of cross products of the columns of $\mathbf{X}$ and the observations $\{y_i\}$.

The estimated regression model is

$$\hat{\mathbf{y}} = \mathbf{X}\hat{\boldsymbol{\beta}} \tag{13-14}$$

In scalar notation, the estimated model is

$$\hat{y}_i = \hat{\beta}_0 + \sum_{j=1}^{k} \hat{\beta}_j x_{ij} \qquad i = 1, 2, \ldots, n$$

The difference between the observation y_i and the estimated value $\hat{y}_i$ is a residual, say $e_i = y_i - \hat{y}_i$. The $(n \times 1)$ vector of residuals is denoted by

$$\mathbf{e} = \mathbf{y} - \hat{\mathbf{y}} \tag{13-15}$$

● **Example 13-1.** A soft drink bottler is analyzing the vending machine service routes in her distribution system. Specifically, she is interested in predicting the amount of time required by the route driver to service the vending machines in an outlet. This service activity includes stocking the machine with beverage products and minor maintenance or housekeeping. The industrial engineer responsible for the study has suggested that the two most important variables affecting the delivery time are the number of cases of product stocked and the distance walked by the route driver. The engineer has collected 25 observations on delivery time, which are shown in Table 13-2. (Note that this is an expansion of the data set used in Example 12-7, where only one independent variable, the number of cases stocked, was considered.)

We will fit the multiple linear regression model

$$y = \beta_0 + \beta_1 x_1 + \beta_2 x_2 + \epsilon$$

to these data. The **X** matrix and **y** vector for this model are

$$\mathbf{X} = \begin{bmatrix} 1 & 2 & 50 \\ 1 & 8 & 110 \\ 1 & 11 & 120 \\ 1 & 10 & 550 \\ 1 & 8 & 295 \\ 1 & 4 & 200 \\ 1 & 2 & 375 \\ 1 & 2 & 52 \\ 1 & 9 & 100 \\ 1 & 8 & 300 \\ 1 & 4 & 412 \\ 1 & 11 & 400 \\ 1 & 12 & 500 \\ 1 & 2 & 360 \\ 1 & 4 & 205 \\ 1 & 4 & 400 \\ 1 & 20 & 600 \\ 1 & 1 & 585 \\ 1 & 10 & 540 \\ 1 & 15 & 250 \\ 1 & 15 & 290 \\ 1 & 16 & 510 \\ 1 & 17 & 590 \\ 1 & 6 & 100 \\ 1 & 5 & 400 \end{bmatrix} \qquad \mathbf{y} = \begin{bmatrix} 9.95 \\ 24.45 \\ 31.75 \\ 35.00 \\ 25.02 \\ 16.86 \\ 14.38 \\ 9.60 \\ 24.35 \\ 27.50 \\ 17.08 \\ 37.00 \\ 41.95 \\ 11.66 \\ 21.65 \\ 17.89 \\ 69.00 \\ 10.30 \\ 34.92 \\ 46.59 \\ 44.88 \\ 54.12 \\ 56.63 \\ 22.13 \\ 21.15 \end{bmatrix}$$

The **X'X** matrix is

$$\mathbf{X'X} = \begin{bmatrix} 1 & 1 \ldots & 1 \\ 2 & 8 \ldots & 5 \\ 50 & 110 \ldots & 400 \end{bmatrix} \begin{bmatrix} 1 & 2 & 50 \\ 1 & 8 & 110 \\ \vdots & \vdots & \vdots \\ 1 & 5 & 400 \end{bmatrix}$$

$$= \begin{bmatrix} 25 & 206 & 8{,}294 \\ 206 & 2{,}396 & 77{,}177 \\ 8{,}294 & 77{,}177 & 3{,}531{,}848 \end{bmatrix}$$

TABLE 13-2 **Delivery Time Data for Example 13-1**

Observation Number	Delivery Time (min) y	Number of Cases x_1	Distance (ft) x_2
1	9.95	2	50
2	24.45	8	110
3	31.75	11	120
4	35.00	10	550
5	25.02	8	295
6	16.86	4	200
7	14.38	2	375
8	9.60	2	52
9	24.35	9	100
10	27.50	8	300
11	17.08	4	412
12	37.00	11	400
13	41.95	12	500
14	11.66	2	360
15	21.65	4	205
16	17.89	4	400
17	69.00	20	600
18	10.30	1	585
19	34.92	10	540
20	46.59	15	250
21	44.88	15	290
22	54.12	16	510
23	56.63	17	590
24	22.13	6	100
25	21.15	5	400

and the $\mathbf{X'y}$ vector is

$$\mathbf{X'y} = \begin{bmatrix} 1 & 1 \ldots & 1 \\ 2 & 8 \ldots & 5 \\ 50 & 110 \ldots & 400 \end{bmatrix} \begin{bmatrix} 9.95 \\ 24.45 \\ \vdots \\ 21.15 \end{bmatrix} = \begin{bmatrix} 725.81 \\ 8,008.37 \\ 274,811.31 \end{bmatrix}$$

The least squares estimators are found from Equation (13-13) as

$$\hat{\boldsymbol{\beta}} = (\mathbf{X'X})^{-1}\mathbf{X'y}$$

or

$$\begin{bmatrix} \hat{\beta}_0 \\ \hat{\beta}_1 \\ \hat{\beta}_2 \end{bmatrix} = \begin{bmatrix} 25 & 206 & 8,294 \\ 206 & 2,396 & 77,177 \\ 8,294 & 77,177 & 3,531,848 \end{bmatrix}^{-1} \begin{bmatrix} 725.81 \\ 8,008.37 \\ 274,811.31 \end{bmatrix}$$

$$= \begin{bmatrix} .214653 & -.007491 & -.000340 \\ -.007491 & & -.000019 \\ -.000340 & -.000019 & .0000015 \end{bmatrix} \begin{bmatrix} 725.81 \\ 8,008.37 \\ 274,811.31 \end{bmatrix}$$

$$= \begin{bmatrix} 2.26379143 \\ 2.74426964 \\ .01252781 \end{bmatrix}$$

Therefore, the estimated regression model is

$$\hat{y} = 2.26379 + 2.74427x_1 + .01253x_2$$

Notice that we have rounded the regression coefficients to five places. Table 13-3 shows the estimated values of y and the residuals. The estimated values and residuals are calculated to the same accuracy as the original data.

The statistical properties of the least squares estimator $\hat{\beta}$ may be easily demonstrated. Consider first bias:

$$\begin{aligned} E(\hat{\beta}) &= E[(\mathbf{X'X})^{-1}\mathbf{X'y}] \\ &= E[(\mathbf{X'X})^{-1}\mathbf{X'}(\mathbf{X\beta} + \boldsymbol{\epsilon})] \\ &= E[(\mathbf{X'X})^{-1}\mathbf{X'X\beta} + (\mathbf{X'X})^{-1}\mathbf{X'\epsilon}] \\ &= \beta \end{aligned}$$

Table 13-3 **Observations, Estimated Values, and Residuals for Example 13-1**

Observation Number	y_i	$\hat{y}_i$	$e_i = y_i - \hat{y}_i$
1	9.95	8.38	1.57
2	24.45	25.60	−1.15
3	31.75	33.95	−2.20
4	35.00	36.60	−1.60
5	25.02	27.91	−2.89
6	16.86	15.75	1.11
7	14.38	12.45	1.93
8	9.60	8.40	1.20
9	24.35	28.21	−3.86
10	27.50	27.98	−.48
11	17.08	18.40	−1.32
12	37.00	37.46	−.46
13	41.95	41.46	.49
14	11.66	12.26	−.60
15	21.65	15.81	5.84
16	17.89	18.25	−.36
17	69.00	64.67	4.33
18	10.30	12.34	−2.04
19	34.93	36.47	−1.54
20	46.59	46.56	.03
21	44.88	47.06	−2.18
22	54.12	52.56	1.56
23	56.63	56.31	.32
24	22.13	19.98	2.15
25	21.15	21.00	.15

since $E(\boldsymbol{\epsilon}) = 0$ and $(\mathbf{X}'\mathbf{X})^{-1}\mathbf{X}'\mathbf{X} = \mathbf{I}$. Thus $\hat{\boldsymbol{\beta}}$ is an unbiased estimator of $\boldsymbol{\beta}$. The variance property of $\hat{\boldsymbol{\beta}}$ is expressed by the covariance matrix

$$\text{Cov}(\hat{\boldsymbol{\beta}}) = E\{[\hat{\boldsymbol{\beta}} - E(\hat{\boldsymbol{\beta}})][\hat{\boldsymbol{\beta}} - E(\hat{\boldsymbol{\beta}})]'\}$$

The covariance matrix of $\hat{\boldsymbol{\beta}}$ is a $(p \times p)$ symmetric matrix whose jjth element is the variance of $\hat{\beta}_j$ and whose (i, j)th element is the covariance between $\hat{\beta}_i$ and $\hat{\beta}_j$. The covariance matrix of $\hat{\boldsymbol{\beta}}$ is

$$\text{Cov}(\hat{\boldsymbol{\beta}}) = \sigma^2(\mathbf{X}'\mathbf{X})^{-1}$$

It is usually necessary to estimate σ^2. To develop this estimator, consider the sum of squares of the residuals, say

$$SS_E = \sum_{i=1}^{n} (y_i - \hat{y}_i)^2$$

$$= \sum_{i=1}^{n} e_i^2$$

$$= \mathbf{e}'\mathbf{e}$$

Substituting $\mathbf{e} = \mathbf{y} - \hat{\mathbf{y}} = \mathbf{y} - \mathbf{X}\hat{\boldsymbol{\beta}}$, we have

$$SS_E = (\mathbf{y} - \mathbf{X}\hat{\boldsymbol{\beta}})'(\mathbf{y} - \mathbf{X}\hat{\boldsymbol{\beta}})$$
$$= \mathbf{y}'\mathbf{y} - \hat{\boldsymbol{\beta}}'\mathbf{X}'\mathbf{y} - \mathbf{y}'\mathbf{X}\hat{\boldsymbol{\beta}} + \hat{\boldsymbol{\beta}}'\mathbf{X}'\mathbf{X}\hat{\boldsymbol{\beta}}$$
$$= \mathbf{y}'\mathbf{y} - 2\hat{\boldsymbol{\beta}}'\mathbf{X}'\mathbf{y} + \hat{\boldsymbol{\beta}}'\mathbf{X}'\mathbf{X}\hat{\boldsymbol{\beta}}$$

Since $\mathbf{X}'\mathbf{X}\hat{\boldsymbol{\beta}} = \mathbf{X}'\mathbf{y}$, this last equation becomes

$$SS_E = \mathbf{y}'\mathbf{y} - \hat{\boldsymbol{\beta}}'\mathbf{X}'\mathbf{y} \tag{13-16}$$

Equation (13-16) is called the error sum of squares, and it has $n - p$ degrees of freedom associated with it. The mean square error is

$$MS_E = \frac{SS_E}{n - p} \tag{13-17}$$

It can be shown that the expected value of MS_E is σ^2, thus an unbiased estimator of σ^2 is given by

$$\hat{\sigma}^2 = MS_E \tag{13-18}$$

● **Example 13-2.** We will estimate the error variance σ^2 for the multiple regression problem in Example 13-1. Using the data in Table 13-2, we find

$$\mathbf{y}'\mathbf{y} = \sum_{i=1}^{25} y_i^2 = 27{,}177.9510$$

and

$$\hat{\boldsymbol{\beta}}'\mathbf{X}'\mathbf{y} = [2.26379143 \quad 2.74426964 \quad .01252781] \begin{bmatrix} 725.82 \\ 8{,}008.37 \\ 274{,}811.31 \end{bmatrix}$$

$$= 27{,}062.7775$$

Therefore, the error sum of squares is

$$SS_E = \mathbf{y'y} - \hat{\boldsymbol{\beta}}'\mathbf{X'y}$$
$$= 27,177.9510 - 27,062.7775$$
$$= 115.1735$$

The estimate of σ^2 is

$$\hat{\sigma}^2 = \frac{SS_E}{n - p} = \frac{115.1735}{25 - 3} = 5.2352$$

13-3   Confidence Intervals in Multiple Linear Regression

It is often necessary to construct confidence interval estimates of the regression coefficients $\{\beta_j\}$. The development of a procedure for obtaining these confidence intervals requires that we assume the errors $\{\epsilon_i\}$ to be normally and independently distributed with mean zero and variance σ^2. Therefore, the observations $\{y_i\}$ are normally and independently distributed with mean $\beta_0 + \sum_{j=1}^{n} \beta_j x_{ij}$ and variance σ^2. Since the least squares estimator $\hat{\boldsymbol{\beta}}$ is a linear combination of the observations, it follows that $\hat{\boldsymbol{\beta}}$ is normally distributed with mean vector $\boldsymbol{\beta}$ and covariance matrix $\sigma^2(\mathbf{X'X})^{-1}$. Then each of the statistics

$$\frac{\hat{\beta}_j - \beta_j}{\sqrt{\hat{\sigma}^2 C_{jj}}} \qquad j = 0, 1, \ldots, k \qquad (13\text{-}19)$$

is distributed as t with $n - p$ degrees of freedom, where C_{jj} is the jjth element of the $(\mathbf{X'X})^{-1}$ matrix, and $\hat{\sigma}^2$ is the estimate of the error variance, obtained from Equation (13-18). Therefore, a $100(1 - \alpha)$ percent confidence interval for the regression coefficient β_j, $j = 0, 1, \ldots, k$, is

$$\hat{\beta}_j - t_{\alpha/2,n-p}\sqrt{\hat{\sigma}^2 C_{jj}} \le \beta_j \le \hat{\beta}_j + t_{\alpha/2,n-p}\sqrt{\hat{\sigma}^2 C_{jj}} \qquad (13\text{-}20)$$

● **Example 13-3.** We will construct a 95 percent confidence interval on the parameter β_1 in Example 13-1. Note that the point estimate of β_1 is $\hat{\beta}_1 = 2.74427$, and that the diagonal element of $(\mathbf{X'X})^{-1}$ corresponding to β_1 is $C_{11} = .001671$. The estimate of σ^2 was obtained in Example 13-2 as 5.2352, and $t_{.025,22} = 2.074$. Therefore, the 95 percent confidence interval on β_1 is computed from Equation (13-20) as

$$2.74427 - (2.074)\sqrt{(5.2352)(.001671)} \le \beta_1 \le 2.74427 + (2.074)\sqrt{(5.2352)(.001671)}$$

which reduces to

$$2.55029 \le \beta_1 \le 2.93825$$

We may also obtain a confidence interval on the mean response at a particular point, say $x_{01}, x_{02}, \ldots, x_{0k}$. To estimate the mean response at this point, define the vector

$$\mathbf{x}_0 = \begin{bmatrix} 1 \\ x_{01} \\ x_{02} \\ \vdots \\ x_{0k} \end{bmatrix}$$

The estimated mean response at this point is

$$\hat{y}_0 = \mathbf{x}_0'\hat{\boldsymbol{\beta}} \tag{13-21}$$

This estimator is unbiased, since $E(\hat{y}_0) = E(\mathbf{x}_0'\hat{\boldsymbol{\beta}}) = \mathbf{x}_0'\boldsymbol{\beta} = E(y_0)$, and the variance of $\hat{y}_0$ is

$$V(\hat{y}_0) = \sigma^2 \mathbf{x}_0'(\mathbf{X}'\mathbf{X})^{-1}\mathbf{x}_0 \tag{13-22}$$

Therefore, a $100(1 - \alpha)$ percent confidence interval on the mean response at the point $x_{01}, x_{02}, \ldots, x_{0k}$ is

$$\hat{y}_0 - t_{\alpha/2, n-p}\sqrt{\hat{\sigma}^2 \mathbf{x}_0'(\mathbf{X}'\mathbf{X})^{-1}\mathbf{x}_0} \le y_0 \le \hat{y}_0 + t_{\alpha/2, n-p}\sqrt{\hat{\sigma}^2 \mathbf{x}_0'(\mathbf{X}'\mathbf{X})^{-1}\mathbf{x}_0} \tag{13-23}$$

Equation (13-23) is a confidence interval about the regression hyperplane. It is the multiple regression generalization of Equation (12-36).

● **Example 13-4.** The soft drink bottler in Example 13-1 would like to construct a 95 percent confidence interval on the mean delivery time for an outlet requiring $x_1 = 8$ cases and where the distance $x_2 = 275$ feet. Therefore,

$$\mathbf{x}_0 = \begin{bmatrix} 1 \\ 8 \\ 275 \end{bmatrix}$$

The estimated mean response at this point is found from Equation (13-21) as

$$\hat{y}_0 = \mathbf{x}_0'\hat{\boldsymbol{\beta}} = \begin{bmatrix} 1 & 8 & 275 \end{bmatrix} \begin{bmatrix} 2.26379 \\ 2.74427 \\ .01253 \end{bmatrix} = 27.66 \text{ minutes}$$

The variance of $\hat{y}_0$ is estimated by

$$\hat{\sigma}^2 \mathbf{x}_0'(\mathbf{X}'\mathbf{X})^{-1}\mathbf{x}_0 = 5.2352 \begin{bmatrix} 1 & 8 & 275 \end{bmatrix} \begin{bmatrix} .214653 & -.007491 & -.000340 \\ -.007491 & .001671 & -.000019 \\ -.000340 & -.000019 & .0000015 \end{bmatrix} \begin{bmatrix} 1 \\ 8 \\ 275 \end{bmatrix}$$

$$= 5.2352(.04444) = .23266$$

Therefore, a 95 percent confidence interval on the mean delivery time at this point is found from Equation (13-23) as

$$27.66 - 2.074\sqrt{.23266} \le y_0 \le 27.66 + 2.074\sqrt{.23266}$$

which reduces to

$$26.66 \le y_0 \le 28.66$$

13-4   Prediction of New Observations

The regression model can be used to predict future observations on y corresponding to particular values of the independent variables, say $x_{01}, x_{02}, \ldots, x_{0k}$. If $\mathbf{x}_0' = [1, x_{01}, x_{02}, \ldots, x_{0k}]$, then a point estimate of the future observation y_0 at the point $x_{01}, x_{02}, \ldots, x_{0k}$ is

$$\hat{y}_0 = \mathbf{x}_0'\hat{\boldsymbol{\beta}} \qquad (13\text{-}24)$$

A $100(1 - \alpha)$ percent prediction interval for this future observation is

$$\hat{y}_0 - t_{\alpha/2, n-p}\sqrt{\hat{\sigma}^2(1 + \mathbf{x}_0'(\mathbf{X}'\mathbf{X})^{-1}\mathbf{x}_0)} \le y_0 \le \hat{y}_0 + t_{\alpha/2, n-p}\sqrt{\hat{\sigma}^2(1 + \mathbf{x}_0'(\mathbf{X}'\mathbf{X})^{-1}\mathbf{x}_0)} \qquad (13\text{-}25)$$

This prediction interval is a generalization of the prediction interval for a future observation in simple linear regression, Equation (12-38).

In predicting new observations and in estimating the mean response at a given point $x_{01}, x_{02}, \ldots, x_{0k}$, one must be careful about extrapolating beyond

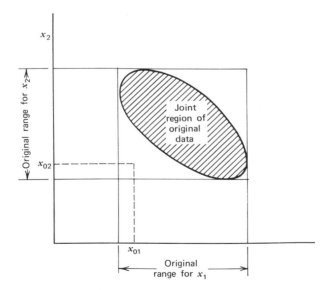

Fig. 13-1. An example of extrapolation in multiple regression.

the region containing the original observations. It is very possible that a model that fits well in the region of the original data will no longer fit well outside of that region. In multiple regression it is often easy to inadvertently extrapolate, since the levels of the variables $(x_{i1}, x_{i2}, \ldots, x_{ik})$, $i = 1, 2, \ldots, n$, jointly define the region containing the data. As an example, consider Fig. 13-1, which illustrates the region containing the observations for a two-variable regression model. Note that the point (x_{01}, x_{02}) lies within the ranges of both independent variables x_1 and x_2, but it is outside the region of the original observations. Thus, either predicting the value of a new observation or estimating the mean response at this point is an extrapolation of the original regression model.

● **Example 13-5.** Suppose that the soft drink bottler in Example 13-1 wishes to construct a 95 percent prediction interval on the delivery time at an outlet where $x_1 = 8$ cases are delivered and the distance walked by the deliveryman is $x_2 = 275$ feet. Note that $\mathbf{x}_0' = [1 \quad 8 \quad 275]$, and the point estimate of the delivery time is $\hat{y}_0 = \mathbf{x}_0' \hat{\boldsymbol{\beta}} = 27.66$ minutes. Also, in Example 13-4 we calculated $\mathbf{x}_0'(\mathbf{X}'\mathbf{X})^{-1}\mathbf{x}_0 = .04444$. Therefore, from Equation (13-25), we have

$$27.66 - 2.074\sqrt{5.2352(1 + .04444)} \leq y_0 \leq 27.66 + 2.074\sqrt{5.2352(1 + .04444)}$$

and the 95 percent prediction interval is

$$22.81 \leq y_0 \leq 32.51$$

13-5 Hypothesis Testing in Multiple Linear Regression

In multiple linear regression problems, certain tests of hypotheses about the model parameters are useful in measuring model adequacy. In this section, we describe several important hypothesis-testing procedures. We continue to require the normality assumption on the errors, which was introduced in the previous section.

13-5.1 Test for Significance of Regression

The test for significance of regression is a test to determine if there is a linear relationship between the dependent variable y and a subset of the independent variables $x_1, x_2, \ldots, x_k$. The appropriate hypotheses are

$$H_0: \beta_1 = \beta_2 = \cdots = \beta_k = 0$$

$$H_1: \beta_j \neq 0 \text{ for at least one } j \quad\quad (13\text{-}26)$$

Rejection of $H_0: \beta_j = 0$ implies that at least one of the independent variables $x_1, x_2, \ldots, x_k$ contributes significantly to the model. The test procedure is a generalization of the procedure used in simple linear regression. The total sum of squares S_{yy} is partitioned into a sum of squares due to regression and a sum

TABLE 13-4   Analysis of Variance for Significance of Regression in Multiple Regression

Source of Variation	Sum of Squares	Degrees of Freedom	Mean Square	F_0
Regression	SS_R	k	MS_R	MS_R/MS_E
Error or residual	SS_E	$n - k - 1$	MS_E	
Total	S_{yy}	$n - 1$		

of squares due to error, say

$$S_{yy} = SS_R + SS_E$$

and if $H_0: \beta_j = 0$ is true, then $SS_R/\sigma^2 \sim \chi_k^2$, where the number of degrees of freedom for χ^2 are equal to the number of independent variables in the model. Also, we can show that $SS_E/\sigma^2 \sim \chi_{n-k-1}^2$, and SS_E and SS_R are independent. The test procedure for $H_0: \beta_j = 0$ is to compute

$$F_0 = \frac{SS_R/k}{SS_E/(n - k - 1)} = \frac{MS_R}{MS_E} \tag{13-27}$$

and to reject H_0 if $F_0 > F_{\alpha,k,n-k-1}$. The procedure is usually summarized in an analysis of variance table such as Table 13-4.

A computational formula for SS_R may be found easily. We have derived a computational formula for SS_E in Equation (13-16), that is,

$$SS_E = \mathbf{y'y} - \hat{\boldsymbol{\beta}}\mathbf{X'y}$$

Now since $S_{yy} = \sum_{i=1}^{n} y_i^2 - (\sum_{i=1}^{n} y_i)^2/n = \mathbf{y'y} - (\sum_{i=1}^{n} y_i)^2/n$, we may rewrite the above equation as

$$SS_E = \mathbf{y'y} - \frac{\left(\sum_{i=1}^{n} y_i\right)^2}{n} - \left[\hat{\boldsymbol{\beta}}'\mathbf{X'y} - \frac{\left(\sum_{i=1}^{n} y_i\right)^2}{n}\right]$$

or

$$SS_E = S_{yy} - SS_R$$

Therefore, the regression sum of squares is

$$SS_R = \hat{\boldsymbol{\beta}}'\mathbf{X'y} - \frac{\left(\sum_{i=1}^{n} y_i\right)^2}{n} \tag{13-28}$$

the error sum of squares is

$$SS_E = \mathbf{y'y} - \hat{\boldsymbol{\beta}}'\mathbf{X'y} \tag{13-29}$$

and the total sum of squares is

$$S_{yy} = \mathbf{y}'\mathbf{y} - \frac{\left(\sum\limits_{i=1}^{n} y_i\right)^2}{n} \tag{13-30}$$

● **Example 13-6.** We will test for significance of regression using the delivery time data from Example 13-1. Some of the numerical quantities required are calculated in Example 13-2. Note that

$$
\begin{aligned}
S_{yy} &= \mathbf{y}'\mathbf{y} - \frac{\left(\sum\limits_{i=1}^{n} y_i\right)^2}{n} \\
&= 27{,}177.9510 - \frac{(725.82)^2}{25} \\
&= 6105.9447
\end{aligned}
$$

$$
\begin{aligned}
SS_R &= \hat{\boldsymbol{\beta}}'\mathbf{X}'\mathbf{y} - \frac{\left(\sum\limits_{i=1}^{n} y_i\right)^2}{n} \\
&= 27{,}062.7775 - \frac{(725.82)^2}{25} \\
&= 5990.7712
\end{aligned}
$$

and

$$
\begin{aligned}
SS_E &= S_{yy} - SS_R \\
&= \mathbf{y}'\mathbf{y} - \hat{\boldsymbol{\beta}}'\mathbf{X}'\mathbf{y} \\
&= 115.1735
\end{aligned}
$$

The analysis of variance is shown in Table 13-5. To test $H_0: \beta_1 = \beta_2 = 0$, we calculate the statistic

$$F_0 = \frac{MS_R}{MS_E} = \frac{2995.3856}{5.2352} = 572.17$$

Since $F_0 > F_{.05,2,22} = 3.44$, we conclude that delivery time is related to both delivery volume and distance. However, we note that this does not necessarily imply that the relationship found is an appropriate one for predicting delivery time as a function of volume and distance. Further tests of model adequacy are required.

TABLE 13-5 Test for Significance of Regression for Example 13-6

Source of Variation	Sum of Squares	Degrees of Freedom	Mean Square	F_0
Regression	5990.7712	2	2995.3856	572.17
Error	115.1735	22	5.2352	
Total	6105.9447	24		

13-5.2   Tests on Individual Regression Coefficients

We are frequently interested in testing hypotheses on the individual regression coefficients. Such tests would be useful in determining the value of each of the independent variables in the regression model. For example, the model might be more effective with the inclusion of additional variables, or perhaps with the deletion of one or more of the variables already in the model.

Adding a variable to a regression model always causes the sum of squares for regression to increase and the error sum of squares to decrease. We must decide whether the increase in the regression sum of squares is sufficient to warrant using the additional variable in the model. Furthermore, adding an unimportant variable to the model can actually increase the mean square error, thereby decreasing the usefulness of the model.

The hypotheses for testing the significance of any individual regression coefficient, say β_j, are

$$H_0: \beta_j = 0$$

$$H_1: \beta_j \neq 0 \tag{13-31}$$

If $H_0: \beta_j = 0$ is not rejected, then this indicates that x_j can be deleted from the model. The test statistic for this hypothesis is

$$t_0 = \frac{\hat{\beta}_j}{\sqrt{\hat{\sigma}^2 C_{jj}}} \tag{13-32}$$

where C_{jj} is the diagonal element of $(\mathbf{X}'\mathbf{X})^{-1}$ corresponding to $\hat{\beta}_j$. The null hypothesis $H_0: \beta_j = 0$ is rejected if $|t_0| > t_{\alpha/2, n-k-1}$. Note that this is really a partial or marginal test, because the regression coefficient $\hat{\beta}_j$ depends on all the other regressor variables $x_i (i \neq j)$ that are in the model. To illustrate the use of this test, consider the data in Example 13-1, and suppose that we want to test

$$H_0: \beta_2 = 0$$

$$H_1: \beta_2 \neq 0$$

The main diagonal element of $(\mathbf{X}'\mathbf{X})^{-1}$ corresponding to $\hat{\beta}_2$ is $C_{22} = .0000015$, so the t statistic in Equation (13-32) becomes

$$t_0 = \frac{\hat{\beta}_2}{\sqrt{\hat{\sigma}^2 C_{22}}} = \frac{.01253}{\sqrt{(5.2352)(.0000015)}} = 4.4767$$

Since $t_{.025, 22} = 2.074$, we reject $H_0: \beta_2 = 0$ and conclude that the variable x_2 (distance) contributes significantly to the model. Note that this test measures the marginal or partial contribution of x_2 *given* that x_1 is in the model.

We may also examine the contribution to the regression sum of squares of a variable, say x_j, given that other variables x_i $(i \neq j)$ are included in the model. The procedure used to do this is called the general regression significance test, or the "extra sum of squares" method. This procedure can also be used to investigate the contribution of a *subset* of the regression variables to the model. Consider the regression model with k independent variables

$$\mathbf{y} = \mathbf{X}\boldsymbol{\beta} + \boldsymbol{\epsilon}$$

where $\mathbf{y}$ is $(n \times 1)$, $\mathbf{X}$ is $(n \times p)$, $\boldsymbol{\beta}$ is $(p \times 1)$, $\boldsymbol{\epsilon}$ is $(n \times 1)$, and $p = k + 1$. We would like to determine if the subset of regressor variables $x_1, x_2, \ldots, x_r$ $(r < k)$ contributes significantly to the regression model. Let the vector of regression coefficients be partitioned as follows:

$$\boldsymbol{\beta} = \begin{bmatrix} \boldsymbol{\beta}_1 \\ \boldsymbol{\beta}_2 \end{bmatrix}$$

where $\boldsymbol{\beta}_1$ is $(r \times 1)$ and $\boldsymbol{\beta}_2$ is $[(p - r) \times 1]$. We wish to test the hypotheses

$$H_0 : \boldsymbol{\beta}_1 = \mathbf{0}$$

$$H_1 : \boldsymbol{\beta}_1 \neq \mathbf{0} \tag{13-33}$$

The model may be written as

$$\mathbf{y} = \mathbf{X}\boldsymbol{\beta} + \boldsymbol{\epsilon} = \mathbf{X}_1\boldsymbol{\beta}_1 + \mathbf{X}_2\boldsymbol{\beta}_2 + \boldsymbol{\epsilon} \tag{13-34}$$

where $\mathbf{X}_1$ represents the columns of $\mathbf{X}$ associated with $\boldsymbol{\beta}_1$ and $\mathbf{X}_2$ represents the columns of $\mathbf{X}$ associated with $\boldsymbol{\beta}_2$.

For the *full* model (including both $\boldsymbol{\beta}_1$ and $\boldsymbol{\beta}_2$), we know that $\hat{\boldsymbol{\beta}} = (\mathbf{X}'\mathbf{X})^{-1}\mathbf{X}'\mathbf{y}$. Also, the regression sum of squares for all variables including the intercept is

$$SS_R(\boldsymbol{\beta}) = \hat{\boldsymbol{\beta}}'\mathbf{X}'\mathbf{y} \qquad (p \text{ degrees of freedom})$$

and

$$MS_E = \frac{\mathbf{y}'\mathbf{y} - \hat{\boldsymbol{\beta}}'\mathbf{X}'\mathbf{y}}{n - p}$$

$SS_R(\boldsymbol{\beta})$ is called the regression sum of squares *due to* $\boldsymbol{\beta}$. To find the contribution of the terms in $\boldsymbol{\beta}_1$ to the regression, fit the model assuming the null hypothesis $H_0 : \boldsymbol{\beta}_1 = \mathbf{0}$ to be true. The *reduced* model is found from Equation (13-34) as

$$\mathbf{y} = \mathbf{X}_2\boldsymbol{\beta}_2 + \boldsymbol{\epsilon} \tag{13-35}$$

The least squares estimator of $\boldsymbol{\beta}_2$ is $\hat{\boldsymbol{\beta}}_2 = (\mathbf{X}_2'\mathbf{X}_2)^{-1}\mathbf{X}_2'\mathbf{y}$, and

$$SS_R(\boldsymbol{\beta}_2) = \hat{\boldsymbol{\beta}}_2'\mathbf{X}_2'\mathbf{y} \qquad (p - r \text{ degrees of freedom}) \tag{13-36}$$

The regression sum of squares due to $\boldsymbol{\beta}_1$ given that $\boldsymbol{\beta}_2$ is already in the model is

$$SS_R(\boldsymbol{\beta}_1 \mid \boldsymbol{\beta}_2) = SS_R(\boldsymbol{\beta}) - SS_R(\boldsymbol{\beta}_2) \tag{13-37}$$

This sum of squares has r degrees of freedom. It is sometimes called the "extra sum of squares" due to $\boldsymbol{\beta}_1$. Note that $SS_R(\boldsymbol{\beta}_1|\boldsymbol{\beta}_2)$ is the increase in the regression sum of squares due to including the variables $x_1, x_2, \ldots, x_r$ in the model. Now $SS_R(\boldsymbol{\beta}_1|\boldsymbol{\beta}_2)$ is independent of MS_E, and the null hypothesis $\boldsymbol{\beta}_1 = \mathbf{0}$ may be tested by the statistic

$$F_0 = \frac{SS_R(\boldsymbol{\beta}_1|\boldsymbol{\beta}_2)/r}{MS_E} \tag{13-38}$$

If $F_0 > F_{\alpha, r, n-p}$ we reject H_0, concluding that at least one of the parameters in $\boldsymbol{\beta}_1$ is not zero and, consequently, at least one of the variables $x_1, x_2, \ldots, x_r$ in $\mathbf{X}_1$ contributes significantly to the regression model. Some authors call the test in Equation (13-38) a *partial F-test*.

The partial F-test is very useful. We can use it to measure the contribution of x_j as if it were the last variable added to the model by computing

$$SS_R(\beta_j|\beta_0, \beta_1, \ldots, \beta_{j-1}, \beta_{j+1}, \ldots, \beta_k)$$

This is the increase in the regression sum of squares due to adding x_j to a model that already includes $x_1, \ldots, x_{j-1}, x_{j+1}, \ldots, x_k$. Note that the partial F-test on a single variable x_j is equivalent to the t-test in Equation (13-32). However, the partial F-test is a more general procedure in that we can measure the effect of sets of variables. In Section 13-11 we will show how the partial F-test plays a major role in *model building*; that is, in searching for the best set of independent variables to use in the model.

- **Example 13-7.** Consider the soft drink delivery time data in Example 13-1. We will investigate the contribution of the variable x_2 (distance) to the model. That is, we wish to test

$$H_0: \beta_2 = 0$$

$$H_1: \beta_2 \neq 0$$

To test this hypothesis, we need the extra sum of squares due to β_2, or

$$SS_R(\beta_2|\beta_1, \beta_0) = SS_R(\beta_1, \beta_2, \beta_0) - SS_R(\beta_1, \beta_0)$$
$$= SS_R(\beta_1, \beta_2|\beta_0) - SS_R(\beta_1|\beta_0)$$

In Example 13-6 we have calculated

$$SS_R(\beta_1, \beta_2|\beta_0) = \hat{\boldsymbol{\beta}}'\mathbf{X}'\mathbf{y} - \frac{\left(\sum_{i=1}^{n} y_i\right)^2}{n} = 5990.7712 \qquad \text{(2 degrees of freedom)}$$

and in Example 12-7, where the model $y = \beta_0 + \beta_1 x_1 + \epsilon$ was estimated, we have calculated

$$SS_R(\beta_1|\beta_0) = \hat{\beta}_1 S_{xy} = 5885.8521 \qquad \text{(1 degree of freedom)}$$

Therefore, we have

$$SS_R(\beta_2 | \beta_1, \beta_0) = 5990.7712 - 5885.8521$$

$$= 104.9191 \qquad \text{(1 degree of freedom)}$$

This is the increase in the regression sum of squares by adding x_2 to a model already containing x_1. To test $H_0: \beta_2 = 0$, form the test statistic

$$F_0 = \frac{SS_R(\beta_2 | \beta_1, \beta_0)/1}{MS_E} = \frac{104.9191/1}{5.2352} = 20.04$$

Note that the MS_E from the *full* model, using both x_1 and x_2, is used in the denominator of the test statistic. Since $F_{.05,1,22} = 4.30$, we reject $H_0: \beta_2 = 0$ and conclude that distance (x_2) contributes significantly to the model.

Since this partial F-test involves a single variable, it is equivalent to the t-test. To see this, recall that the t-test on $H_0: \beta_2 = 0$ resulted in the test statistic $t_0 = 4.4767$. Furthermore, the square of a t random variable with ν degrees of freedom is an F random variable with one and ν degrees of freedom, and we note that $t_0^2 = (4.4767)^2 = 20.04 = F_0$.

13-6 Measures of Model Adequacy

A number of techniques can be used to measure the adequacy of a multiple regression model. This section will present several of these techniques. Model validation is an important part of the multiple regression model building process. A good paper on this subject is Snee (1977).

13-6.1 The Coefficient of Multiple Determination

The coefficient of multiple determination R^2 is defined as

$$R^2 = \frac{SS_R}{S_{yy}} = 1 - \frac{SS_E}{S_{yy}} \qquad (13\text{-}39)$$

R^2 is a measure of the amount of reduction in the variability of y obtained by using the regressor variables $x_1, x_2, \ldots, x_k$. As in the simple linear regression case, we must have $0 \le R^2 \le 1$. However, a large value of R^2 does not necessarily imply that the regression model is a good one. Adding a variable to the model will always increase R^2, regardless of whether the additional variable is statistically significant or not. Thus it is possible for models that have large values of R^2 to yield poor predictions of new observations or estimates of the mean response.

The positive square root of R^2 is the multiple correlation coefficient between y and the set of regressor variables $x_1, x_2, \ldots, x_k$. That is, R is a measure of the linear association between y and $x_1, x_2, \ldots, x_k$. When $k = 1$, this becomes the simple correlation between y and x.

● **Example 13-8.** The coefficient of multiple determination for the regression model estimated in Example 13-1 is

$$R^2 = \frac{SS_R}{S_{yy}} = \frac{5990.7712}{6105.9447} = .981137$$

That is, about 98.11 percent of the variability in delivery time y has been explained when the two independent variables delivery volume (x_1) and distance (x_2) are used. In Example 12-7, a model relating y to x_1 only was developed. The value of R^2 for this model is $R^2 = .963954$. Therefore, adding the variable x_2 to the model has increased R^2 from .963954 to .981137.

13-6.2 Residual Analysis

The residuals from the estimated multiple regression model, defined by $e_i = y_i - \hat{y}_i$, play an important role in judging model adequacy just as they do in simple linear regression. As noted in Section 12-5.1, there are several residual plots that are often useful. These are illustrated in Example 13-9. It is also helpful to plot the residuals against variables not presently in the model that are possible candidates for inclusion. Patterns in these plots, similar to those in Fig. 12-5, indicate that the model may be improved by adding the candidate variable.

● **Example 13-9.** The residuals for the model estimated in Example 13-1 are shown in Table 13-3. These residuals are plotted on normal probability paper in Fig. 13-2. No severe deviations from normality are obviously apparent, although the two largest residuals ($e_{15} = 5.88$ and $e_{17} = 4.33$) do not fall extremely close to a straight line drawn through the remaining residuals. However, the standardized residuals, $5.88/\sqrt{5.2352} = 2.57$ and $4.33/\sqrt{5.2352} = 1.89$, do not seem excessively large. Inspection of the data does not reveal any error in collecting observations 15 and 17, or any other reason to discard or modify these two points.

The residuals are plotted against $\hat{y}$ in Fig. 13-3, and against x_1 and x_2 in Fig. 13-4 and 13-5, respectively. The two largest residuals e_{15} and e_{17} are apparent. In Fig. 13-4 there is some indication that the model underpredicts the time at outlets with small delivery volumes ($x_1 \leq 6$ cases) and large delivery volumes ($x_1 \geq 15$ cases), and over predicts the time at outlets with intermediate delivery volumes ($7 \leq x_1 \leq 14$ cases). The same impression is obtained from Fig. 13-3. Possibly the relationship between time and delivery volume is not linear (requiring that a term involving x_1^2, say, be added to the model), or other regressor variables not presently in the model affect the response. We will see subsequently that a third regressor variable is required to adequately model this data.

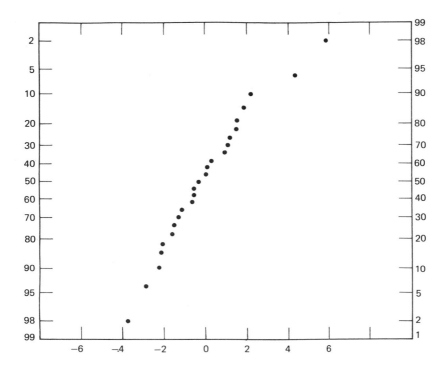

Fig. 13-2. Normal probability plot of residuals.

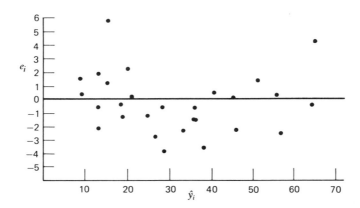

Fig. 13-3. Plot of residuals against $\hat{y}$.

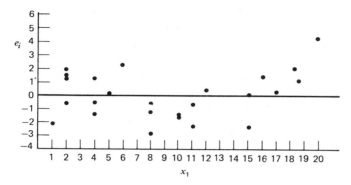

Fig. 13-4. Plot of residuals against x_1.

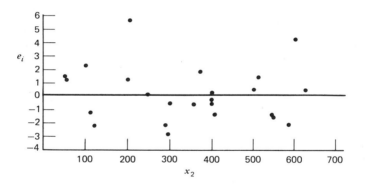

Fig. 13-5. Plot of residuals against x_2.

13-6.3 Estimation of Pure Error from Near Neighbors

In Section 12-5.2 we described a test for lack of fit in simple linear regression. The procedure involved partitioning the error or residual sum of squares into a component due to pure error and a component due to lack of fit, say

$$SS_E = SS_{PE} + SS_{LOF}$$

The pure error sum of squares SS_{PE} is computed from the responses obtained by repeated observations at the same level of x.

This general procedure can, in principle, be extended to multiple regression. The calculation of SS_{PE} requires repeated observations on y at the same set of levels on the regressor variables $x_1, x_2, \ldots, x_k$. That is, some of the *rows* of the **X** matrix must be the same. However, the occurrence of repeated observations is relatively unlikely in multiple regression, and the procedure described in Section 12-5.2 is not often useful.

Daniel and Wood (1971) have suggested a method for measuring pure error in cases where there are no exact repeat points. The procedure searches for points in the x-space that are "near neighbors," that is, sets of observations that have been taken with nearly identical levels of $x_1, x_2, \ldots, x_k$. The responses y_i from such near neighbors can be considered as repeat points and used to obtain an estimate of pure error. As a measure of the distance between any two points $x_{i1}, x_{i2}, \ldots, x_{ik}$ and $x_{i'1}, x_{i'2}, \ldots, x_{i'k}$, Daniel and Wood propose the weighted sum of squared distance

$$D_{ii'}^2 = \sum_{j=1}^{k} \left[\frac{\hat{\beta}_j(x_{ij} - x_{i'j})}{\sqrt{MS_E}} \right]^2 \tag{13-40}$$

Pairs of points that have small values of $D_{ii'}^2$ are "near neighbors"; that is, they are relatively close together in x-space. Pairs of points for which $D_{ii'}^2$ is large ($D_{ii'}^2 >> 1$, say) are widely separated in x-space. The residuals of two points with a small value of $D_{ii'}^2$, can be used to obtain an estimate of pure error. The estimate is obtained from the range of the residuals at the points i and i', say

$$E = |e_i - e_{i'}|$$

There is a relationship between the range of a sample from a normal population and the population standard deviation. For samples of size 2, this relationship is

$$\hat{\sigma} = (1.128)^{-1} E = .886E$$

The quantity $\hat{\sigma}$ so obtained is an estimate of the standard deviation of pure error.

An efficient algorithm may be used to compute this estimate. First, arrange the data points $x_{i1}, x_{i2}, \ldots, x_{ik}$ in order of increasing $\hat{y}_i$. Note that points with very different values of $\hat{y}_i$ cannot be near neighbors, but those with similar values of $\hat{y}_i$ could be neighbors (or they could be near the same contour of constant $\hat{y}$ but far apart in some x coordinates). Then

1. Compute the values of $D_{ii'}^2$ for all $n-1$ pairs of points with adjacent values of $\hat{y}$. Repeat this calculation for the pairs of points separated by 1, 2, and 3 intermediate $\hat{y}$ values. This will produce $4n - 10$ values of $D_{ii'}^2$.

2. Arrange the $4n - 10$ values of $D_{ii'}^2$ found in (1) in ascending order. Let E_u, $u = 1, 2, \ldots, 4n - 10$, be the range of the residuals at these points.

3. For the first m values of E_u, calculate an estimate of the standard deviation of pure error as

$$\hat{\sigma} = \frac{.886}{m} \sum_{u=1}^{m} E_u$$

TABLE 13-6 Calculation of $D_{ii'}^2$ for the Soft Drink Delivery Time Data

Near-Neighbor Calculations

Delta Residuals and the Weighted Standardized Squared Distances of Near Neighbors

Observation	Ordered Fitted Y	Residual	Adjacent Delta	Adjacent $D_{ii'}^2$	R^a	1 Apart Delta	1 Apart $D_{ii'}^2$	R	2 Apart Delta	2 Apart $D_{ii'}^2$	R	3 Apart Delta	3 Apart $D_{ii'}^2$	R
1	8.380	1.570	.370	.1199E−03	1	2.170	.2881E+01		3.610	.1002E+02		.360	.3167E+01	
8	8.400	1.200	1.800	.2844E+01		3.240	.9955E+01		.730	.3128E+01		.090	.6411E+01	
14	12.260	−.600	1.440	.2956E+01		2.530	.6745E−02	6	1.710	.6522E+01		6.440	.6474E+01	
18	12.340	−2.040	3.970	.2761E+01		3.150	.1739E+02		7.880	.1728E+02		1.680	.1397E+02	
7	12.450	1.930	.820	.6672E+01		3.910	.6621E+01		2.290	.5773E+01		3.250	.5795E+01	
6	15.750	1.110	4.730	.7495E−03	2	1.470	.1199E+01	11	2.430	.1347E+01	13	1.040	.6054E+01	
15	15.810	5.840	6.200	.1140E+01	10	7.160	.1285E+01	12	3.690	.6085E+01		5.690	.2578E+01	
16	18.250	−.360	.960	.4317E−02	5	2.510	.8452E+01		.510	.1439E+01	14	.790	.2554E+02	
11	18.400	−1.320	3.470	.8672E+01		1.470	.1443E+01	15	.170	.2575E+02		1.570	.2343E+02	
24	19.980	2.150	2.000	.4137E+01		3.300	.5757E+01		5.040	.6894E+01		2.630	.6953E+01	
25	21.000	.150	1.300	.1547E+02		3.040	.1328E+02		.630	.1325E+02		4.010	.2571E+02	
2	25.600	−1.150	1.740	.1026E+01	8	.670	.1082E+01	9	2.710	.1442E+02		1.050	.1295E+02	
5	27.910	−2.890	2.410	.7495E−03	3	.970	.2578E+01		.690	.1386E+02		1.350	.7554E+01	
10	27.980	−.480	3.380	.2638E+01		1.720	.1392E+02		1.060	.7481E+01		1.120	.7628E+01	
9	28.210	−3.860	1.660	.5766E+01		2.320	.7242E+01		2.260	.7509E+01		3.400	.8452E+01	
3	33.950	−2.200	.660	.6727E+01		.600	.6982E+01		1.740	.2350E+01		2.690	.5768E+01	
19	36.470	−1.540	.060	.2998E−02	4	1.080	.2026E+01		2.030	.5802E+01		1.570	.3848E+02	
4	36.600	−1.600	1.140	.2113E+01		2.090	.5829E+01		1.630	.3866E+02		.580	.3799E+02	
12	37.460	−.460	.950	.1738E+01		.490	.2369E+02		1.720	.2338E+02		2.020	.3633E+02	
13	41.460	.490	.460	.1482E+02		2.670	.1427E+02		1.070	.2302E+02		.170	.3621E+02	
20	46.560	.030	2.210	.4797E−01	7	1.530	.3465E+01		.290	.9220E+01		4.300	.3964E+02	
21	47.060	−2.180	3.740	.2890E+01		2.500	.8452E+01		6.510	.3884E+02				
22	52.560	1.560	1.240	.1630E+01		2.770	.2326E+02							
23	56.310	.320	4.010	.1295E+02										
17	64.670	4.330												

aColumn R gives the rank order of the 15 smallest values for the weighted standardized squared distances.

415

TABLE 13-7 **Calculation of σ̂ from Residuals of Observations that are Near Neighbors**

| | | Standard Deviation Estimated from Residuals of Neighboring Observations | | | |
| | | Ordered by Weighted Standardized Squared Distance | | | |
Number	Cumulative Standard Deviation	$D_{ii'}^2$	Observation	Observation	Delta Residual
1	.3278E+00	.1199E−03	1	8	.3700
2	.2259E+01	.7495E−03	6	15	4.7300
3	.2218E+01	.7495E−03	5	10	2.4100
4	.1677E+01	.2998E−02	19	4	.0600
5	.1512E+01	.4317E−02	16	11	.9600
6	.1633E+01	.6745E−02	14	7	2.5300
7	.1680E+01	.4797E−01	20	21	2.2100
8	.1662E+01	.1026E+01	2	5	1.7400
9	.1544E+01	.1082E+01	2	10	.6700
10	.1939E+01	.1140E+01	15	16	6.2000
11	.1881E+01	.1199E+01	6	16	1.4700
12	.2253E+01	.1285E+01	15	11	7.1600
13	.2245E+01	.1347E+01	6	11	2.4300
14	.2117E+01	.1439E+01	16	25	.5100
15	.2136E+01	.1442E+01	2	9	2.7100
16	.2084E+01	.1443E+01	11	25	1.4700

17	.2026E + 01	.1630E + 01	22	23	1.2400
18	.1960E + 01	.1738E + 01	12	13	.9500
19	.1907E + 01	.2026E + 01	19	12	1.0800
20	.1862E + 01	.2113E + 01	4	12	1.1400
21	.1847E + 01	.2350E + 01	3	12	1.7400
22	.1802E + 01	.2578E + 01	5	9	.9700
23	.1943E + 01	.2578E + 01	15	25	5.6900
24	.1987E + 01	.2638E + 01	10	9	3.3800
25	.2048E + 01	.2761E + 01	18	7	3.9700
26	.2031E + 01	.2844E + 01	8	14	1.8000
27	.2027E + 01	.2881E + 01	1	14	2.1700
28	.2073E + 01	.2890E + 01	21	22	3.7400
29	.2045E + 01	.2956E + 01	14	18	1.4400
30	.1999E + 01	.3128E + 01	8	7	.7300
31	.1944E + 01	.3167E + 01	1	7	.3600
32	.1926E + 01	.3465E + 01	20	22	1.5300
33	.1921E + 01	.4137E + 01	24	25	2.0000
34	.1951E + 01	.5757E + 01	24	2	3.3000
35	.1937E + 01	.5766E + 01	9	3	1.6600
36	.1949E + 01	.5768E + 01	3	13	2.6900
37	.1952E + 01	.5773E + 01	7	16	2.2900
38	.1976E + 01	.5795E + 01	7	11	3.2500
39	.1971E + 01	.5802E + 01	19	13	2.0300
40	.1968E + 01	.5829E + 01	4	13	2.0900

Note that $\hat{\sigma}$ is based on the average range of the residuals associated with the m smallest values of $D_{ii'}^2$. The value of m must be chosen after inspecting the values of $D_{ii'}^2$. One should not include values of E_u in the calculations for which the weighted sum of squared distance is too large.

● **Example 13-10.** We will use this three-step procedure to calculate an estimate of the standard deviation of pure error for the soft drink delivery time data in Example 13-1. Table 13-6 displays the calculation of $D_{ii'}^2$ for pairs of points that, in terms of $\hat{y}$, are adjacent, one apart, two apart, and three apart. The columns labeled "R" in this table identify the 15 smallest values of $D_{ii'}^2$. The residuals at these pairs of points are used to estimate σ. The calculation of $\hat{\sigma}$ is shown in Table 13-7. The value of $\hat{\sigma} = 2.136$ is reasonably close to $\sqrt{MS_E} = \sqrt{5.2352} = 2.288$, from Table 13-5. Since $\hat{\sigma} \approx \sqrt{MS_E}$, we would conclude that there is no strong indication of lack of fit.

13-7 Polynomial Regression

The linear model $y = X\beta + \epsilon$ is a general model that can be used to fit any relationship that is *linear* in the unknown parameters β. This includes the important class of polynomial regression models. For example, the second-degree polynomial in one variable

$$y = \beta_0 + \beta_1 x + \beta_{11} x^2 + \epsilon \tag{13-41}$$

and the second-degree polynomial in two variables

$$y = \beta_0 + \beta_1 x_1 + \beta_2 x_2 + \beta_{11} x_1^2 + \beta_{22} x_2^2 + \beta_{12} x_1 x_2 + \epsilon \tag{13-42}$$

are linear regression models.

Polynomial regression models are widely used in cases where the response is curvilinear, because the general principles of multiple regression can be applied. The following example illustrates some of the types of analyses that can be performed.

● **Example 13-11.** The data shown below gives the average cost per unit for a product (y) and the production lot size (x). The scatter diagram, shown in Fig. 13-6, indicates that a second-order polynomial may be appropriate.

y	1.81	1.70	1.65	1.55	1.48	1.40	1.30	1.26	1.24	1.21	1.20	1.18
x	20	25	30	35	40	50	60	65	70	75	80	90

We will fit the model

$$y = \beta_0 + \beta_1 x + \beta_{11} x^2 + \epsilon$$

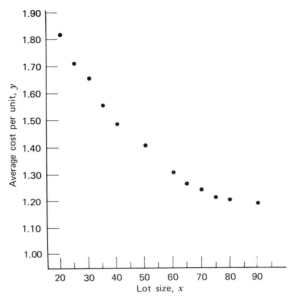

Fig. 13-6. Data for Example 13-11.

The **y** vector, **X** matrix, and **β** vector are as follows:

$$
\mathbf{y} = \begin{bmatrix} 1.81 \\ 1.70 \\ 1.65 \\ 1.55 \\ 1.48 \\ 1.40 \\ 1.30 \\ 1.26 \\ 1.24 \\ 1.21 \\ 1.20 \\ 1.18 \end{bmatrix}
\qquad
\mathbf{X} = \begin{bmatrix} 1 & 20 & 400 \\ 1 & 25 & 625 \\ 1 & 30 & 900 \\ 1 & 35 & 1225 \\ 1 & 40 & 1600 \\ 1 & 50 & 2500 \\ 1 & 60 & 3600 \\ 1 & 65 & 4225 \\ 1 & 70 & 4900 \\ 1 & 75 & 5625 \\ 1 & 80 & 6400 \\ 1 & 90 & 8100 \end{bmatrix}
\qquad
\boldsymbol{\beta} = \begin{bmatrix} \beta_0 \\ \beta_1 \\ \beta_{11} \end{bmatrix}
$$

Solving the normal equations $\mathbf{X}'\mathbf{X}\hat{\boldsymbol{\beta}} = \mathbf{X}'\mathbf{y}$ gives the estimated model

$$\hat{y} = 2.19826629 - .02252236x + .00012507x^2$$

The test for significance of regression is shown in Table 13-8. Since $F_0 = 2171.07$ is significant at 1 percent, we conclude that at least one of the parameters β_1 and β_{11} is not zero. Furthermore, the standard tests for model adequacy do not reveal any unusual behavior.

TABLE 13-8 **Test for Significance of Regression for the Second-Order Model in Example 13-11**

Source of Variation	Sum of Squares	Degrees of Freedom	Mean Square	F_0
Regression	.5254	2	.262700	2171.07
Error	.0011	9	.000121	
Total	.5265	11		

In fitting polynomials, we generally like to use the lowest-degree model consistent with the data. In this example, it would seem logical to investigate dropping the quadratic term from the model. That is, we would like to test

$$H_0: \beta_{11} = 0$$
$$H_1: \beta_{11} \neq 0$$

The general regression significance test can be used to test this hypothesis. We need to determine the "extra sum of squares" due to β_{11}, or

$$SS_R(\beta_{11} \mid \beta_1, \beta_0) = SS_R(\beta_1, \beta_{11} \mid \beta_0) - SS_R(\beta_1 \mid \beta_0)$$

The sum of squares $SS_R(\beta_1, \beta_{11} \mid \beta_0) = .5254$, from Table 13-8. To find $SS_R(\beta_1 \mid \beta_0)$, we fit a simple linear regression model to the original data, yielding

$$\hat{y} = 1.90036320 - .00910056x$$

It can be easily verified that the regression sum of squares for this model is

$$SS_R(\beta_1 \mid \beta_0) = .4942$$

Therefore, the extra sum of squares due to β_{11}, given that β_1 and β_0 are in the model, is

$$SS_R(\beta_{11} \mid \beta_0, \beta_1) = SS_R(\beta_1, \beta_{11} \mid \beta_0) - SS_R(\beta_1 \mid \beta_0)$$
$$= .5254 - .4942$$
$$= .0312$$

The analysis of variance, with the test of $H_0: \beta_{11} = 0$ incorporated into the procedure, is displayed in Table 13-9. Note that the quadratic term contributes significantly to the model.

TABLE 13.9 **Analysis of Variance for Example 13-11, Showing the Test for** $H_0: \beta_{11} = 0$

Source of Variation	Sum of Squares	Degrees of Freedom	Mean Square	F_0
Regression	$SS_R(\beta_1, \beta_{11} \mid \beta_0) = .5254$	2	.262700	2171.07
Linear	$SS_R(\beta_1 \mid \beta_0) \quad = .4942$	1	.494200	4084.30
Quadratic	$SS_R(\beta_{11} \mid \beta_0, \beta_1) = .0312$	1	.031200	258.18
Error	.0011	9	.000121	
Total	.5265	11		

13-8 Indicator Variables

The regression models presented in previous sections have been based on *quantitative* variables, that is, variables that are measured on a numerical scale. For example, variables such as temperature, pressure, distance, and age are quantitative variables. Occasionally, we need to incorporate *qualitative* variables in a regression model. For example, suppose that one of the variables in a regression model is the operator who is associated with each observation y_i. Assume that only two operators are involved. We may wish to assign different levels to the two operators to account for the possibility that each operator may have a different effect on the response.

The usual method of accounting for the different levels of a qualitative variable is by using indicator variables. For example, to introduce the effect of two different operators into a regression model, we could define an indicator variable as follows:

$$x = 0 \text{ if the observation is from operator 1}$$
$$x = 1 \text{ if the observation is from operator 2}$$

In general, a qualitative variable with t levels is represented by $t - 1$ indicator variables, which are assigned the values either 0 or 1. Thus, if there were *three* operators, the different levels would be accounted for by *two* indicator variables defined as follows:

x_1	x_2	
0	0	if the observation is from operator 1
1	0	if the observation is from operator 2
0	1	if the observation is from operator 3

Indicator variables are also referred to as *dummy* variables. The following example illustrates some of the uses of indicator variables. For other applications, see Draper and Smith (1966).

● **Example 13-12.** A mechanical engineer is investigating the surface finish of metal parts produced on a lathe and its relationship to the speed (in RPM) of the lathe. The data are shown in Table 13-10. Note that the data have been collected using two different types of cutting tool. Since it is likely that the type of cutting tool affects the surface finish, we will fit the model

$$y = \beta_0 + \beta_1 x_1 + \beta_2 x_2 + \epsilon$$

where y is the surface finish, x_1 is the lathe speed in RPM, and x_2 is an indicator variable denoting the type of cutting tool used; that is,

$$x_2 = \begin{cases} 0, \text{ for tool type 302} \\ 1, \text{ for tool type 416} \end{cases}$$

The parameters in this model may be easily interpreted. If $x_2 = 0$, then the model becomes

$$y = \beta_0 + \beta_1 x_1 + \epsilon$$

which is a straight-line model with slope β_1 and intercept β_0.
However, if $x_2 = 1$, then the model becomes

$$y = \beta_0 + \beta_1 x_1 + \beta_2(1) + \epsilon = \beta_0 + \beta_2 + \beta_1 x_1 + \epsilon$$

Table 13-10 **Surface Finish Data for Example 13-12**

Observation Number, i	Surface Finish, y_i	rpm	Type of Cutting Tool
1	45.44	225	302
2	42.03	200	302
3	50.10	250	302
4	48.75	245	302
5	47.92	235	302
6	47.79	237	302
7	52.26	265	302
8	50.52	259	302
9	45.58	221	302
10	44.78	218	302
11	33.50	224	416
12	31.23	212	416
13	37.52	248	416
14	37.13	260	416
15	34.70	243	416
16	33.92	238	416
17	32.13	224	416
18	35.47	251	416
19	33.49	232	416
20	32.29	216	416

which is a straight-line model with slope β_1 and intercept $\beta_0 + \beta_2$. Thus, the model $y = \beta_0 + \beta_1 x_1 + \beta_2 x_2 + \epsilon$ implies that surface finish is linearly related to lathe speed and that the slope β_1 does not depend on the type of cutting tool used. However, the type of cutting tool does affect the intercept, and β_2 indicates the change in the intercept associated with a change in tool type from 302 to 416.

The **X** matrix and **y** vector for this problem are as follows:

$$
\mathbf{X} = \begin{bmatrix}
1 & 225 & 0 \\
1 & 200 & 0 \\
1 & 250 & 0 \\
1 & 245 & 0 \\
1 & 235 & 0 \\
1 & 237 & 0 \\
1 & 265 & 0 \\
1 & 259 & 0 \\
1 & 221 & 0 \\
1 & 218 & 0 \\
1 & 224 & 1 \\
1 & 212 & 1 \\
1 & 248 & 1 \\
1 & 260 & 1 \\
1 & 243 & 1 \\
1 & 238 & 1 \\
1 & 224 & 1 \\
1 & 251 & 1 \\
1 & 232 & 1 \\
1 & 216 & 1
\end{bmatrix}
\qquad
\mathbf{y} = \begin{bmatrix}
45.44 \\
42.03 \\
50.10 \\
48.75 \\
47.92 \\
47.79 \\
52.26 \\
50.52 \\
45.58 \\
44.78 \\
33.50 \\
31.23 \\
37.52 \\
37.13 \\
34.70 \\
33.92 \\
32.13 \\
35.47 \\
33.49 \\
32.29
\end{bmatrix}
$$

The estimated model is

$$\hat{y} = 14.27620 + .11415 x_1 - 13.28020 x_2$$

The analysis of variance for this model is shown in Table 13-11. Note that the hypothesis $H_0 : \beta_1 = \beta_2 = 0$ (significance of regression) is rejected. This table also contains the sums of squares

$$
\begin{aligned}
SS_R &= SS_R(\beta_1, \beta_2 \mid \beta_0) \\
&= SS_R(\beta_1 \mid \beta_0) + SS_R(\beta_2 \mid \beta_1, \beta_0)
\end{aligned}
$$

so that a test of the hypothesis $H_0 : \beta_2 = 0$ can be made. This hypothesis is also rejected, so we conclude that tool type has an effect on surface finish.

It is also possible to use indicator variables to investigate whether tool type affects *both* the slope *and* intercept. Let the model be

$$y = \beta_0 + \beta_1 x_1 + \beta_2 x_2 + \beta_3 x_1 x_2 + \epsilon$$

TABLE 13-11 **Analysis of Variance for Example 13-12**

Source of Variation	Sum of Squares	Degrees of Freedom	Mean Square	F_0
Regression	1012.0595	2	506.0297	1103.69[a]
$SS_R(\beta_1 \mid \beta_0)$	(130.6091)	(1)	130.6091	284.87[a]
$SS_R(\beta_2 \mid \beta_1, \beta_0)$	(881.4504)	(1)	881.4504	1922.52[a]
Error	7.7943	17	.4508	
Total	1019.8538	19		

[a]Significant at 1 percent.

where x_2 is the indicator variable. Now if tool type 302 is used, $x_2 = 0$, and the model is

$$y = \beta_0 + \beta_1 x_1 + \epsilon$$

If tool type 416 is used, $x_2 = 1$, and the model becomes

$$y = \beta_0 + \beta_1 x_1 + \beta_2 + \beta_3 x_1 + \epsilon$$
$$= (\beta_0 + \beta_2) + (\beta_1 + \beta_3) x_1 + \epsilon$$

Note that β_2 is the change in the intercept, and β_3 is the change in slope produced by a change in tool type.

Another method of analyzing this data set is to fit separate regression models to the data for each tool type. However, the indicator variable approach has several advantages. First, only one regression model must be estimated. Second, by pooling the data on both tool types, more degrees of freedom for error are obtained. Third, tests of hypotheses on the parameters β_2 and β_3 are just special cases of the general regression significance test.

13-9 The Correlation Matrix

Suppose we wish to estimate the parameters in the model

$$y_i = \beta_0 + \beta_1 x_{i1} + \beta_2 x_{i2} + \epsilon_i, \quad i = 1, 2, \ldots, n \tag{13-43}$$

We may rewrite this model with a transformed intercept β_0' as

$$y_i = \beta_0' + \beta_1(x_{i1} - \bar{x}_1) + \beta_2(x_{i2} - \bar{x}_2) + \epsilon_i \tag{13-44}$$

or, since $\hat{\beta}_0' = \bar{y}$,

$$y_i - \bar{y} = \beta_1(x_{i1} - \bar{x}_1) + \beta_2(x_{i2} - \bar{x}_2) + \epsilon_i \tag{13-45}$$

The $\mathbf{X'X}$ matrix for this model is

$$\mathbf{X'X} = \begin{bmatrix} S_{11} & S_{12} \\ S_{12} & S_{22} \end{bmatrix} \tag{13-46}$$

where

$$S_{kj} = \sum_{i=1}^{n} (x_{ik} - \bar{x}_k)(x_{ij} - \bar{x}_j) \qquad k, j = 1, 2 \qquad (13\text{-}47)$$

It is possible to express this $\mathbf{X'X}$ matrix in correlation form. Let

$$r_{kj} = \frac{S_{kj}}{(S_{kk}S_{jj})^{1/2}} \qquad k, j = 1, 2 \qquad (13\text{-}48)$$

and note that $r_{11} = r_{22} = 1$. Then the correlation form of the $\mathbf{X'X}$ matrix, Equation (13-47), is

$$\mathbf{R} = \begin{bmatrix} 1 & r_{12} \\ r_{12} & 1 \end{bmatrix} \qquad (13\text{-}49)$$

The quantity r_{12} is the simple correlation between x_1 and x_2. We may also define the simple correlation between x_j and y as

$$r_{jy} = \frac{S_{jy}}{(S_{jj}S_{yy})^{1/2}} \qquad j = 1, 2 \qquad (13\text{-}50)$$

where

$$S_{jy} = \sum_{u=1}^{n} (x_{uj} - \bar{x}_j)(y_u - \bar{y}) \qquad j = 1, 2 \qquad (13\text{-}51)$$

is the corrected sum of cross products between x_j and y, and S_{yy} is the usual total corrected sum of squares of y.

These transformations result in a new regression model

$$y_i^* = b_1 z_{i1} + b_2 z_{i2} + \epsilon_i^* \qquad (13\text{-}52)$$

in the new variables

$$y_i^* = \frac{y_i - \bar{y}}{S_{yy}^{1/2}}$$

$$z_{ij} = \frac{x_{ij} - \bar{x}_j}{S_{jj}^{1/2}} \qquad j = 1, 2$$

The relationship between the parameters b_1 and b_2 in the new model, Equation (13-52), and the parameters β_0, β_1, and β_2 in the original model, Equation (13-43), is as follows:

$$\beta_1 = b_1 \left(\frac{S_{yy}}{S_{11}} \right)^{1/2} \qquad (13\text{-}53)$$

$$\beta_2 = b_2 \left(\frac{S_{yy}}{S_{22}} \right)^{1/2} \qquad (13\text{-}54)$$

$$\beta_0 = \bar{y} - \beta_1 \bar{x}_1 - \beta_2 \bar{x}_2 \qquad (13\text{-}55)$$

The least squares normal equations for the transformed model, Equation (13-52), are

$$\begin{bmatrix} 1 & r_{12} \\ r_{12} & 1 \end{bmatrix} \begin{bmatrix} b_1 \\ b_2 \end{bmatrix} = \begin{bmatrix} r_{1y} \\ r_{2y} \end{bmatrix} \tag{13-56}$$

The solution to Equation (13-56) is

$$\begin{bmatrix} \hat{b}_1 \\ \hat{b}_2 \end{bmatrix} = \begin{bmatrix} 1 & r_{12} \\ r_{12} & 1 \end{bmatrix}^{-1} \begin{bmatrix} r_{1y} \\ r_{2y} \end{bmatrix}$$

$$= \frac{1}{1 - r_{12}^2} \begin{bmatrix} 1 & -r_{12} \\ -r_{12} & 1 \end{bmatrix} \begin{bmatrix} r_{1y} \\ r_{2y} \end{bmatrix}$$

or

$$\hat{b}_1 = \frac{r_{1y} - r_{12}r_{2y}}{1 - r_{12}^2} \tag{13-57a}$$

$$\hat{b}_2 = \frac{r_{2y} - r_{12}r_{1y}}{1 - r_{12}^2} \tag{13-57b}$$

The regression coefficients, Equations (13-57), are usually called *standardized regression coefficients*. Many multiple regression computer programs use this transformation to reduce roundoff errors in the $(\mathbf{X'X})^{-1}$ matrix. These roundoff errors may be very serious if the original variables differ considerably in magnitude. Some of these computer programs also display both the original regression coefficients and the standardized coefficients. The standardized regression coefficients are dimensionless, and this may make it easier to compare regression coefficients in situations where the original variables x_j differ considerably in their units of measurement. In interpreting these standardized regression coefficients, however, we must remember that they are still partial regression coefficients (i.e., b_j shows the effect of z_j given that other z_i, $i \neq j$, are in the model). Furthermore, the $\hat{b}_j$ are affected by the spacing of the levels of the x_j. Consequently, we should not use the magnitude of the $\hat{b}_j$ as a measure of the importance of the independent variables.

While we have explicitly treated only the case of two independent variables, the results generalize. If there are k independent variables $x_1, x_2, \ldots, x_k$, one may write the $\mathbf{X'X}$ matrix in correlation form as

$$\mathbf{R} = \begin{bmatrix} 1 & r_{12} & r_{13} \ldots r_{1k} \\ r_{12} & 1 & r_{23} \ldots r_{2k} \\ r_{13} & r_{23} & 1 \ldots r_{3k} \\ & & \vdots \\ r_{1k} & r_{2k} & r_{3k} \ldots 1 \end{bmatrix} \tag{13-58}$$

where $r_{ij} = S_{ij}/(S_{ii}S_{jj})^{1/2}$ is the simple correlation between x_i and x_j and $S_{ij} = \sum_{u=1}^{n} (x_{ui} - \bar{x}_i)(x_{uj} - \bar{x}_j)$. The correlations between x_j and y are

$$g = \begin{bmatrix} r_{1y} \\ r_{2y} \\ \vdots \\ r_{ky} \end{bmatrix} \tag{13-59}$$

where $r_{jy} = \sum_{u=1}^{n} (x_{uj} - \bar{x}_j)(y_u - \bar{y})$. The vector of standardized regression coefficients $\hat{\mathbf{b}}' = [b_1, b_2, \ldots, b_u]$ is

$$\hat{\mathbf{b}} = \mathbf{R}^{-1}\mathbf{g} \tag{13-60}$$

The relationship between the standardized regression coefficients and the original regression coefficients is

$$\hat{\beta}_j = \hat{b}_j \left(\frac{S_{yy}}{S_{jj}}\right)^{1/2} \qquad j = 1, 2, \ldots, k \tag{13-61}$$

● **Example 13-13.** For the data in Example 13-1, we find

$$S_{yy} = 6105.9447 \qquad S_{11} = 698.5600$$
$$S_{1y} = 2027.7132 \qquad S_{22} = 780,230.5600$$
$$S_{2y} = 34,018.6668 \qquad S_{12} = 8834.4400$$

Therefore,

$$r_{12} = \frac{S_{12}}{(S_{11}S_{22})^{1/2}} = \frac{8834.4400}{\sqrt{(698.5600)(780,230.5600)}} = .378413$$

$$r_{1y} = \frac{S_{1y}}{(S_{11}S_{yy})^{1/2}} = \frac{2027.7132}{\sqrt{(698.5600)(6105.9447)}} = .981812$$

$$r_{2y} = \frac{S_{2y}}{(S_{22}S_{yy})^{1/2}} = \frac{34,018.6668}{\sqrt{(780,230.5600)(6105.9447)}} = .492867$$

and the correlation matrix for this problem is

$$\begin{bmatrix} 1 & .378413 \\ .378413 & 1 \end{bmatrix}$$

From Equation (13-56), the normal equations in terms of the standardized regression coefficients are

$$\begin{bmatrix} 1 & .378413 \\ .378413 & 1 \end{bmatrix} \begin{bmatrix} \hat{b}_1 \\ \hat{b}_2 \end{bmatrix} = \begin{bmatrix} .981812 \\ .492867 \end{bmatrix}$$

Consequently, the standardized regression coefficients are

$$\begin{bmatrix} \hat{b}_1 \\ \hat{b}_2 \end{bmatrix} = \begin{bmatrix} 1 & .378413 \\ 378413 & 1 \end{bmatrix}^{-1} \begin{bmatrix} .981812 \\ .492867 \end{bmatrix}$$

$$= \begin{bmatrix} 1.16713 & -.44166 \\ -.44166 & 1.16713 \end{bmatrix} \begin{bmatrix} .981812 \\ .492867 \end{bmatrix}$$

$$= \begin{bmatrix} .928223 \\ .141615 \end{bmatrix}$$

These standardized regression coefficients could also have been computed directly from either Equation (13-57) or Equation (13-61). Note that although $\hat{b}_1 > \hat{b}_2$, we should be cautious about concluding that the number of cases delivered (x_1) is more important than distance (x_2), since $\hat{b}_1$ and $\hat{b}_2$ are still *partial* regression coefficients.

13-10   Problems in Multiple Regression

There are a number of problems often encountered in the use of multiple regression. In this section, we briefly discuss three of these problem areas: the effect of multicollinearity on the regression model, the effect of outlying points in the x-space on the regression coefficients, and autocorrelation in the errors.

13-10.1   Multicollinearity

In most multiple regression problems, the independent or regressor variables x_j are intercorrelated. In situations where this intercorrelation is very large, we say that *multicollinearity* exists. Multicollinearity can have serious effects on the estimates of the regression coefficients and on the general applicability of the estimated model.

The effects of multicollinearity may be easily demonstrated. Consider a regression model with two regressor variables x_1 and x_2, and suppose that x_1 and x_2 have been "standardized" as in Section 13-9, so that the $\mathbf{X}'\mathbf{X}$ matrix is in correlation form, as in Equation (13-49). The model is

$$y_i = \beta_0 + \beta_1 x_{i1} + \beta_2 x_{i2} + \epsilon_i \qquad i = 1, 2, \ldots, n$$

The $(\mathbf{X}'\mathbf{X})^{-1}$ matrix for this model is

$$\mathbf{C} = (\mathbf{X}'\mathbf{X})^{-1} = \begin{bmatrix} 1/(1 - r_{12}^2) & -r_{12}/(1 - r_{12}^2) \\ -r_{12}/(1 - r_{12}^2) & 1/(1 - r_{12}^2) \end{bmatrix}$$

and the estimators of the parameters are

$$\hat{\beta}_1 = \frac{\mathbf{x}_1'\mathbf{y} - r_{12}\mathbf{x}_2'\mathbf{y}}{1 - r_{12}^2}$$

$$\hat{\beta}_2 = \frac{\mathbf{x}_2'\mathbf{y} - r_{12}\mathbf{x}_1'\mathbf{y}}{1 - r_{12}^2}$$

where r_{12} is the simple correlation between x_1 and x_2, and $\mathbf{x}_1'\mathbf{y}$ and $\mathbf{x}_2'\mathbf{y}$ are the elements of the $\mathbf{X}'\mathbf{y}$ vector.

Now, if multicollinearity is present, x_1 and x_2 are highly correlated, and $|r_{12}| \rightarrow 1$. In such a situation, the variances and covariances of the regression coefficients become very large, since $V(\hat{\beta}_j) = C_{jj}\sigma^2 \rightarrow \infty$ as $|r_{12}| \rightarrow 1$, and $\text{Cov}(\hat{\beta}_1, \hat{\beta}_2) = C_{12}\sigma^2 \rightarrow \pm\infty$ depending on whether $r_{12} \rightarrow \pm 1$. The large variances for $\hat{\beta}_j$ imply that the regression coefficients are very poorly estimated. Note that the effect of multicollinearity is to introduce a "near" linear dependency in the columns of the $\mathbf{X}$ matrix. As $r_{12} \rightarrow \pm 1$, this linear dependency becomes exact. Furthermore, if we assume that $\mathbf{x}_1'\mathbf{y} \rightarrow \mathbf{x}_2'\mathbf{y}$ as $|r_{12}| \rightarrow \pm 1$, then the estimates of the regression coefficients become equal in magnitude but opposite in sign; that is, $\hat{\beta}_1 = -\hat{\beta}_2$, *regardless* of the true values of β_1 and β_2.

Similar problems occur when multicollinearity is present and there are more than two regressor variables. In general, the diagonal elements of the matrix $\mathbf{C} = (\mathbf{X}'\mathbf{X})^{-1}$ can be written as

$$C_{jj} = \frac{1}{(1 - R_j^2)} \qquad j = 1, 2, \ldots, k \tag{13-62}$$

where R_j^2 is the coefficient of multiple determination resulting from regressing x_j on the other $k - 1$ regressor variables. Clearly, the stronger the linear dependency of x_j on the remaining regressor variables (and hence the stronger the multicollinearity), the larger the value of R_j^2 will be. We say that the variance of $\hat{\beta}_j$ is "inflated" by the quantity $(1 - R_j^2)^{-1}$. Consequently, we usually call

$$VIF(\hat{\beta}_j) = \frac{1}{(1 - R_j^2)} \qquad j = 1, 2, \ldots, k \tag{13-63}$$

the *variance inflation factor* for $\hat{\beta}_j$. Note that these factors are the main diagonal elements of the inverse of the correlation matrix. They are an important measure of the extent to which multicollinearity is present.

Although the estimates of the regression coefficients are very imprecise when multicollinearity is present, the estimated equation may still be useful. For example, suppose we wish to predict new observations. If these predictions are required in the region of the x-space where the multicollinearity is in effect, then often satisfactory results will be obtained because, while individual β_j may be poorly estimated, the function $\sum_{j=1}^{k} \beta_j x_{ij}$ may be estimated quite well. On the other hand, if the prediction of new observations requires extrapolation, then generally we would expect to obtain poor results. Extrapolation usually requires good estimates of the individual model parameters.

Multicollinearity arises for several reasons. It will occur when the analyst collects the data such that a constraint of the form $\sum_{j=1}^{k} a_j x_j = 0$ holds among the columns of the **X** matrix (the a_j are constants, not all zero). For example, if four regressor variables are the components of a mixture, then such a constraint will always exist because the sum of the components is always constant. Usually, these constraints do not hold exactly, and the analyst does not know that they exist.

There are several ways to detect the presence of multicollinearity. Some of the more important of these will be briefly discussed.

1. The variance inflation factors, defined in Equation (13-63), are very useful measures of multicollinearity. The larger the variance inflation factor, the more severe the multicollinearity. Some authors have suggested that if any variance inflation factors exceed 10, then multicollinearity is a problem. Other authors consider this value too liberal and suggest that the variance inflation factors should not exceed 4 or 5.

2. The determinant of the correlation matrix may also be used as a measure of multicollinearity. The value of this determinant can range between 0 and 1. When the value of the determinant is 1, the columns of the **X** matrix are orthogonal (i.e., there is no intercorrelation between the regression variables), and when the value is 0, there is an exact linear dependency among the columns of **X**. The smaller the value of the determinant, the greater the degree of multicollinearity.

3. The eigenvalues or characteristic roots of the correlation matrix provide a measure of multicollinearity. If **X'X** is in correlation form, then the eigenvalues of **X'X** are the roots of the equation

$$|\mathbf{X'X} - \lambda \mathbf{I}| = 0$$

One or more eigenvalues near zero implies that multicollinearity is present. If $\lambda_{\max}$ and $\lambda_{\min}$ denote the largest and smallest eigenvalues of **X'X**, then the ratio $\lambda_{\max}/\lambda_{\min}$ can also be used as a measure of multicollinearity. The larger the value of this ratio, the greater the degree of multicollinearity. Generally, if the ratio $\lambda_{\max}/\lambda_{\min}$ is less than 10, there is little problem with multicollinearity.

4. Sometimes inspection of the individual elements of the correlation matrix can be helpful in detecting multicollinearity. If an element $|r_{ij}|$ is close to one, then x_i and x_j may be strongly multicollinear. However, when more than two regressor variables are involved in a multicollinear

not always enable us to detect the presence of multicollinearity.
fashion, the individual r_{ij} are not necessarily large. Thus, this method will
5. If the F-test for significance of regression is significant, but tests on the individual regression coefficients are not significant, then multicollinearity may be present.

Several remedial measures have been proposed for resolving the problem of multicollinearity. Augmenting the data with new observations specifically designed to break up the approximate linear dependencies that currently exist is often suggested. However, sometimes this is impossible for economic reasons, or because of physical constraints that relate the x_j. Another possibility is to delete certain variables from the model. This suffers from the disadvantage of discarding the information contained in the deleted variables.

Since multicollinearity primarily affects the stability of the regression coefficients, it would seem that estimating these parameters by some method that is less sensitive to multicollinearity than ordinary least squares would be helpful. Several methods have been suggested for this. Hoerl and Kennard (1970*a,b*) have proposed ridge regression as an alternative to ordinary least squares. In ridge regression, the parameter estimates are obtained by solving

$$\boldsymbol{\beta}^*(l) = (\mathbf{X'X} + l\mathbf{I})^{-1}\mathbf{X'y} \tag{13-64}$$

where $l \geq 0$ is a constant. Generally, values of l in the interval $0 \leq l \leq 1$ are appropriate. The ridge estimator $\boldsymbol{\beta}^*(l)$ is not an unbiased estimator of $\boldsymbol{\beta}$, as is the ordinary least squares estimator $\hat{\boldsymbol{\beta}}$, but the mean square error of $\boldsymbol{\beta}^*(l)$ will be smaller than the mean square error of $\hat{\boldsymbol{\beta}}$. Thus ridge regression seeks to find a set of regression coefficients that is more "stable," in the sense of having a small mean square error. Since multicollinearity usually results in ordinary least squares estimators that may have extremely large variances, ridge regression is suitable for situations where the multicollinearity problem exists.

To obtain the ridge regression estimator from Equation (13-64), one must specify a value for the constant l. Generally, there is an "optimum" l for any problem, but the most simple approach is to solve Equation (13-64) for several values of l in the interval $0 \leq l \leq 1$. Then a plot of the values of $\boldsymbol{\beta}^*(l)$ against l is constructed. This display is called the *ridge trace*. The appropriate value of l is chosen subjectively by inspection of the ridge trace. Typically, a value for l is chosen such that relatively stable parameter estimates are obtained. In general, the variance of $\boldsymbol{\beta}^*(l)$ is a decreasing function of l, while the squared bias $[\boldsymbol{\beta} - \boldsymbol{\beta}^*(l)]^2$ is an increasing function of l. Choosing the value of l involves involves trading off these two properties of $\boldsymbol{\beta}^*(l)$.

A good discussion of the practical use of ridge regression is in Marquardt and Snee (1975). Also, there are several other biased estimation techniques that have been proposed for dealing with multicollinearity. Several of these are discussed by Hocking, Speed, and Lynn (1976).

TABLE 13-12 **Data for Example 13-14**

Observation Number	y	z_1	z_2	z_3	z_4
1	28.25	10	31	5	45
2	24.80	12	35	5	52
3	11.86	5	15	3	24
4	36.60	17	42	9	65
·5	15.80	8	6	5	19
6	16.23	6	17	3	25
7	29.50	12	36	6	55
8	28.75	10	34	5	50
9	43.20	18	40	10	70
10	38.47	23	50	10	80
11	10.14	16	37	5	61
12	38.92	20	40	11	70
13	36.70	15	45	8	68
14	15.31	7	22	2	30
15	8.40	9	12	3	24

● **Example 13-14.** The heat generated in calories per gram for a particular type of cement as a function of the quantities of four additives (z_1, z_2, z_3, and z_4) is shown in Table 13-12. We wish to fit a multiple linear regression model to these data.

The data will be coded by defining a new set of regressor variables as

$$x_{ij} = \frac{z_{ij} - \bar{z}_j}{\sqrt{S_{jj}}} \qquad i = 1, 2, \ldots, 15 \qquad j = 1, 2, 3, 4$$

where $S_{jj} = \sum_{i=1}^{n}(z_{ij} - \bar{z}_j)^2$ is the corrected sum of squares of the levels of z_j. The coded data are shown in Table 13-13. This transformation makes the intercept orthogonal to the other regression coefficients, since the first column of the **X** matrix consists of ones. Therefore, the intercept in this model will always be estimated by $\bar{y}$. The (4×4) **X'X** matrix for the four coded variables is the correlation matrix

$$\mathbf{X'X} = \begin{bmatrix} 1.00000 & .84894 & .91412 & .93367 \\ .84894 & 1.00000 & .76899 & .97567 \\ .91412 & .76899 & 1.00000 & .86784 \\ .93367 & .97567 & .86784 & 1.00000 \end{bmatrix}$$

This matrix contains several large correlation coefficients, and this may

TABLE 13-13   **Coded Data for Example 13-14**

Observation Number	y	x_1	x_2	x_3	x_4
1	28.25	−.12515	.00405	−.09206	−.05538
2	24.80	−.02635	.08495	−.09206	.03692
3	11.86	−.37217	−.31957	−.27617	−.33226
4	36.60	.22066	.22653	+.27617	.20832
5	15.80	−.22396	−.50161	−.09206	−.39819
6	16.23	−.32276	−.27912	−.27617	−.31907
7	29.50	−.02635	.10518	.00000	.07647
8	28.75	−.12515	.06472	−.09206	.01055
9	43.20	.27007	.18608	.36823	.27425
10	38.47	.51709	.38834	.36823	.40609
11	10.14	.17126	.12540	−.09206	.15558
12	38.92	.36887	.18608	.46029	.27425
13	36.70	.12186	.28721	.18411	.24788
14	15.31	−.27336	−.17799	−.36823	−.25315
15	8.40	−.17456	−.38025	−.27617	−.33226

indicate significant multicollinearity. The inverse of $X'X$ is

$$(X'X)^{-1} = \begin{bmatrix} 20.769 & 25.813 & -.608 & -44.042 \\ 25.813 & 74.486 & 12.597 & -107.710 \\ -.608 & 12.597 & 8.274 & -18.903 \\ -44.042 & -107.710 & -18.903 & 163.620 \end{bmatrix}$$

The variance inflation factors are the main diagonal elements of this matrix. Note that three of the variance inflation factors exceed 10, a good indication that multicollinearity is present. The eigenvalues of $X'X$ are $\lambda_1 = 3.657$, $\lambda_2 = .2679$, $\lambda_3 = .07127$, and $\lambda_4 = .004014$. Two of the eigenvalues, λ_3 and λ_4, are relatively close to zero. Also, the ratio of the largest to the smallest eigenvalue is

$$\frac{\lambda_{max}}{\lambda_{min}} = \frac{3.657}{.004014} = 911.06$$

which is considerably larger than 10. Therefore, since examination of the variance inflation factors and the eigenvalues indicates potential problems with multicollinearity, we will use ridge regression to estimate the model parameters.

Equation (13-64) was solved for various values of l, and the results are summarized in Table 13-14. The ridge trace is shown in Fig. 13-7. The

TABLE 13-14 **Ridge Regression Estimates for Example 13-14**

l	$\beta_1^*(l)$	$\beta_2^*(l)$	$\beta_3^*(l)$	$\beta_4^*(l)$
.000	−28.3318	65.9996	64.0479	−57.2491
.001	−31.0360	57.0244	61.9645	−44.0901
.002	−32.6441	50.9649	60.3899	−35.3088
.004	−34.1071	43.2358	58.0266	−24.3241
.008	−34.3195	35.1426	54.7018	−13.3348
.016	−31.9710	27.9534	50.0949	−4.5489
.032	−26.3451	22.0347	43.8309	1.2950
.064	−18.0566	17.2202	36.0743	4.7242
.128	−9.1786	13.4944	27.9363	6.5914
.256	−1.9896	10.9160	20.8028	7.5076
.512	2.4922	9.2014	15.3197	7.7224

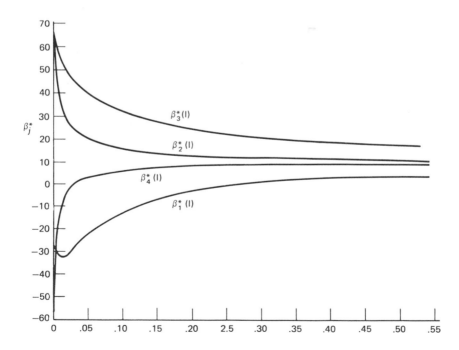

Fig. 13-7. Ridge Trace for Example 13-14.

instability of the least squares estimates $\beta_j^*(l = 0)$ is evident from inspection of the ridge trace. It is often difficult to choose a value of l from the ridge trace that simultaneously stabilizes the estimates of all regression coefficients. We will choose $l = .064$, which implies that the regression model is

$$\hat{y} = 25.53 - 18.0566x_1 + 17.2202x_2 + 36.0743x_3 + 4.7242x_4$$

using $\hat{\beta}_0 = \bar{y} = 25.53$. Converting the model to the original variables z_j, we have

$$\hat{y} = 2.9913 - .8920z_1 + .3483z_2 + 3.3209z_3 - .0623z_4$$

13-10.2   The Effect of Outermost Points in x-Space

When using multiple regression, we occasionally find that some of the observations have been taken under extreme conditions. That is, a few observations are relatively far away from the vicinity where the rest of the data were collected. A hypothetical situation for two variables is depicted in Fig. 13-8, where one observation in x-space is remote from the rest of the data. The disposition of points in the x-space is important in determining the properties of the model. For example, the point (x_{i1}, x_{i2}) in Fig. 13-8 may be very influential in determining the estimates of the regression coefficients, the value of R^2, and the value of MS_E.

We would like to examine data points that are remote in x-space to determine if they control many model properties. If these points taken under extreme conditions are "bad" points, or are erroneous in any way, then they should be eliminated. On the other hand, there may be nothing wrong with these remote points, but at least we would like to determine whether or not they produce results consistent with the rest of the data. In any event, even if the remote point is a valid one, if it controls important model properties, we

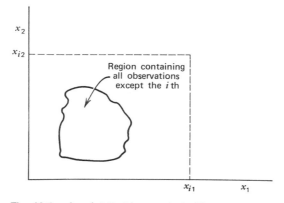

Fig. 13-8. A point that is remote in X-space.

would like to know this, since it could have an impact on the use of the model.

Usually, points remote in x-space cannot be easily located by examining the residuals. In fact, if a remote point does exert a great deal of influence over the model, its residual will usually be small. Daniel and Wood (1971) have suggested that the weighted sum of the squared distance of the ith point from the center of the data, say

$$D_i^2 = \sum_{j=1}^{k} \left[\frac{\hat{\beta}_j(x_{ij} - \bar{x}_j)}{\sqrt{MS_E}} \right]^2 \qquad i = 1, 2, \ldots, n \qquad (13\text{-}65)$$

be used to locate points that are remote in x-space. The general procedure is to fit the model, calculate D_i^2, $i = 1, 2, \ldots, n$, and then rank the observations in order of increasing values of D_i^2. Points that have large values of D_i^2 are

TABLE 13-15 Values of D_i^2 for the Soft Drink Delivery Time Data of Example 13-1

Observation No., i	D_i^2
10	.11
5	.12
2	1.56
9	2.44
19	5.76
4	5.88
24	8.83
12	11.10
3	12.30
25	15.24
13	21.19
16	26.00
11	26.05
15	26.34
6	26.38
14	56.04
7	56.07
8	58.38
1	58.39
21	65.79
20	65.94
18	77.33
22	87.58
23	112.40
17	201.10

remote in x-space. Once a remote point thas been located, its effect may be studied by refitting the model with the point deleted. If several points are remote, then several additional runs may be required to investigate their effects.

- **Example 13-15.** Table 13-15 lists the values of D_i^2 for the 25 data points for the soft drink delivery time study in Example 13-1. To illustrate the calculations, consider the first observation:

$$D_1^2 = \sum_{j=1}^{2} \left[\frac{\hat{\beta}_j(x_{1j} - \bar{x}_j)}{\sqrt{MS_E}} \right]^2$$
$$= \left[\frac{2.74427(2 - 8.24)}{\sqrt{5.2352}} \right]^2 + \left[\frac{.01253(50 - 331.76)}{\sqrt{5.2352}} \right]^2$$
$$= 58.39$$

The two most distant observations are 17 and 23, and 17 is nearly twice as remote as 23. This is confirmed by inspection of the data, from which we see that, for both points 17 and 23, x_1 and x_2 are at or near the upper limits of their range. Note that observation number 17 has one of the two largest residuals (see Table 13-3 and Example 13-9), and this increases our suspicions about this point. To investigate the effect of these two points on the model, three additional analyses were performed—one deleting observation 17, a second deleting observation 23, and the third deleting both 23 and 17. The results of these additional runs are shown in the following table.

	Regression Coefficients				
Run	$\hat{\beta}_0$	$\hat{\beta}_1$	$\hat{\beta}_2$	MS_E	R^2
17 and 23 in	2.26379	2.74427	.01253	5.2352	.981137
17 out	3.07952	2.65899	.01148	4.2767	.979781
23 out	2.30853	2.74043	.01244	5.4784	.978345
17 and 23 out	3.39961	2.62993	.01092	4.3268	.975613

Deleting observation 17 produces only minor changes in $\hat{\beta}_1$ and $\hat{\beta}_2$, but results in approximately a 36 percent change in $\hat{\beta}_0$. This could indicate that observation 17 lies somewhat off the plane that passes through the remaining 24 points, but it does not exert much influence over the orientation of the plane in the $x_1 - x_2$ space. Deleting observation 23 has no appreciable effect on the model, while deleting both 17 and 23 produces effects similar to those observed by deleting 17 alone. Thus, while point 17 is remote in x-space, it seems to exert little effect on the model's properties. This is somewhat confirmed by the large residual of point 17—if the point controlled model properties, its residual would likely be small. If observation 17 were deleted, a smaller mean square error would result. However, there are no legitimate reasons for its removal, such as an error in recording the data, and so observation 17 is retained. Observation 23 is retained also.

13-10.3 Autocorrelation

The regression models developed thus far have assumed that the model error components ϵ_i are uncorrelated random variables. Many applications of regression analysis involve data for which this assumption may be inappropriate. In regression problems where the dependent and independent variables are time-oriented or are time series data, the assumption of uncorrelated errors is often untenable. For example, suppose we regressed the quarterly sales of a product against the quarterly point-of-sale advertising expenditures. Both variables are time series, and if they are positively correlated with other factors such as disposable income and population size, which are not included in the model, then it is likely that the error terms in the regression model are positively correlated over time. Variables that exhibit correlation over time are referred to as *autocorrelated* variables. Many regression problems in economics, business, and agriculture involve autocorrelated errors.

The occurrence of positively autocorrelated errors has several potentially serious consequences. The ordinary least squares estimators of the parameters are affected in that they are no longer minimum variance estimators, although they are still unbiased. Furthermore, the mean square error MS_E may underestimate the error variance σ^2. Also, confidence intervals and tests of hypothesis, which are developed assuming uncorrelated errors, are not valid if autocorrelation is present.

There are several statistical procedures that can be used to determine if the error terms in the model are uncorrelated. We will describe one of these, the Durbin-Watson test. This test assumes that the data are generated by the *first-order autoregressive model*

$$y_t = \beta_0 + \beta_1 x_t + \epsilon_t \qquad t = 1, 2, \ldots, n \qquad (13\text{-}66)$$

where t is the index of time and the error terms are generated according to the process

$$\epsilon_t = \rho \epsilon_{t-1} + a_t \qquad (13\text{-}67)$$

where $|\rho| < 1$ is an unknown parameter and a_t is a NID$(0, \sigma^2)$ random variable. Equation (13-66) is a simple linear regression model, except for the errors, which are generated from Equation (13-67). The parameter ρ in Equation (13-67) is the autocorrelation coefficient. The Durbin-Watson test

TABLE 13-16 **Critical Values of the Durbin-Watson Statistic**

Sample Size	Probability in Lower Tail (Significance Level = α)	k = Number of Regressors (Excluding the Intercept)									
		1		2		3		4		5	
		D_L	D_U	D_L	D_U	D_L	D_U	D_L	D_U	D_L	D_U
15	.01	.81	1.07	.70	1.25	.59	1.46	.49	1.70	.39	1.96
	.025	.95	1.23	.83	1.40	.71	1.61	.59	1.84	.48	2.09
	.05	1.08	1.36	.95	1.54	.82	1.75	.69	1.97	.56	2.21
20	.01	.95	1.15	.86	1.27	.77	1.41	.63	1.57	.60	1.74
	.025	1.08	1.28	.99	1.41	.89	1.55	.79	1.70	.70	1.87
	.05	1.20	1.41	1.10	1.54	1.00	1.68	.90	1.83	.79	1.99
25	.01	1.05	1.21	.98	1.30	.90	1.41	.83	1.52	.75	1.65
	.025	1.13	1.34	1.10	1.43	1.02	1.54	.94	1.65	.86	1.77
	.05	1.20	1.45	1.21	1.55	1.12	1.66	1.04	1.77	.95	1.89
30	.01	1.13	1.26	1.07	1.34	1.01	1.42	.94	1.51	.88	1.61
	.025	1.25	1.38	1.18	1.46	1.12	1.54	1.05	1.63	.98	1.73
	.05	1.35	1.49	1.28	1.57	1.21	1.65	1.14	1.74	1.07	1.83
40	.01	1.25	1.34	1.20	1.40	1.15	1.46	1.10	1.52	1.05	1.58
	.025	1.35	1.45	1.30	1.51	1.25	1.57	1.20	1.63	1.15	1.69
	.05	1.44	1.54	1.39	1.60	1.34	1.66	1.29	1.72	1.23	1.79
50	.01	1.32	1.40	1.28	1.45	1.24	1.49	1.20	1.54	1.16	1.59
	.025	1.42	1.50	1.38	1.54	1.34	1.59	1.30	1.64	1.26	1.69
	.05	1.50	1.59	1.46	1.63	1.42	1.67	1.38	1.72	1.34	1.77
60	.01	1.38	1.45	1.35	1.48	1.32	1.52	1.28	1.56	1.25	1.60
	.025	1.47	1.54	1.44	1.57	1.40	1.61	1.37	1.65	1.33	1.69
	.05	1.55	1.62	1.51	1.65	1.48	1.69	1.44	1.73	1.41	1.77
80	.01	1.47	1.52	1.44	1.54	1.42	1.57	1.39	1.60	1.36	1.62
	.025	1.54	1.59	1.52	1.62	1.49	1.65	1.47	1.67	1.44	1.70
	.05	1.61	1.66	1.59	1.69	1.56	1.72	1.53	1.74	1.51	1.77
100	.01	1.52	1.56	1.50	1.58	1.48	1.60	1.45	1.63	1.44	1.65
	.025	1.59	1.63	1.57	1.65	1.55	1.67	1.53	1.70	1.51	1.72
	.05	1.65	1.69	1.63	1.72	1.61	1.74	1.59	1.76	1.57	1.78

Source: Adapted from *Econometrics*, by R. J. Wonnacott and T. H. Wonnacott, John Wiley & Sons, New York, 1970, with permission of the publisher.

can be applied to the hypotheses

$$H_0 : \rho = 0$$

$$H_1 : \rho > 0 \tag{13-68}$$

Note that if $H_0 : \rho = 0$ is not rejected, we are implying that there is no auto-correlation in the errors, and the ordinary linear regression model is appropriate.

To test $H_0: \rho = 0$, first fit the regression model by ordinary least squares. Then, calculate the Durbin-Watson test statistic

$$D = \frac{\sum_{t=2}^{n} (e_t - e_{t-1})^2}{\sum_{t=1}^{n} e_t^2} \qquad (13\text{-}69)$$

where e_t is the tth residual. For a suitable value of α, obtain the critical values $D_{\alpha, U}$ and $D_{\alpha, L}$ from Table 13-16. If $D > D_{\alpha, U}$, do not reject $H_0: \rho = 0$; but if $D < D_{\alpha, L}$, reject $H_0: \rho = 0$ and conclude that the errors are positively autocorrelated. If $D_{\alpha, L} \le D \le D_{\alpha, U}$, the test is inconclusive. When the test is inconclusive, the implication is that more data must be collected. In many problems this is difficult to do.

To test for *negative* autocorrelation, that is, if the alternative hypothesis in Equation (13-68) is $H_1: \rho < 0$, then use $D' = 4 - D$ as the test statistic, where D is defined in Equation (13-69). If a two-sided alternative is specified, then use both of the one-sided procedures, noting that the type I error for the two-sided test is 2α, where α is the type I error for the one-sided tests.

The only effective remedial measure when autocorrelation is present is to build a model that accounts explicitly for the autocorrelative structure of the errors. For an introductory treatment of these models, refer to Neter and Wasserman (1974).

13-11 Selection of Variables in Multiple Regression

13-11.1 The Model-Building Problem

An important problem in many applications of regression analysis is the selection of the set of independent or regressor variables to be used in the model. Sometimes previous experience or underlying theoretical considerations can help the analyst specify the set of independent variables. Usually, however, the problem of selecting an appropriate set of independent variables is more complex. Analysts will generally have a set of *candidate* independent variables that quite likely includes all of the important variables, but they are not sure that *all* of these candidate variables are necessary to adequately model the dependent variable y.

In such a situation, we are interested in screening the candidate variables to obtain a regression model that contains the "best" subset of independent variables. We would like for the final model to contain enough independent variables so that in the intended use of the model (prediction, for example) it will perform satisfactorily. On the other hand, to keep model maintenance costs to a minimum, we would like the model to use as few independent variables as possible. The compromise between these conflicting objectives is often called finding the "best" regression equation. However, in most problems, there is no single regression model that is "best" in terms of the various evaluation criteria that have been proposed. A great deal of judgment and experience with the system being modeled is usually necessary to select an appropriate set of independent variables for a regression equation.

There is no algorithm that will always produce a good solution to the variable selection problem. Most of the currently available procedures are search techniques. To perform satisfactorily, they require interaction with and judgment by the analyst. We now briefly discuss some of the more popular variable selection techniques.

13-11.2   Computational Procedures for Variable Selection

We assume that there are k candidate variables, $x_1, x_2, \ldots, x_k$, and a single dependent variable y. All models will include an intercept term β_0, so that the model with *all* variables included would have $k + 1$ terms. Furthermore, the functional form of each candidate variable (for example, $x_1 = 1/x_1$, $x_2 = \ln x_2$, etc.) is correct.

All Possible Regressions.   This approach requires that the analyst fit all of the regression equations involving one candidate variable, all regression equations involving two candidate variables, and so on. Then these equations are evaluated according to some suitable criteria to select the "best" regression model. If there are k candidate variables, there are 2^k total equations to be examined. For example, if $k = 4$, there are $2^4 = 16$ possible regression equations; while if $k = 10$, there are $2^{10} = 1024$ possible regression equations. Hence, the number of equations to be examined increases rapidly as the number of candidate variables increases.

There are a number of criteria that may be used for evaluating and comparing the different regression models obtained. Perhaps the most commonly used criterion is based on the coefficient of multiple determination. Let R_p^2 denote the coefficient of determination for a regression model with p terms, that is $p - 1$ candidate variables and an intercept term (note that $p \leq k + 1$). Computationally, we have

$$R_p^2 = \frac{SS_R(p)}{S_{yy}} = 1 - \frac{SS_E(p)}{S_{yy}} \tag{13-70}$$

where $SS_R(p)$ and $SS_E(p)$ denote the regression sum of squares and the error sum of squares, respectively, for the p-variable equation. Now R_p^2 increases as p increases and is a maximum when $p = k + 1$. Therefore, the analyst uses this criterion by adding variables to the model up to the point where an

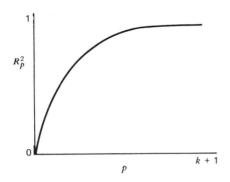

Fig. 13-9. Plot of R_p^2 against p.

additional variable is not useful in that it gives only a small increase in R_p^2. The general approach is illustrated in Fig. 13-9, which gives a hypothetical plot of R_p^2 against p. Typically, one examines a display such as this and chooses the number of variables in the model as the point at which the "knee" in the curve becomes apparent. Clearly, this requires judgment on the part of the analyst.

A second criterion is to consider the mean square error for the p-variable equation, say $MS_E(p) = SS_E(p)/(n - p)$. Generally, $MS_E(p)$ decreases as p increases, but this is not necessarily so. If the addition of a variable to the model with $p - 1$ terms does not reduce the error sum of squares in the new p term model by an amount equal to the error mean square in the old $p - 1$ term model, $MS_E(p)$ will *increase*, because of the loss of one degree of freedom for error. Therefore, a logical criteria is to select p as the value that minimizes $MS_E(p)$, or since $MS_E(p)$ is usually relatively flat in the vicinity of the minimum, we could choose p such that adding more variables to the model produces only very small reductions in $MS_E(p)$. The general procedure is illustrated in Fig. 13-10.

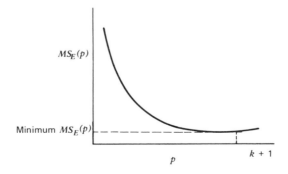

Fig. 13-10. Plot of $MS_E(p)$ against p.

A third criterion is the C_p statistic, which is a measure of the total mean square error for the regression model. We define the total standardized mean squared error as

$$\Gamma_p = \frac{1}{\sigma^2} \sum_{i=1}^{n} E(y_i - \hat{y}_i)^2$$

$$= \frac{1}{\sigma^2}\left[\sum_{i=1}^{n} \{E(y_i) - E(\hat{y}_i)\}^2 + \sum_{i=1}^{n} V(\hat{y}_i) \right]$$

$$= \frac{1}{\sigma^2}[(\text{bias})^2 + \text{variance}]$$

We use the mean square error from the *full $k + 1$ term* model as an estimate of σ^2; that is, $\hat{\sigma}^2 = MS_E(k + 1)$. An estimator of Γ_p is

$$C_p = \frac{SS_E(p)}{\hat{\sigma}^2} - n + 2p \tag{13-71}$$

If the p-term model has negligible bias, then it can be shown that

$$E(C_p | \text{zero bias}) = p$$

Therefore, the values of C_p for each regression model under consideration should be plotted against p. The regression equations that have negligible bias will have values of C_p that fall near the line $C_p = p$, while those with significant bias will have values of C_p that plot above this line. One then chooses as the "best" regression equation either a model with minimum C_p or a model with a slightly larger C_p that does not contain as much bias (i.e., $C_p \simeq p$) as the minimum.

Another criterion is based on a modification of R_p^2 that accounts for the number of variables in the model. This statistic is called the *adjusted R_p^2* defined as

$$\bar{R}_p^2 = 1 - \frac{n-1}{n-p}(1 - R_p^2) \tag{13-72}$$

Note that $\bar{R}_p^2$ may decrease as p increases if the decrease in $(n - 1)(1 - R_p^2)$ is not compensated for by the loss of one degree of freedom in $n - p$. The experimenter would usually select the regression model that has the maximum value of $\bar{R}_p^2$. However, note that this is equivalent to the model that minimizes $MS_E(p)$, since

$$\bar{R}_p^2 = 1 - \left(\frac{n-1}{n-p}\right)(1 - R_p^2)$$

$$= 1 - \left(\frac{n-1}{n-p}\right)\frac{SS_E(p)}{S_{yy}}$$

$$= 1 - \left(\frac{n-1}{S_{yy}}\right)MS_E(p)$$

● **Example 13-16.** The data in Table 13-17 are an expanded set of data for the soft drink delivery time study in Example 13-1. There are now four candidate variables, delivery volume (x_1), distance (x_2), the number of vending machines at the outlet (x_3), and the number of different machine locations (x_4).

Table 13-18 presents the results of running all possible regressions (except the trivial model with only an intercept) on these data. The values of R_p^2, $\bar{R}_p^2$, $SS_R(p)$, $SS_E(p)$, $MS_E(p)$, and C_p are also shown in this table. A plot of R_p^2 against p is shown in Fig. 13-11. In terms of R^2 improvement, there is very little gain in going from a two-variable model to a three-variable model. A plot of the minimum $MS_E(p)$ for each subset of size p is shown in Fig. 13-12. Several models have good values of $MS_E(p)$. The best two-variable model is either (x_1, x_2) or (x_1, x_4), and the best three-variable model is (x_1, x_2, x_4). The

TABLE 13-17 Soft Drink Delivery Time Data for Example 13-16

Observation	Delivery Time y	Number of Cases x_1	Distance x_2	Number of Machines x_3	Number of Machine Locations x_4
1	9.95	2	50	1	1
2	24.45	8	110	1	1
3	31.75	11	120	2	1
4	35.00	10	550	2	2
5	25.02	8	295	1	1
6	16.86	4	200	1	1
7	14.38	2	375	1	1
8	9.60	2	52	1	1
9	24.35	9	100	1	1
10	27.50	8	300	2	1
11	17.08	4	412	2	2
12	37.00	11	400	3	2
13	41.95	12	500	3	3
14	11.66	2	360	1	1
15	21.65	4	205	2	2
16	17.89	4	400	2	1
17	69.00	20	600	4	4
18	10.30	1	585	1	1
19	34.93	10	540	2	1
20	46.59	15	250	3	2
21	44.88	15	290	3	1
22	54.12	16	510	3	3
23	56.63	17	590	2	2
24	22.13	6	100	2	1
25	21.15	5	400	1	1

TABLE 13-18 All Possible Regressions for the Data in Example 13-16

Number of Variables in Model	p	Variables in model	R_p^2	$SS_R(p)$	$SS_E(p)$	$MS_E(p)$	$\bar{R}_p^2$	C_p
1	2	x_2	.24291747	1483.2406	4622.7041	200.9871	.2100	1479.93
1	2	x_4	.56007492	3419.7865	2686.1582	116.7895	.5409	851.16
1	2	x_3	.69830969	4263.8404	1842.1043	80.0915	.6852	577.11
1	2	x_1	.96395437	5885.8521	220.0926	9.5692	.9624	50.46
2	3	x_2, x_4	.57158123	3490.0434	2615.9013	118.9046	.5326	830.35
2	3	x_2, x_3	.72162460	4406.1999	1699.7448	77.2611	.6963	532.88
2	3	x_3, x_4	.72232683	4410.4877	1695.4570	77.0662	.6971	531.49
2	3	x_1, x_3	.97220237	5936.2139	169.7308	7.7150	.9697	36.11
2	3	x_1, x_2	.98113748	5990.7712	115.1735	5.2352	.9794	18.40
2	3	x_1, x_4	.98302963	6002.3246	103.6201	4.7100	.9815	14.64
3	4	x_2, x_3, x_4	.73299076	4475.6010	1630.3437	77.6354	.6948	512.35
3	4	x_1, x_3, x_4	.98327895	6003.8469	102.0978	4.8618	.9809	16.15
3	4	x_1, x_2, x_3	.98517507	6015.4245	90.5203	4.3105	.9831	12.39
3	4	x_1, x_2, x_4	.98961493	6042.5340	63.4107	3.0196	.9881	3.59
4	5	x_1, x_2, x_3, x_4	.98991196	6044.3477	61.5970	3.0798	.9879	5

445

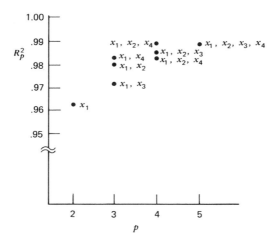

Fig. 13-11. The R_p^2 plot for Example 13-16.

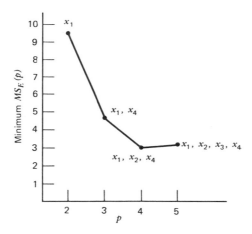

Fig. 13-12. The $MS_E(p)$ plot for Example 13-16.

minimum values of $MS_E(p)$ occur for the three-variable model (x_1, x_2, x_4). While there are several other models that have relatively small values of $MS_E(p)$, such as (x_1, x_2, x_3) and (x_1, x_2), the model (x_1, x_2, x_4) is definitely superior with respect to the $MS_E(p)$ criterion. Note that, as expected, this model also maximizes the adjusted $\bar{R}_p^2$. A C_p plot is shown in Fig. 13-13. The only model with $C_p \leq p$ is the three-variable equation in (x_1, x_2, x_4). To

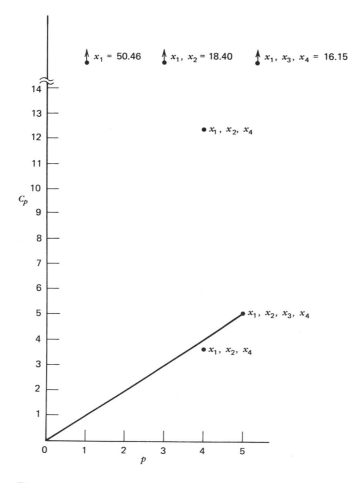

Fig. 13-13. The C_p plot for Example 13-16.

illustrate the calculations, for this equation we would find that

$$C_p = \frac{SS_E(p)}{\hat{\sigma}^2} - n + 2p$$

$$= \frac{63.4107}{3.0799} - 25 + 2(4) = 3.59$$

noting that $\hat{\sigma}^2 = 3.0799$ is obtained from the *full* equation (x_1, x_2, x_3, x_4). Since all other models contain substantial bias, we would conclude on the basis of the C_p criterion that the best subset of the regressor variable is (x_1, x_2, x_4). Since this model also results in a minimum $MS_E(p)$ and a high R_p^2, we would select it as the "best" regression equation. The final model is

$$\hat{y} = 1.36707 + 2.53492x_1 + .00852x_2 + 2.59928x_4$$

The all-possible regressions approach requires considerable computational effort, even when k is moderately small. However, if the analyst is willing to look at something less than the estimated model and all of its associated statistics, it is possible to devise algorithms for all possible regressions that produce less information about each model but which are more efficient computationally. For example, suppose that we could efficiently calculate only the MS_E for each model. Since models with large MS_E are not likely to be selected as the best regression equations, we would only then have to examine in detail the models with small values of MS_E. There are several approaches to developing a computationally efficient algorithm for all possible regressions; for example, see Furnival and Wilson (1974). Even with computational refinements, however, currently only problems with up to about 30 candidate variables can be investigated by this approach.

Directed Search on t.   This variable selection procedure is based on using the t statistic for an individual variable, Equation (13-32), from the *full* model containing all k candidate variables. In general, the t statistic is a good indicator of the contribution of an individual variable to the regression sum of squares. The general procedure consists of ranking the candidate variables in order of decreasing $|t|$. Then variables are added to the model, one at a time in order of decreasing $|t|$ until $|t|$ is less than some predetermined value. The C_p statistic is generally used to evaluate each model. This will produce a subset of the original candidate variables that are to be automatically included in the model. Then a search is carried out on all combinations of the remaining variables. This procedure often leads to several regression models, all of which adequately fit the data. When there is one "best" equation, the directed search on t will generally find it with much less computation than would be required by the all-possible regressions approach.

● **Example 13-17.**   To illustrate the directed search on t method for the soft drink delivery time data in Table 13-17, we must first fit the full model in all four variables. This gives the following results.

Variable	Regression Coefficient	t Statistic
x_1	2.48423	22.57
x_2	.00855	3.63
x_3	.62237	.77
x_4	2.29392	3.06

Since the largest values of t (in descending order) are for variables x_1, x_2, and x_4, this would imply that we need to build three additional regression models: one with x_1, one with x_1 and x_2, and one with x_1, x_2, and x_4. This results in the following data.

Variables in Model	C_p
x_1	50.46
x_1, x_2	18.40
x_1, x_2, x_4	3.59

Note that the third equation gives the minimum value of C_p. If we consider x_1, x_2, and x_4 to be the set of variables automatically included in the model, then the search of all possible combinations of the remaining variables (x_3 only) is trivial, and we would conclude that the three-variable model in x_1, x_2, and x_4 is the best regression equation. Note that this equation has been found by fitting only four equations. Even if we had only automatically retained x_1, only eight additional runs would have been required to investigate all possible combinations of x_2, x_3, and x_4. This would have produced the same final equation.

Stepwise Regression.   This is probably the most widely used variable selection technique. The procedure iteratively constructs a sequence of regression models by adding or removing variables at each step. The criterion for adding or removing a variable at any step is usually expressed in terms of a partial F-test. Let F_{in} be the value of the F statistic for adding a variable to the model, and let F_{out} be the value of the F statistic for removing a variable from the model. We must have $F_{in} \geq F_{out}$, and usually $F_{in} = F_{out}$.

Stepwise regression begins by forming a one-variable model using the regressor variable that has the highest correlation with the response variable y. This will also be the variable producing the largest F statistic. If no F statistic exceeds F_{in}, the procedure terminates. For example, suppose that at this step, x_1 is selected. At the second step, the remaining $k - 1$ candidate variables are examined, and the variable for which the statistic

$$F_j = \frac{SS_R(\beta_j | \beta_1, \beta_0)}{MS_E(x_j, x_1)} \tag{13-73}$$

is a maximum is added to the equation, provided that $F_j > F_{in}$. In Equation (13-73), $MS_E(x_j, x_1)$ denotes the mean square for error for the model containing both x_1 and x_j. Suppose that this procedure now indicates that x_2 should be added to the model. Now the stepwise regression algorithm determines whether the variable x_1 added at the first step should be removed. This is done by calculating the F statistic

$$F_1 = \frac{SS_R(\beta_1 | \beta_2, \beta_0)}{MS_E(x_1, x_2)} \tag{13-74}$$

If $F_1 < F_{out}$, the variable x_1 is removed.

In general, at each stage the set of remaining candidate variables are examined and the variable with the largest partial F statistic is entered, provided that the observed value of F exceeds F_{in}. Then the partial F statistic for each variable in the model is calculated, and the variable with the smallest observed value of F is deleted if the observed $F < F_{out}$. The procedure continues until no other variables can be added to or removed from the model.

Stepwise regression is usually performed using a computer program. The analyst exercises control over the procedure by the choice of F_{in} and F_{out}. Some stepwise regression computer programs require that numerical values be specified for F_{in} and F_{out}. Since the number of degrees of freedom on MS_E depend on the number of variables in the model, which changes from step to step, a fixed value of F_{in} and F_{out} causes the type I and type II error rates to vary. Some computer programs allow the analyst to specify the type I error levels for F_{in} and F_{out}. However, the "advertised" significance level is not the true level, because the variable selected is the one that maximizes the partial F statistic at that stage. Sometimes it is useful to experiment with different values of F_{in} and F_{out} (or different advertised type I error rates) in several runs to see if this substantially affects the choice of the final model.

● **Example 13-18.** We will apply stepwise regression to the soft drink delivery time data in Table 13-17. Instead of specifying numerical values of F_{in} and F_{out}, we use an advertised type I error of $\alpha = .10$. The first step consists of building a simple linear regression model using the variable that gives the largest F statistic. This is x_1, and since

$$F_1 = \frac{SS_R(\beta_1 | \beta_0)}{MS_E(x_1)} = \frac{5885.8521}{9.5692} = 615.08 > F_{in} = F_{.10,1,23} = 2.94$$

x_1 is entered into the model.

The second step begins by finding the variable x_j that has the largest partial F statistic, given that x_1 is in the model. This is x_4, and since

$$F_4 = \frac{SS_R(\beta_4 | \beta_1, \beta_0)}{MS_E(x_4, x_1)} = \frac{116.4725}{4.7100} = 29.73 > F_{in} = F_{.10,1,22} = 2.95$$

x_4 is added to the model. Now the procedure evaluates whether or not x_1

should be retained, given that x_4 is in the model. This involves calculating

$$F_1 = \frac{SS_R(\beta_1 \mid \beta_4, \beta_0)}{MS_E(x_4, x_1)} = \frac{2582.5381}{4.7100} = 548.31 > F_{\text{out}} = F_{.10,1,22} = 2.95$$

Therefore x_1 should be retained. Step 2 terminates with both x_1 and x_4 in the model.

The third step finds the next variable for entry as x_2. Since

$$F_2 = \frac{SS_R(\beta_2 \mid \beta_1, \beta_4, \beta_0)}{MS_E(x_2, x_4, x_1)} = \frac{40.2094}{3.0196} = 13.32 > F_{\text{in}} = F_{.10,1,21} = 2.96$$

x_2 is added to the model. Partial F-tests on x_1 (given x_2 and x_4) and x_4 (given x_2 and x_1) indicate that these variables should be retained. Therefore, the third step concludes with the variables x_1, x_2, and x_4 in the model.

At the fourth step, the procedure attempts to add x_3. However, the partial F statistic for x_3 is $F_3 = .59$, which is less than $F_{\text{in}} = F_{.10,1,20} = 2.97$; therefore, since this variable cannot be added and since there are no other candidate variables to consider, the procedure terminates.

The stepwise regression procedure would conclude that the three-variable model in x_1, x_2, and x_4 is the best regression equation. However, the usual checks of model adequacy, such as residual analysis and C_p plots, should be applied to the equation. Note also that this is the same regression model found by both the all-possible regressions method and the directed search on t methods.

Forward Selection. This variable selection procedure is based on the principle that variables should be added to the model one at a time until there are no remaining candidate variables that produce a significant increase in the regression sum of squares. That is, variables are added one at a time as long as $F > F_{\text{in}}$. Forward selection is a simplification of stepwise regression that omits the partial F-test for deleting variables from the model that have been added at previous steps. This is a potential weakness of forward selection; the procedure does not explore the effect that adding a variable at the current step has on variables added at earlier steps.

● **Example 13-19.** Application of the forward selection algorithm to the soft drink delivery time data in Table 13-17 would begin by adding x_1 to the model. Then, the variable that induces the largest partial F-test, given that x_1 is in the model, is added—this is variable x_4. The third step enters x_2, which produces the largest partial F statistic, given that x_1 and x_4 are in the model. Since the partial F statistic for x_3 is not significant, the procedure terminates. Note that forward selection leads to the same final model as stepwise regression. This is not always the case.

Backward Elimination. This algorithm begins with all k candidate variables in the model. Then, the variable with the smallest partial F statistic is deleted, if this F statistic is insignificant, that is, if $F < F_{out}$. Next, the model with $k - 1$ variables is estimated, and the next variable for potential elimination is found. The algorithm terminates when no further variables can be deleted.

● **Example 13-20.** To apply backward elimination to the data in Table 13-17, we begin by estimating the full model in all four variables. This model is

$$\hat{y} = 1.06813 + 2.48423x_1 + .00855x_2 + .62237x_3 + 2.29392x_4$$

The partial F-tests for each variable are:

$$F_1 = \frac{SS_R(\beta_1 \mid \beta_2, \beta_3, \beta_4, \beta_0)}{MS_E} = 509.36$$

$$F_2 = \frac{SS_R(\beta_2 \mid \beta_1, \beta_3, \beta_4, \beta_0)}{MS_E} = 13.15$$

$$F_3 = \frac{SS_R(\beta_3 \mid \beta_1, \beta_2, \beta_4, \beta_0)}{MS_E} = .59$$

$$F_4 = \frac{SS_R(\beta_4 \mid \beta_1, \beta_2, \beta_3, \beta_0)}{MS_E} = 9.39$$

Clearly x_3 does not contribute significantly to the fit and should be deleted. The three-variable model in (x_1, x_2, x_4) has all variables significant by the partial F-test criterion and, consequently, the algorithm terminates. Note that backward elimination has resulted in the same model that was found by forward selection and stepwise regression. This may not always happen.

Some Comments on Final Model Selection. We have illustrated several different approaches to the selection of variables in multiple linear regression. The final model obtained from any model-building procedure should be subjected to the usual adequacy checks, such as residual analysis, lack-of-fit testing, and examination of the effects of outermost points. The analyst may also consider augmenting the original set of candidate variables with cross products, polynomial terms, or other transformations of the original variables that might improve the model.

A major criticism of variable selection methods, such as stepwise regression, is that the analyst may conclude that there is one "best" regression equation. This generally is not the case, because there are often several equally good regression models that can be used. One way to avoid this problem is to use several different model-building techniques and see if different models result. For example, we have found the same model for the soft drink delivery time data by using stepwise regression, forward selection, and backward elimination. This is a good indication that the three-variable model is the best regression equation. The same model was also found from all possible regressions and the directed search on t. The directed search on t method often works well in identifying good alternative models. Furthermore, there are variable selection techniques that are designed to find the best one-variable model, the best two-variable model, and so forth. For a good discussion of these methods, and the variable selection problem in general, see Hocking (1976).

13-12   Summary

This chapter has introduced multiple linear regression, including least squares estimation of the parameters, interval estimation, prediction of new observations, and methods for hypothesis testing. Various tests of model adequacy, including residual plots, have been discussed. An extension of the lack-of-fit test to multiple regression, using pairs of points that are near neighbors to obtain an estimate of pure error, was given. It was shown that polynomial regression models can be handled by the usual multiple linear regression methods. Indicator variables were introduced for dealing with qualitative variables. It also was observed that the problem of multicollinearity, or intercorrelation between the regressor variables, can greatly complicate the regression problem and often leads to a regression model that may not predict new observations well. Several causes and remedial measures of this problem including biased estimation techniques, were discussed. Finally, the variable selection problem in multiple regression was introduced. A number of model-building procedures, including all possible regressions, directed search on t, stepwise regression, forward selection, and backward elimination, were illustrated.

13-3   Exercises

13-1   Consider the soft drink delivery time data in Table 13-17.
  (a) Fit a regression model using x_1 (delivery volume) and x_4 (number of machine locations) to these data.
  (b) Test for significance of regression.
  (c) Compute the residuals from this model. Analyze these residuals using the methods discussed in Section 13-6.2.
  (d) How does this two-variable model compare with the two-variable model in x_1 and x_2 from Example 13-1?

13-2.   Consider the soft drink delivery time data in Table 13-17.
  (a) Fit a regression model using x_1 (delivery volume), x_2 (distance), and x_3 (number of machines) to these data.

(b) Test for significance of regression.

(c) Compute the residuals from this model. Analyze these residuals using the methods discussed in Section 13-6.2.

13-3. Using the results of Exercise 13-1, find a 95 percent confidence interval on β_4.

13-4. Using the results of Exercise 13-2, find a 95 percent confidence interval on β_3.

13-5. The tensile strength of a synthetic fiber is influenced by the drying time (x_1), the drying temperature (x_2), and the percentage of cotton in the fiber (x_3). The data are shown in the following table.

y	x_1	x_2	x_3
213	2.0	145	13
220	2.3	140	15
216	2.3	140	15
234	2.5	146	18
230	3.0	138	20
235	3.4	135	19
238	3.4	135	19
230	3.4	135	19
236	4.0	141	16
231	4.0	141	16
243	4.1	145	17

(a) Fit a multiple regression model to the data.

(b) Test for significance of regression and lack of fit.

(c) Construct a 90 percent confidence interval on the mean response at $x_1 = 2.5$, $x_2 = 140$, and $x_3 = 16$.

(d) Construct a 95 percent confidence interval on β_1.

(e) Construct a 90 percent prediction interval on a new observation of strength at the point $x_1 = 2.5$, $x_2 = 140$, and $x_3 = 16$.

13-6. The time required for an operator to move a workpiece is related to the distance of the move (x_1) and the weight of the workpiece (x_2). The industrial engineering department has obtained the following data.

y(s)	x_1(ft)	x_2(lb)
2.03	1.50	.85
2.30	2.25	.69
2.40	2.46	1.90
1.96	1.50	.85
2.16	2.25	1.86
2.43	2.40	.98
2.21	2.40	.98
2.15	1.75	1.02
2.07	1.75	1.02

(a) Fit a multiple regression model to the data.

(b) Test for significance of regression and lack of fit.

(c) Use the t statistic to test $H_0: \beta_1 = 0$.

(d) Use the partial F statistic to test $H_0: \beta_1 = 0$. Also use the partial F statistic to test $H_0: \beta_2 = 0$. Are both variables necessary in this model?

(e) Construct a 95 percent confidence interval on β_1.

13-7. The time to failure of a machine component is related to the operating voltage (x_1), the motor speed in revolutions per minute (x_2), and the operating temperature (x_3). A designed experiment is run in the research and development laboratory, and the following data are obtained:

y	x_1	x_2	x_3
2145	110	750	140
2155	110	750	180
2220	110	1000	140
2225	110	1000	180
2260	120	750	140
2266	120	750	180
2334	120	1000	140
2340	120	1000	180

(a) Fit a multiple regression model to these data.

(b) Test for significance of regression.

(c) Do all three independent variables contribute significantly to the model?

(d) Find a 95 percent confidence interval estimate of β_1.

13-8. Suppose that in Exercise 13-7 the following additional data had been collected.

y	x_1	x_2	x_3
2175	115	875	160
2186	115	875	160
2177	115	875	160
2170	115	875	160
2181	115	875	160

(a) How do these additional runs change the estimates of the regression coefficients in Exercise 13-7?

(b) Incorporate a test for lack of fit into the analysis.

13-9. The electric power consumed each month by a chemical plant is thought to be related to the average ambient temperature (x_1), the number of days in the month (x_2), the average product purity (x_3), and the tons of product produced (x_4). The past year's historical data is available and is presented in the following table.

y	x_1	x_2	x_3	x_4
240	25	24	91	100
236	31	21	90	95
290	45	24	88	110
274	60	25	87	88
301	65	25	91	94
316	72	26	94	99
300	80	25	87	97
296	84	25	86	96
267	75	24	88	110
276	60	25	91	105
288	50	25	90	100
261	38	23	89	98

(a) Fit a multiple regression model to these data.
(b) Test for significance of regression.
(c) Use partial F statistics to test $H_0: \beta_3 = 0$ and $H_0: \beta_4 = 0$.
(d) Compute the residuals from this model. Analyze the residuals using the methods discussed in Section 13-6.2.

13-10. Consider the data shown in the following table.

y	x_1	x_2	x_3	x_4
45.16	8	3	2	5
43.20	7	5	1	6
40.75	4	8	2	10
46.53	9	2	1	4
63.05	15	2	4	3
54.45	12	4	3	2
86.89	18	10	5	1
41.68	5	7	2	4
63.31	10	10	3	1
38.26	8	2	3	2
31.88	7	1	2	1
29.14	2	4	3	8
56.73	6	12	1	7
29.08	4	3	4	5
52.94	12	2	2	3
39.17	6	5	1	2
76.45	18	4	3	1
94.26	20	10	6	4

(a) Fit a multiple regression model to these data.
(b) Test for significance of regression.
(c) Test the hypothesis $H_0: \beta_4 = 0$ using the partial F-test.
(d) Test the hypothesis $H_0: \beta_4 = 0$ using the t statistic.

(e) What is the increase in the regression sum of squares associated with adding the variables x_3 and x_4 to a model that already contains x_1 and x_2?

(f) Test the hypothesis $H_0: \beta_3 = \beta_4 = 0$ using a partial F-test.

(g) Construct a 95 percent confidence interval estimate for β_1.

13-11. The data in this problem are adapted from Hald (1952). The data represents the heat evolved in calories per gram of cement (y) for various amounts of four ingredients (x_1, x_2, x_3, x_4).

Observation Number	y	x_1	x_2	x_3	x_4
1	78.5	7	26	6	60
2	74.3	1	29	15	52
3	104.3	11	56	8	20
4	87.6	11	31	8	47
5	95.9	7	52	6	33
6	109.2	11	55	9	22
7	102.7	3	71	17	6
8	72.5	1	31	22	44
9	93.1	2	54	18	22
10	115.9	21	47	4	26
11	83.8	1	40	23	34
12	113.3	11	66	9	12
13	109.4	10	68	8	12

(a) Fit a multiple regression model to these data.

(b) Test for significance of regression.

(c) Test the hypothesis $\beta_4 = 0$ using the partial F-test.

(d) Compute the t statistics for each independent variable. What conclusions can you draw?

(e) Test the hypothesis $\beta_2 = \beta_3 = \beta_4 = 0$ using the partial F-test.

(f) Construct a 95 percent confidence interval estimate for β_2.

13-12. Consider the data shown in the following table.

y	x_1	x_2
2.60	1.0	1.0
2.40	1.0	1.0
17.32	1.5	4.0
15.60	1.5	4.0
16.12	1.5	4.0
5.36	.5	2.0
6.19	1.5	2.0
10.17	.5	3.0
2.62	1.0	1.5
2.98	.5	1.5
6.92	1.0	2.5
7.06	.5	2.5

(a) Fit the second-order polynomial model $y = \beta_0 + \beta_1 x_1 + \beta_2 x_2 + \beta_{11} x_1^2 + \beta_{22} x_2^2 + \beta_{12} x_1 x_2 + \epsilon$.

(b) Test for significance of regression and lack of fit.

(c) Test the hypothesis $H_0: \beta_{11} = \beta_{22} = \beta_{12} = 0$.

13-13. Consider the following data.

y	−4.42	−1.39	−1.55	−1.89	−2.43	−3.15	−4.05	−5.15	−6.43	−7.89
x	.25	.50	.75	1.00	1.25	1.50	1.75	2.00	2.25	2.50

(a) Fit a second-order polynomial to the data.

(b) Test for significance of regression.

(c) Test the hypothesis that $\beta_{11} = 0$.

13-14. Consider the following data.

y	24.60	24.71	23.90	39.50	39.60	57.12	67.11	67.24	67.15	77.87	80.11	84.67
x	4.0	4.0	4.0	5.0	5.0	6.0	6.5	6.5	6.75	7.0	7.1	7.3

(a) Fit a second-order model to the data.

(b) Test for significance of regression and lack of fit.

(c) Test the hypothesis that $\beta_{11} = 0$.

13-15. Consider the data in Example 13-12. Test the hypothesis that two different regression models (with different slopes and intercepts) are required to adequately model the data.

13-16. **Piecewise Linear Regression (I).** Suppose that y is piecewise linearly related to x. That is, different linear relationships are appropriate over the intervals $-\infty < x \leq x^*$ and $x^* < x < \infty$. Show how indicator variables can be used to fit such a piecewise linear regression model, assuming that the point x^* is known.

13-17. **Piecewise Linear Regression (II).** Consider the piecewise linear regression model described in Exercise 13-16. Suppose that at the point x^* a discontinuity occurs in the regression function. Show how indicator variables can be used to incorporate the discontinuity into the model.

13-18. **Piecewise Linear Regression (III).** Consider the piecewise linear regression model described in Exercise 13-16. Suppose that the point x^* is not known with certainty and must be estimated. Develop an approach that could be used to fit the piecewise linear regression model.

13-19. Calculate the standardized regression coefficients for the regression model developed in Exercise 13-1.

13-20. Calculate the standardized regression coefficients for the regression model developed in Exercise 13-2.

13-21. Find the variance inflation factors for the regression model developed in Example 13-1. Do they indicate that multicollinearity is a problem in this model?

13-22. Find the variance inflation factors for the four-variable regression model discussed in Example 13-18. Do they indicate that multicollinearity is a problem in this model?

13-23. Analyze the data in Exercise 13-10 using the following variable selection techniques:
(*a*) All possible regressions
(*b*) Directed search on *t*
(*c*) Stepwise regression
(*d*) Forward selection
(*e*) Backward elimination

13-24. Analyze the data in Exercise 13-11 using the following variable selection techniques:
(*a*) All possible regressions
(*b*) Directed search on *t*
(*c*) Stepwise regression
(*d*) Forward selection
(*e*) Backward elimination

13-25. Consider the data in Exercise 13-11. Fit a regression model of the form $y = \beta_0 + \beta_1 x_1 + \beta_2 x_2 + \epsilon$ to these data. Compute the value of D_i^2 for each observation, and determine if there are any points remote in x-space that exert unusual influence over the model.

13-26. Consider the data in Exercise 13-11. Fit a regression model of the form $y = \beta_0 + \beta_1 x_1 + \beta_2 x_2 + \epsilon$ to these data. Using data points that are near neighbors, calculate an estimate of the standard deviation of pure error. Does the regression model demonstrate any obvious lack of fit?

Chapter 14
Design of Experiments

Experiments are performed by investigators in all disciplines. The objective of an experiment is either to explore the relationships between the factors that affect a process or to confirm some hypothesis. Many of the hypothesis-testing procedures discussed in Chapter 10 can be thought of as methods for analyzing simple comparative experiments. For example, in comparing the equality of two means, experimenters must decide whether to pair or to completely randomize the observations. They must also determine the number of observations required to achieve a certain power of the test. In other words, experimenters will often have certain aspects of the experiment under their control. They must plan or *design* the experiment to ensure that the data collected are relevant and to obtain as much information as possible.

The literature of experimental design is extensive. In this chapter we will introduce some of the more useful design techniques. For a more complete treatment of the subject, see Montgomery (1976).

14-1 The Need for Designed Experiments

Statisticians are often asked for advice in drawing conclusions from the results of experiments. However, since the results will depend in part on how the experiment was conducted, statisticians should be familar with the experimental procedure. From this information they may be able to assist the experimenter in finding answers to those questions that prompted the experiment. Unfortunately, however, the experiment may have been performed in such a way that some (or perhaps all) important questions cannot be answered. In this latter situation usually all the statistician can do is to point out ways to avoid this outcome in future work. Because of the possibility of performing an ill-advised set of experiments we must consider questions of design at the start of an experimental program, before the data are collected.

When designing an experiment, we must remember two basic considerations: (1) statistical accuracy and (2) cost. Statistical accuracy involves the proper selection of the response to be measured, determination of the number of factors that influence the response, the selection of the subset of these factors to be studied in the experiment being planned, the number of times the basic experiment should be repeated (replicated), and the form of the analysis to be conducted. This aspect of experimental design is reasonably well understood by statisticians.

Cost is less often emphasized, but is of equal importance. To minimize the cost of an experimental investigation, the statistician usually attempts to choose the simplest experimental design possible and to utilize the smallest sample size consistent with satisfactory results. Fortunately, most simple experimental designs are both statistically efficient and economical.

14-2   Factorial Experiments

Factorial experiments are employed to simultaneously study the effects of two or more factors. By a *factorial* experiment we mean that in each complete trial or *replicate* of the experiment all possible combinations of the levels of the factors are investigated. Thus, if there are a levels of factor A and b levels of factor B, then each replicate contains all ab treatment combinations.

The effect of a factor is defined as the change in response produced by a change in the level of the factor. This is called a *main effect* because it refers to the primary factors in the study. For example, consider the data in Table 14-1. The main effect of factor A is the difference between the average response at the first level of A and the average response at the second level of A, or

$$A = \frac{30 + 40}{2} - \frac{10 + 20}{2} = 20$$

TABLE 14-1 **A Factorial Experiment with Two Factors**

Factor A	Factor B	
	B_1	B_2
A_1	10	20
A_2	30	40

TABLE 14-2 **A Factorial Experiment with Interaction**

Factor A	Factor B	
	B_1	B_2
A_1	10	20
A_2	30	0

That is, changing factor A from level 1 to level 2 causes an average response increase of 20 units. Similarly, the main effect of B is

$$B = \frac{20 + 40}{2} - \frac{10 + 30}{2} = 10$$

In some experiments, the difference in response between the levels of one factor is not the same at all levels of the other factors. When this occurs, there is an *interaction* between the factors. For example, consider the data in Table 14-2. At the first level of factor B, the A effect is

$$A = 30 - 10 = 20$$

and at the second level of factor B, the A effect is

$$A = 0 - 20 = -20$$

Since the effect of A depends on the level chosen for factor B, there is interaction between A and B.

When an interaction is large, the corresponding main effects have little meaning. For example, by using the data in Table 14-2, we find the main effect of A as

$$A = \frac{30 + 0}{2} - \frac{10 + 20}{2} = 0$$

and we would be tempted to conclude that there is no A effect. However, when we examined the effects of A at *different levels of factor B*, we saw that this was not the case. The effect of factor A depends on the levels of factor B. Thus, knowledge of the AB interaction is more useful than knowledge of the main effect. A significant interaction can mask the significance of main effects.

The concept of interaction can be illustrated graphically. Figure 14-1 plots the data in Table 14-1 against the levels of A for both levels of B. Note that the B_1 and B_2 lines are roughly parallel, indicating that factors A and B do not

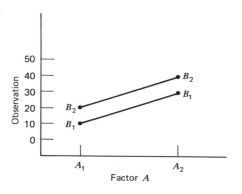

Fig. 14-1. Factorial experiment, no interaction.

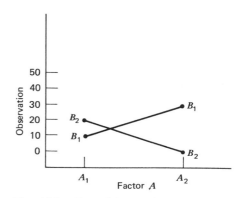

Fig. 14-2. Factorial experiment, with interaction.

interact significantly. Figure 14-2 plots the data in Table 14-2. In this graph, the B_1 and B_2 lines are not parallel, indicating the interaction between factors A and B. Such graphical displays are often useful in presenting the results of experiments.

14-3 Two-Factor Factorial Experiments

The simplest type of factorial experiment involves only two factors, say A and B. There are a levels of factor A and b levels of factor B. The two-factor factorial is shown in Table 14-3. Note that there are n *replicates* of the experiment and that each replicate contains all ab treatment combinations.

TABLE 14-3 **Data Arrangement for a Two-Factor Factorial Design**

		Factor B			
		1	2	$\cdots$	b
	1	$y_{111}, y_{112},$ $\ldots, y_{11n}$	$y_{121}, y_{122},$ $\ldots, y_{12n}$		$y_{1b1}, y_{1b2},$ $\ldots, y_{1bn}$
Factor A	2	$y_{211}, y_{212},$ $\ldots, y_{21n}$	$y_{221}, y_{222},$ $\ldots, y_{22n}$		$y_{2b1}, y_{2b2},$ $\ldots, y_{2bn}$
	$\vdots$				
	a	$y_{a11}, y_{a12},$ $\ldots, y_{a1n}$	$y_{a21}, y_{a22},$ $\ldots, y_{a2n}$		$y_{ab1}, y_{ab2},$ $\ldots, y_{abn}$

The observation in the *ij*th cell in the *k*th replicate is denoted by y_{ijk}. In collecting the data, the *abn* observations would be run in *random* order. Thus, like the single-factor experiments studied in Chapter 11, the two-factor factorial is a *completely randomized design*.

The observations may be described by the linear statistical model

$$y_{ijk} = \mu + \tau_i + \beta_j + (\tau\beta)_{ij} + \epsilon_{ijk} \quad \left\{ \begin{array}{l} i = 1, 2, \ldots, a \\ j = 1, 2, \ldots, b \\ k = 1, 2, \ldots, n \end{array} \right. \qquad (14\text{-}1)$$

where μ is the overall mean effect, τ_i is the effect of the *i*th level of factor A, β_j is the effect of the *j*th level of factor B, $(\tau\beta)_{ij}$ is the effect of the interaction between A and B, and ϵ_{ijk} is a NID$(0, \sigma^2)$ random error component. We are interested in testing the hypotheses of no significant factor A effect, no significant factor B effect, and no significant AB interaction. As with the single-factor experiments of Chapter 11, the analysis of variance will be used to test these hypotheses. Since there are two factors under study, the procedure used is known as the two-way classification analysis of variance.

14-3.1 Statistical Analysis of the Fixed Effects Model

Suppose that factors A and B are fixed. That is, the a levels of factor A and the b levels of factor B are specifically chosen by the experimenter, and inferences are confined to these levels only. In this model, it is customary to define the effects τ_i, β_j, and $(\tau\beta)_{ij}$ as deviations from the mean, so that $\Sigma_{i=1}^a \tau_i = 0$, $\Sigma_{j=1}^b \beta_j = 0$, $\Sigma_{i=1}^a (\tau\beta)_{ij} = 0$, and $\Sigma_{j=1}^b (\tau\beta)_{ij} = 0$.

Let $y_{i..}$ denote the total of the observations under the *i*th level of factor A, $y_{.j.}$ denote the total of the observations under the *j*th level of factor B, $y_{ij.}$ denote the total of the observations in the *ij*th cell of Table 14-3, B, and $y_{...}$ denote the grand total of all the observations. Define $\bar{y}_{i..}$, $\bar{y}_{.j.}$, $\bar{y}_{ij.}$, and $\bar{y}_{...}$ as the corresponding row, column, cell and grand averages. That is,

$$y_{i..} = \sum_{j=1}^b \sum_{k=1}^n y_{ijk} \qquad \bar{y}_{i..} = \frac{y_{i..}}{bn} \qquad i = 1, 2, \ldots, a$$

$$y_{.j.} = \sum_{i=1}^a \sum_{k=1}^n y_{ijk} \qquad \bar{y}_{.j.} = \frac{y_{.j.}}{an} \qquad j = 1, 2, \ldots, b$$

$$y_{ij.} = \sum_{k=1}^n y_{ijk} \qquad \bar{y}_{ij.} = \frac{y_{ij.}}{n} \qquad \begin{array}{l} i = 1, 2, \ldots, a \\ j = 1, 2, \ldots, b \end{array}$$

$$y_{...} = \sum_{i=1}^a \sum_{j=1}^b \sum_{k=1}^n y_{ijk} \qquad \bar{y}_{...} = \frac{y_{...}}{abn} \qquad (14\text{-}2)$$

The total corrected sum of squares may be written as

$$\sum_{i=1}^{a}\sum_{j=1}^{b}\sum_{k=1}^{n}(y_{ijk}-\bar{y}_{...})^2$$

$$=\sum_{i=1}^{a}\sum_{j=1}^{b}\sum_{k=1}^{n}[(\bar{y}_{i..}-\bar{y}_{...})+(\bar{y}_{.j.}-\bar{y}_{...})+(\bar{y}_{ij.}-\bar{y}_{i..}-\bar{y}_{.j.}+\bar{y}_{...})+(y_{ijk}-\bar{y}_{ij.})]^2$$

$$=bn\sum_{i=1}^{a}(\bar{y}_{i..}-\bar{y}_{...})^2+an\sum_{j=1}^{b}(\bar{y}_{.j.}-\bar{y}_{...})^2+n\sum_{i=1}^{a}\sum_{j=1}^{b}(\bar{y}_{ij.}-\bar{y}_{i..}-\bar{y}_{.j.}+\bar{y}_{...})^2$$

$$+\sum_{i=1}^{a}\sum_{j=1}^{b}\sum_{k=1}^{n}(y_{ijk}-\bar{y}_{ij.})^2 \qquad (14\text{-}3)$$

Thus, the total sum of squares is partitioned into a sum of squares due to "rows" or factor A (SS_A), a sum of squares due to "columns" or factor B (SS_B), a sum of squares due to the interaction between A and B (SS_{AB}), and a sum of squares due to error (SS_E). Notice that there must be at least two replicates to obtain a nonzero error sum of squares.

The sum of squares identity in Equation (14-3) may be written symbolically as

$$SS_T = SS_A + SS_B + SS_{AB} + SS_E \qquad (14\text{-}4)$$

There are $abn - 1$ total degrees of freedom. The main effects A and B have $a - 1$ and $b - 1$ degrees of freedom, while the interaction effect AB has $(a - 1)(b - 1)$ degrees of freedom. Within each of the ab cells in Table 14-3, there are $n - 1$ degrees of freedom between the n replicates, and observations in the same cell can differ only due to random error. Therefore, there are $ab(n - 1)$ degrees of freedom for error. The ratio of each sum of squares on the right-hand side of Equation (14-3) to its degrees of freedom is a *mean square*.

Assuming that factors A and B are fixed, the expected values of the mean squares are

$$E(MS_A) = E\left(\frac{SS_A}{a-1}\right) = \sigma^2 + \frac{bn\sum_{i=1}^{a}\tau_i^2}{a-1}$$

$$E(MS_B) = E\left(\frac{SS_B}{b-1}\right) = \sigma^2 + \frac{an\sum_{j=1}^{b}\beta_j^2}{b-1}$$

$$E(MS_{AB}) = E\left(\frac{SS_{AB}}{(a-1)(b-1)}\right) = \sigma^2 + \frac{n\sum_{i=1}^{a}\sum_{j=1}^{b}(\tau\beta)_{ij}^2}{(a-1)(b-1)}$$

TABLE 14-4 **The Analysis of Variance Table for the Two-Way Classification, Fixed Effects Model**

Source of Variation	Sum of Squares	Degrees of Freedom	Mean Square	F_0
A treatments	SS_A	$a - 1$	$MS_A = \dfrac{SS_A}{a - 1}$	$F_0 = \dfrac{MS_A}{MS_E}$
B treatments	SS_B	$b - 1$	$MS_B = \dfrac{SS_B}{b - 1}$	$F_0 = \dfrac{MS_B}{MS_E}$
Interaction	SS_{AB}	$(a - 1)(b - 1)$	$MS_{AB} = \dfrac{SS_{AB}}{(a - 1)(b - 1)}$	$F_0 = \dfrac{MS_{AB}}{MS_E}$
Error	SS_E	$ab(n - 1)$	$MS_E = \dfrac{SS_E}{ab(n - 1)}$	
Total	SS_T	$abn - 1$		

and

$$E(MS_E) = E\left(\frac{SS_E}{ab(n - 1)}\right) = \sigma^2$$

Therefore, to test $H_0: \tau_i = 0$ (no row factor effects), $H_0: \beta_j = 0$ (no column factor effects), and $H_0: (\tau\beta)_{ij} = 0$ (no interaction effects), we would divide the corresponding mean square by mean square error. Each of these ratios will follow an F distribution with numerator degrees of freedom equal to the number of degrees of freedom for the numerator mean square and $ab(n - 1)$ denominator degrees of freedom, and the critical region will be located in the upper tail. The test procedure is arranged in an analysis of variance table, such as shown in Table 14-4.

Computational formulas for the sums of squares in Equation (14-4) may be obtained easily. The total sum of squares is computed from

$$SS_T = \sum_{i=1}^{a} \sum_{j=1}^{b} \sum_{k=1}^{n} y_{ijk}^2 - \frac{y_{...}^2}{abn} \tag{14-5}$$

The sums of squares for main effects are

$$SS_A = \sum_{i=1}^{a} \frac{y_{i..}^2}{bn} - \frac{y_{...}^2}{abn} \tag{14-6}$$

and

$$SS_B = \sum_{j=1}^{b} \frac{y_{.j.}^2}{an} - \frac{y_{...}^2}{abn} \qquad (14\text{-}7)$$

We usually calculate the SS_{AB} in two steps. First, we compute the sum of squares between the ab cell totals, called the sum of squares due to "subtotals."

$$SS_{\text{subtotals}} = \sum_{i=1}^{a} \sum_{j=1}^{b} \frac{y_{ij.}^2}{n} - \frac{y_{...}^2}{abn}$$

This sum of squares also contains SS_A and SS_B. Therefore, the second step is to compute SS_{AB} as

$$SS_{AB} = SS_{\text{subtotals}} - SS_A - SS_B \qquad (14\text{-}8)$$

The error sum of square is found by subtraction as either

$$SS_E = SS_T - SS_{AB} - SS_A - SS_B \qquad (14\text{-}9a)$$

or

$$SS_E = SS_T - SS_{\text{subtotals}} \qquad (14\text{-}9b)$$

● **Example 14-1.** An experiment is performed to determine the ability of three different chemicals to prevent rust from forming on steel specimens. The chemicals may be applied by either dipping or spraying. Three clean specimens are treated by each chemical, using each method of application. The specimens are then exposed to a salt spray, and the amount of rust after 10 days is noted. The data are shown in Table 14-5. The circled numbers in the cells are the cell totals $y_{ij.}$.

TABLE 14-5 **Data for Example 14-1**

	Application Method		
Chemical Type	Dipping	Spraying	$y_{i..}$
1	4.0, 4.5, 4.3 (12.8)	5.4, 4.9, 5.6 (15.9)	28.7
2	5.6, 4.9, 5.4 (15.9)	5.8, 6.1, 6.3 (18.2)	34.1
3	3.8, 3.7, 4.0 (11.5)	5.5, 5.0, 5.0 (15.5)	27.0
$y_{.j.}$	40.2	49.6	$89.8 = y_{...}$

The sums of squares are computed as follows:

$$SS_T = \sum_{i=1}^{a} \sum_{j=1}^{b} \sum_{k=1}^{n} y_{ijk}^2 - \frac{y_{...}^2}{abn}$$

$$= (4.0)^2 + (4.5)^2 + \cdots + (5.0)^2 - \frac{(89.8)^2}{18} = 10.72$$

$$SS_{types} = \sum_{i=1}^{a} \frac{y_{i..}^2}{bn} - \frac{y_{...}^2}{abn}$$

$$= \frac{(28.7)^2 + (34.1)^2 + (27.0)^2}{6} - \frac{(89.8)^2}{18} = 4.91$$

$$SS_{methods} = \sum_{j=1}^{b} \frac{y_{.j.}^2}{an} - \frac{y_{...}^2}{abn}$$

$$= \frac{(40.2)^2 + (49.6)^2}{9} - \frac{(89.8)^2}{18} = 4.58$$

$$SS_{interaction} = \sum_{i=1}^{a} \sum_{j=1}^{b} \frac{y_{ij.}^2}{n} - \frac{y_{...}^2}{abn} - SS_{types} - SS_{methods}$$

$$= \frac{(12.8)^2 + (15.9)^2 + (11.5)^2 + (15.9)^2 + (18.2)^2 + (15.5)^2}{3}$$

$$- \frac{(89.8)^2}{18} - 4.91 - 4.58 = .24$$

and

$$SS_E = SS_T - SS_{types} - SS_{method} - SS_{interaction}$$

$$= 10.72 - 4.91 - 4.58 - .24 = .99$$

The analysis of variance is summarized in Table 14-6. Since $F_{.05, 1, 12} = 4.75$, we conclude that the main effects of chemical type and application method affect rust formation and, since $F_{.05, 2, 12} = 3.89$, there is no indication of interaction between these factors.

TABLE 14-6   **Analysis of Variance for Example 14-1**

Source of Variation	Sum of Squares	Degrees of Freedom	Mean Square	F_0
Chemical types	4.91	2	2.46	30.75
Application methods	4.58	1	4.58	57.25
Interaction	.24	2	.12	1.5
Error	.99	12	.08	
Total	10.72	17		

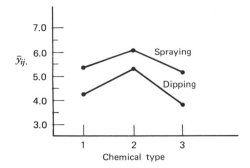

Fig. 14-3. Graph of average rust formation versus chemical types for Example 14-1.

A graph of the cell averages $\{\bar{y}_{ij.}\}$ versus the levels of chemical type for each application method is shown in Fig. 14-3. The absence of interaction is evident by the parallelism of the two lines. Furthermore, since a large response indicates greater rust formation, we observe that dipping is a superior application method and that chemical type 3 is most effective.

When both factors are fixed, comparisons between the individual means of either factor may be made using Duncan's multiple range test. When there is no interaction, these comparisons may be made using either the row averages $\bar{y}_{i..}$ or the column averages $\bar{y}_{.j.}$. However, when interaction is significant, comparisons between the means of one factor (say A) may be obscured by the AB interaction. In this case, we may apply Duncan's multiple range test to the means of factor A, with factor B set at a particular level.

14-3.2 Statistical Analysis of the Random Effects Model

Now consider the situation in which the levels of both factors are selected at random from larger populations of factor levels, and we wish to extend our conclusions about the sampled population of factor levels. The observations are represented by the model

$$y_{ijk} = \mu + \tau_i + \beta_j + (\tau\beta)_{ij} + \epsilon_{ijk} \quad \begin{cases} i = 1, 2, \ldots, a \\ j = 1, 2, \ldots, b \\ k = 1, 2, \ldots, n \end{cases} \quad (14\text{-}10)$$

where the parameters τ_i, β_j, $(\tau\beta)_{ij}$, and ϵ_{ijk} are random variables. Specifically, we assume that τ_i is NID$(0, \sigma_\tau^2)$, β_j is NID$(0, \sigma_\beta^2)$, $(\tau\beta)_{ij}$ is NID$(0, \sigma_{\tau\beta}^2)$, and ϵ_{ijk} is NID$(0, \sigma^2)$. The variance of any observation is

$$V(y_{ijk}) = \sigma_\tau^2 + \sigma_\beta^2 + \sigma_{\tau\beta}^2 + \sigma^2$$

and σ_τ^2, σ_β^2, $\sigma_{\tau\beta}^2$, and σ^2 are called *variance components*. The hypotheses that we are interested in testing are $H_0: \sigma_\tau^2 = 0$, $H_0: \sigma_\beta^2 = 0$, and $H_0: \sigma_{\tau\beta}^2 = 0$. Notice the similarity to the one-way classification random effects model.

The basic analysis of variance remain unchanged; that is, SS_A, SS_B, SS_{AB}, SS_T, and SS_E are all calculated as in the fixed effects case. To construct the test statistics, we must examine the expected mean squares. They are:

$$E(MS_A) = \sigma^2 + n\sigma_{\tau\beta}^2 + bn\sigma_\tau^2$$
$$E(MS_B) = \sigma^2 + n\sigma_{\tau\beta}^2 + an\sigma_\beta^2$$
$$E(MS_{AB}) = \sigma^2 + n\sigma_{\tau\beta}^2 \qquad (14\text{-}11)$$

and

$$E(MS_E) = \sigma^2$$

Note from the expected mean squares that the appropriate statistic for testing $H_0: \sigma_{\tau\beta}^2 = 0$ is

$$F_0 = \frac{MS_{AB}}{MS_E} \qquad (14\text{-}12)$$

since under H_0 both numerator and denominator of F_0 have expectation σ^2, and only if H_0 is false is $E(MS_{AB})$ greater than $E(MS_E)$. The ratio F_0 is distributed as $F_{(a-1)(b-1),ab(n-1)}$. Similarly, for testing $H_0: \sigma_\tau^2 = 0$ we would use

$$F_0 = \frac{MS_A}{MS_{AB}} \qquad (14\text{-}13)$$

which is distributed as $F_{a-1,(a-1)(b-1)}$, and for testing $H_0: \sigma_\beta^2 = 0$ the statistic is

$$F_0 = \frac{MS_B}{MS_{AB}} \qquad (14\text{-}14)$$

which is distributed as $F_{b-1,(a-1)(b-1)}$. These are all upper-tail, one-tail tests. Notice that these test statistics are not the same as those used if both factors A and B are fixed. The expected mean squares are always used as a guide to test statistic construction.

The variance components may be estimated by equating the observed mean squares to their expected values and solving for the variance components. This yields

$$\hat{\sigma}^2 = MS_E$$
$$\hat{\sigma}_{\tau\beta}^2 = \frac{MS_{AB} - MS_E}{n}$$
$$\hat{\sigma}_\beta^2 = \frac{MS_B - MS_{AB}}{an}$$
$$\hat{\sigma}_\tau^2 = \frac{MS_A - MS_{AB}}{bn} \qquad (14\text{-}15)$$

TABLE 14-7   **Analysis of Variance for Example 14-2**

Source of Variation	Sum of Squares	Degrees of Freedom	Mean Square	F_0
Chemical types	4.91	2	2.46	20.50
Application methods	4.58	1	4.58	38.17
Interaction	.24	2	.12	1.5
Error	.99	12	.08	
Total	10.72	17		

● **Example 14-2.** Suppose that in Example 14-1, a large number of chemicals could be used to inhibit rust, and several application methods could be used. Three chemicals, say 1, 2, and 3, were selected at random as were the two application methods. The analysis of variance assuming the random effects model is shown in Table 14-7.

Notice that the first four columns in the analysis of variance table are exactly as in Example 14-1. Now, however, the F ratios are computed according to Equations (14-12) and (14-14). Since $F_{.05, 2, 12} = 3.89$, we conclude that interaction is not significant. Also, since $F_{.05, 2, 2} = 19.0$ and $F_{.05, 1, 2} = 18.5$, we conclude that both chemical types and application methods significantly affect rust formation. The variance components may be estimated using Equation (14-15) as follows:

$$\hat{\sigma}^2 = .08$$

$$\hat{\sigma}^2_{\tau\beta} = \frac{.12 - .08}{3} = .0133$$

$$\hat{\sigma}^2_{\tau} = \frac{2.46 - .12}{6} = .39$$

$$\hat{\sigma}^2_{\beta} = \frac{4.58 - .12}{9} = .50$$

Clearly, the two largest variance components are for chemical types ($\hat{\sigma}^2_{\tau} = .39$) and application methods ($\hat{\sigma}^2_{\beta} = .50$).

14-3.3   Statistical Analysis of the Mixed Model

Now suppose that one of the factors, A, is fixed and the other, B, is random. This is called the *mixed model* analysis of variance. The linear model is

$$y_{ijk} = \mu + \tau_i + \beta_j + (\tau\beta)_{ij} + \epsilon_{ijk} \begin{cases} i = 1, 2, \ldots, a \\ j = 1, 2, \ldots, b \\ k = 1, 2, \ldots, n \end{cases} \qquad (14\text{-}16)$$

In this model, τ_i is a fixed effect defined such that $\sum_{i=1}^{a} \tau_i = 0$, β_j is a random effect, the interaction term $(\tau\beta)_{ij}$ is a random effect, and ϵ_{ijk} is a NID$(0, \sigma^2)$ random error. It is also customary to assume that β_j is NID$(0, \sigma_\beta^2)$ and that the interaction elements $(\tau\beta)_{ij}$ are normal random variables with mean zero and variance $[(a-1)/a]\sigma_{\tau\beta}^2$. The interaction elements are not all independent.

The expected mean squares in this case are

$$E(MS_A) = \sigma^2 + n\sigma_{\tau\beta}^2 + \frac{bn\sum_{i=1}^{a}\tau_i^2}{a-1}$$

$$E(MS_B) = \sigma^2 + an\sigma_\beta^2$$

$$E(MS_{AB}) = \sigma^2 + n\sigma_{\tau\beta}^2$$

and

$$E(MS_E) = \sigma^2 \tag{14-17}$$

Therefore, the appropriate test statistic for testing $H_0: \tau_i = 0$ is

$$F_0 = \frac{MS_A}{MS_{AB}} \tag{14-18}$$

which is distributed as $F_{a-1,(a-1)(b-1)}$. For testing $H_0: \sigma_\beta^2 = 0$, the test statistic is

$$F_0 = \frac{MS_B}{MS_E} \tag{14-19}$$

which is distributed as $F_{b-1,ab(n-1)}$. Finally, for testing $H_0: \sigma_{\tau\beta}^2 = 0$ we would use

$$F_0 = \frac{MS_{AB}}{MS_E} \tag{14-20}$$

which is distributed as $F_{(a-1)(b-1),ab(n-1)}$.

The variance components σ_β^2, $\sigma_{\tau\beta}^2$, and σ^2 may be estimated by eliminating the first equation from Equation (14-17), leaving three equations in three unknowns, whose solution is

$$\hat{\sigma}_\beta^2 = \frac{MS_B - MS_E}{an}$$

$$\hat{\sigma}_{\tau\beta}^2 = \frac{MS_{AB} - MS_E}{n}$$

and

$$\hat{\sigma}^2 = MS_E \tag{14-21}$$

This general approach can be used to estimate the variance components in *any* mixed model. After eliminating the mean squares containing fixed factors, there will always be a set of equations remaining that can be solved for

TABLE 14-8 **Analysis of Variance for the Two-Factor Mixed Model**

Source of Variation	Sum of Squares	Degrees of Freedom	Expected Mean Square
Rows (A)	SS_A	$a - 1$	$\sigma^2 + n\sigma_{\tau\beta}^2 + bn \sum \tau_i^2/(a - 1)$
Columns (B)	SS_B	$b - 1$	$\sigma^2 + an\sigma_{\beta}^2$
Interaction	SS_{AB}	$(a - 1)(b - 1)$	$\sigma^2 + n\sigma_{\tau\beta}^2$
Error	SS_E	$ab(n - 1)$	σ^2
Total	SS_T	$abn - 1$	

the variance components. Table 14-8 summarizes the analysis of variance for the two-factor mixed model.

14-4   General Factorial Experiments

Many experiments involve more than two factors. In this section we consider the case where there are a levels of factor A, b levels of factor B, c levels of factor C, and so on, arranged in a factorial experiment. In general, there will be $abc \ldots n$ total observations, if there are n replicates of the complete experiment.

For example, consider the three-factor analysis of variance model

$$y_{ijkl} = \mu + \tau_i + \beta_j + \gamma_k + (\tau\beta)_{ij} + (\tau\gamma)_{ik} + (\beta\gamma)_{jk}$$

$$+ (\tau\beta\gamma)_{ijk} + \epsilon_{ijkl} \left\{ \begin{array}{l} i = 1, 2, \ldots, a \\ j = 1, 2, \ldots, b \\ k = 1, 2, \ldots, c \\ l = 1, 2, \ldots, n \end{array} \right. \quad (14\text{-}22)$$

Assuming that A, B, and C are fixed, the analysis of variance is shown in Table 14-9. Note that there must be at least two replicates ($n \geq 2$) to obtain an error sum of squares. The F-tests on main effects and interactions follow directly from the expected mean squares.

Computing formulas for the sums of squares in Table 14-9 are easily obtained. The total sum of squares is

$$SS_T = \sum_{i=1}^{a} \sum_{j=1}^{b} \sum_{k=1}^{c} \sum_{l=1}^{n} y_{ijkl}^2 - \frac{y_{\ldots}^2}{abcn} \quad (14\text{-}23)$$

The sum of squares for the main effects are computed from the totals for

TABLE 14-9 The Analysis of Variance Table for the Three-Factor Fixed Effects Model

Source of Variation	Sum of Squares	Degrees of Freedom	Mean Square	Expected Mean Squares	F_0
A	SS_A	$a-1$	MS_A	$\sigma^2 + \dfrac{bcn \sum \tau_i^2}{a-1}$	$F_0 = \dfrac{MS_A}{MS_E}$
B	SS_B	$b-1$	MS_B	$\sigma^2 + \dfrac{acn \sum \beta_j^2}{b-1}$	$F_0 = \dfrac{MS_B}{MS_E}$
C	SS_C	$c-1$	MS_C	$\sigma^2 + \dfrac{abn \sum \gamma_k^2}{c-1}$	$F_0 = \dfrac{MS_C}{MS_E}$
AB	SS_{AB}	$(a-1)(b-1)$	MS_{AB}	$\sigma^2 + \dfrac{cn \sum \sum (\tau\beta)_{ij}^2}{(a-1)(b-1)}$	$F_0 = \dfrac{MS_{AB}}{MS_E}$
AC	SS_{AC}	$(a-1)(c-1)$	MS_{AC}	$\sigma^2 + \dfrac{bn \sum \sum (\tau\gamma)_{ik}^2}{(a-1)(c-1)}$	$F_0 = \dfrac{MS_{AC}}{MS_E}$
BC	SS_{BC}	$(b-1)(c-1)$	MS_{BC}	$\sigma^2 + \dfrac{an \sum \sum (\beta\gamma)_{jk}^2}{(b-1)(c-1)}$	$F_0 = \dfrac{MS_{BC}}{MS_E}$
ABC	SS_{ABC}	$(a-1)(b-1)(c-1)$	MS_{ABC}	$\sigma^2 + \dfrac{n \sum \sum \sum (\tau\beta\gamma)_{ijk}^2}{(a-1)(b-1)(c-1)}$	$F_0 = \dfrac{MS_{ABC}}{MS_E}$
Error	SS_E	$abc(n-1)$	MS_E	σ^2	
Total	SS_T	$abcn-1$			

factors $A(y_{i..})$, $B(y_{.j.})$, and $C(y_{..k})$ as follows:

$$SS_A = \sum_{i=1}^{a} \frac{y_{i..}^2}{bcn} - \frac{y_{....}^2}{abcn} \tag{14-24}$$

$$SS_B = \sum_{j=1}^{b} \frac{y_{.j.}^2}{acn} - \frac{y_{....}^2}{abcn} \tag{14-25}$$

$$SS_C = \sum_{k=1}^{c} \frac{y_{..k}^2}{abn} - \frac{y_{....}^2}{abcn} \tag{14-26}$$

To compute the two-factor interaction sums of squares, the totals for the $A \times B$, $A \times C$, and $B \times C$ cells are needed. It may be helpful to collapse the original data table into three two-way tables in order to compute these totals. The sums of squares are

$$SS_{AB} = \sum_{i=1}^{a} \sum_{j=1}^{b} \frac{y_{ij.}^2}{cn} - \frac{y_{....}^2}{abcn} - SS_A - SS_B$$
$$= SS_{\text{subtotals }(AB)} - SS_A - SS_B \tag{14-27}$$

$$SS_{AC} = \sum_{i=1}^{a} \sum_{k=1}^{c} \frac{y_{i.k}^2}{bn} - \frac{y_{....}^2}{abcn} - SS_A - SS_C$$
$$= SS_{\text{subtotals }(AC)} - SS_A - SS_C \tag{14-28}$$

and

$$SS_{BC} = \sum_{j=1}^{b} \sum_{k=1}^{c} \frac{y_{.jk}^2}{an} - \frac{y_{....}^2}{abcn} - SS_B - SS_C$$
$$= SS_{\text{subtotals }(BC)} - SS_B - SS_C \tag{14-29}$$

The three-factor interaction sum of squares is computed from the three-way cell totals $\{y_{ijk.}\}$ as

$$SS_{ABC} = \sum_{i=1}^{a} \sum_{j=1}^{b} \sum_{k=1}^{c} \frac{y_{ijk.}^2}{n} - \frac{y_{....}^2}{abcn} - SS_A - SS_B - SS_C - SS_{AB} - SS_{AC} - SS_{BC} \tag{14-30a}$$

$$= SS_{\text{subtotals }(ABC)} - SS_A - SS_B - SS_C - SS_{AB} - SS_{AC} - SS_{BC} \tag{14-30b}$$

The error sum of squares may be found by subtracting the sum of squares for each main effect and interaction from the total sum of squares, or by

$$SS_E = SS_T - SS_{\text{subtotals }(ABC)} \tag{14-31}$$

● **Example 14-3.** A mechanical engineer is studying the surface finish of a part produced in a metal-cutting operation. Three factors, feed rate (A), depth of cut (B), and tool angle (C), are of interest. All three factors have been

TABLE 14-10 **Coded Surface Finish Data for Example 14-3**

	Depth of Cut (B)				
	.025 inch		.040 inch		
Feed Rate (A)	Tool Angle (C)		Tool Angle (C)		$y_{i...}$
	15°	25°	15°	25°	
20 in./min	9 7 ⑯	11 10 ㉑	9 11 ⑳	10 8 ⑱	
30 in./min	10 12 ㉒	10 13 ㉓	12 15 ㉗	16 14 ㉚	102
B × C totals $y_{.jk.}$	38	44	47	48	177 = $y_{....}$

A × B Totals $y_{ij..}$ B				A × C Totals $y_{i.k.}$ C		
A	.025	.040		A	15	25
20	37	38		20	36	39
30	45	57		30	49	53
$y_{.j..}$	82	95		$y_{..k.}$	85	92

assigned two levels, and two replicates of a factorial design are run. The coded data are shown in Table 14-10. The three-way cell totals $\{y_{ijk.}\}$ are circled in this table.

The sums of squares are calculated as follows, using Equations (14-23) to (14-31):

$$SS_T = \sum_{i=1}^{a} \sum_{j=1}^{b} \sum_{k=1}^{c} \sum_{l=1}^{n} y_{ijkl}^2 - \frac{y_{....}^2}{abcn} = 2051 - \frac{(177)^2}{16} = 92.9375$$

$$SS_A = \sum_{i=1}^{a} \frac{y_{i...}^2}{bcn} - \frac{y_{....}^2}{abcn}$$
$$= \frac{(75)^2 + (102)^2}{8} - \frac{(177)^2}{16} = 45.5625$$

$$SS_B = \sum_{j=1}^{b} \frac{y_{.j..}^2}{acn} - \frac{y_{....}^2}{abcn}$$

$$= \frac{(82)^2 + (95)^2}{8} - \frac{y_{....}^2}{abcn} = 10.5625$$

$$SS_C = \sum_{k=1}^{c} \frac{y_{..k.}^2}{abn} - \frac{y_{....}^2}{abcn}$$

$$= \frac{(85)^2 + (92)^2}{8} - \frac{(177)^2}{16} = 3.0625$$

$$SS_{AB} = \sum_{i=1}^{a} \sum_{j=1}^{b} \frac{y_{ij..}^2}{cn} - \frac{y_{....}^2}{abcn} - SS_A - SS_B$$

$$= \frac{(37)^2 + (38)^2 + (45)^2 + (57)^2}{4} - \frac{(177)^2}{16} - 45.5625 - 10.5625$$

$$= 7.5625$$

$$SS_{AC} = \sum_{i=1}^{a} \sum_{k=1}^{c} \frac{y_{i.k.}^2}{bn} - \frac{y_{....}^2}{abcn} - SS_B - SS_C$$

$$= \frac{(36)^2 + (39)^2 + (49)^2 + (53)^2}{4} - \frac{(177)^2}{16} - 45.5625 - 3.0625$$

$$= .0625$$

$$SS_{BC} = \sum_{j=1}^{b} \sum_{k=1}^{c} \frac{y_{.jk.}^2}{an} - \frac{y_{....}^2}{abcn} - SS_B - SS_C$$

$$= \frac{(38)^2 + (44)^2 + (47)^2 + (48)^2}{4} - \frac{(177)^2}{16} - 10.5625 - 3.0625$$

$$= 1.5625$$

$$SS_{ABC} = \sum_{i=1}^{a} \sum_{j=1}^{b} \sum_{k=1}^{c} \frac{y_{ijk.}^2}{n} - \frac{y_{....}^2}{abcn} - SS_A - SS_B - SS_C - SS_{AB} - SS_{AC} - SS_{BC}$$

$$= \frac{(16)^2 + (21)^2 + (20)^2 + (18)^2 + (22)^2 + (23)^2 + (27)^2 + (30)^2}{2} - \frac{(177)^2}{16}$$

$$- 45.5625 - 10.5625 - 3.0625 - 7.5625 - .0625 - 1.5625$$

$$= 73.4573 - 45.5625 - 10.5625 - 3.0625 - 7.5625 - .0625 - 1.5625$$

$$= 5.0625$$

$$SS_E = SS_T - SS_{\text{subtotals }(ABC)}$$

$$= 92.9375 - 73.4375 = 19.5000$$

The analysis of variance is summarized in Table 14-11. Feed rate has a significant effect on surface finish ($\alpha < .01$), as does the depth of cut ($.05 < \alpha < .10$). There is some indication of a mild interaction between these factors, as the F-test for AB is just less than the 10 percent critical value.

TABLE 14-11 **Analysis of Variance for Example 14-3**

Source of Variation	Sum of Squares	Degrees of Freedom	Mean Square	F_0
Feed Rate (A)	45.4625	1	45.4625	18.69[a]
Depth of Cut (B)	10.5625	1	10.5625	4.33[b]
Tool Angle (C)	3.0625	1	3.0625	1.26
AB	7.5625	1	7.5625	3.10
AC	.0625	1	.0625	.03
BC	1.5625	1	1.5625	.64
ABC	5.0625	1	5.0625	2.08
Error	19.5000	8	2.4375	
Total	92.9375	15		

[a] Significant at 1 percent.
[b] Significant at 10 percent.

If a factorial experiment with three or more factors involves a random or mixed model, then the expected mean squares must be examined in order to determine the appropriate test statistics. We will give a set of rules for constructing the expected mean squares for any balanced multifactor design. The rules will be illustrated for the two-factor factorial mixed model.

1. Each effect has either a variance component (random effect) or a fixed factor (fixed effect) associated with it. If an interaction contains at least one random effect, the entire interaction is considered as random. A variance component has Greek letters as subscripts to identify the particular random effect. Thus, in a two-way mixed model with factor A fixed and factor B random, the variance component for B is σ_β^2, and the variance component for AB is $\sigma_{\tau\beta}^2$. A fixed effect is always represented by the sum of squares of the model components associated with that factor, divided by its degrees of freedom. In our example, the effect of A is $\sum_{i=1}^a \tau_i^2/(a-1)$.

2. Enclose all the subscripts on the error term except the replication subscript in parentheses. Thus ϵ_{ijk} becomes $\epsilon_{(ij)k}$.

3. Prepare a table with a row for each model component (mean square) and a column for each subscript. Over each subscript write the number of levels of the factor associated with that subscript, and whether the factor is fixed (F) or random (R). Replicates are always considered random.

   (a) In each row, write 1 if any of the subscripts in parentheses in the row component matches the subscript in the column.

Factor	F a i	R b j	R n k
τ_i			
β_j			
$(\tau\beta)_{ij}$			
$\epsilon_{(ij)k}$	1	1	

(b) In each row, if any of the subscripts on the row component match the subscript in the column, write 0 if the column is headed by a fixed factor and 1 if the column is headed by a random factor.

Factor	F a i	R b j	R n k
τ_i	0		
β_j		1	
$(\tau\beta)_{ij}$	0	1	
$\epsilon_{(ij)k}$	1	1	1

(c) In the remaining empty row positions, write the number of levels shown above the column heading.

Factor	F a i	R b j	R n k
τ_i	0	b	n
β_j	a	1	n
$(\tau\beta)_{ij}$	0	1	n
$\epsilon_{(ij)k}$	1	1	1

4. To generate the expected mean square for any model component, first cover all columns headed by subscripts that are present in the component and not contained in parentheses on that component. Then, in each row that contains at least the same subscripts as those on the component being considered, take the product of the visible numbers and multiply by the appropriate fixed or random factor from Rule 1. The sum of these quantities is the expected mean square of the model component being

TABLE 14-12 **Expected Mean Square Derivation for the Two-Factor Mixed Model**

Factor	F a i	R b j	R n k	Expected Mean Square
τ_i	0	b	n	$\sigma^2 + n\sigma^2_{\tau\beta} + \dfrac{bn\sum \tau_i^2}{a-1}$
β_j	a	1	n	$\sigma^2 + an\sigma^2_\beta$
$(\tau\beta)_{ij}$	0	1	n	$\sigma^2 + n\sigma^2_{\tau\beta}$
$\epsilon_{(ij)k}$	1	1	1	σ^2

considered. To find $E(MS_A)$, for example, cover column i. The product of the visible numbers in the rows that contain at least subscript i are bn (row 1), n (row 3), and 1 (row 4). Note that i is missing in row 2. Therefore, the expected mean square is

$$E(MS_A) = \sigma^2 + n\sigma^2_{\tau\beta} + \frac{bn\sum_{i=1}^{a} \tau_i^2}{a-1}$$

The complete table of expected mean squares for this design is shown in Table 14-12.

● **Example 14-4.** Consider the three-factor factorial design model in Equation (14-22). Assuming that all three factors are random, the expected mean squares are derived in Table 14-13 using the algorithm presented above.

TABLE 14-13 **Expected Mean Square Derivation for the Three-Way Random Model**

Factor	R a i	R b j	R c k	R n l	Expected Mean Squares
τ_i	1	b	c	n	$\sigma^2 + cn\sigma^2_{\tau\beta} + bn\sigma^2_{\tau\gamma} + n\sigma^2_{\tau\beta\gamma} + bcn\sigma^2_\tau$
β_j	a	1	c	n	$\sigma^2 + cn\sigma^2_{\tau\beta} + an\sigma^2_{\beta\gamma} + n\sigma^2_{\tau\beta\gamma} + acn\sigma^2_\beta$
γ_k	a	b	1	n	$\sigma^2 + bn\sigma^2_{\tau\gamma} + an\sigma^2_{\beta\gamma} + n\sigma^2_{\tau\beta\gamma} + abn\sigma^2_\gamma$
$(\tau\beta)_{ij}$	1	1	c	n	$\sigma^2 + n\sigma^2_{\tau\beta\gamma} + cn\sigma^2_{\tau\beta}$
$(\tau\gamma)_{ik}$	1	b	1	n	$\sigma^2 + n\sigma^2_{\tau\beta\gamma} + bn\sigma^2_{\tau\gamma}$
$(\beta\gamma)_{jk}$	a	1	1	n	$\sigma^2 + n\sigma^2_{\tau\beta\gamma} + an\sigma^2_{\beta\gamma}$
$(\tau\beta\gamma)_{ijk}$	1	1	1	n	$\sigma^2 + n\sigma^2_{\tau\beta\gamma}$
$\epsilon_{(ijk)l}$	1	1	1	1	σ^2

From examining the expected mean squares in Table 14-13, we note that there is no appropriate test statistic for the main effects A, B, and C, unless one or more of the two-factor interactions is negligible. For example, if MS_{AB}/MS_{ABC} indicates that $\sigma_{\tau\beta}^2 = 0$, then we could set $\sigma_{\tau\beta}^2 = 0$ in $E(MS_A)$ and test $H_0: \sigma_\tau^2 = 0$ with the ratio MS_A/MS_{AC}. If certain interactions are not negligible, and we must make inferences about those effects for which tests do not exist, a procedure developed by Satterthwaite (1946) can be used. This method uses linear combinations of mean squares, say,

$$MS' = MS_r + \cdots + MS_s \tag{14-32}$$

and

$$MS'' = MS_u + \cdots + MS_v \tag{14-33}$$

where MS' and MS'' are chosen so that $E(MS') - E(MS'')$ is equal to the effect in the null hypothesis. The test statistic would be

$$F_0 = \frac{MS'}{MS''} \sim F_{p,q} \tag{14-34}$$

where

$$p = \frac{(MS_r + \cdots + MS_s)^2}{MS_r^2/f_r + \cdots + MS_s^2/f_s} \tag{14-35}$$

and

$$q = \frac{(MS_u + \cdots + MS_v)^2}{MS_u^2/f_u + \cdots + MS_v^2/f_v} \tag{14-36}$$

In p and q, f_i is the number of degrees of freedom associated with the mean square MS_i. For example, in the three-way classification random effects model (Table 14-13), we note that an appropriate test statistic for $H_0: \sigma_\tau^2 = 0$ would be $F_0 = MS'/MS''$, with

$$MS' = MS_A + MS_{ABC}$$

and

$$MS'' = MS_{AB} + MS_{AC}$$

14-5   The Randomized Complete Block Design

The paired t-test is a procedure for comparing two treatment means when all observations cannot be run under homogeneous conditions. The paired t-test reduces the error variance by removing or blocking out the extraneous effect, but at the same time reducing the error degrees of freedom for the test. The randomized block design is a design for investigating the effects of one or more factors when the entire experiment cannot be run under homogeneous conditions.

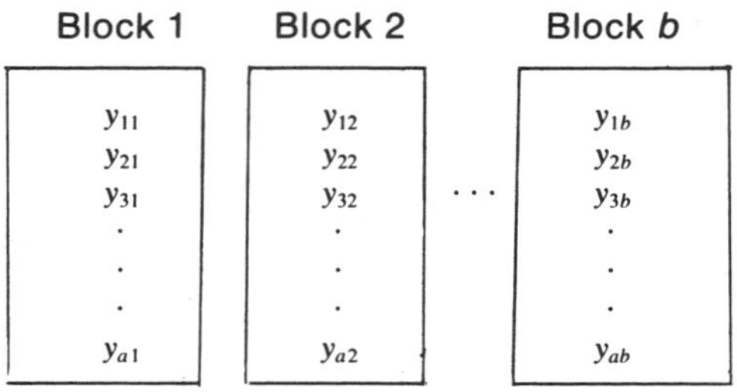

Block 1   Block 2   Block b

Fig. 14-4. The randomized complete block design.

As an example, suppose that we wish to compare the effect of four different chemicals on the strength of a particular fabric. It is known that the effect of these chemicals varies considerably from one fabric specimen to another. In this example, we have only one factor, chemical type. Therefore, we might select three pieces of fabric and compare all four chemicals within the relatively homogeneous conditions provided by each piece of fabric. This would remove any variation due to the fabric.

The general procedure for a randomized complete block design consists of selecting b blocks and running a complete replicate of the experiment in each block. If the blocks are too small to hold a complete replicate, a variation of this design, called the randomized incomplete block, must be used. A randomized complete block design for investigating a single factor with a levels would appear as in Fig. 14-4. There will be a observations (one per factor level) in each block, and the order in which these observations are run is randomly assigned within the block.

We will now describe the statistical analysis for a randomized block design. Suppose that a single factor with a levels is of interest, and the experiment is run in b blocks, as shown in Fig. 14-4. The observations may be represented by the linear statistical model

$$y_{ij} = \mu + \tau_i + \beta_j + \epsilon_{ij} \begin{cases} i = 1, 2, \ldots, a \\ j = 1, 2, \ldots, b \end{cases} \tag{14-37}$$

where μ is an overall mean, τ_i is the effect of the ith treatment, β_j is the effect of the jth block, and ϵ_{ij} is the usual NID(0, σ^2) random error term. Treatments and blocks will be considered initially as fixed factors. Furthermore, the treatment and block effects are defined as deviations from the overall mean, so that $\sum_{i=1}^{a} \tau_i = 0$ and $\sum_{j=1}^{b} \beta_j = 0$. We are interested in testing the equality of the treatment effects. That is,

$$H_0: \tau_1 = \tau_2 = \cdots = \tau_a = 0$$
$$H_1: \tau_i \neq 0 \text{ at least one } i$$

Let $y_{i.}$ be the total of all observations taken under treatment i, $y_{.j}$ be the total of all observations in block j, $y_{..}$ be the grand total of all observations, and $N = ab$ be the total number of observations. Similarly, $\bar{y}_{i.}$ is the average of the observations taken under treatments i, $\bar{y}_{.j}$ is the average of the observations in block j, and $\bar{y}_{..}$ is the grand average of all observations. The total corrected sum of squares is

$$\sum_{i=1}^{a} \sum_{j=1}^{b} (y_{ij} - \bar{y}_{..})^2 = \sum_{i=1}^{a} \sum_{i=1}^{b} [(\bar{y}_{i.} - \bar{y}_{..}) + (\bar{y}_{.j} - \bar{y}_{..}) + (y_{ij} - \bar{y}_{i.} - \bar{y}_{.j} + \bar{y}_{..})]^2$$

(14-38)

Expanding the right-hand side of Equation (14-38) yields

$$\sum_{i=1}^{a} \sum_{j=1}^{b} (y_{ij} - \bar{y}_{..})^2 = b \sum_{i=1}^{a} (\bar{y}_{i.} - \bar{y}_{..})^2 + a \sum_{j=1}^{b} (\bar{y}_{.j} - \bar{y}_{..})^2 + \sum_{i=1}^{a} \sum_{j=1}^{b} (y_{ij} - \bar{y}_{.j} - \bar{y}_{i.} + \bar{y}_{..})^2$$

(14-39)

or, symbolically,

$$SS_T = SS_{\text{treatments}} + SS_{\text{blocks}} + SS_E$$

(14-40)

The degree of freedom breakdown corresponding to Equation (14-40) is

$$ab - 1 = (a - 1) + (b - 1) + (a - 1)(b - 1)$$

(14-41)

The null hypothesis of no treatment effects ($H_0 : \tau_i = 0$) is tested by the F ratio, $MS_{\text{treatments}}/MS_E$. The analysis of variance is summarized in Table 14-14. Computing formulas for the sums of squares are also shown in this table. The same test procedure is used in cases where treatments and/or blocks are random.

TABLE 14-14   **Analysis of Variance for Randomized Complete Block Design**

Source of Variation	Sum of Squares	Degrees of Freedom	Mean Square	F_0
Treatments	$\sum_{i=1}^{a} \dfrac{y_{i.}^2}{b} - \dfrac{y_{..}^2}{ab}$	$a - 1$	$\dfrac{SS_{\text{treatments}}}{a - 1}$	$\dfrac{MS_{\text{treatments}}}{MS_E}$
Blocks	$\sum_{j=1}^{b} \dfrac{y_{.j}^2}{a} - \dfrac{y_{..}^2}{ab}$	$b - 1$	$\dfrac{SS_{\text{blocks}}}{b - 1}$	
Error	SS_E (by subtraction)	$(a - 1)(b - 1)$	$\dfrac{SS_E}{(a - 1)(b - 1)}$	
Total	$\sum_{i=1}^{a} \sum_{j=1}^{b} y_{ij}^2 - \dfrac{y_{..}^2}{ab}$	$ab - 1$		

TABLE 14-15 **Data for Example 14-5**

Chemical Type	Fabric Specimen			
	1	2	3	$y_{i.}$
1	1.3	1.6	.5	3.4
2	2.2	2.4	.4	5.0
3	1.8	1.7	.1	3.6
4	3.9	4.4	2.2	10.5
$y_{.j}$	9.2	10.1	3.2	22.5 = $y_{..}$

● **Example 14-5.** An experiment was performed to study the effect of four different chemicals on the strength of a particular type of fabric. Three fabric specimens were selected, and a randomized block design was run by testing all four chemicals in random order on each fabric specimen. The data are shown in Table 14-15.

The sums of squares are computed as follows:

$$SS_T = \sum_{i=1}^{a} \sum_{j=1}^{b} y_{ij}^2 - \frac{y_{..}^2}{ab}$$

$$= 60.81 - \frac{(22.5)^2}{12} = 18.62$$

$$SS_{\text{treatments}} = \sum_{i=1}^{a} \frac{y_{i.}^2}{b} - \frac{y_{..}^2}{ab}$$

$$= \frac{(3.4)^2 + (5.0)^2 + (3.6)^2 + (.5)^2}{3} - \frac{(22.5)^2}{12}$$

$$= 11.07$$

$$SS_{\text{blocks}} = \sum_{j=1}^{b} \frac{y_{.j}^2}{a} - \frac{y_{..}^2}{ab}$$

$$= \frac{(9.2)^2 + (10.1)^2 + (3.2)^2}{4} - \frac{(22.5)^2}{12} = 7.03$$

$$SS_E = SS_T - SS_{\text{treatments}} - SS_{\text{blocks}}$$

$$= 18.62 - 11.07 - 7.03 = .52$$

The analysis of variance is summarized in Table 14-16. We conclude that chemical types have a significant effect on fabric strength.

We note from Equation (14-37) that the randomized complete block design

TABLE 14-16   **Analysis of Variance for Example 14-5**

Source of Variation	Sum of Squares	Degrees of Freedom	Mean Square	F_0
Chemical types	11.07	3	3.69	41.00[a]
Fabric	7.03	2	3.52	
Error	.52	6	.09	
Total	18.62	11		

[a] Significant at 1 percent.

is very similar to the two-way analysis of variance model with no interaction. However, since each treatment appears only once in each block, there is only one replicate. If there is no interaction, the residual sum of squares can only be attributable to random error. On the other hand, if we have incorrectly assumed the absence of interaction, then the test on treatments can be adversely affected.

Suppose an experiment is conducted as a randomized block design, and blocking was not really necessary. There are ab observations and $(a - 1)(b - 1)$ degrees of freedom for error. If the experiment had been run as a completely randomized single factor design with b replicates, we would have had $a(b - 1)$ degrees of freedom for error. Therefore, blocking has cost $a(b - 1) - (a - 1)(b - 1) = b - 1$ degrees of freedom for error. Thus, since the loss in error degrees of freedom is usually small, if there is a reasonable chance that block effects may be important, the experimenter should use the randomized block design. For example, consider the experiment described in Example 14-5 as a one-way classification analysis of variance. We would have eight degrees of freedom for error. In the randomized block design there are six degrees of freedom for error. Therefore, blocking has cost only two degrees of freedom, a very small loss considering the possible gain in information that would be achieved if block effects are really important. As a general rule, when in doubt as to the importance of block effects, the experimenter should block and gamble that the block effect does exist. If the experimenter is wrong, the slight loss in the degrees of freedom for error will have a negligible effect, unless the number of degrees of freedom is very small.

14-6   The 2^k Factorial Design

There are certain special types of factorial experiments that are very useful. One of these is a factorial design with k factors, each at two levels. Because each complete replicate of the design has 2^k treatment combinations, the

TABLE 14-17 **Treatment Combinations in the 2^2 Design**

Factor A	Factor B	
	Low	High
Low	(1)	b
High	a	ab

arrangement is called a 2^k factorial design. These designs have a greatly simplified statistical analysis, and they also form the basis of many other useful designs.

The simplest type of 2^k design is the 2^2; that is, two factors A and B, each at two levels. We usually think of these levels as the "low" and "high" levels of the factor. The 2^2 design is shown in Table 14-17. Note that a special notation is used to represent the treatment combinations. In general a treatment combination is represented by a series of lowercase letters. If a letter is present, then the corresponding factor is run at the high level in that treatment combination; if it is absent, the factor is run at its low level. For example, treatment combination a indicates that factor A is at the high level and factor B is at the low level. The treatment combination with both factors at the low level is represented by (1). This notation is used throughout the 2^k design series. For example, the treatment combination in a 2^4 with A and C at the high level and B and D at the low level is denoted by ac.

The effects of interest in the 2^2 design are the main effects A and B and their interaction AB. Let the letters (1), a, b, and ab also represent the totals of all n observations taken at these design points. There is a single-degree-of-freedom contrast in these treatment totals associated with each effect. These contrasts are:

$$\text{Contrast}_A = a + ab - b - (1)$$
$$\text{Contrast}_B = b + ab - a - (1)$$
$$\text{Contrast}_{AB} = ab + (1) - a - b$$

Note that the contrast for A is just the sum of the treatment combinations where A is at the high level minus the sum of the treatment combinations where A is at the low level. Similarly, the contrast for B is the sum of the treatment combinations where B is at the high level minus the sum of the treatment combinations where B is at the low level. The contrast for the AB interaction is the difference in the diagonal treatment totals in Table 14-17.

To obtain the sums of squares for A, B, and AB, we can use Equation (11-18), which expresses the relationship between a single degree of freedom contrast and its sum of squares:

$$SS = \frac{(\text{contrast})^2}{n \sum (\text{contrast coefficients})^2} \tag{14-42}$$

Therefore, the sums of squares for A, B, and AB are

$$SS_A = \frac{[a + ab - b - (1)]^2}{4n}$$

$$SS_B = \frac{[b + ab - a - (1)]^2}{4n}$$

$$SS_{AB} = \frac{[ab + (1) - a - b]^2}{4n}$$

The contrast coefficients are always either $+1$ or -1. A table of plus and minus signs, such as Table 14-18, can be used to determine the sign on each treatment combination for a particular contrast. The column headings for Table 14-18 are the main effects A and B, the AB interaction, and I, which represents the total. The row headings are the treatment combinations. Note that the signs in the AB column are the product of signs from columns A and B. To generate a contrast from this table, multiply the signs in the appropriate column of Table 14-18 by the treatment combinations listed in the rows and add. The analysis of variance for the 2^2 design is completed by computing SS_T (with $4n - 1$ degrees of freedom), as usual, and obtaining SS_E [with $4(n - 1)$ degrees of freedom] by subtraction.

The same general procedure is used for analyzing other designs in the 2^k series. For example, consider the 2^3 design. The eight treatment combinations in this design are denoted by (1), a, b, ab, c, ac, bc, and abc. This design allows three main effects A, B, and C to be computed, along with three two-factor interactions AB, AC, and BC, and the three-factor interaction ABC. The table of

TABLE 14-18   **Signs for Effects in the 2^2 Design**

Treatment Combination	Factorial Effect			
	I	A	B	AB
(1)	+	−	−	+
a	+	+	−	−
b	+	−	+	−
ab	+	+	+	+

TABLE 14-19   **Signs for Effects in the 2^3 Design**

Treatment Combination	Factorial Effect							
	I	A	B	AB	C	AC	BC	ABC
(1)	+	−	−	+	−	+	+	−
a	+	+	−	−	−	−	+	+
b	+	−	+	−	−	+	−	+
ab	+	+	+	+	−	−	−	−
c	+	−	−	+	+	−	−	+
ac	+	+	−	−	+	+	−	−
bc	+	−	+	−	+	−	+	−
abc	+	+	+	+	+	+	+	+

plus and minus signs for generating the contrasts is shown in Table 14-19. The contrast for any effect is obtained by appending the signs in the appropriate column of Table 14-19 to the treatment combinations in the rows and adding. Then the sums of squares for the effects are computed as

$$SS = \frac{(\text{contrast})^2}{n2^3} \tag{14-43}$$

The total sum of squares SS_T (with $8n - 1$ degrees of freedom) is calculated as usual, and the error sum of squares SS_E with $8(n - 1)$ degrees of freedom is found by subtraction.

Instead of using the table of plus and minus signs to obtain the contrasts for the sums of squares, a simple tabular algorithm devised by Yates can be employed. To use Yates' algorithm, construct a table with the treatment combinations and the corresponding treatment totals recorded in *standard order*. By standard order, we mean that each factor is introduced one at a time by combining it with all factor levels above it. Thus for a 2^2, the standard order is (1), a, b, ab, while for a 2^3 it is (1), a, b, ab, c, ac, bc, abc, and for a 2^4 it is (1), a, b, ab, c, ac, bc, abc, d, ad, bd, abd, cd, acd, bcd, $abcd$. Then follow this three-step procedure:

1. Label the adjacent column (1). Compute the entries in the top half of this column by adding the observations in adjacent pairs. Compute the entries in the bottom half of this column by changing the sign of the first entry in each pair of the original observations and adding the adjacent pairs.
2. Label the adjacent column (2). Construct column (2) using the entries in column (1). Follow the same procedure employed to generate column (1).

TABLE 14-20   **Yates' Algorithm for the 2^3 Design in Example 14-3**

Treatment Combinations	Response	(1)	(2)	(3)	Effect	Sum of Squares $(3)^2/n2^3$
(1)	16	38	85	177	Total	—
a	22	47	92	27	A	45.5625
b	20	44	13	13	B	10.5625
ab	27	48	14	11	AB	7.5625
c	21	6	9	7	C	3.0625
ac	23	7	4	1	AC	.0625
bc	18	2	1	−5	BC	1.5625
abc	30	12	10	9	ABC	5.0625

Continue this process until k columns have been constructed. Column (k) contains the contrasts designated in the rows.

3. Calculate the sums of squares for the effects by squaring the entries in column k and dividing by $n2^k$.

● **Example 14-6.** Consider the data in Example 14-3. This is a 2^3 design with $n = 2$ replicates. The analysis of this data using Yates' algorithm is illustrated in Table 14-20. Note that the sums of squares computed from Yates' algorithm agree with the results obtained by the general method of analysis for factorials in Example 14-3.

Frequently, only a single replicate ($n = 1$) of the 2^k design is run. This is often done when the number of factors is moderately large, say $k \geq 4$ or 5. In these cases, the usual practice is to combine higher-order interactions that the experimenter feels are negligible to obtain an estimate of error. For further details, see Montgomery (1976).

14-7   Confounding in the 2^k Design

It is often impossible to run a complete replicate of a factorial design under homogeneous experimental conditions. *Confounding* is a design technique for running a factorial experiment in blocks, where the block size is smaller than the number of treatment combinations in one complete replicate. The technique causes certain interaction effects to be indistinguishable from or *confounded* with blocks. We will illustrate confounding in the 2^k factorial design in 2^p blocks, where $p < k$.

Consider the 2^2 design. Each of the $2^2 = 4$ treatment combinations requires four hours of laboratory analysis. Thus, two days are required to perform the

Block 1   Block 2

```
┌─────────┐   ┌─────────┐
│  (1)    │   │  ⌐      │
│  ab     │   │  b      │
│         │   │         │
└─────────┘   └─────────┘
```
Fig. 14-5. The 2^2 design in two blocks.

experiment. If days are considered as blocks, then we must assign two of the four treatment combinations to each day.

Consider the design shown in Fig. 14-5. Notice that block 1 contains the treatment combinations (1) and ab, and that block 2 contains a and b. The contrasts for estimating the main effects A and B are

$$\text{Contrast}_A = ab + a - b - (1)$$
$$\text{Contrast}_B = ab + b - a - (1)$$

Note that these contrasts are unaffected by blocking since in each contrast there is one plus and one minus treatment combination from each block. That is, any difference between block 1 and block 2 will cancel out. The contrast for the AB interaction is

$$\text{Contrast}_{AB} = ab + (1) - a - b$$

Since the two treatment combinations with the plus sign, ab and (1), are in block 1 and the two with the minus sign, a and b, are in block 2, the block effect and the AB interaction are identical. That is, AB is confounded with blocks.

The reason for this is apparent from the table of plus and minus signs for the 2^2 design, shown in Table 14-18. From this table, we see that all treatment combinations that have a plus on AB are assigned to block 1, while all treatment combinations that have a minus sign on AB are assigned to block 2.

This scheme can be used to confound any 2^k design in two blocks. As a second example, consider a 2^3 design, run in two blocks. Suppose we wish to confound the three-factor interaction ABC with blocks. From the table of plus and minus signs, shown in Table 14-19, we assign the treatment combinations that are minus on ABC to block 1 and those that are plus on ABC to block 2. The resulting design is shown in Fig. 14-6.

There is a more general method of constructing the blocks. The method

Block 1   Block 2

```
┌─────────┐   ┌─────────┐
│  (1)    │   │  a      │
│  ab     │   │  b      │
│  ac     │   │  c      │
│  bc     │   │  abc    │
└─────────┘   └─────────┘
```
Fig. 14-6. The 2^3 design in two blocks, ABC confounded.

employs a *defining contrast*, say

$$L = \alpha_1 x_1 + \alpha_2 x_2 + \cdots + \alpha_k x_k \qquad (14\text{-}44)$$

where x_i is the level of the ith factor appearing in a treatment combination and α_i is the exponent appearing on the ith factor in the effect to be confounded. For the 2^k system, we have either $\alpha_i = 0$ or 1, and either $x_i = 0$ (low level) or $x_i = 1$ (high level). Treatment combinations that produce the same value of L modulus (2) will be placed in the same block. Since the only possible values of L (mod 2) are 0 and 1, this will assign the 2^k treatment combinations to exactly two blocks.

As an example consider a 2^3 design with ABC confounded with blocks. Here x_1 corresponds to A, x_2 to B, x_3 to C, and $\alpha_1 = \alpha_2 = \alpha_3 = 1$. Thus, the defining contrast for ABC is

$$L = x_1 + x_2 + x_3$$

To assign the treatment combinations to the two blocks, we substitute the treatment combinations into the defining contrast as follows:

$$(1): \ L = 1(0) + 1(0) + 1(0) = 0 = 0 \ (\text{mod } 2)$$
$$a: \ L = 1(1) + 1(0) + 1(0) = 1 = 1 \ (\text{mod } 2)$$
$$b: \ L = 1(0) + 1(1) + 1(0) = 1 = 1 \ (\text{mod } 2)$$
$$ab: \ L = 1(1) + 1(1) + 1(0) = 2 = 0 \ (\text{mod } 2)$$
$$c: \ L = 1(0) + 1(0) + 1(1) = 1 = 1 \ (\text{mod } 2)$$
$$ac: \ L = 1(1) + 1(0) + 1(1) = 2 = 0 \ (\text{mod } 2)$$
$$bc: \ L = 1(0) + 1(1) + 1(1) = 2 = 0 \ (\text{mod } 2)$$
$$abc: \ L = 1(1) + 1(1) + 1(1) = 3 = 1 \ (\text{mod } 2)$$

Thus (1), ab, ac, and bc are run in block 1, and a, b, c, and abc are run in block 2. This is the same design shown in Figure 14-6.

Another method may be used to construct these designs. The block containing the treatment combination (1) is called the *principal block*. Any element [except (1)] in the principal block may be generated by multiplying two other elements in the principal block modulus 2. For example, consider the principal block of the 2^3 design with ABC confounded, shown in Fig. 14-6. Note that

$$ab \cdot ac = a^2 bc = bc$$
$$ab \cdot bc = ab^2 c = ac$$
$$ac \cdot bc = abc^2 = ab$$

Treatment combinations in the other block (or blocks) may be generated by multiplying one element in the new block by each element in the principal block modulus 2. For the 2^3 with ABC confounded, since the principal block is (1), ab, ac, and bc, we know that b is in the other block. Thus, the elements of this second block are

$$b \cdot (1) \qquad = b$$
$$b \cdot ab = ab^2 = a$$
$$b \cdot ac \qquad = abc$$
$$b \cdot bc = b^2c = c$$

● **Example 14-7.** An experiment is performed to study the effect of four factors on the target error of a shoulder-fired ground-to-air missile. The four factors are propellant type (A), seeker calibration (B), target altitude (C), and target range (D). Each factor has two levels. Two different gunners are used in the flight test and, since there may be differences between individuals, it was decided to conduct the 2^4 design in two blocks with $ABCD$ confounded. Thus, the defining contrast is

$$L = x_1 + x_2 + x_3 + x_4$$

The resulting design is

Block 1	Block 2
(1) = 3	a = 7
ab = 7	b = 5
ac = 6	c = 6
bc = 8	d = 4
ad = 10	abc = 6
bd = 4	bcd = 7
cd = 8	acd = 9
$abcd$ = 9	abd = 12

The analysis of the design by Yates' algorithm is shown in Table 14-21. The analysis of variance is summarized in Table 14-22, assuming that all three-factor interactions are negligible. The main effects of A (propellant type) and D (target range) are significant, as are the AC and AD interactions.

It is possible to confound the 2^k design in four blocks of 2^{k-2} observations each. To construct the design, two effects are chosen to confound with blocks and their defining contrasts obtained. A third effect, the generalized interaction of the two initially chosen, is also confounded with blocks. The generalized interaction of two effects is found by multiplying them together modulus 2.

For example, consider the 2^4 design in four blocks. Let AC and BD be confounded, along with their generalized interaction $(AC)(BD) = ABCD$. The

TABLE 14-21   **Yates' Algorithm for the 2^4 Design in Example 14-7**

Treatment Combination	Response	(1)	(2)	(3)	(4)	Effect	Sum of Squares
(1)	3	10	22	48	111	Total	—
a	7	12	26	63	21	A	27.5625
b	5	12	30	4	5	B	1.5625
ab	7	14	33	17	−1	AB	.0625
c	6	14	6	4	7	C	3.0625
ac	6	16	−2	1	−19	AC	22.5625
bc	8	17	14	−4	−3	BC	.5625
abc	6	16	3	3	−1	ABC	.0625
d	4	4	2	4	15	D	14.0625
ad	10	2	2	3	13	AD	10.5625
bd	4	0	2	−8	−3	BD	.5625
abd	12	−2	−1	−11	7	ABD	3.0625
cd	8	6	−2	0	−1	CD	.0625
acd	9	8	−2	−3	−3	ACD	.5625
bcd	7	1	2	0	−3	BCD	.5625
abcd	9	2	1	−1	−1	ABCD	.0625

TABLE 14-22   **Analysis of Variance for Example 14-7**

Source of Variation	Sum of Squares	Degrees of Freedom	Mean Square	F_0
Blocks (ABCD)	.0625	1	27.5625	25.94
A	27.5625	1	27.5625	25.94
B	1.5625	1	1.5625	1.47
C	3.0625	1	3.0625	2.88
D	14.0625	1	14.0625	13.24
AB	.0625	1	.0625	.06
AC	22.5625	1	22.5625	21.24
AD	10.5625	1	10.4525	9.94
BC	.5625	1	.5625	.53
BD	.5625	1	.5625	.53
CD	.0625	1	.0625	.06
Error (ABC + ABD + ACD + BCD)	4.2500	4	1.0625	
Total	84.9375	15		

design is constructed by using the defining contrasts for AC and BD:

$$L_1 = x_1 + x_3$$
$$L_2 = x_2 + x_4$$

It is easy to verify that the four blocks are

Block 1 $L_1 = 0, L_2 = 0$	Block 2 $L_1 = 1, L_2 = 0$	Block 3 $L_1 = 0, L_2 = 1$	Block 4 $L_1 = 1, L_2 = 1$
(1)	a	b	ab
ac	c	abc	bc
bd	abd	d	ad
abcd	bcd	acd	cd

This general procedure can be extended to confounding the 2^k design in 2^k blocks, where $p < k$. Select p effects to be confounded, such that no effect chosen is a generalized interaction of the others. The blocks can be constructed from the p defining contrasts $L_1, L_2, \ldots, L_p$ associated with these effects. In addition, exactly $2^p - p - 1$ other effects are confounded with blocks, these being the generalized interaction of the original p effects chosen. Care should be taken so as not to confound effects of potential interest.

14-8 Fractional Replication of the 2^k Design

As the number of factors in a 2^k increases, the number of runs required increases rapidly. For example, a 2^5 requires 32 runs. In this design, only 5 degrees of freedom correspond to main effects and 10 degrees of freedom correspond to two-factor interactions. If we can assume that certain high-order interactions are negligible, then a fractional factorial design can be used to obtain information on the main effects and low-order interactions. In this section, we will introduce fractional replication of the 2^k design. For a more complete treatment, see Montgomery (1976).

A one-half fraction of the 2^k design contains 2^{k-1} runs and is often called a 2^{k-1} fractional factorial design. As an example, consider the 2^{3-1} design; that is, a one-half fraction of the 2^3. The table of plus and minus signs for the 2^3 design is shown in Table 14-23. Suppose we select the four treatment combinations a, b, c, and abc as our one-half fraction. These treatment combinations are shown in the top half of Table 14-23. We will use both the conventional notation $(a, b, c, \ldots)$ and the plus and minus notation for the treatment combinations. The equivalence between the two notations is as follows:

Notation 1	Notation 2
a	$+--$
b	$-+-$
c	$--+$
abc	$+++$

Notice that the 2^{3-1} design is formed by selecting only those treatment combinations that yield a plus on the ABC effect. Thus ABC is called the *generator* of this particular fraction. Furthermore, the identity element I is also always plus, so we call

$$I = ABC$$

the *defining relation* for the design.

The treatment combinations in the 2^{3-1} design yield three degrees of freedom associated with the main effects. From Table 14-23, we obtain the contrasts for the main effects as

$$\text{Contrast}_A = a - b - c + abc$$
$$\text{Contrast}_B = -a + b - c + abc$$
$$\text{Contrast}_C = -a - b + c + abc$$

It is also easy to verify that the contrast for the two-factor interactions are

$$\text{Contrast}_{BC} = a - b - c + abc$$
$$\text{Contrast}_{AC} = -a + b - c + abc$$
$$\text{Contrast}_{AB} = -a - b + c + abc$$

TABLE 14-23   **Plus and Minus Signs for the 2^3 Factorial Design**

Treatment Combination	Factorial Effect							
	I	A	B	C	AB	AC	BC	ABC
a	$+$	$+$	$-$	$-$	$-$	$-$	$+$	$+$
'b	$+$	$-$	$+$	$-$	$-$	$+$	$-$	$+$
c	$+$	$-$	$-$	$+$	$+$	$-$	$-$	$+$
abc	$+$	$+$	$+$	$+$	$+$	$+$	$+$	$+$
ab	$+$	$+$	$+$	$-$	$+$	$-$	$-$	$-$
ac	$+$	$+$	$-$	$+$	$-$	$+$	$-$	$-$
bc	$+$	$-$	$+$	$+$	$-$	$-$	$+$	$-$
(1)	$+$	$-$	$-$	$-$	$+$	$+$	$+$	$-$

Thus we cannot differentiate between A and BC, B and AC, and C and AB. In fact, we can show that the effects A, B, and C really estimate $A + BC$, $B + AC$, and $C + AB$. Two or more effects that have this property are called aliases. In our example, A and BC are aliases, B and AC are aliases, and C and AB are aliases.

The alias structure for this design is found by using the defining relation $I = ABC$. Multiplying any effect by the defining relation modulus 2 yields the aliases for that effect. In our example, this yields as the alias of A

$$A = A \cdot ABC = A^2 BC = BC$$

since $A \cdot I = A$ and $A^2 = I$. The aliases of B and C are

$$B = B \cdot ABC = AB^2 C = AC$$

and

$$C = C \cdot ABC = ABC^2 = AB$$

Now suppose that we had chosen the other one-half fraction; that is, the treatment combinations in Table 14-23 associated with minus on ABC. The defining relation for this design is $I = -ABC$. The aliases are $A = -BC$, $B = -AC$, and $C = -AB$. Thus the effects A, B, and C with this particular fraction really estimate $A - BC$, $B - AC$, and $C - AB$.

In general, a one-half fraction of a 2^k design may be constructed by first partitioning the full 2^k design into two blocks, using the highest-order interaction as the defining contrast. Each block is then a 2^{k-1} fractional factorial design. The defining relation of the 2^{k-1} is $I = \pm ABC \ldots K$, where the sign on the generator depends on the fraction chosen.

A 2^{k-1} design may also be generated by writing down the treatment combination for a full 2^{k-1} factorial and then adding the kth factor by identifying its plus and minus levels with the plus and minus signs of the highest-order interaction $ABC \ldots (K - 1)$. Therefore, the 2^{3-1} fractional factorial is obtained by writing down the full 2^2 factorial and then equating factor C to the AB interaction. Thus,

Full 2^2		2^{3-1}, $I = ABC$		
A	B	A	B	$C = AB$
$-$	$-$	$-$	$-$	$+$
$+$	$-$	$+$	$-$	$-$
$-$	$+$	$-$	$+$	$-$
$+$	$+$	$+$	$+$	$+$

Now consider the one-quarter fraction of the 2^k design. This design contains 2^{k-p} runs and is often called a 2^{k-2} fractional factorial design. This design may be constructed by selecting two effects or *generators* and partitioning the full 2^k factorial into four blocks. Any one of these four blocks is a 2^{k-2} fractional factorial design. The complete defining relation used to generate the alias structure consists of the two originally chosen effects and their generalized interaction. For example, consider the 2^{6-2} design. Choose $I = ABCE$ and $I = ACDF$ as the generators. Then, the generalized interaction of $ABCE$ and $ACDF$ is $(ABCE)(ACDF) = BDEF$, and the complete defining relation is

$$I = ABCE = ACDF = BDEF$$

To find the alias of any effect, simply multiply the effect by each *word* in the above defining relation. The complete alias structure is shown here.

$$
\begin{aligned}
A &= BCE = CDF &= ABDEF \\
B &= ACE = DEF &= ABCDF \\
C &= ABE = ADF &= BCDEF \\
D &= ACF = BEF &= ABCDE \\
E &= ABC = BDF &= ACDEF \\
F &= ACD = BDE &= ABCEF \\
AB &= CE = BCDF &= ADEF \\
AC &= BE = DF &= ABCDEF \\
AD &= CF = BCDE &= ABEF \\
AE &= BC = CDEF &= ABDF \\
AF &= CD = BCEF &= ABDE \\
BD &= EF = ACDE &= ABCF \\
BF &= DE = ABCD &= ACEF \\
ABF &= CEF = BCD &= ADE \\
CDE &= ABD = AEF &= CBF
\end{aligned}
$$

To construct this design, we could form the four blocks of the 2^6 design, with $ABCE$ and $ACDF$ confounded, and then choose the block with treatment combinations that are positive on $ABCE$ and $ACDF$. This will produce the treatment combinations (1), *aef, be, abf, cef, ac, bcf, abce, df, ade, bdef, abd, cde, acdf, bcd,* and *abcdef.* This design could also be constructed by first writing down a full $2^{6-2} = 2^4$ in the factors $A, B, C,$ and D, and setting $E = ABC$ and $F = ACD$ to produce the levels for these two additional factors.

14-9 Summary

This chapter has introduced the design and analysis of experiments, concentrating on factorial designs and the randomized block design. Fixed, random, and mixed models were considered. The F-tests for main effects and interactions depends on whether the factors are fixed or random. An algorithm for determining the expected mean squares was given. These expected mean squares are a guide to test statistic construction. The 2^k factorial design was also introduced. These are very useful designs in which all k factors appear at two levels. They have a greatly simplified method of statistical analysis. In situations where the design cannot be run under homogeneous conditions, the 2^k design can be easily confounded in 2^p blocks. This requires that certain interactions be confounded with blocks. The 2^k design also lends itself to fractional replication, in which only a particular subset of the 2^k treatment combinations are run. In fractional replication, each effect is aliased with one or more other effects.

14-10 Exercises

14-1. The following data represent the yield of a chemical process under several predetermined operating conditions. Analyze the data and draw appropriate conclusions.

	Pressure		
Temperature	240	260	280
Low	40.4 40.2	40.7 40.6	40.2 40.4
Medium	40.1 40.3	40.5 40.6	40.0 40.1
High	40.5 40.7	40.8 40.9	40.3 40.1

14-2. The life of a particular type of storage battery is thought to be influenced by the material used in the plates and the temperature in the location at which the battery is installed. An experiment is run in the laboratory for three materials and three temperatures, and the results are shown in the following table. Analyze the data and draw appropriate conclusions.

	Temperature (°F)					
Material	50		65		80	
1	130	155	34	40	20	70
	74	180	80	75	82	58
2	150	188	136	122	25	70
	159	126	106	115	58	45
3	138	110	174	120	96	104
	168	160	150	139	82	60

14-3. An engineer suspects that the surface finish of a metal part is influenced by the type of paint used and the drying time. He selects three drying times—20, 25, and 30 minutes—and randomly chooses two types of paint from several that are available. He conducts an experiment and obtains the data shown here. Analyze the data and draw conclusions. Estimate the variance components.

Paint	Drying Time (min) 20	25	30
1	74	73	78
	64	61	85
	50	44	92
2	92	98	66
	86	73	45
	68	88	85

14-4. Suppose that in Exercise 14-3 paint types were fixed effects. Compute a 95 percent interval estimate of the mean difference between the responses for paint type 1 and paint type 2.

14-5. The factors that influence the breaking strength of cloth are being studied. Four machines and three operators are chosen at random and an experiment is run using cloth from the same one-yard segment. The results are as follows.

Operator	Machine 1	2	3	4
A	109	110	108	110
	110	115	109	116
B	111	110	111	114
	112	111	109	112
C	109	112	114	111
	111	115	109	112

Test for interaction and main effects at the 5 percent level. Estimate the components of variance.

14-6. Suppose that in Exercise 14-5 the operators were chosen at random, but only four machines were available for the test. Does this influence the analysis or your conclusions?

14-7. A company employs two time-study engineers. Their supervisor wishes to determine whether the standards set by them are influenced by any interaction between engineers and operators. She selects three operators at random and conducts an experiment in which the engineers set standard times for the same job. She obtains the data shown here. Analyze the data and draw conclusions.

		Operator	
Engineer	1	2	3
1	2.59	2.38	2.40
	2.78	2.49	2.72
2	2.15	2.85	2.66
	2.86	2.72	2.87

14-8. Analyze the data shown in the following table, assuming that both row and column factors are fixed.

Row Factor	Column Factor		
	1	2	3
1	580	1090	1392
	570	1085	1386
2	530	1070	1328
	579	1000	1299
3	546	1045	1355
	599	1066	1368

14-9. Analyze the data shown in the following table, assuming that both row and column factors are fixed.

Row Factor	Column Factor		
	A	B	C
1	565	1080	510
	583	1043	590
2	528	988	526
	547	1026	538

14-10. Analyze the following data, where row and column treatments are both fixed effects, and it is known that no interaction is present.

Row Factor	Column Factor		
	1	2	3
A	36	38	37
B	18	19	22
C	30	38	33

14-11. For the data shown here, assume that machines are fixed and operators are random, and conduct an analysis of variance to determine whether operators or machines significantly affect the response. Assume that no interaction exists.

	Machine		
Operator	A	B	C
1	9.60	11.28	9.00
2	9.69	10.10	9.57
3	8.43	11.01	9.03
4	9.98	10.44	9.80

14-12. For the data in Exercise 14-11, suppose both operators and machines are chosen at random. Assuming no interaction, conduct the analysis of variance. How do your conclusions differ, if at all, from the conclusions of Exercise 14-11.

14-13. The percentage of hardwood concentration in raw pulp, the freeness, and the cooking time of pulp are being investigated for their effects on the strength of

Percentage of Hardwood Concentration	Cooking Time 1.5 hours			Cooking Time 2.0 hours		
	Freeness			Freeness		
	400	500	650	400	500	650
10	96.6	97.7	99.4	98.4	99.6	100.6
	96.0	96.0	99.8	98.6	100.4	100.9
15	98.5	96.0	98.4	97.5	98.7	99.6
	97.2	96.9	97.6	98.1	98.0	99.0
20	97.5	95.6	97.4	97.6	97.0	98.5
	96.6	96.2	98.1	98.4	97.8	99.8

paper. Analyze the data shown in the following table, assuming that all three factors are fixed.

14-14. The quality control department of a fabric finishing plant is studying the effect of several factors on the dyeing of cotton-synthetic cloth, which is used to manufacture men's shirts. Analyze the data shown in the following table, assuming that all three factors are fixed.

	250°F			275°F		
	Operator			Operator		
Vat	A	B	C	A	B	C
1	23	37	31	24	38	34
	34	28	32	33	36	36
	35	37	35	38	35	39
2	36	34	33	37	34	34
	35	38	34	39	38	36
	36	39	35	30	39	31
3	28	35	26	26	36	28
	24	35	27	29	37	26
	27	34	25	30	34	24

14-15. Use the algorithm in the text to derive the expected mean squares for a three-factor factorial design where factors A and B are fixed and factor C is random. How would the F ratios be found?

14-16. Use the algorithm in the test to derive expected mean squares for a three-factor factorial design where factor A is fixed and factors B and C are random. How would the F ratios be formed?

14-17. Analyze the data in Exercise 14-13, assuming that freeness is a random factor and that hardwood concentrations and cooking times are fixed.

14-18. Consider the three-factor model

$$y_{ijk} = \mu + \alpha_i + \beta_j + \gamma_k + (\alpha\beta)_{ij} + (\beta\gamma)_{jk} + \epsilon_{ijk}$$

for $i = 1, 2, \ldots, a, j = 1, 2, \ldots, b,$ and $k = 1, 2, \ldots, c.$ Assuming all treatments to be random effects, write down the analysis of variance table, including the expected mean squares. Do exact tests exist for all main effects? If not, propose appropriate statistics.

14-19. A company wishes to test the effect of four different chemical agents on a particular type of paper. Because there might be variation from one roll of paper to another, four rolls are selected and all four chemicals are applied to each. The discoloration of each sample is then measured and coded as shown here. Analyze the data and draw conclusions.

		Bolt		
Chemical	1	2	3	4
1	73	68	74	71
2	73	67	75	72
3	73	68	75	73
4	75	72	78	75

14-20. The owner of a fleet of cars used by salesmen wishes to study the wear characteristics of four brands of tires. Because driver characteristics, driving conditions, and vehicle condition have a considerable effect on tire performance, he decides to place one tire of each brand on each vehicle. The actual wheel location is chosen at random. Five cars are selected at random and the wear after 10,000 km is measured. The results are shown in the following table. Analyze the data and draw conclusions.

			Car		
Brand	1	2	3	4	5
A	17	13	13	13	13
B	14	13	13	8	13
C	12	10	10	9	12
D	13	11	11	9	10

14-21. An engineer is interested in the effect of cutting speed (A), metal hardness (B), and cutting angle (C) on the life of a cutting tool. Two levels of each factor are chosen, and two replicates of a 2^3 factorial design are run. The results are shown in the following tables. Analyze the data from this experiment.

Treatment Combination	Replicate	
	I	II
(1)	22	31
a	32	43
b	35	34
ab	55	47
c	44	45
ac	40	37
bc	60	50
abc	39	41

14-22. Data from two replicates of a 2^4 factorial design are shown below. Analyze the data and draw conclusions.

Treatment Combination	Replicate I	Replicate II	Treatment Combination	Replicate I	Replicate II
(1)	190	193	d	198	195
a	174	178	ad	172	176
b	181	185	bd	187	183
ab	183	180	abd	185	186
c	177	178	cd	199	190
ac	181	180	acd	179	175
bc	188	182	bcd	187	184
abc	173	170	abcd	180	180

14-23. Consider the data in Exercise 14-13. Delete the 500 level of freeness and the hardwood concentration of 10 percent. Analyze the data as a 2^3 factorial design.

14-24. The data shown here represent a single replicate of a 2^5 design that is used in an experiment to study the compressive strength of concrete. The factors are mix (A), time (B), laboratory (C), temperature (D), and drying time (E). Analyze the data, assuming that three-factor and higher interactions are negligible.

(1) = 700	d = 1000	e = 800	de = 1000
a = 900	ad = 1100	ae = 1200	ade = 1500
b = 3400	bd = 3000	be = 3500	bde = 4000
ab = 5500	abd = 6100	abe = 6200	abde = 6500
c = 600	cd = 800	ce = 500	cde = 1500
ac = 1000	acd = 1100	ace = 1200	acde = 2000
bc = 3000	bcd = 3400	bce = 2500	bcde = 3400
abc = 5060	abcd = 6000	abce = 5500	abcde = 6500

14-25. Consider the data from the first replicate of Exercise 14-21. Suppose that these observations could not all be run under the same conditions. Set up a design to run these observations in two blocks of four observations, each with ABC confounded. Analyze the data.

14-26. Consider the data from the first replicate of Exercise 14-22. Construct a design with two blocks of eight observations each, with $ABCD$ confounded. Analyze the data.

14-27. Repeat Exercise 14-26, assuming that four blocks are required. Confound ABD and ABC (and consequently CD) with blocks.

14-28. Suppose that in Exercise 14-22 it was only possible to run a one-half fraction of the 2^4 design. Construct the design and perform the statistical analysis, using the data from replicate I.

14-29. Suppose that in Exercise 14-24 only a one-half fraction of the 2^5 design could be run. Construct the design and perform the analysis.

14-30. Consider the data in Exercise 14-24. Suppose that only a one-quarter fraction of the 2^5 design could be run. Construct the design and analyze the data.

Chapter 15

Quality Control and Reliability

Many important industrial applications of probability and statistics involve quality control and reliability engineering. The field of statistical quality control consists of two general types of techniques: *control charts*, for monitoring the performance of a production process, and *acceptance sampling*. These techniques will be discussed in this chapter. We will also discuss the use of probability models for reliability analysis, as well as some aspects of life testing.

15-1 Statistical Quality Control

All manufacturing processes, however good, are characterized by a certain amount of random variation that cannot be completely eliminated. When this variability is confined to *chance variation* only, the process is said to be in a state of *statistical control*. However, another situation may exist in which the process variability is also affected by some *assignable cause*, such as a faulty machine setting, operator error, unsatisfactory raw material, worn machine components, and so on. These assignable causes of variation usually have an adverse effect on product quality, so it is important to have some systematic technique for detecting serious departures from a state of statistical control as soon as possible after they occur. Control charts are principally used for this purpose.

We distinguish between control charts for *measurements* and control charts for *attributes*, depending on whether the observations of the quality characteristic are measurements or enumeration data. For example, we may choose to measure the diameter of a shaft, say with a micrometer, and utilize these data in conjunction with a control chart for measurements. On the other hand, we may judge each unit of product as either defective or nondefective, and

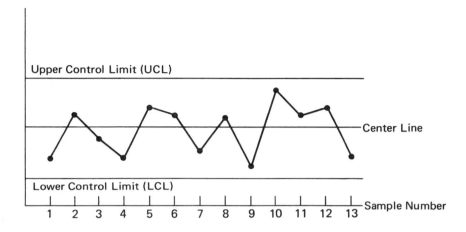

Fig. 15-1. A typical control chart.

use the total count of defectives in conjunction with a control chart for
attributes. Obviously, certain products and quality characteristics lend them-
selves to analysis by either method, and a clear-cut choice between the two
methods may be difficult.

A control chart, whether for measurements or attributes, consists of a
centerline corresponding to the average quality at which the process should
perform when statistical control is exhibited, and two *control limits*, called the
upper and lower control limits. A typical control chart is shown in Fig. 15-1.
The control limits are chosen so that values falling between them can be
attributed to chance variation, while values falling beyond them can be taken
to indicate a lack of statistical control. The general approach consists of
periodically taking a random sample from the process, computing some
appropriate quantity, and plotting that quantity on the control chart. When a
sample value falls outside the control limits, we search for some assignable
cause of variation. However, even if a sample value falls between the control
limits, a trend or some other systematic pattern may indicate that some action
is necessary, usually to avoid more serious trouble. The samples should be
selected in such a way that each sample is as homogeneous as possible, and at
the same time maximizes the opportunity for variations due to assignable
cause to be present. This is usually called the *rational subgroup* concept.
Order of production, and source (if more than one source exists), are
commonly used bases for obtaining rational subgroups.

The ability to interpret control charts accurately is usually acquired with
experience. It is necessary that the user be thoroughly familiar with both the

statistical foundation of control charts and the nature of the production process itself.

15-2 Control Charts for Measurements

When dealing with a quality characteristic that can be expressed as a measurement, it is customary to exercise control over both the average value of the quality characteristic and its variability. Control over the average quality is exercised by the control chart for means, usually called the $\bar{X}$ chart. Process variability can be controlled by either an R chart or a σ chart, depending on how the population standard deviation is estimated. We will discuss only the R chart.

Suppose that the process mean and standard deviation, say μ and σ, are known, and, furthermore, that we can assume the quality characteristic follows the normal distribution. Let $\bar{X}$ be the sample mean based on a random sample of size n from this process. Then the probability is $1 - \alpha$ that the mean of such random samples will fall between $\mu + Z_{\alpha/2}(\sigma/\sqrt{n})$ and $\mu - Z_{\alpha/2}(\sigma/\sqrt{n})$. Therefore, we could use these two values as the upper and lower control limits, respectively. However, we usually do not know μ and σ and they must be estimated. Furthermore, we may not be able to make the normality assumption. For these reasons, the probability limit $1 - \alpha$ is seldom used in practice. Usually $Z_{\alpha/2}$ is replaced by 3, and "three-sigma" control limits are used.

When μ and σ are unknown, we usually estimate them on the basis of preliminary samples, taken when the process is thought to be in control. Often 20 to 25, or more, preliminary samples are used. Suppose k preliminary samples are available, each of size n. Typically n will be 4, 5, or 6; these relatively small sample sizes are widely used and often arise from the construction of rational subgroups. Let the sample mean for the ith sample be $\bar{X}_i$. Then we estimate the mean of the population, μ, by the grand mean

$$\bar{\bar{X}} = \frac{1}{k} \sum_{i=1}^{k} \bar{X}_i \qquad (15\text{-}1)$$

Thus, we may take $\bar{\bar{X}}$ as the centerline on the $\bar{X}$ control chart.

We may estimate σ from either the standard deviations or the ranges of the k samples. Since it is more frequently used in practice, we confine our discussion to the range method. The sample size is relatively small, so there is little loss in efficiency in estimating σ from the sample ranges. The relationship between the range, R, of a sample from a normal population with known parameters and the standard deviation of that population is needed. Since R is a random variable, the quantity $W = R/\sigma$, called the relative range, is also a random variable. The parameters of the distribution of W have been determined for any sample size n. The mean of the distribution of W is called d_2, and a table of d_2 for various n is given in Table XIII of the Appendix. Let R_i be the range of the ith sample, and let

$$\bar{R} = \frac{1}{k} \sum_{i=1}^{k} R_i \qquad (15\text{-}2)$$

be the average range. Then an estimate of σ would be

$$\hat{\sigma} = \frac{\bar{R}}{d_2} \qquad (15\text{-}3)$$

Therefore, we may use as our upper and lower control limits for the $\bar{X}$ chart

$$\text{UCL} = \bar{\bar{X}} + \frac{3}{d_2 \sqrt{n}} \bar{R}$$

$$\text{LCL} = \bar{\bar{X}} - \frac{3}{d_2 \sqrt{n}} \bar{R} \qquad (15\text{-}4)$$

We note that the quantity

$$A_2 = \frac{3}{d_2 \sqrt{n}}$$

is a constant depending on the sample size, so it is possible to rewrite Equation (15-4) as

$$\text{UCL} = \bar{\bar{X}} + A_2 \bar{R}$$
$$\text{LCL} = \bar{\bar{X}} - A_2 \bar{R} \qquad (15\text{-}5)$$

The constant A_2 is tabulated for various sample sizes in Table XIII of the Appendix.

The parameters of the R chart may also be easily determined. The centerline will obviously be $\bar{R}$. To determine the control limits, we need an estimate of σ_R, the standard deviation of R. Once again, assuming the process is in control, the distribution of the relative range, W, will be useful. The standard deviation of W, say σ_W, is a function of n, which has been determined. Thus, since

$$R = W\sigma$$

we may obtain the standard deviation of R as

$$\sigma_R = \sigma_W \sigma$$

As σ is unknown, we may estimate σ_R as

$$\hat{\sigma}_R = \sigma_W \frac{\bar{R}}{d_2}$$

and we would use as the upper and lower control limits on the R chart

$$\text{UCL} = \bar{R} + \frac{3\sigma_W}{d_2} \bar{R}$$

$$\text{LCL} = \bar{R} - \frac{3\sigma_W}{d_2} \bar{R} \tag{15-6}$$

Setting $D_3 = 1 - 3\sigma_W/d_2$ and $D_4 = 1 + 3\sigma_W/d_2$, we may rewrite Equation (15-6) as

$$\text{UCL} = D_4\bar{R}$$

$$\text{LCL} = D_3\bar{R} \tag{15-7}$$

where D_3 and D_4 are tabulated in Table XIII of the Appendix.

When preliminary samples are used to construct limits for control charts, it is customary to treat these limits as trial values. Therefore, the k sample means and ranges should be plotted on the appropriate charts, and any points that exceed the control limits should be investigated. If assignable causes for these points are discovered, they should be eliminated and new limits for the control charts determined. In this way, the process may eventually be brought into statistical control and its inherent capabilities assessed. Other changes in process centering and dispersion may then be contemplated.

● **Example 15-1.** Suppose that we desire to construct $\bar{X}$ and R control charts for the diameters of bearings. The data on 20 preliminary samples, each of size 5, are shown below. The values given are the last three digits of the dimension reading, that is, 31.6 should be .50316.

Sample	$\bar{X}$	R	Sample	$\bar{X}$	R
1	31.6	4	11	29.8	4
2	33.0	3	12	34.0	4
3	35.0	4	13	33.0	10
4	32.2	4	14	34.8	4
5	33.8	2	15	35.6	7
6	38.4	3	16	30.8	6
7	31.6	4	17	33.0	5
8	36.8	10	18	31.6	3
9	35.0	15	19	28.2	9
10	34.0	6	20	33.8	6

We may compute

$$\sum_{i=1}^{20} \bar{X}_i = 666.0 \qquad \bar{\bar{X}} = 33.3 \qquad \sum_{i=1}^{20} R_i = 113 \qquad \bar{R} = 5.65$$

Notice that even though $\bar{X}_i$, $\bar{X}$, R, and $\bar{R}$ are now realizations of random variables, we have still written them as uppercase letters. This is the usual convention in quality control, and it will always be clear from the context what the notation implies. The trial control limits are, for the $\bar{X}$ chart,

or

$$\bar{\bar{X}} \pm A_2\bar{R} = 33.3 \pm (.577)(5.65) = 33.3 \pm 3.26$$

$$\text{UCL} = 36.56$$

$$\text{LCL} = 30.04$$

For the R chart, the trial control limits are

$$\text{UCL} = D_4\bar{R} = (2.115)(5.65) = 11.95$$

$$\text{LCL} = D_3\bar{R} = (0)(5.65) = 0$$

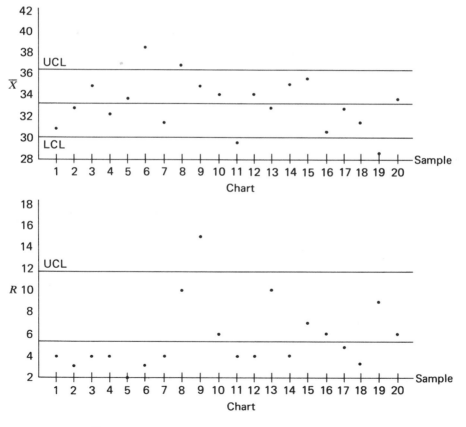

Fig. 15-2. The $\bar{X}$ and R control charts for bearing diameters.

The $\bar{X}$ and R control charts with these trial limits are shown in Fig. 15-2. We see that samples 6, 8, 9, 11, and 19 fall outside the control limits. Suppose that assignable causes of variation can be found for these five samples. Discarding the five samples we may now recompute the control limits to be, for the $\bar{X}$ chart,

$$\text{UCL} = \bar{\bar{X}} + A_2\bar{R} = 33.19 + (.577)(4.8) = 35.96$$
$$\text{LCL} = \bar{\bar{X}} - A_2\bar{R} = 33.19 - (.577)(4.8) = 30.42$$

and, for the R chart,

$$\text{UCL} = D_4\bar{R} = (2.115)(4.8) = 10.15$$
$$\text{LCL} = D_3\bar{R} = (0)(4.8) = 0$$

As all remaining points now fall within these limits, we assume the process can be controlled at this level of quality.

15-3 Control Charts for Attributes

Often it is desirable to classify a product as either defective or nondefective on the basis of comparison with a standard. This is usually done to achieve economy and simplicity in the inspection operation. For example, the diameter of a ball bearing may be checked by determining whether it will pass through a gauge consisting of circular holes cut in a template. This would be much simpler than measuring the diameter with a micrometer. Control charts for attributes are used in these situations. However, attribute control charts require a considerably larger sample size than do their measurements counterparts. We will discuss the fraction defective chart, or p chart, and the defects per unit chart, or c chart. Note that it is possible for a unit to have many defects, yet it is either defective or nondefective. In many applications a unit can have several defects, yet be classified as nondefective.

Suppose D is the number of defective units in a random sample of size n. We assume that D is a binomial random variable with unknown parameter p. Now the sample fraction defective is an estimator of p, that is

$$\hat{p} = \frac{D}{n} \tag{15-8}$$

Furthermore, the variance of the binomial distribution with parameter p is

$$\sigma_p^2 = \frac{p(1-p)}{n}$$

so we may estimate σ_p^2 as

$$\hat{\sigma}_p^2 = \frac{\hat{p}(1-\hat{p})}{n} \tag{15-9}$$

The centerline and control limits for the fraction defective control chart may now be easily determined. Suppose k preliminary samples are available, each of size n, and D_i is the number of defectives in the ith sample. Then we may take

$$\hat{p} = \frac{\sum\limits_{i=1}^{k} D_i}{kn} \tag{15-10}$$

as the centerline, and

$$\mathrm{UCL} = \hat{p} + 3\sqrt{\frac{\hat{p}(1-\hat{p})}{n}}$$

$$\mathrm{LCL} = \hat{p} - 3\sqrt{\frac{\hat{p}(1-\hat{p})}{n}} \tag{15-11}$$

as the upper and lower control limits, respectively. These control limits are based on the normal approximation to the binomial distribution. When p is small, the normal approximation may not always be adequate. In such cases, it is best to use control limits obtained directly from a table of binomial probabilities or, perhaps, from the Poisson approximation to the binomial distribution. If p is small, the lower control limit may be a negative number. If this should occur, it is customary to consider zero as the lower control limit.

● **Example 15-2.** Suppose we wish to construct a fraction defective control chart for a transistor production line. We have 20 preliminary samples, each of size 100; the number of defectives in each sample are shown in the following table. Assume that the samples are numbered in the sequence of production.

Sample	Number of Defectives	Sample	Number of Defectives
1	44	11	36
2	48	12	52
3	32	13	35
4	50	14	41
5	29	15	42
6	31	16	30
7	46	17	46
8	52	18	38
9	44	19	26
10	48	20	30

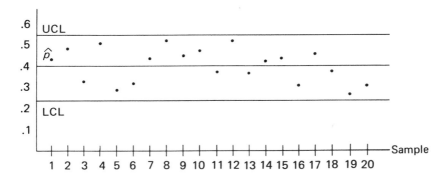

Fig. 15-3.   The p chart for a transistor production line.

Note that $\hat{p} = (800/2000) = .40$, and therefore the trial parameters for the control chart are

$$\text{Centerline} = .40$$

$$\text{UCL} = .40 + 3\sqrt{\frac{(.40)(.60)}{100}} = .55$$

$$\text{LCL} = .40 - 3\sqrt{\frac{(.40)(.60)}{100}} = .25$$

The control chart is shown in Fig. 15-3. All samples are in control. If they were not, we would search for assignable causes of variation and revise the limits accordingly.

● **Example 15-3.**   The advantage of measurement control charts relative to the p chart with respect to size of sample may be easily illustrated. Suppose that a normally distributed quality characteristic has a standard deviation of 4 and specification limits of 52 and 68. The process is centered at 60, which results in a fraction defective of .0454. Let the process mean shift to 56. Now the fraction defective is .1610. If the probability of detecting the shift on the first sample following the shift is to be .50, then the sample size must be such that the lower 3-sigma limit will be at 56. This implies

$$60 - \frac{3(4)}{\sqrt{n}} = 56$$

whose solution is $n = 9$. For a p chart, using the normal approximation to the binomial, we must have

$$.0454 + 3\sqrt{\frac{(.0454)(.9546)}{n}} = .1601$$

whose solution is $n = 30$. Thus, unless the cost of measurement inspection is more than three times as costly as attributes inspection, the measurement control chart is cheaper to operate.

In some situations it may be necessary to control the number of defects in a unit of product, rather than the fraction defective. In these situations we may use the defects per unit control chart, or c chart. Suppose that in the production of cloth it is necessary to control the number of defects per yard, or that in assembling an aircraft wing the number of missing rivets must be controlled. Many defects-per-unit situations can be modeled by the Poisson distribution.

Let c be the number of defects per unit, where c is a Poisson random variable with parameter α. Now the mean and variance of this distribution are both α. Therefore, if k units are available and c_i is the number of defects in unit i, the centerline of the control chart is

$$\bar{c} = \frac{1}{k} \sum_{i=1}^{k} c_i \qquad (15\text{-}12)$$

and

$$\text{UCL} = \bar{c} + 3\sqrt{\bar{c}}$$
$$\text{LCL} = \bar{c} - 3\sqrt{\bar{c}} \qquad (15\text{-}13)$$

are the upper and lower control limits, respectively.

● **Example 15-4.** In the manufacture of carpet the number of defects per 100 feet should be controlled. From 20 randomly selected 100-foot lengths we obtain the following data.

Unit	Number of Defects	Unit	Number of Defects
1	6	11	9
2	4	12	15
3	8	13	8
4	10	14	10
5	9	15	8
6	12	16	2
7	16	17	7
8	2	18	1
9	3	19	7
10	10	20	13

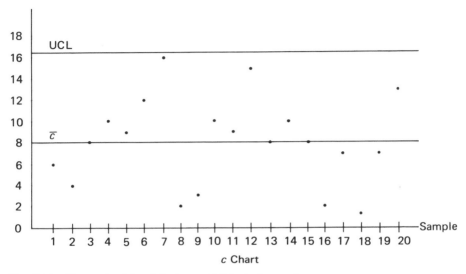

Fig. 15-4. The *c* chart for defects per 100 feet of carpet.

Now $\bar{c} = (160/20) = 8$, and therefore

$$UCL = 8 + 3\sqrt{8} = 16.484$$
$$LCL = 8 - 3\sqrt{8} = 0$$

From the control chart in Fig. 15-4, we see that the process is in control.

15-4 Acceptance Sampling

Inspection of product is an integral part of all production processes. The general situation in which products, grouped in lots, are sampled and the results of the sample are used to draw inferences concerning the quality of the product (or lot) is called *acceptance sampling*. Acceptance sampling plans can be applied either to suppliers' goods, prior to their introduction into another production process, or to the output of a company's own production processes. The purpose of acceptance sampling is to estimate the pertinent quality characteristic of each lot of product and indicate whether the lot should be accepted or rejected.

While acceptance sampling is usually classified as a quality control technique, we must point out that often no direct control over process quality is exercised. This is especially true when we are sampling from suppliers' goods and rejected lots are not returned. Obviously, some indirect control will always be achieved through communication with the manufacturer.

We first consider a procedure known as a single sampling plan. The procedure consists of drawing a random sample of size n from a lot composed of N items. Let d be the number of defective items in this random sample. Then if d is less than or equal to some *acceptance number, c*, the lot is accepted. Since N is fixed, the sampling plan is completely specified by the parameters n and c. The procedure is called a single sampling plan because a decision is made based on the results of one sample. If d is greater than c, we reject the lot and several possibilities exist. We may return a rejected lot to the manufacturer, in which case the sampling plan is said to be *nonrectifying*: the average quality entering the production process is the same as the average quality leaving the manufacturer.

On the other hand, we may screen (inspect 100 percent) the rejected lots, either replacing all defective items with good ones, or simply removing them. This alternative is called *rectifying* inspection: the average quality entering the production process is superior to the average quality leaving the manufacturer. Thus, for rectifying inspection we think of the *average outgoing quality* from the inspection process. The average outgoing quality will be high both when the incoming quality is high (few high-quality lots are rejected) and when the incoming quality is low (many low-quality lots are rejected and screened). It can be shown that the average outgoing quality has a lower limit, called the average outgoing quality limit (AOQL). Thus, no matter how bad the incoming quality becomes, the average outgoing quality will never be worse than the AOQL. This process is illustrated in Figs. 15-5 and 15-6.

Any acceptance sampling plan may be described in terms of its operating characteristic curve. A typical operating characteristic curve for a single sampling plan is shown in Fig. 15-7. This curve relates the true fraction defective for the lot to the probability of acceptance. Some standard ter-

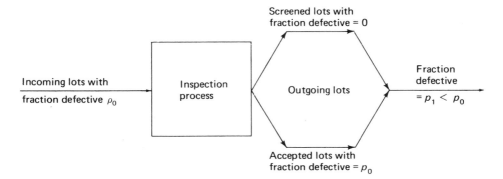

Fig. 15-5. Rectifying inspection where the quality characteristic is the lot fraction defective.

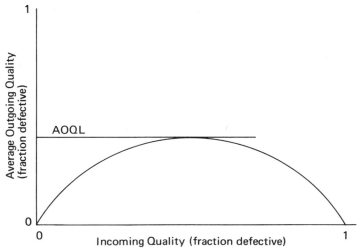

Fig. 15-6. Average outgoing quality (lot fraction defective) for rectifying inspection.

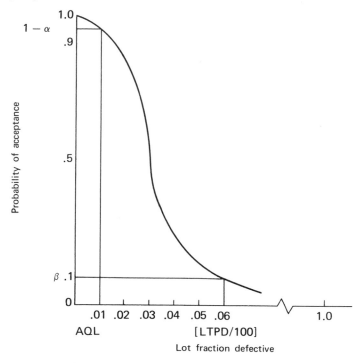

Fig. 15-7. Operating characteristic curve.

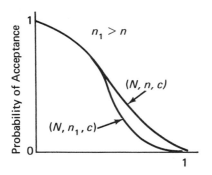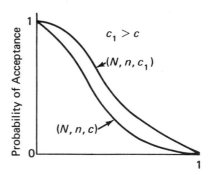

Fig. 15-8. Effect of varying n and c in a single sampling plan.

minology has been adopted for certain features of a sampling plan, shown in Fig. 15-7. The level of quality that is considered "good" and that is desired to accept most of the time is called the acceptable quality level (AQL). The level that is considered "bad" and that should be rejected most of the time is called the lot tolerance percent defective (LTPD). The probability that a sampling plan will reject AQL lots is called the producer's risk (α), and the probability that a plan will accept LTPD lots is called the consumer's risk (β). Any operating characteristic curve may be defined by choosing the points (AQL, $1 - \alpha$) and (LTPD/100, β). The operating characteristic curve essentially gives the probabilities of the type I and the type II errors associated with the sampling plan.

The effect of n and c on the operating characteristic curve of a single sampling plan is shown in Fig. 15-8. For N and n constant we see that increasing c shifts the operating characteristic curve to the right; that is, the plan becomes less selective. For N and c constant, increasing n causes the operating characteristic curve to become steeper. If the lot size N is large relative to the sample size n, the operating characteristic curve is essentially independent of the lot size.

Single sampling plans may be constructed for either measurement or attribute quality characteristics. Since any sampling plan is defined by its operating characteristic curve, we may design a plan by specifying two points on the curve, say (AQL, $1 - \alpha$) and (LTPD/100, β), and find the corresponding parameters of the plan using the appropriate probability model. For attributes data, the probability model is the binomial distribution, although probabilities are often computed using the Poisson or normal approximation.[1] Single

---

[1]Technically, we are considering a "stream" of lots. Thus, the population we are sampling is infinite, and the binomial distribution is appropriate. If we are buying only isolated lots, or if the quality of individual lots is important, then the hypergeometric distribution must be used.

sampling plans for measurements require that a sample size and either a single or double specification limit for the sample mean be specified. The normal distribution is the probability model usually chosen.

● **Example 15-5.** Consider a single sampling plan for attributes. If a lot of quality p_1 is submitted, the probability of acceptance should be $1 - \alpha$. If a lot of quality p_2 is submitted, the probability acceptance should be β. Thus,

$$1 - \alpha = \sum_{i=0}^{c} \binom{n}{i} p_1^i (1 - p_1)^{n-i}$$

$$\beta = \sum_{j=0}^{c} \binom{n}{j} p_2^j (1 - p_2)^{n-j}$$

It may be difficult to solve these equations for n and c. However, if np is small we may use the Poisson approximation to the binomial, or

$$1 - \alpha = \sum_{j=0}^{c} \frac{e^{-np_1}(np_1)^j}{j!}$$

$$\beta = \sum_{j=0}^{c} \frac{e^{-np_2}(np_2)^j}{j!}$$

These equations may still be difficult to solve analytically. However, Duncan (1974) describes a simple procedure that produces an approximate solution. Suppose we wish to design a single sampling plan that requires that $p_1 = .01$, $p_2 = .06$, $\alpha = .05$, and $\beta = .10$. It may be easily verified that $n = 89$ and $c = 2$ is an approximate solution to the latter pair of equations. The operating characteristic curve for this plan is shown in Fig. 15-7.

Sometimes, smaller average sample sizes (and hence reductions in the cost of sampling) can be achieved without loss of protection by the use of *double* or *multiple* sampling plans. We briefly consider the attributes case. A double sampling plan requires that a random sample of size n_1 be taken from the lot, and the number of defectives, say d_1, is noted. If $d_1 \leq c_1$, the lot is accepted without further sampling. If $c_1 < d_1 \leq c_2$, a second random sample of size n_2 is taken, and the number of defectives d_2 is noted. Now if $d_1 + d_2 \leq c_2$, the lot is accepted; otherwise it is rejected. A multiple sampling plan is similar in nature to a double sampling plan, but it involves more than two stages. The advantage of these plans is that lots of very high quality will be accepted on the first sample with a high probability, and lots of very low quality will be rejected quickly, thereby reducing the average amount of inspection required. On the other hand, lots that are "intermediate" in quality can actually require more inspection than would be needed with a single sampling plan. Furthermore, the design and administration of these plans are more complicated than for simple sampling plans.

The concept of multiple sampling can be generalized to *sequential sampling*, in which a decision is made to accept, reject, or continue sampling after each observation (that is, all sample sizes are one). The construction of sequential sampling plans requires that two sequences of numbers, say a_n and r_n, be generated, where n is the number of observations. The procedure would then reject the lot as soon as the number of defectives exceeds r_n for some n, and accept the lot as soon as the number of defectives is less than a_n for some n. Sampling continues as long as the number of defectives out of n observations falls between a_n and r_n.

To facilitate the design and use of acceptance sampling plans, several tables of standard plans have been published. Among the most widely used are the *Military Standard 105D Tables* (1963) for attribute sampling and the *Military Standard 414 Tables* (1957) for measurements sampling, both published by the Department of Defense.

15-5   Tolerance Limits

In most production processes we wish to compare a product to a set of specifications. These specifications, usually called *tolerance limits*, are determined either by the designer or the customer. Sometimes a product is made without prior specifications, and then we speak of "natural" tolerance limits for the process. In any case, tolerance limits are simply a set of limits between which we can expect to find any given proportion, say P, of the population.

If the distribution underlying the quality characteristic in question and its parameters are known, say from long experience, then tolerance limits may be easily established. For example, if we know that a dimension is normally distributed with mean μ and variance σ^2, then tolerance limits can be constructed for any P. If $P = .95$, we see that the tolerance limits are $\mu \pm 1.96\sigma$, by using the tables of the cumulative normal distribution.

If μ and σ^2 are not known, they must be estimated from a random sample, say by $\bar{X}$ and S^2. Then it is possible to determine a constant K such that we can assert with a degree of confidence $1 - \alpha$ that the proportion of the population contained between $\bar{X} - KS$ and $\bar{X} + KS$ is at least P. A brief table of K for random samples from normal populations is given in Table XIV of the Appendix.

● **Example 15-6.** A manufacturer takes a random sample of size 100 from a large population of shafts. From this he computes $\bar{x} = 1.407$ and $s = .001$ inch. Choosing $1 - \alpha = .99$ and $P = .95$, we obtain the tolerance limits

$$1.407 \pm (2.355)(.001)$$

Thus, the manufacturer can assert with a degree of confidence of .99 that at least 95 percent of the shafts have diameters from 1.404 to 1.410 inches. Notice that the lower tolerance limit is rounded down and the upper tolerance limit is rounded up.

It is possible to construct *nonparametric tolerance limits* which are based on the extreme values in a random sample of size n from any continuous population. If P is the minimum proportion of the population contained between the largest and smallest observation with confidence $1 - \alpha$, then it can be shown that

$$nP^{n-1} - (n-1)P^n = \alpha$$

and n is approximately

$$n = \frac{1}{2} + \frac{1+P}{1-P} \cdot \frac{\chi^2_{\alpha,4}}{4} \tag{15-14}$$

Thus, in order to be 95 percent certain that at least 90 percent of the population will be included between the extreme values of the sample, we require a sample of

$$n \approx \frac{1}{2} + \frac{1.9}{.1} \cdot \frac{9.488}{4} \approx 46$$

observations.

Note that there is a fundamental difference between confidence limits and tolerance limits. Confidence limits are used to estimate a parameter of a population, while tolerance limits are used to indicate the limits between which we can expect to find a proportion of a population. As n approaches infinity the length of a confidence interval approaches zero, while tolerance limits approach the corresponding values for the population. Thus, in Table XI of the Appendix as n becomes large for $P = .90$, say, K approaches 1.645.

15-6 Applications to Reliability

One of the challenging engineering endeavors of the past two decades has been the design and development of large-scale systems for space exploration, military use, and commercial operation. System design, as differentiated from equipment design, involves the broader aspects of organizing composite equipment, operating schedules, maintenance schedules, and the skills required to assure system performance as a unified entity. As a general rule, complex systems perform a number of functions, cost a great deal of money, and require substantial support facilities. System performance is of primary concern, and the consequences of failure must be evaluated carefully. Human life has been frequently involved, for example, in space exploration or in operating an automobile on an interstate highway. The emphasis historically placed on equipment reliability by the military community stems largely from increasing ratios of maintenance costs to procurement costs and the strategic and tactical implications of system failure. In the area of consumer product manufacture, high reliability has come to be expected as much as conformance to other important quality characteristics.

Reliability engineering encompasses several activities, one of which is reliability modeling. Essentially, the system survival probability is expressed as a function of subsystem or component reliabilities (survival probabilities). Usually, these models are time dependent, but there are some situations where this is not the case. A second important activity is that of life testing and reliability estimation.

Let us consider a basic component that has just been manufactured. It is to be operated at a stated "stress level" or within some range of stress such as temperature, shock, and so on. The random variable T will be defined as time to failure, and the *reliability* of the component (or subsystem or system) at time t is $R(t) = P[T > t]$. R is called the *reliability function*. The failure process is usually complex, consisting of at least three types of failures: initial failures, wearout failures, and those that fail between these. A hypothetical composite distribution of time to failure is shown in Fig. 15-9. This is a mixed distribution, and

$$p(0) + \int_0^\infty g(t)\, dt = 1 \qquad (15\text{-}15)$$

Since for many components (or systems) the initial failures or time zero failures are removed during test, the random variable T is conditioned on the

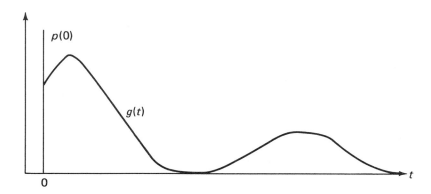

Fig. 15-9. A composite failure distribution.

event that $T > 0$, so that the failure density is

$$f(t) = \frac{g(t)}{1 - p(0)} \qquad t > 0$$

$$= 0 \qquad \text{otherwise} \qquad (15\text{-}16)$$

Thus, in terms of f, the reliability function, R, is

$$R(t) = 1 - F(t) = \int_t^\infty f(x) \, dx \qquad (15\text{-}17)$$

The term *interval failure rate* denotes the rate of failure on a particular interval of time $[t_1, t_2]$ and the terms *failure rate, instantaneous failure rate,* and *hazard* will be used synonymously as a limiting form of the interval failure rate as $t_2 \to t_1$. The interval failure rate $FR(t_1, t_2)$ is as follows:

$$FR(t_1, t_2) = \left[\frac{R(t_1) - R(t_2)}{R(t_1)}\right] \cdot \left[\frac{1}{(t_2 - t_1)}\right] \qquad (15\text{-}18)$$

The first bracketed term is simply

$$P\{\text{Failure during } [t_1, t_2] \,|\, \text{Survival to time } t_1\} \qquad (15\text{-}19)$$

The second term is for the dimensional characteristic, so that we may express the conditional probability of Equation (15-19) on a per-unit time basis.

We will develop the instantaneous failure rate (as a function of t). Let $h(t)$ be the hazard function. Then

$$h(t) = \lim_{\Delta t \to 0} \frac{R(t) - R(t + \Delta t)}{R(t)} \frac{1}{\Delta t}$$

$$= -\lim_{\Delta t \to 0} \frac{R(t + \Delta t) - R(t)}{\Delta t} \cdot \frac{1}{R(t)}$$

or

$$h(t) = \frac{-R'(t)}{R(t)} = \frac{f(t)}{R(t)} \qquad (15\text{-}20)$$

since $R(t) = 1 - F(t)$ and $-R'(t) = f(t)$. A typical hazard function is shown in Fig. 15-10. Note that $h(t) \cdot dt$ might be thought of as the instantaneous probability of failure at t, given survival to t.

A useful result is that the reliability function, R, may be easily expressed in terms of h as

$$R(t) = e^{-\int_0^t h(x) \, dx} = e^{-H(t)} \qquad (15\text{-}21)$$

where

$$H(t) = \int_0^t h(x) \, dx.$$

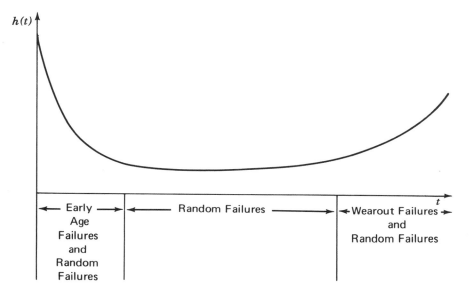

Fig. 15-10. A typical hazard function.

Equation (15-21) results from the definition

$$h(t) = \frac{f(t)}{R(t)} = -\frac{R'(t)}{R(t)}$$

and the integration of both sides

$$\int_0^t h(x) \, dx = -\int_0^t \frac{R'(x)}{R(x)} \, dx = -\log_e R(x) \Big|_0^t$$

so that

$$\int_0^t h(x) \, dx = -\log_e R(t) + \log_e R(0)$$

Since $F(0) = 0$, $\log_e R(0) = 0$, and

$$e^{-\int_t^\infty h(x) \, dx} = e^{\log_e R(t)} = R(t)$$

The mean time to failure (MTTF) is

$$E[T] = \int_0^\infty t \cdot f(t) \, dt$$

A useful alternate form is

$$E[T] = \int_0^\infty R(t) \, dt \tag{15-22}$$

Most complex system modeling assumes that only random component failures need be considered. This is equivalent to stating that the time to failure density is exponential, that is,

$$f(t) = \lambda e^{-\lambda t} \qquad t \geq 0$$
$$= 0 \qquad \text{otherwise}$$

so that

$$h(t) = \frac{f(t)}{R(t)} = \frac{\lambda e^{-\lambda t}}{e^{-\lambda t}} = \lambda$$

a constant. When all early-age failures have been removed by *burn in*, and the time to occurrence of wearout failures is very great (as with electronic parts), then this assumption is reasonable.

The normal distribution is most generally used to model wearout failure or stress failure (where the random variable is stress level rather than time). In situations where most failures are due to wear, the normal distribution may very well be appropriate.

The lognormal distribution has been found to be applicable in describing time to failure for some types of components, and the literature seems to indicate an increased utilization of this density for this purpose.

The Weibull distribution has been extensively used to represent time to failure, and its nature is such that it may be made to approximate closely the observed phenomena. When a system is composed of a number of components and failure is due to the most serious of a large number of defects or possible defects, the Weibull distribution seems to do particularly well as a model.

The gamma distribution frequently results from modeling standby redundancy where components have an exponential time to failure density.

15-7   Exponential Failure Law

In this section, it will be assumed that the time to failure density is exponential; that is, only "random failures" are considered. The density, reliability function, and hazard functions are given in Equations (15-23) through (15-25), and they are shown in Fig. 15-11.

$$f(t) = e^{-\lambda t} \qquad\qquad t \geq 0$$
$$= 0 \qquad\qquad \text{otherwise} \qquad (15\text{-}23)$$

$$R(t) = P[T > t] = e^{-\lambda t} \qquad t \geq 0$$
$$= 0 \qquad\qquad \text{otherwise} \qquad (15\text{-}24)$$

$$h(t) = \frac{f(t)}{R(t)} = \lambda \qquad\qquad t \geq 0$$
$$= 0 \qquad\qquad \text{otherwise} \qquad (15\text{-}25)$$

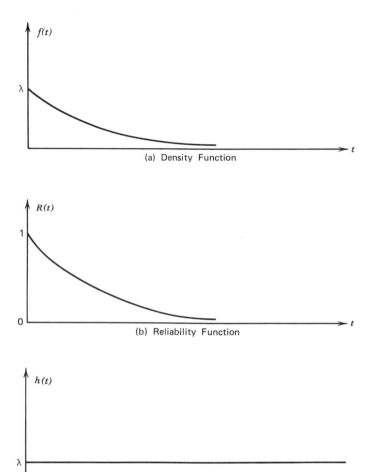

$f(t)$

λ

(a) Density Function

$R(t)$

1

0

(b) Reliability Function

$h(t)$

λ

(c) Hazard Function

Fig. 15-11.   Density, reliability function, and hazard function for the exponential failure law.

The constant hazard is interpreted to mean that the failure process has no memory; that is,

$$P\{t \le T \le t + \Delta t \,|\, T > t\} = \frac{e^{-\lambda t} - e^{-\lambda(t+\Delta t)}}{e^{-\lambda t}} = 1 - e^{-\lambda \Delta t} \qquad (15\text{-}26)$$

a quantity which is independent of t. Thus if a component is functioning at time t, it is as good as new. The remaining life has the same density as f.

● **Example 15-7.** A certain type diode has a rated failure rate of 2.3×10^{-8} failures per hour. However, under an increased temperature stress, it is felt that the rate is about 1.5×10^{-5} failures per hour. The time to failure is exponentially distributed, so that we have

$$f(t) = (1.5 \times 10^{-5})e^{-(1.5 \times 10^{-5})t} \qquad t \geq 0$$
$$= 0 \qquad \text{otherwise}$$
$$R(t) = e^{-(1.5 \times 10^{-5})t} \qquad t \geq 0$$
$$= 0 \qquad \text{otherwise}$$

and

$$h(t) = 1.5 \times 10^{-5} \qquad t \geq 0$$
$$= 0 \qquad \text{otherwise}$$

To determine the reliability at $t = 10^4$ and $t = 10^5$, we evaluate $R(10^4) = e^{-0.15} = .86071$, and $R(10^5) = e^{-1.5} = .22313$.

15-8   Simple Serial Systems

A simple serial system is shown in Fig. 15-12. In order for the system to function, all components must function, and it is assumed that the components function *independently*. We let T_j be the time to failure for component c_j for $j = 1, 2, \ldots, n$, and T represents system time to failure. The reliability model is thus

$$R(t) = P[T > t] = P(T_1 > t) \cdot P(T_2 > t) \cdot \ldots \cdot P(T_n > t)$$

or

$$R(t) = R_1(t) \cdot R_2(t) \cdot \ldots \cdot R_n(t) \qquad (15\text{-}27)$$

where

$$P[T_j > t] = R_j(t)$$

● **Example 15-8.** Three components must all function for a simple system to function. The random variables T_1, T_2, and T_3 representing time to failure for

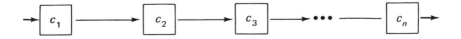

Fig. 15-12.   A simple serial system.

the components are independent with the following distributions:

$$T_1 \sim N(2 \times 10^3, 4 \times 10^4) \qquad -$$

$$T_2 \sim \text{Weibull} \left(\gamma = 0, \delta = 1, \beta = \frac{1}{7} \right)$$

$$T_3 \sim \text{lognormal} (\mu = 10, \sigma^2 = 4)$$

It follows that

$$R_1(t) = 1 - \Phi\left(\frac{t - 2 \times 10^3}{200}\right)$$

$$R_2(t) = e^{-t^{(1/7)}}$$

and

$$R_3(t) = 1 - \Phi\left(\frac{\log_e t - 10}{2}\right)$$

so that

$$R(t) = \left[1 - \Phi\left(\frac{t - 2 \times 10^3}{200}\right)\right] \cdot \left[e^{-t^{(1/7)}}\right] \cdot \left[1 - \Phi\left(\frac{\log_e t - 10}{2}\right)\right]$$

For example, if $t = 2187$ hours, then

$$R(2187) = [1 - \Phi(.937)][e^{-3}][1 - \Phi(-1.154)]$$
$$= [.174][.0498][.876]$$
$$\simeq .0076$$

For the simple serial system, system reliability may be calculated using the product of the component reliability functions as demonstrated; however, when all components have an exponential distribution, the calculations are greatly simplified since

$$R(t) = e^{-\lambda_1 t} \cdot e^{-\lambda_2 t} \ldots e^{-\lambda_n t} = e^{-(\lambda_1 + \lambda_2 + \cdots + \lambda_n)t}$$

or

$$R(t) = e^{-\lambda_s t} \qquad (15\text{-}28)$$

where $\lambda_s = \sum_{j=1}^{n} \lambda_j$ represents the *system failure rate*. We also note that the system reliability function is of the same form as the component reliability functions. The system failure rate is simply the sum of the component failure rates, and this makes application very easy.

● **Example 15-9.** Consider an electronic circuit with 3 silicon transistors, 12 silicon diodes, 8 ceramic capacitors, and 15 composition resistors. Suppose under given stress levels of temperature, shock, and so on, each component has failure rates, as shown in the following table, and the component failures are independent.

	Failures per Hour
Transistors	1.3×10^{-5}
Diodes	1.7×10^{-7}
Capacitors	1.2×10^{-7}
Resistors	6.1×10^{-6}

Therefore,

$$\lambda_s = 3(.013 \times 10^{-7}) + 12(1.7 \times 10^{-7}) + 8(1.2 \times 10^{-7}) + 15(.61 \times 10^{-7})$$
$$= 3.9189 \times 10^{-6}$$

and

$$R(t) = e^{-(3.9189 \times 10^{-6})t}$$

The circuit mean time to failure is

$$\text{MTTF} = E[T] = \frac{1}{\lambda_s} = \frac{1}{3.9189} \times 10^6 = 2.56 \times 10^5 \text{ hours}$$

If we wish to determine, say $R(10^4)$, we get $R(10^4) = e^{-.039189} \simeq .96$.

15-9 Simple Active Redundancy

An active redundant configuration is shown in Fig. 15-13. The assembly functions if k or more of the components function ($k \leq n$). All components begin operation at time zero, thus the term "active" is used to describe the redundancy. Again, independence is assumed.

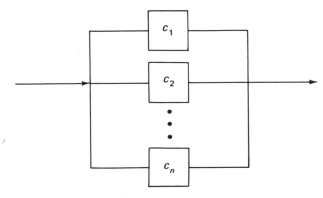

Fig. 15-13. An active redundant configuration.

A general formulation is not convenient to work with, and in most cases it is not necessary to utilize a general formulation. When all components have the same reliability function, as is the case when the components are the same type, we let $R_j(t) = r(t)$ for $j = 1, 2, \ldots, n$, so that

$$R(t) = \sum_{x=k}^{n} \binom{n}{x} [r(t)]^x [1 - r(t)]^{n-x}$$

$$= 1 - \sum_{x=0}^{k-1} \binom{n}{x} [r(t)]^x [1 - r(t)]^{n-x} \tag{15-29}$$

Equation (15-29) is derived from the definition of reliability.

● **Example 15-10.** Three identical components are arranged in active redundancy, operating independently. In order for the assembly to function, at least two of the components must function ($k = 2$). The reliability function for the system is thus

$$R(t) = \sum_{x=2}^{3} \binom{3}{x} [r(t)]^x [1 - r(t)]^{n-x}$$

$$= 3[r(t)]^2 [1 - r(t)] + [r(t)]^3$$

$$= [r(t)]^2 [3 - 2r(t)]$$

It is noted that R is a function of time, t.

When only one of the n components is required, as is often the case, and the components are not identical, we obtain

$$R(t) = 1 - \prod_{j=1}^{n} [1 - R_j(t)] \tag{15-30}$$

The product is the probability that all components fail, and, obviously, if they do not all fail the system survives. When the components are identical and only one is required, Equation (15-30) reduces to

$$R(t) = 1 - [1 - r(t)]^n \tag{15-31}$$

where $r(t) = R_j(t)$, $j = 1, 2, \ldots, n$.

When the components have exponential failure laws, we will consider two cases. First, when the components are identical with failure rate λ and at least k components are required for the assembly to operate, Equation (15-29) becomes

$$R(t) = \sum_{x=k}^{n} \binom{n}{x} [e^{-\lambda t}]^x [1 - e^{-\lambda t}]^{n-x} \tag{15-32}$$

The second case is considered for the situation where the components have identical exponential failure densities and where only one component must function for the assembly to function. Using Equation (15-31), we get

$$R(t) = 1 - [1 - e^{-\lambda t}]^n \qquad (15\text{-}33)$$

● **Example 15-11.** In Example 15-10, where three identical components were arranged in an active redundancy, and at least two were required for system operation, we found

$$R(t) = [r(t)]^2[3 - 2r(t)]$$

If the component reliability functions are

$$r(t) = e^{-\lambda t}$$

then

$$R(t) = e^{-2\lambda t}[3 - 2e^{-\lambda t}]$$
$$= 3e^{-2\lambda t} - 2e^{-3\lambda t}$$

If two components are arranged in an active redundancy as described, and only one must function for the assembly to function, and, furthermore, if the time to failure densities are exponential with failure rate λ, then from Equation (15-33), we obtain

$$R(t) = 1 - [1 - e^{-\lambda t}]^2 = 2e^{-\lambda t} - e^{-2\lambda t}$$

15-10 Standby Redundancy

A common form of redundancy called standby redundancy is shown in Fig. 15-14. The unit labeled DS is a decision switch that we will assume has reliability of 1 for all t. The operating rules are as follows. Component 1 is initially "on-line," and when this component fails, the decision switch switches in component 2, which remains on-line until it fails. Standby units are not subject to failure until activated. The time to failure for the assembly is

$$T = T_1 + T_2 + \cdots + T_n$$

where T_i is the time to failure for the ith component and $T_1, T_2, \ldots, T_n$ are independent random variables. The most common value for n in practice is two, so the central limit theorem is of little value. However, we know from the property of linear combinations that

$$E[T] = \sum_{i=1}^{n} E(T_i)$$

and

$$V[T] = \sum_{i=1}^{n} V(T_i)$$

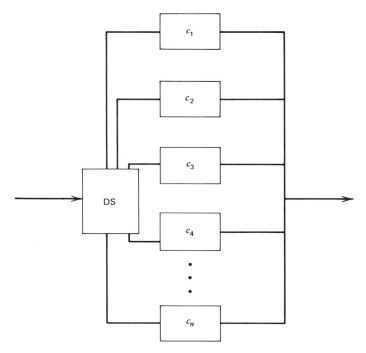

Fig. 15-14. Standby redundancy.

We must know the distributions of the random variables T_i in order to find the distribution of T. The most common case occurs when the components are identical and the time to failure densities are assumed to be exponential. In this case, T has a gamma density

$$f(t) = \frac{\lambda}{(n-1)!}(\lambda t)^{n-1}e^{-\lambda t} \qquad t > 0$$
$$= 0 \qquad\qquad\qquad \text{otherwise}$$

so that the reliability function is

$$R(t) = \sum_{k=0}^{n-1} e^{-\lambda t}(\lambda t)^k/k! \qquad t > 0 \qquad (15\text{-}34)$$

The parameter λ is the component failure rate; that is, $E(T_i) = 1/\lambda$. The mean time to failure and variance are

$$\text{MTTF} = E[T] = n/\lambda \qquad\qquad (15\text{-}35)$$

and

$$V[T] = n/\lambda^2 \qquad\qquad (15\text{-}36)$$

respectively.

● **Example 15-12.** Two identical components are assembled in a standby redundant configuration with perfect switching. The component lives are identically distributed, independent random variables having an exponential density with failure rate 100^{-1}. The mean time to failure is

$$\text{MTTF} = 2/100^{-1} = 200$$

and the variance is

$$V[T] = 2/(100^{-1})^2 = 20,000$$

The reliability function R is

$$R(t) = \sum_{k=0}^{1} e^{-100^{-1}t}(100^{-1}t)^k/k!$$

or

$$R(t) = e^{-t/100}[1 + t/100]$$

15-11 Life Testing

Life tests are conducted for different purposes. Sometimes, n units are placed on test and aged until all or most units have failed; the purpose is to test a hypothesis about the form of the time to failure density with certain parameters. Both formal statistical tests and probability plotting are widely used in life testing.

A second objective in life testing is to estimate reliability. Suppose, for example, that a manufacturer is interested in estimating $R(1000)$ for a particular component or system. One approach to this problem would be to place n units on test and count the number of failures, r, occurring before 1000 hours of operation. Failed units are not to be replaced in this example. An estimate of unreliability is $\hat{p} = r/n$, and an estimate of reliability is

$$\hat{R}(1000) = 1 - \frac{r}{n} \tag{15-37}$$

A $(1 - \alpha)100$ lower confidence limit on $R(1000)$ is given by [1 − upper limit on p], where p is the unreliability. This upper limit on p may be determined using a table of the binomial distribution. In the case where n is large, an estimate of the upper limit on p is

$$\hat{p} + Z_{1-\alpha} \sqrt{\frac{\hat{p}(1 - \hat{p})}{n}} \tag{15-38}$$

● **Example 15-13.** One hundred units are placed on life test, and the test is run for 1000 hours. There are two failures during test, so $\hat{p} = \frac{2}{100} = .02$, and $\hat{R}(1000) = .98$. Using a table of the binomial distribution, an upper 95 percent confidence limit on p is .06, so that a lower limit on $R(1000)$ is given by .94.

15-12   Reliability Estimation with a Known Density Form

In the case where the form of the reliability function is assumed known and there is only one parameter, the maximum likelihood estimator for $R(t)$ is $\hat{R}(t)$, which is formed by substituting $\hat{\theta}$ for the parameter θ in the expression for $R(t)$, where $\hat{\theta}$ is the maximum likelihood estimator of θ. The case where there are multiple parameters requires the use of approximation techniques such as those described by Lloyd and Lipow (1962).

15-13   Estimation with the Exponential Time to Failure Density

The most common case for the one-parameter situation is where the time to failure density is exponential, $R(t) = e^{-t/\theta}$. The parameter $\theta = E[T]$ is called the mean time to failure and the estimator of R is $\hat{R}(t)$, where

$$\hat{R}(t) = e^{-t/\hat{\theta}}$$

and $\hat{\theta}$ is the maximum likelihood estimator of θ.

Epstein (1960) developed the maximum likelihood estimators for θ under a number of different conditions and, furthermore, showed that a $100(1-\alpha)$ percent confidence interval on $R(t)$ is given by

$$[e^{-t/\hat{\theta}_L};\ e^{t/\hat{\theta}_U}] \tag{15-39}$$

for the two-sided case or

$$[e^{-t/\hat{\theta}_L};\ 1] \tag{15-40}$$

for the lower, one-sided interval. In these cases, the values $\hat{\theta}_L$ and $\hat{\theta}_U$ are the lower and upper confidence limits on θ.

The following symbols will be used:

n = number of units placed on test at $t = 0$

Q = total test time in unit hours

t^* = time at which the test is terminated

r = number of failures accumulated to time t

r^* = preassigned number of failures

$1 - \alpha$ = confidence level

$\chi^2_{\alpha,k}$ = read from the chi-square table with k degrees of freedom

There are four situations to consider, according to whether the test is stopped after a preassigned time or after a preassigned number of failures and whether failed items are replaced or not replaced during test.

For replacement test, the total test time in unit hours is $Q = nt^*$, and for nonreplacement test,

$$Q = \sum_{i=1}^{r} t_i + (n - r)t^* \tag{15-41}$$

If items are censored (withdrawn items which have not failed), and failures are replaced while censored items are not replaced, then

$$Q = \sum_{j=1}^{c} t_j + (n - c)t^* \tag{15-42}$$

where c represents the number of censored items and t_j is the time of the jth censorship. If neither censored items nor failed items are replaced, then

$$Q = \sum_{i=1}^{r} t_i + \sum_{j=1}^{c} t_j + (n - r - c)t^* \tag{15-43}$$

The development of the maximum likelihood estimators for θ is rather straightforward. In the case where the test is nonreplacement, and the test is discontinued after a fixed number of items have failed, the likelihood function is

$$L = \prod_{i=1}^{r} f(t_i) \cdot \prod_{i=r}^{n} R(t^*)$$

$$= \frac{1}{\theta^r} e^{-(1/\theta) \sum_{i=1}^{r} t_i} \cdot e^{-(n-r)t^*/\theta} \tag{15-44}$$

Then

$$l = \log_e L = -r \log_e \theta - \frac{1}{\theta} \sum_{i=1}^{r} t_i - (n - r)t^*/\theta$$

and solving $(\partial l/\partial \theta) = 0$ yields the estimator

$$\hat{\theta} = \frac{\sum_{i=1}^{r} t_i + (n - r)t^*}{r} = \frac{Q}{r} \tag{15-45}$$

It turns out that

$$\hat{\theta} = Q/r \tag{15-46}$$

is the maximum likelihood estimator of θ for all cases considered for the test design and operation.

The quantity $2r\hat{\theta}/\theta$ has a chi-square distribution with $2r$ degrees of freedom in the case where test is terminated after a fixed number of failures. For fixed termination time, t^*, the degrees of freedom become $2r + 2$.

Since the expression $2r\hat{\theta}/\theta = 2Q/\theta$, confidence limits on θ may be expressed as indicated in Table 15-1. The results presented in the table may be used

TABLE 15-1   Confidence Limits on θ

Nature of Limit	Fixed Number of Failures $r*$	Fixed Termination Time $t*$
Two-sided limits	$\left[\dfrac{2Q}{\chi^2_{\alpha/2,\,2r}}; \dfrac{2Q}{\chi^2_{1-\alpha/2,\,2r}}\right]$	$\left[\dfrac{2Q}{\chi^2_{\alpha/2,\,2r+2}}; \dfrac{2Q}{\chi^2_{1-\alpha/2,\,2r}}\right]$
Lower, one-sided limit	$\left[\dfrac{2Q}{\chi^2_{\alpha,\,2r}}; \infty\right]$	$\left[\dfrac{2Q}{\chi^2_{\alpha,\,2r+2}}; \infty\right]$

directly with Equations (15-39) and (15-40) to establish confidence limits on $R(t)$. It should be noted that this testing procedure does not require that the test be run for the time at which a reliability estimate is required. For example, 100 units may be placed on a nonreplacement test for 200 hours, the parameter θ estimated, and $\hat{R}(1000)$ calculated. In the case of the binomial testing mentioned earlier, it would have been necessary to run the test for 1000 hours.

The results are, however, dependent on the assumption that the distribution is exponential.

It is sometimes necessary to estimate the time t_R for which the reliability will be R. For the exponential model, this estimate is

$$\hat{t}_R = \hat{\theta} \cdot \log_e \frac{1}{R} \tag{15-47}$$

and confidence limits on t_R are given in Table 15-2.

● **Example 15-14.** Twenty items are placed on a replacement test that is to be operated until ten failures occur. The tenth failure occurs at 80 hours, and the reliability engineer wishes to estimate the mean time to failure, 95 percent two-sided limits on θ, $R(100)$, and 95 percent two-sided limits on $R(100)$. Finally, she wishes to estimate the time for which the reliability will be .8 with point and 95 percent two-sided confidence interval estimates.

TABLE 15-2   Confidence Limits on t_R

Nature of Limit	Fixed Number of Failures $r*$	Fixed Termination Time $t*$
Two-sided limits	$\left[\dfrac{2Q\log_e(1/R)}{\chi^2_{\alpha/2,\,2r}}; \dfrac{2Q\log_e(1/R)}{\chi^2_{1-\alpha/2,\,2r}}\right]$	$\left[\dfrac{2Q\log_e(1/R)}{\chi^2_{\alpha/2,\,2r+2}}; \dfrac{2Q\log_e(1/R)}{\chi^2_{1-\alpha/2,\,2r}}\right]$
Lower, one-sided limit	$\left[\dfrac{2Q\log_e(1/R)}{\chi^2_{\alpha,\,2r}}; \infty\right]$	$\left[\dfrac{2Q\log_e(1/R)}{\chi^2_{\alpha,\,2r+2}}; \infty\right]$

Using Equation (15-46) and the results presented in Tables 15-1 and 15-2,

$$\hat{\theta} = \frac{nt^*}{r} = \frac{20(80)}{10} = 160 \text{ hours}$$

$$Q = nt^* = 1600 \text{ unit hours}$$

$$\left[\frac{2Q}{\chi^2_{.025,\,20}}; \frac{2Q}{\chi^2_{.975,\,20}}\right] = \left[\frac{3200}{34.17}; \frac{3200}{9.591}\right]$$
$$= [93.65; 333.65]$$

$$\hat{R}(100) = e^{-100/\hat{\theta}} = e^{-100/160} = .535$$

Using Equation (15-39), the confidence interval on $R(100)$ is

$$[e^{-100/93.65}; e^{-100/333.65}] = [.344; .741]$$

$$\hat{t}_{.80} = \hat{\theta} \log_e \frac{1}{R} = 160 \log_e \frac{1}{.8} = 35.70 \text{ hours}$$

The two-sided 95 percent confidence limit is determined from Table 14-2 as

$$\left[\frac{2(1600)(.22314)}{34.17}; \frac{2(1600)(.22314)}{9.591}\right] = [20.9; 74.45]$$

15-14   Demonstration and Acceptance Testing

It is not uncommon for a purchaser to test incoming products to assure that the vendor is conforming to reliability specifications. These tests are destructive tests and, in the case of attribute measurement, the test design follows that of acceptance sampling discussed earlier in this chapter.

A special set of sampling plans which assume an exponential time to failure density has been presented in a Department of Defense handbook (DOD H-108), and these plans are in wide use.

15-15   Summary

This chapter has presented several widely used methods for statistical quality control. Control charts were introduced and their use as process surveillance devices discussed. The $\bar{X}$ and R control charts are used for measurements data. When the quality characteristic is an attribute, either the p chart for fraction defective or the c chart for defects may be used. Acceptance sampling was also introduced as a technique for estimating lot quality and providing guidelines for lot disposition.

The use of probability as a modeling technique in reliability analysis was also discussed. The exponential distribution is widely used as the distribution of time to failure, although other plausible models include the normal, lognormal, Weibull and gamma distributions. System reliability analysis methods were presented for serial systems, as well as systems having active or standby redundancy. Life testing and reliability estimation was also briefly introduced.

15-16 Exercises

15-1. The following are $\bar{X}$ and R values for 20 samples of size 5. Specifications are given as $.5037 \pm .0010$. The values given are the last three decimals of the measurement; that is, 34.2 should read .50342.

Sample	$\bar{X}$	R	Sample	$\bar{X}$	R
1	34.2	3	11	35.4	8
2	31.6	4	12	34.0	6
3	31.8	4	13	36.0	4
4	33.4	5	14	37.2	7
5	35.0	4	15	35.2	3
6	32.1	2	16	33.4	10
7	32.6	7	17	35.0	4
8	33.8	9	18	34.4	7
9	34.8	10	19	33.9	8
10	38.6	4	20	34.0	4

(a) Set up the $\bar{X}$ and R charts, revising the trial control limits if necessary, assuming assignable causes can be found.

(b) What percentage of defectives is being produced by this process?

15-2. Suppose a process is in control, and 3-sigma control limits are in use on the $\bar{X}$ chart. Let the mean shift by 1.5σ. What is the probability that this shift will remain undetected for 3 consecutive samples? What would this probability be if 2-sigma control limits are used? The sample size is 4.

15-3. Twenty-five samples of size 5 are drawn from a process at regular intervals, and the following data are obtained:

$$\sum_{i=1}^{25} \bar{X}_i = 362.75 \qquad \sum_{i=1}^{25} R_i = 8.60$$

(a) Compute the control limits for the $\bar{X}$ and R chart.

(b) Assuming the process is in control and specification limits are $14.50 \pm .40$, what conclusions can you draw about the ability of the process to operate within these limits? Estimate the percentage of defective items that will be produced.

15-4. Suppose an $\bar{X}$ chart for a process is in control with 3-sigma limits. Samples of size 4 are drawn every 15 minutes, on the quarter-hour. Now suppose the process mean shifts out of control by 1.5σ 10 minutes after the hour. If D is the expected number of defectives produced per quarter-hour in this out-of-control state, find the expected loss (in terms of defective units) that results from this control procedure.

15-5. The following table gives measurements of inside diameters multiplied by 10^4.

Sample	Observation 1	2	3	4	Sample	Observation 1	2	3	4
1	15	11	6	9	11	13	6	9	5
2	14	4	10	6	12	10	15	7	10
3	7	10	9	11	13	8	12	14	9
4	8	6	9	13	14	15	12	4	6
5	14	8	9	12	15	13	16	9	5
6	9	10	7	13	16	14	7	8	12
7	15	10	12	12	17	8	10	16	9
8	14	16	11	10	18	7	4	10	9
9	11	7	16	10	19	13	15	10	8
10	11	14	11	12	20	9	7	15	8

Set up the $\bar{X}$ and R charts. Is this process in control?

15-6. The following are the number of defective shaft and washer assemblies found during successive samples of 1000 each.

Day	Number of Defectives	Day	Number of Defectives
1	106	11	42
2	116	12	37
3	164	13	25
4	89	14	88
5	99	15	101
6	40	16	64
7	112	17	51
8	36	18	74
9	69	19	71
10	74	20	43
—	—	21	80

Construct a fraction defective chart. Is this process in control?

15-7. The process fraction has been shown to be .04. Your control chart calls for daily samples of 800 items. What is the probability that you would detect a shift to .08 on the first sample taken after the shift? On the first or second sample after the shift?

15-8. Suppose we know the process fraction defective to be .08. We wish to detect a shift to .12 with probability .5 on the first sample after the shift occurs. What sample size for the p chart is necessary?

15-9. Consider a process whose specifications on a quality characteristic are 100 ± 10. We know that the standard deviation of this quality characteristic is 5. Where should we center the process to minimize the fraction defective produced? Now suppose the mean shifts to 105 and we are using a sample size of 4 on an $\bar{X}$ chart. What is the probability that such a shift will be detected on the first sample following the shift? What sample size would be needed on a p chart to obtain a similar degree of protection?

15-10. Suppose the following fraction defective had been found in successive samples of size 100:

.09	.03	.12
.10	.05	.14
.13	.13	.06
.08	.10	.05
.14	.14	.14
.09	.07	.11
.10	.06	.09
.15	.09	.13
.13	.08	.12
.06	.11	.09

Is the process in control with respect to its fraction defective?

15-11. The following represent the number of defects per radio assembly: 7, 6, 8, 10, 24, 6, 5, 4, 8, 11, 15, 8, 4, 16, 11, 12, 8, 6, 5, 9, 7, 14, 8, 21. Can we conclude that the process is in control? If not, assume assignable causes can be found and revise the control limits.

15-12. The following represent the number of defects per 1000 feet in rubber-covered wire: 1, 1, 3, 7, 8, 10, 5, 13, 0, 19, 24, 6, 9, 11, 15, 8, 3, 6, 7, 4, 9, 20, 11, 7, 18, 10, 6, 4, 0, 9, 7, 3, 1, 8, 12. Do the data come from a controlled process?

15-13. Suppose the number of defects per unit is known to be 9. If the number of defects per unit shifts to 16, what is the probability that it will be detected by the c chart on the first sample following the shift?

15-14. Suppose we are inspecting radios for defects per unit, and it is known that there are an average of 6 defects per unit. If we decided to make our inspection unit for the c chart three radios, and control the total number of defects per inspection unit, describe the new control chart.

15-15. A single sampling plan for attributes has $n = 50$ and $c = 3$. Draw the operating characteristic curve. What is the AQL at a producer's risk of .05?

15-16. A single sampling plan for attributes has $n = 40$ and $c = 2$. Suppose the AQL $= .03$ and LTPD $= 15$ percent. Find the producer's and consumer's risks.

15-17. A single sampling plan for attributes requires a sample size of $n = 100$. If the AOQ is $.03$, find the producer's risk for $c = 0, 1,$ and 2.

15-18. Consider a single sampling plan for attributes with rectifying inspection. Assuming defective items are replaced with good ones, derive an equation for the average outgoing quality. Assume that n is small relative to N.

15-19. In a random sample of size $n = 40$ from a process which produces pipe, the mean thickness was .1264 inch and the standard deviation was .0003 inch. Assume the thickness is normally distributed.

    (*a*) Between what limits can we say with 95 percent confidence that 95 percent of the thicknesses produced by this process should be found?

    (*b*) Construct a 95 percent confidence interval estimate of the true mean thickness. How does this interval differ from the one constructed in (*a*)?

15-20. Suppose the weights of 25 cartons of soap powder are obtained, yielding a sample mean of 24.32 ounces and standard deviation of .02 ounce. Assuming the weight to be normally distributed, construct a tolerance interval for $\alpha = .05$ and $P = .90$. Interpret this tolerance interval.

15-21. A time to failure distribution is given by a uniform density

$$f(t) = \frac{1}{\beta - \alpha} \qquad \alpha \leq t \leq \beta$$
$$= 0 \qquad \qquad \text{otherwise}$$

    (*a*) Determine the reliability function.

    (*b*) Show that

$$\int_0^\infty R(t)\, dt = \int_0^\infty t f(t)\, dt$$

    (*c*) Determine the hazard function.

    (*d*) Show that

$$R(t) = e^{-H(t)}$$

    where H is defined as in Equation (15-21).

15-22. Three units which operate and fail independently form a series configuration, as shown in the following figure.

The time to failure density for each unit is exponential with the failure rates as indicated.

    (*a*) Find $R(60)$ for the system.

    (*b*) What is the mean time to failure for this system?

15-23. Four identical units are arranged in an active redundancy to form a subsystem. Unit failures are independent, and at least two of the units must survive 1000 hours for the subsystem to perform its mission.

(*a*) If the units have exponential time to failure densities with failure rate .001, what is the subsystem reliability?

(*b*) What is the reliability if only one unit is required?

15-24. If the units described in the previous exercise are operated in a standby redundancy with a perfect decision switch and only one unit is required for subsystem survival, determine the subsystem reliability.

15-25. One hundred units are placed on test and aged until all units have failed. The following results are obtained, and a mean life of $\bar{t} = 160$ hours is calculated from the serial data.

Time Interval	Number of Failures
0–100 hours	50
100–200	18
200–300	17
300–400	8
400–500	4
After 500 hours	3

Use the chi-square goodness-of-fit test to determine whether you consider the exponential distribution to represent a reasonable time to failure model for these data.

15-26. Fifty units are placed on a life test for 1000 hours. Ten units fail during that period. Estimate $R(1000)$ for these units. Determine a lower 95 percent confidence interval on $R(1000)$.

15-27. In Section 15-12 it was noted that for one-parameter reliability functions, $R(t; \theta), \hat{R}(t; \theta) = R(t; \hat{\theta})$, where $\hat{\theta}$ and $\hat{R}$ are the maximum likelihood estimators. Prove this statement for the case

$$R(t; \theta) = e^{-t/\theta} \qquad t \geq 0$$
$$= 0 \qquad \text{otherwise}$$

Hint: Express the density function f in terms of R.

15-28. For a nonreplacement test that is terminated after 100 hours of operation, it is noted that failures occur at the following times: 9, 21, 40, 55, and 85 hours. The units are assumed to have an exponential time to failure density, and 50 units were on test initially.

(*a*) Estimate the mean time to failure.

(*b*) Construct a 95 percent lower confidence limit on the mean time to failure.

15-29. Use the statements in Exercise 15-28.

(*a*) Estimate $R(200)$ and construct a 95 percent lower confidence limit on $R(200)$

(*b*) Estimate the time for which the reliability will be .9, and construct a 95 percent lower limit on $t_{.9}$.

Chapter 16

Stochastic Processes and Queueing

16-1   Introduction

The term *stochastic process* is frequently used in connection with observations from a time oriented, physical process that is controlled by a random mechanism. More precisely, a stochastic process is a sequence of random variables $\{X_t\}$, where $t \in T$ is a time or sequence index. The range space for X_t may be discrete or continuous; however, in this chapter we will consider only the case where at a particular time t the process is in exactly one of $m + 1$ mutually. exclusive and exhaustive *states*. The states are labeled 0, 1, 2, 3, ..., m.

The variables $X_1, X_2, \ldots$, might represent the number of customers awaiting service at a ticket booth at times 1 minute, 2 minutes, and so on, after the booth opens. Another example would be daily demands for a certain product on successive days. X_0 represents the initial state of the process.

This chapter will introduce a special type of stochastic process called a *Markov process*. We will also discuss the *Chapman–Kolmogorov equations*, various special properties of *Markov chains*, the *birth–death equations*, and some applications to waiting line, or *queueing*, and interference problems.

In the study of stochastic processes, certain assumptions are required about the joint probability distribution of the random variables $X_1, X_2, \ldots$. In the case of Bernoulli trials, presented in Chapter 5, recall that these variables were defined to be independent and the range space (state space) consisted of two values (0, 1). Here we will first consider discrete-time Markov chains, the case where time is discrete and the independence assumption is relaxed to allow for a one-stage dependence.

16-2 Discrete-Time Markov Chains

A stochastic process exhibits the Markovian property if

$$P\{X_{t+1} = j | X_t = i\} = P\{X_{t+1} = j | X_t = i, X_{t-1} = i_1, X_{t-2} = i_2, \ldots, X_0 = i_t\} \qquad (16\text{-}1)$$

for $t = 0, 1, 2, \ldots$, and every sequence $j, i, i_1, \ldots, i_t$. This is equivalent to stating that the probability of an event at time $t + 1$ *given* only the outcome at time t is equal to the probability of the event at time $t + 1$ *given* the entire state history of the system. In other words, the probability of the event at $t + 1$ is not dependent on the state history prior to time t.

The conditional probabilities

$$p\{X_{t+1} = j | X_t = i\} = p_{ij} \qquad (16\text{-}2)$$

are called one-step transition probabilities, and they are said to be *stationary* if

$$P\{X_{t+1} = j | X_t = i\} = P\{X_1 = j | X_0 = i\} \qquad \text{for } t = 0, 1, 2, \ldots \qquad (16\text{-}3)$$

so that the transition probabilities remain unchanged through time. These values may be displayed in a matrix $\mathbf{P} = [p_{ij}]$, called the one-step transition matrix. The matrix $\mathbf{P}$ has $m + 1$ rows and $m + 1$ columns, and

$$0 \le p_{ij} \le 1$$

while

$$\sum_{j=0}^{m} p_{ij} = 1 \qquad \text{for } i = 0, 1, 2, \ldots, m$$

That is, each element of the $\mathbf{P}$ matrix is a probability, and each row of the matrix sums to one.

The existence of the one-step, stationary transition probabilities implies that

$$p_{ij}^{(n)} = P\{X_{t+n} = j | X_t = i\} = P\{X_n = j | X_0 = i\} \qquad (16\text{-}4)$$

for all $t = 0, 1, 2, \ldots$. The values $p_{ij}^{(n)}$ are called n-step transition probabilities, and they may be displayed in an n-step transition matrix

$$\mathbf{P}^{(n)} = [p_{ij}^{(n)}]$$

where

$$0 \le p_{ij}^{(n)} \le 1 \qquad n = 0, 1, 2, \ldots \qquad i = 0, 1, 2, \ldots, m \qquad j = 0, 1, 2, \ldots, m$$

and

$$\sum_{j=0}^{m} p_{ij}^{(n)} = 1 \qquad n = 0, 1, 2, \ldots \qquad i = 0, 1, 2, \ldots, m$$

The 0-step transition matrix is the identity matrix.

A *finite-state Markov chain* is defined as a stochastic process having a finite number of states, the Markovian property, stationary transition probabilities, and an initial set of probabilities $\mathbf{A} = [a_0^{(0)} a_1^{(0)} a_2^{(0)} \ldots a_m^{(0)}]$, where $a_i^{(0)} = P\{X_0 = i\}$.

The *Chapman–Kolmogorov* equations are useful in computing n-step transition probabilities. These equations are

$$p_{ij}^{(n)} = \sum_{l=0}^{m} p_{il}^{(v)} \cdot p_{lj}^{(n-v)} \qquad \begin{array}{l} i = 0, 1, 2, \ldots, m \\ j = 0, 1, 2, \ldots, m \\ 0 \leq v \leq n \end{array} \qquad (16\text{-}5)$$

and they indicate that in passing from state i to state j in n steps the process will be in some state, say l, after exactly v steps ($v \leq n$). Therefore $p_{il}^{(v)} \cdot p_{lj}^{(n-v)}$ is the conditional probability that given state i as the starting state, the process goes to state l in v steps and from l to j in $(n - v)$ steps. When summed over l, the sum of the products yields $p_{ij}^{(n)}$.

By setting $v = 1$ or $v = n - 1$, we obtain

$$p_{ij}^{(n)} = \sum_{l=0}^{m} p_{il} p_{lj}^{(n-1)} = \sum_{l=0}^{m} p_{il}^{(n-1)} \cdot p_{lj}, \qquad \begin{array}{l} i = 0, 1, 2, \ldots, m \\ j = 0, 2, 3, \ldots, m \\ n = 0, 1, 2, \ldots \end{array}$$

It follows that the n-step transition probabilities, $\mathbf{P}^{(n)}$, may be obtained from the one-step probabilities, and

$$\mathbf{P}^{(n)} = \mathbf{P}^n \qquad (16\text{-}6)$$

The unconditional probability of being in state j at time $t = n$ is

$$\mathbf{A}^{(n)} = [a_0^{(n)} a_1^{(n)} \ldots a_m^{(n)}] \qquad (16\text{-}7)$$

where

$$a_j^{(n)} = P\{X_n = j\} = \sum_{i=0}^{m} a_i^{(0)} \cdot p_{ij}^{(n)} \qquad j = 0, 1, 2, \ldots, m$$

$$n = 1, 2, \ldots$$

We note that the rule for matrix multiplication solves the total probability law of Theorem 1-9, so that $\mathbf{A}^{(n)} = \mathbf{A}^{(n-1)} \cdot \mathbf{P}$.

● **Example 16-1.** In a computing system, the probability of an error on each cycle depends on whether or not it was preceeded by an error. We will define 0 as the error state and 1 as the nonerror state. Suppose the probability of an error if preceded by an error is .75, the probability of an error if preceded by a nonerror is .50, the probability of a nonerror if preceded by an error is .25, and the probability of nonerror if preceded by nonerror is .50. Thus,

$$\mathbf{P} = \begin{bmatrix} .75 & .25 \\ .50 & .50 \end{bmatrix}$$

Two-step, three-step, . . . , seven-step transition matrices are shown below:

$$\mathbf{P}^2 = \begin{bmatrix} 688 & .312 \\ .625 & .375 \end{bmatrix} \qquad \mathbf{P}^3 = \begin{bmatrix} .672 & .328 \\ .656 & .344 \end{bmatrix}$$

$$\mathbf{P}^4 = \begin{bmatrix} .668 & .332 \\ .664 & .336 \end{bmatrix} \qquad \mathbf{P}^5 = \begin{bmatrix} .667 & .333 \\ .666 & .334 \end{bmatrix}$$

$$\mathbf{P}^6 = \begin{bmatrix} .667 & .333 \\ .667 & .333 \end{bmatrix} \qquad \mathbf{P}^7 = \begin{bmatrix} .667 & .333 \\ .667 & .333 \end{bmatrix}$$

If we know that initially the system is in the nonerror state, then $a_1^{(0)} = 1$, $a_2^{(0)} = 0$, and $\mathbf{A}^{(n)} = [a_j^{(n)}] = \mathbf{A} \cdot \mathbf{P}^{(n)}$. Thus, for example, $\mathbf{A}^{(7)} = [.667, .333]$.

An alternate approach would be to perform the above calculations as $\mathbf{A}^{(1)} = \mathbf{A} \cdot \mathbf{P}$, $\mathbf{A}^{(2)} = \mathbf{A}^{(1)} \cdot \mathbf{P}, \dots, \mathbf{A}^{(n)} = \mathbf{A}^{(n-1)} \cdot \mathbf{P}$. Then $\mathbf{A}^{(7)} = \mathbf{A}^{(6)} \cdot \mathbf{P}$. This leads to an alternate development of the results presented in Equation (16-7) as:

$$\mathbf{A}^{(1)} = \mathbf{A} \cdot \mathbf{P}$$
$$\mathbf{A}^{(2)} = \mathbf{A}^{(1)} \cdot \mathbf{P} = \mathbf{A} \cdot \mathbf{P} \cdot \mathbf{P} = \mathbf{A} \cdot \mathbf{P}^2$$
$$\vdots$$
$$\mathbf{A}^{(n)} = \mathbf{A} \cdot \mathbf{P}^n$$

16-3 Classification of States and Chains

We will first consider the notion of *first passage times*. The length of time (number of steps in discrete-time systems) for the process to go from state i to state j for the first time is called the first passage time. If $i = j$, then this is the number of steps needed for the process to return to state i for the first time, and this is termed the *first return time* or *recurrence time* for state i.

First passage times under certain conditions are random variables with an associated probability distribution. We let $f_{ij}^{(n)}$ denote the probability that the first passage time from state i to j is equal to n, where it can be shown directly from Theorem 1-5 that

$$f_{ij}^{(1)} = p_{ij}^{(1)} = p_{ij}$$
$$f_{ij}^{(2)} = p_{ij}^{(2)} - f_{ij}^{(1)} \cdot p_{jj}$$
$$\vdots$$
$$f_{ij}^{(n)} = p_{ij}^{(n)} - f_{ij}^{(1)} \cdot p_{jj}^{(n-1)} - f_{ij}^{(2)} \cdot p_{jj}^{(n-2)} - \cdots - f_{ij}^{(n-1)} p_{jj} \qquad (16\text{-}8)$$

Thus, recursive computation from the one-step transition probabilities yields the probability of n for given i, j.

● **Example 16-2.** Using the one-step transition probabilities presented in Example 16-1, the distribution of the passage time index n is determined as indicated below for $i = 0$, $j = 1$.

$f_{01}^{(1)} = p_{01} = .250$

$f_{01}^{(2)} = (.312) - (.25)(.5) = .187$

$f_{01}^{(3)} = (.328) - (.25)(.375) - (.187)(.5) = .141$

$f_{01}^{(4)} = (.332) - (.25)(.344) - (.187)(.375) - (.141)(.5) = .105$

$\vdots$

There are four such distributions corresponding to i, j values $(0, 0), (0, 1)$, $(1, 0), (1, 1)$.

If i and j are fixed, $\sum_{n=1}^{\infty} f_{ij}^{(n)} \leq 1$. When the sum is equal to one, the values $f_{ij}^{(n)}$, for $n = 1, 2, 3, \ldots$, represent the probability distribution of first passage time for specific i, j. In the case where a process in state i may never reach state j, $\sum_{n=1}^{\infty} f_{ij}^{(n)} < 1$.

Where $i = j$, and $\sum_{n=1}^{\infty} f_{ii}^{(n)} = 1$, the state i is termed a *recurrent state*, since given that the process is in state i it will always return to i.

If $p_{ii} = 1$ for some state i, then that state is called an *absorbing state*, and the process will never leave it after it is entered.

The state i is called a *transient state* if

$$\sum_{n=1}^{\infty} f_{ii}^{(n)} < 1$$

since there is a positive probability that given the process is in state i, it will never return to this state. It is not always easy to classify a state as transient or recurrent, since it is generally not possible to calculate first passage time probabilities for all n as was the case in Example 16-2. Although calculating the $f_{ij}^{(n)}$ for all n may be difficult, the expected first passage time is

$$E(n|i, j) = \mu_{ij} = \begin{cases} \infty, & \sum_{n=1}^{\infty} f_{ij}^{(n)} < 1 \\ \sum_{n=1}^{\infty} n \cdot f_{ij}^{(n)}, & \sum_{n=1}^{\infty} f_{ij}^{(n)} = 1 \end{cases} \tag{16-9}$$

and if $\sum_{n=1}^{\infty} f_{ij}^{(n)} = 1$, it can be shown that

$$\mu_{ij} = 1 + \sum_{l \neq j} p_{il} \cdot \mu_{lj} \tag{16-10}$$

If we take $i = j$, the expected first passage time is called the *expected recurrence time*. If $\mu_{ii} = \infty$ for a recurrent state, it is called *null*; if $\mu_{ii} < \infty$, it is called *nonnull* or *positive recurrent*.

There are no null recurrent states in a finite-state Markov chain. All of the states in such chains are either positive recurrent or transient.

A state is called *periodic* with period $\tau > 1$ if a return is possible only in $\tau, 2\tau, 3\tau, \ldots$, steps; so that $p_{ii}^{(n)} = 0$ for all values of n that are not divisible by $\tau > 1$, and τ is the smallest integer having this property.

A state j is termed *accessible* from state i if $p_{ij}^{(n)} > 0$ for some $n = 1, 2, \ldots$. In our example of the computing system, each state, 0 and 1, is accessible from the other since $p_{ij}^{(n)} > 0$ for all i, j and all n. If state j is accessible from i and state i is accessible from j, then the states are said to *communicate*. This is the case in Example 16-1. We note that any state communicates with itself. If state i communicates with j, j also communicates with i. Also, if i communicates with l and l communicates with j, then i also communicates with j.

If the state space is partitioned into disjoint sets (called classes) of states, where communicating states belong to the same class, then the Markov chain may consist of one or more classes. If there is only one class so that all states communicate, the Markov chain is said to be *irreducible*. The chain represented by Example 16-1 is thus also irreducible. For finite-state Markov chains, the states of a class are either all positive recurrent or all transient. In many applications, the states will all communicate. This is the case if there is a value of n for which $p_{ij}^{(n)} > 0$ for all values of i and j.

If state i in a class is aperiodic (not periodic), and if the state is also positive recurrent, then the state is said to be *ergodic*. An irreducible Markov chain is ergodic if all of its states are ergodic. In the case of such Markov chains the distribution

$$\mathbf{A}^{(n)} = \mathbf{A} \cdot \mathbf{P}^n$$

converges as $n \to \infty$, and the limiting distribution is independent of the initial probabilities, $\mathbf{A}$. In Example 16-1, this was clearly observed to be the case; and after five steps ($n > 5$), $P(X_n = 0) = .667$, and $P\{X_n = 1\} = .333$ when three significant figures are used.

In general, for irreducible, ergodic Markov chains,

$$\lim_{n \to \infty} p_{ij}^{(n)} = \lim_{n \to \infty} a_j^{(n)} = p_j$$

and, furthermore, these values p_j are independent of i. These "steady state" probabilities, p_j, satisfy the following *state equations*:

1. $p_j > 0$

2. $\displaystyle\sum_{j=0}^{m} p_j = 1$

3. $p_j = \displaystyle\sum_{i=0}^{m} p_i \cdot p_{ij} \qquad j = 0, 1, 2, \ldots, m$ $\qquad\qquad$ (16-11)

Since there are $m+2$ equations in (2) and (3) above and there are $m+1$ unknowns, one of the equations is redundant. Therefore, we will use m of the $m+1$ equations in (3) with equation (2).

● **Example 16-3.** In the case of the computing system presented in Example 16-1, we have from Equation (16-11) equations (2) and (3).

$$1 = p_0 + p_1$$
$$p_0 = p_0 \cdot (.75) + p_1(.50)$$

or

$$p_0 = 2/3 \quad \text{and} \quad p_1 = 1/3$$

which agrees with the emerging result as $n > 5$ in Example 16-1.

The steady state probabilities and the mean recurrence time for irreducible ergodic Markov chains have a reciprocal relationship

$$\mu_{jj} = \frac{1}{p_j} \quad j = 0, 1, 2, \ldots, m \tag{16-12}$$

In Example 16-3 note that $\mu_{00} = 1/p_0 = 1/(2/3) = 1.5$ and $\mu_{11} = 1/p_1 = 1/(1/3) = 3$.

● **Example 16-4.** The mood of a corporate president is observed over a period of time by a psychologist in the operations research department. Being inclined toward mathematical modeling, the psychologist classifies mood into three states as follows:

0: Good (cheerful)
1: Fair (so-so)
2: Poor (glum and depressed)

The psychologist observes that mood changes occur only overnight; thus, the data allow estimation of the following transition probabilities.

$$\mathbf{P} = \begin{bmatrix} .6 & .2 & .2 \\ .3 & .4 & .3 \\ .0 & .3 & .7 \end{bmatrix}$$

The following equations are solved simultaneously

$$p_0 = .6p_0 + .3p_1 + 0p_2$$
$$p_1 = .2p_0 + .4p_1 + .3p_2$$
$$1 = p_0 + p_1 + p_2$$

for the steady state probabilities

$$p_0 = 3/13$$
$$p_1 = 4/13$$
$$p_2 = 6/13$$

Given that the president is in a bad mood, that is, state 2, the mean time required to return to that state is μ_{22}, where

$$\mu_{22} = \frac{1}{p_2} = \frac{13}{6} \text{ days}$$

As noted earlier, if $p_{kk} = 1$, state k is called an *absorbing state*, and the process remains in state k once that state is reached. In this case, b_{ik} is called the absorption probability, which is the conditional probability of absorption into state k given state i. Mathematically, we have

$$b_{ik} = \sum_{j=0}^{m} p_{ij} \cdot b_{jk} \qquad i = 0, 1, 2, \ldots, m \qquad (16\text{-}13)$$

where

$$b_{kk} = 1$$

and

$$b_{ik} = 0 \qquad \text{for } i \text{ recurrent}, i \neq k$$

16-4   Continuous-Time Markov Chains

If the time parameter is continuous rather than a discrete index, as assumed in the previous sections, the Markov chain is called a *continuous-parameter* chain. It is customary to use a slightly different notation for continuous-parameter Markov chains, namely $X(t) = X_t$, where $\{X(t)\}, t \geq 0$, will be considered to have states $0, 1, \ldots, m + 1$. The discrete nature of the state space [range space for $X(t)$] is thus maintained, and

$$p_{ij}(t) = P[X(t+s) = j | X(s) = i] \qquad \begin{matrix} i = 0, 1, 2, \ldots, m+1 \\ j = 0, 1, 2, \ldots, m+1 \\ s \geq 0, t \geq 0 \end{matrix}$$

is the stationary transition probability function. It is noted that these probabilities are not dependent on s but only on t for a specified i, j pair of states. Furthermore, at time $t = 0$, the function is continuous with

$$\lim_{t \to 0} p_{ij}(t) = \begin{cases} 0 & i \neq j \\ 1 & i = j \end{cases}$$

There is a direct correspondence between the discrete-time and continuous-time models. The Chapman–Kolmogorov equations become

$$p_{ij}(t) = \sum_{l=0}^{m} p_{il}(v) \cdot p_{lj}(t - v) \tag{16-14}$$

for $0 \le v \le t$, and for the specified state pair i, j, and time, t. If there are times t_1 and t_2 such that $p_{ij}(t_1) > 0$ and $p_{ji}(t_2) > 0$, then states i and j are said to communicate. Once again states that communicate form a class and, where the chain is irreducible (all states form a single class),

$$p_{ij}(t) > 0 \qquad t > 0$$

for each state pair i, j.

We also have the property that

$$\lim_{t \to \infty} p_{ij}(t) = p_j$$

where p_j exists and is independent of the initial state probability vector $\mathbf{A}$. The values p_j are again called the steady state, state probabilities and they satisfy

$$p_j > 0 \qquad j = 0, 1, 2, \ldots, m$$

$$\sum_{j=0}^{m} p_j = 1$$

$$p_j = \sum_{i=0}^{m} p_i \cdot p_{ij}(t) \qquad j = 0, 1, 2, \ldots, m, t \ge 0.$$

The *intensity of transition*, given that the state is j, is defined as

$$u_j = \lim_{\Delta t \to 0} \left\{ \frac{1 - p_{jj}(\Delta t)}{\Delta t} \right\} = -\frac{d}{dt} p_{jj}(t)|_{t=0} \tag{16-15}$$

where the limit exists and is finite. Likewise, the *intensity of passage* from state i to state j, given that the system is in state i, is

$$u_{ij} = \lim_{\Delta t \to 0} \left\{ \frac{p_{ij}(\Delta t)}{\Delta t} \right\} = \frac{d}{dt} p_{ij}(t)|_{t=0} \tag{16-16}$$

again where the limit exists and is finite. The interpretation of the intensities is that they represent an instantaneous rate of transition from state i to j. For a small Δt, $p_{ij}(\Delta t) = u_{ij} \Delta t + o(\Delta t)$, where $o(\Delta t)/\Delta t \to 0$ as $\Delta t \to 0$, so that u_{ij} is a proportionality constant by which $p_{ij}(\Delta t)$ is proportional to Δt as $\Delta t \to 0$. The transition intensities also satisfy

$$p_j \cdot u_j = \sum_{i \ne j} p_i \cdot u_{ij} \qquad j = 0, 1, 2, \ldots, m \tag{16-17}$$

● **Example 16-5.** An electronic control mechanism for a chemical process is constructed with two identical modules, operating as a parallel, active redundant pair. The function of at least one module is necessary for the mechanism to operate. The maintenance shop has two identical repair stations for these modules and, furthermore, when a module fails and enters the shop, other work is moved aside and repair work is immediately initiated. The "system" here consists of the mechanism and repair facility and the states are:

0: Both modules operating

1: One unit operating and one unit in repair

2: Two units in repair (mechanism down)

The random variable representing time to failure for a module has an exponential density, say

$$f(t) = \lambda e^{-\lambda t} \qquad t \geq 0$$
$$= 0 \qquad t < 0$$

and the random variable describing repair time at a repair station also has an exponential density, say

$$r(t) = \mu e^{-\mu t} \qquad t \geq 0$$
$$= 0 \qquad t < 0$$

Interfailure and interrepair times are independent, and $\{X(t)\}$ can be shown to be a continuous-parameter, irreducible Markov chain with transitions only from a state to its neighbor states: $0 \to 1$, $1 \to 0$, $1 \to 2$, $2 \to 1$. Of course, there may be no state change.

The transition intensities are:

$$u_0 = 2\lambda \qquad\qquad u_1 = (\lambda + \mu)$$
$$u_{01} = 2\lambda \qquad\qquad u_{12} = \lambda$$
$$u_{02} = 0 \qquad\qquad u_{20} = 0$$
$$u_{10} = \mu \qquad\qquad u_{21} = 2\mu$$
$$u_2 = 2\mu$$

Using Equation (16-17),

$$2\lambda p_0 = \mu p_1$$
$$(\lambda + \mu)p_1 = 2\lambda p_0 + 2\mu p_2$$
$$2\mu p_2 = \lambda p_1$$

and, since $p_0 + p_1 + p_2 = 1$,

$$p_0 = \frac{\mu^2}{(\lambda + \mu)^2}$$

$$p_1 = \frac{2\lambda\mu}{(\lambda + \mu)^2}$$

$$p_2 = \frac{\lambda^2}{(\lambda + \mu)^2}$$

The system availability (probability that the mechanism is up) in the steady state condition is thus

$$\text{Availability} = 1 - \frac{\lambda^2}{(\lambda + \mu)^2}$$

The matrix of transition probabilities for time increment Δt may be expressed as

$$\mathbf{P} = \{p_{ij}(\Delta t)\} = \begin{bmatrix} (1 - u_0\Delta t) & (u_{01}\Delta t) & \ldots (u_{0j}\Delta t) \ldots (u_{0m}\Delta t) \\ (u_{10}\Delta t) & (1 - u_1\Delta t) & \ldots (u_{1j}\Delta t) \ldots (u_{1m}\Delta t) \\ \vdots & & \\ (u_{i0}\Delta t) & (u_{i1}\Delta t) & \ldots (u_{ij}\Delta t) \ldots (u_{im}\Delta t) \\ \vdots & & \\ (u_{m0}\Delta t) & (u_{m1}\Delta t) & \ldots (u_{mj}\Delta t) \ldots (1 - u_m\Delta t) \end{bmatrix} \quad (16\text{-}18)$$

and

$$p_j(t + \Delta t) = \sum_{i=0}^{m} p_i(t) \cdot p_{ij}(\Delta t) \qquad j = 0, 1, 2, \ldots, m \quad (16\text{-}19)$$

where

$$p_j(t) = P[X(t) = j]$$

For the jth equation in the $m + 1$ equations of Equation (16-19),

$$p_j(t + \Delta t) = p_0(t) \cdot u_{0j}\Delta t + \cdots + p_i(t) \cdot u_{ij}\Delta t + \cdots + p_j(t)[1 - u_j\Delta t] + \cdots \\ + p_m(t) \cdot u_{mj} \cdot \Delta t$$

which may be rewritten as

$$\frac{d}{dt} \cdot p_j^{(t)} = \lim_{\Delta t \to 0} \left[\frac{p_j(t + \Delta t) - p_j(t)}{\Delta t} \right] = -u_j \cdot p_j(t) + \sum_{i \neq j} u_{ij} \cdot p_i(t) \quad (16\text{-}20)$$

The resulting system of differential equations is

$$p_j'(t) = -u_j \cdot p_j(t) + \sum_{i \neq j} u_{ij} \cdot p_i(t) \qquad j = 0, 1, 2, \ldots, m \quad (16\text{-}21)$$

which may be solved when m is finite, given initial conditions (probabilities) $\mathbf{A} = [a_0^{(0)} a_1^{(0)} \ldots a_m^{(0)}]$, and using the result that $\sum_{j=0}^{m} p_j(t) = 1$. The solution

$$[p_0(t)p_1(t) \ldots p_m(t)] = \mathbf{P}(t) \quad (16\text{-}22)$$

presents the state probabilities as a function of time in the same manner that $p_j^{(n)}$ presented state probabilities as a function of the number of transitions, n, given an initial condition vector $\mathbf{A}$ in the discrete-time model. The solution to Equations (16-21) may be somewhat difficult to obtain and, in general practice, transformation techniques are employed.

16-5   The Birth-Death Process in Queueing

The major application of the birth-death process that we will study is in *queueing* or *waiting-line theory*. Therefore birth will refer to an *arrival* and death to a *departure* from a physical system, as shown in Fig. 16-1.

Queueing theory is the mathematical study of queues or waiting lines. These waiting lines occur in a variety of problem environments. There is an input process or "calling population" and a queueing *system*, which in Fig. 16-1 consists of the queue and service facility. The calling population may be finite or infinite. Arrivals occur in a probabilistic manner. A common assumption is that the interarrival times are exponentially distributed. The queue is generally classified according to whether its capacity is infinite or finite, and the service discipline refers to the order in which the customers in the queue are served. The service mechanism consists of one or more servers, and the elapsed service time is commonly called the holding time.

The following notation will be employed:

$X(t)$ = Number of customers in system at time t

States = $0, 1, 2, \ldots, j, j+1, \ldots$

s = Number of servers

$p_j(t) = P\{X(t) = j | A\}$

$p_j = \lim_{t \to \infty} p_j(t)$

λ_n = Mean arrival rate given that n customers are in the system

μ_n = Mean service rate given that n customers are in the system

The birth-death process can be used to describe how $X(t)$ changes through time. It will be assumed here that when $X(t) = j$, the probability distribution

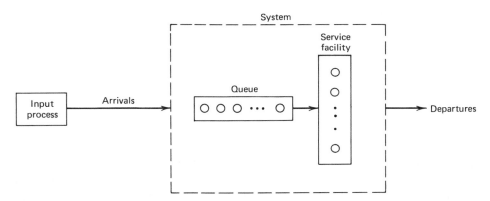

Fig. 16-1. A simple queueing system.

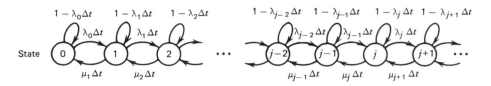

Fig. 16-2.   Transition diagram for the birth-death process.

of the time to the next birth (arrival) is exponential with parameter λ_j, $j = 0, 1, 2, \ldots$. Furthermore, given $X(t) = j$, the remaining time to the next service completion is taken to be exponential with parameter μ_j, $j = 1, 2, \ldots$. Poisson type postulates are assumed to hold, so that the probability of more than one birth or death at the same instant is zero.

A transition diagram is shown in Fig. 16-2. The transition matrix corresponding to Equation (16-18) is

$$\mathbf{P} = \begin{bmatrix} 1 - \lambda_0 \Delta t & \lambda_0 \Delta t & 0 & \cdots & 0 & 0 & \cdots \\ \mu_1 \Delta t & 1 - (\lambda_1 + \mu_1)\Delta t & \lambda_1 \Delta t & \cdots & 0 & 0 & \cdots \\ 0 & \mu_2 \Delta t & 1 - (\lambda_2 + \mu_2)\Delta t & \cdots & 0 & 0 & \cdots \\ 0 & 0 & \mu_3 \Delta t & \cdots & 0 & 0 & \cdots \\ \vdots & \vdots & \vdots & \cdots & \vdots & \vdots & \\ 0 & 0 & 0 & \cdots & \lambda_{j-2}\Delta t & 0 & \cdots \\ \vdots & \vdots & \vdots & & 1 - (\lambda_{j-1} + \mu_{j-1})\Delta t & \lambda_{j-1}\Delta t & \cdots \\ & & & & \mu_j \Delta t & 1 - (\lambda_j + \mu_j)\Delta t & \cdots \\ 0 & 0 & 0 & \cdots & 0 & \mu_{j+1}\Delta t & \cdots \\ \vdots & \vdots & \vdots & & \vdots & \vdots & \cdots \\ 0 & 0 & 0 & \cdots & 0 & 0 & \cdots \end{bmatrix}$$

$$(16\text{-}23)$$

We note that $p_{ij}(\Delta t) = 0$ for $j < i - 1$ or $j > i + 1$. Furthermore, the transition intensities and intensities of passage shown in Equation (16-17) are

$$u_0 = \lambda_0$$
$$u_j = (\lambda_j + \mu_j) \qquad \text{for } j = 1, 2, \ldots$$
$$u_{ij} = \lambda_i \qquad \text{for } j = i + 1$$
$$\quad = \mu_i \qquad \text{for } j = i - 1$$
$$\quad = 0 \qquad \text{for } j < i - 1, j > i + 1$$

The fact that the transition intensities and intensities of passage are constant with time is important in the development of this model. The nature of transition can be viewed to be specified by assumption, or it may be considered as a result of the prior assumption about the distribution of time between occurrences (births and deaths).

The assumptions of independent, exponentially distributed service times and independent, exponentially distributed interarrival times yield transition intensities that are constant in time. This was also observed in the development of the Poisson and exponential distributions in Chapters 5 and 6.

The methods used in Equations (16-19) to (16-21) may be used to formulate an infinite set of differential state equations from the transition matrix of Equation (16-22). Thus, the time-dependent behavior is described in the following equation.

$$p_0'(t) = -\lambda_0 p_0(t) + \mu_1 p_1(t)$$
$$p_j'(t) = -(\lambda_j + \mu_j) \cdot p_j(t) + \lambda_{j-1} p_{j-1}(t) + \mu_{j+1} \cdot p_{j+1}(t) \qquad j = 1, 2, \ldots \qquad (16\text{-}24)$$

$$\sum_{j=0}^{\infty} p_j(t) = 1 \qquad \text{and} \qquad \mathbf{A} = [a_0^{(0)} a_1^{(0)} \ldots a_j^{(0)} \ldots]$$

In the steady state $(t \to \infty)$, we have $p_j'(t) = 0$, so the steady state equations are obtained from Equations (16-23)

$$\mu_1 p_1 = \lambda_0 p_0$$
$$\lambda_0 p_0 + \mu_2 p_2 = (\lambda_1 + \mu_1) \cdot p_1$$
$$\lambda_1 p_1 + \mu_3 p_3 = (\lambda_2 + \mu_2) \cdot p_2$$
$$\vdots$$
$$\lambda_{j-2} \cdot p_{j-2} + \mu_j p_j = (\lambda_{j-1} + \mu_{j-1}) \cdot p_{j-1}$$
$$\lambda_{j-1} p_{j-1} + \mu_{j+1} p_{j+1} = (\lambda_j + \mu_j) \cdot p_j$$
$$\vdots \qquad (16\text{-}25)$$

and $\Sigma_{j=0}^{\infty} p_j = 1$.

Equations (16-25) could have also been determined by the direct application of Equation (16-17) which provides a "rate balance" or "intensity balance." Solving Equations (16-25) we obtain

$$p_1 = \frac{\lambda_0}{\mu_1} \cdot p_0$$

$$p_2 = \frac{\lambda_1}{\mu_2} \cdot p_1 = \frac{\lambda_1 \lambda_0}{\mu_2 \mu_1} \cdot p_0$$

$$p_3 = \frac{\lambda_2}{\mu_3} \cdot p_2 = \frac{\lambda_2 \lambda_1 \lambda_0}{\mu_3 \mu_2 \mu_1} \cdot p_0$$

$$\vdots$$

$$p_j = \frac{\lambda_{j-1}}{\mu_j} \cdot p_{j-1} = \frac{\lambda_{j-1} \lambda_{j-2} \ldots \lambda_0}{\mu_j \mu_{j-1} \ldots \mu_1} \cdot p_0$$

$$p_{j+1} = \frac{\lambda_j}{\mu_{j+1}} \cdot p_j = \frac{\lambda_j \lambda_{j-1} \ldots \lambda_0}{\mu_{j+1} \mu_j \ldots \mu_1} \cdot p_0 \qquad (16\text{-}26)$$

If we let

$$C_j = \frac{\lambda_{j-1}\lambda_{j-2}\ldots\lambda_0}{\mu_j\mu_{j-1}\ldots\mu_1}$$

then

$$p_j = C_j \cdot p_0 \qquad j = 1, 2, 3, \ldots$$

and since

$$\sum_{j=0}^{\infty} p_j = 1 \qquad \text{or} \qquad p_0 + \sum_{j=1}^{\infty} p_j = 1$$

$$p_0 = \frac{1}{1 + \sum_{j=1}^{\infty} C_j} \tag{16-27}$$

These steady state results assume that the λ_j, μ_j values are such that a steady state can be reached. This will be true if $\lambda_j = 0$ for $j > k$, so that there is a finite number of states. It is also true if $\rho = \dfrac{\lambda}{s\mu} < 1$, where λ and μ are constant. The steady state will not be reached if $\Sigma_{j=1}^{\infty} C_j = \infty$.

16-6 Considerations in Queueing Models

When the arrival rate λ_j is constant for all j, the constant is denoted as λ. Similarly, when the mean service rate per busy server is constant, it will be denoted as μ, so that $\mu_j = s\mu$ if $j \geq s$ and $\mu_j = j \cdot \mu$ if $j < s$. The exponential distributions

$$f_a(t) = \lambda e^{-\lambda t} \qquad t \geq 0$$
$$= 0 \qquad t < 0$$
$$f_s(t) = \mu e^{-\mu t} \qquad t \geq 0$$
$$= 0 \qquad t < 0$$

for interarrival times and service times in a busy channel produce rates λ and μ, which are constant. The mean interarrival time is $1/\lambda$, and the mean time for a busy channel to complete service is $1/\mu$.

A special set of notation has been widely employed in the analysis of queueing systems. This notation is given in the following list.

$L = \Sigma_{j=0}^{\infty} j \cdot p_j =$ expected number of customers in the queueing system
$L_q = \Sigma_{j=0}^{\infty} (j - s) \cdot p_j =$ expected queue length
$W =$ Expected waiting time in the system
$W_q =$ Expected waiting time in the queue (excluding service time)

If λ is constant for all j, then it has been shown that

$$L = \lambda W \qquad (16\text{-}28)$$

and

$$L_q = \lambda W_q$$

If the λ_j are not equal, $\bar{\lambda}$ replaces λ, where

$$\bar{\lambda} = \sum_{j=0}^{\infty} \lambda_j \cdot p_j \qquad (16\text{-}29)$$

The system utilization coefficient $\rho = \lambda/s\mu$ is the fraction of time that the servers are busy. In the case where the mean service time is $1/\mu$ for all $j \geq 1$,

$$W = W_q + \frac{1}{\mu} \qquad (16\text{-}30)$$

The birth-death process rates, $\lambda_0, \lambda_1, \ldots, \lambda_j, \ldots$, and $\mu_1, \mu_2, \ldots, \mu_j, \ldots$, may be assigned any positive values so long as the assignment leads to a steady state solution. This allows considerable flexibility in using the results given in Equation (16-27). The specific models presented will differ in the manner in which λ_j and μ_j vary as a function of j.

16-7 Basic Single Server Model with Constant Rates

We will now consider the case where $s = 1$, that is, a single server. We will also assume an unlimited queue length with exponential interarrivals having a constant parameter λ, so that $\lambda_0 = \lambda_1 = \cdots = \lambda_{j-1} = \lambda$. Furthermore, service times will be assumed to be independent and exponentially distributed with $\mu_1 = \mu_2 = \mu_3 = \cdots = \mu_j = \mu$. We will assume $\lambda < \mu$. As a result of Equations (16-26), we have

$$C_j = \left(\frac{\lambda}{\mu}\right)^j = \rho^j \qquad j = 1, 2, 3, \ldots \qquad (16\text{-}31)$$

and from Equation (16-27),

$$p_j = \rho^j p_0 \qquad j = 1, 2, 3, \ldots$$

$$p_0 = \frac{1}{1 + \sum_{j=1}^{\infty} \rho^j} = 1 - \rho \qquad (16\text{-}32)$$

Thus, the steady state, state equations are

$$p_j = (1 - \rho)\rho^j \qquad j = 0, 1, 2, \ldots \qquad (16\text{-}33)$$

Note that the probability that there are j customers in the system p_j is given by a geometric distribution with parameter ρ. The mean number of customers in the system, L, is determined as

$$L = \sum_{j=0}^{\infty} j \cdot (1 - \rho)\rho^j$$

$$= (1 - \rho) \cdot \rho \sum_{j=0}^{\infty} \frac{d}{d\rho} (\rho^j)$$

$$= (1 - \rho) \cdot \rho \frac{d}{d\rho} \sum_{j=0}^{\infty} \rho^j$$

$$= \frac{\rho}{1 - \rho} \tag{16-34}$$

And the expected queue length is

$$L_q = \sum_{j=1}^{\infty} (j - 1) \cdot p_j$$

$$= L - (1 - p_0)$$

$$= \frac{\lambda^2}{\mu(\mu - \lambda)} \tag{16-35}$$

Using Equations (16-28) and (16-29), we find that the expected waiting time in the system is

$$W = \frac{L}{\lambda} = \frac{\rho}{\lambda(1 - \rho)} = \frac{1}{\mu - \lambda} \tag{16-36}$$

and the expected waiting time in the queue is

$$W_q = \frac{L_q}{\lambda} = \frac{\lambda^2}{\mu(\mu - \lambda) \cdot \lambda} = \frac{\lambda}{\mu(\mu - \lambda)} \tag{16-37}$$

These results could have been developed directly from the distributions of time in the system and time in the queue, respectively. Since the exponential distribution reflects a memoryless process, an arrival finding j units in the system will wait through $j + 1$ services, including its own, and thus its waiting time T_{j+1} is the sum of $j + 1$ independent, exponentially distributed random variables. This random variable was shown in Chapter 6 to have a gamma distribution. This is a conditional density given that the arrival finds j units in the system. Thus, if S represents time in the system,

$$P(S > w) = \sum_{j=0}^{\infty} p_j \cdot P(T_{j+1} > w)$$

$$= \sum_{j=0}^{\infty} (1 - \rho)\rho^j \cdot P(T_{j+1} > w)$$

$$= e^{-\mu(1-\rho)w} \qquad w \geq 0$$

$$= 0 \qquad w < 0 \tag{16-38}$$

which is seen to be the complement of the distribution function for an exponential random variable with parameter $\mu(1-\rho)$. The value $W = 1/\mu(1-\rho) = 1/\mu - \lambda$ follows directly.

If we let S_q represent time in the queue, excluding service time, then

$$P(S_q = 0) = p_0 = 1 - \rho$$

If we take T_j as the sum of j service times, T_j will again have a gamma distribution. Then,

$$
\begin{aligned}
P(S_q > w_q) &= \sum_{j=1}^{\infty} p_j \cdot P(T_j > w_q) \\
&= \sum_{j=1}^{\infty} (1-\rho)\rho^j \cdot P(T_j > w_q) \\
&= \rho e^{-\mu(1-\rho)w_q} \qquad w_q > 0 \\
&= 0 \qquad\qquad\quad w_q < 0
\end{aligned}
\qquad (16\text{-}39)
$$

and we find the distribution of time in the queue $g(w_q)$, for $w_q > 0$, to be

$$g(w_q) = \frac{d}{dw_q}[1 - \rho e^{-\mu(1-\rho)w_q}] = \rho(1-\rho)\mu e^{-\mu(1-\rho)w_q} \qquad w_q > 0$$

Thus, the probability distribution is

$$
\begin{aligned}
g(w_q) &= 1 - \rho \qquad\qquad\quad w_q = 0 \\
&= \lambda(1-\rho)e^{-(\mu-\lambda)w_q} \qquad w_q > 0
\end{aligned}
\qquad (16\text{-}40)
$$

The expected waiting time in the queue W_q could be determined directly from this distribution as

$$W_q = (1-\rho) \cdot 0 + \int_0^{\infty} w_q \cdot \lambda(1-\rho)e^{-(\mu-\lambda)w_q} \, dw_q$$

$$= \frac{\lambda}{\mu(\mu-\lambda)} \qquad (16\text{-}41)$$

When $\lambda \geq \mu$, the summation of the terms p_j in Equation (16-32) diverges. In this case, there is no steady state solution since the steady state is never reached. That is, the queue would grow without bound.

16-8   Single Server with Limited Queue Length

If the queue is limited so that at most N units can be in the system, and if the exponential service time and exponential interarrival times are retained from the prior model, we have

$$\lambda_0 = \lambda_1 = \cdots = \lambda_{N-1} = \lambda$$

$$\lambda_j = 0 \qquad j \geq N$$

and

$$\mu_1 = \mu_2 = \cdots = \mu_N = \mu$$

It follows from Equation (16-26) that

$$C_j = \left(\frac{\lambda}{\mu}\right)^j \quad j \leq N$$

$$= 0 \qquad j > N \tag{16-42}$$

Thus,

$$p_j = \left(\frac{\lambda}{\mu}\right)^j p_0 \quad j = 0, 1, 2, \ldots, N$$

$$p_0 \sum_{j=0}^{N} \rho^j = 1$$

and

$$p_0 = \frac{1}{1 + \sum_{j=1}^{N} \rho^j} = \frac{1 - \rho}{1 - \rho^{N+1}} \tag{16-43}$$

As a result, the steady state, state equations are given by

$$p_j = \rho^j \left[\frac{1 - \rho}{1 - \rho^{N+1}}\right] \quad j = 0, 1, 2, \ldots, N \tag{16-44}$$

The mean number of customers in the system in this case is

$$L = \sum_{j=0}^{N} j \cdot \rho^j \left[\frac{1 - \rho}{1 - \rho^{N+1}}\right]$$

$$= \rho \left[\frac{1 - (N+1)\rho^N + N\rho^{N+1}}{(1 - \rho)(1 - \rho^{N+1})}\right] \tag{16-45}$$

The mean number of customers in the queue is

$$L_q = \sum_{j=1}^{N} (j - 1) \cdot p_j$$

$$= \sum_{j=0}^{N} j p_j - \sum_{j=1}^{N} p_j$$

$$= L - (1 - p_0) \tag{16-46}$$

The mean time in the system is found as

$$W = \frac{L}{\lambda} \tag{16-47}$$

and the mean time in the queue is

$$W_q = \frac{L_q}{\lambda} = \frac{L - 1 + p_0}{\lambda} \tag{16-48}$$

where L is given by Equation (16-45).

16-9 Multiple Servers with an Unlimited Queue

We now consider the case where there are multiple servers. We also assume that the queue is unlimited and that exponential assumptions hold for interarrival times and service times. In this case, we have

$$\lambda_0 = \lambda_1 = \cdots = \lambda_j = \cdots = \lambda \tag{16-49}$$

and

$$\mu_j = j \cdot \mu \qquad \text{for } j \leq s$$
$$= s\mu \qquad \text{for } j > s$$

Thus,

$$C_j = \frac{\lambda^j}{j! \cdot \mu^j} = \frac{\left(\dfrac{\lambda}{\mu}\right)^j}{j!} \qquad j \leq s$$

$$= \frac{\lambda^j}{s! \cdot s^{j-s}\mu^j} = \frac{\left(\dfrac{\lambda}{\mu}\right)^j}{s!s^{j-s}} \qquad j > s \tag{16-50}$$

It follows from Equation (16-27) that the state equations are developed as

$$p_j = \frac{\left(\dfrac{\lambda}{\mu}\right)^j}{j!} \cdot p_0 \qquad j < s$$

$$= \frac{\left(\dfrac{\lambda}{\mu}\right)^j}{s!s^{j-s}} \cdot p_0 \qquad j \geq s$$

$$p_0 = \frac{1}{1 + \displaystyle\sum_{j=1}^{s} \frac{(\lambda/\mu)^j}{j!} + \sum_{j=s+1}^{\infty} \frac{(\lambda/\mu)^j}{s!s^{j-s}}}$$

$$= \frac{1}{\displaystyle\sum_{j=0}^{s-1} \frac{\phi^j}{j!} + \frac{\phi^j}{s!}\left(\frac{1}{1-\rho}\right)} \tag{16-51}$$

where $\phi = \lambda/\mu$, and $\rho = \lambda/s\mu = \phi/s$ is the utilization coefficient, assuming $\rho < 1$.

The value L_q, representing the mean number of units in the queue, is developed as follows:

$$L_q = \sum_{j=s}^{\infty} (j - s)p_j = \sum_{j=0}^{\infty} j \cdot p_{s+j}$$

$$= \left[\frac{\phi^s}{s!} \cdot \rho \frac{d}{d\rho}\left(\sum_{j=0}^{\infty} \rho^j\right)\right] \cdot p_0$$

$$= \left[\frac{\phi^s}{s!(1-\rho)^2}\right] \cdot p_0 \tag{16-52}$$

Then

$$W_q = \frac{L_q}{\lambda} \tag{16-53}$$

and

$$W = W_q + \frac{1}{\mu} \tag{16-54}$$

so that

$$L = \left(W_q + \frac{1}{\mu} \right) = L_q + \phi \tag{16-55}$$

16-10 Other Queueing Models

There are numerous other queueing models that can be developed from the birth-death process. In addition, it is also possible to develop queueing models for situations involving nonexponential distributions. A good reference is White, Schmidt, and Bennet (1975). One useful result, given without development, is for a single server system having exponential interarrivals and arbitrary service time distribution with mean $1/\mu$ and variance σ^2. If $\rho = \lambda/\mu < 1$, the steady state measures are given by Equations (16-56).

$$p_0 = 1 - \rho$$
$$L_q = \frac{\lambda^2 \sigma^2 + \rho^2}{2(1 - \rho)}$$
$$L = \rho + L_q$$
$$W_q = \frac{L_q}{\lambda}$$
$$W = W_q + \frac{1}{\mu} \tag{16-56}$$

In the case where service times are constant at $1/\mu$, the above relationships yield the measures of system performance if the variance $\sigma^2 = 0$.

16-11 Summary

This chapter introduced the notion of discrete-state space stochastic processes for discrete-time and continuous-time orientations. The Markov process was developed along with the presentation of state properties and characteristics. This was followed by a presentation of the birth-death process and several important applications to queueing models for the description of waiting time phenomena.

16-12 Exercises

16-1. Weather data are analyzed for a particular locality, and a Markov chain is employed as a model for weather change as follows. The conditional probability of change from rain to clear weather in one day is .3. Likewise, the conditional probability of transition from clear to rain in one day is .1. The model is to be a discrete-time model, with transitions occurring only between days.

(a) Determine the matrix **P** of one-step transition probabilities.

(b) Find the steady state, state probabilities.

(c) If today is clear, find the probability that it will be clear exactly three days hence.

(d) Find the probability that the first passage from a clear day to a rainy day occurs in exactly two days, given a clear day is the initial state.

(e) What is the mean recurrence time for the rainy day state?

16-2. Consider a two-component active redundancy where the components are identical and time to failure distributions are exponential. When both units are operating, each carries load $L/2$ and each has failure rate λ. However, when one unit fails, the load carried by the other component is L, and its failure rate under this load is $(1.5)\lambda$. There is only one repair facility available, and repair time is exponentially distributed with mean $1/\mu$. The system is considered failed when both components are in the failed state. Both components are initially operating. Assume that $\mu > (1.5)\lambda$. Let the states be

0: No components are failed.

1: One component is failed and is in repair.

2: Two components are failed, one is in repair, one is waiting, and the system is in the failed condition.

(a) Determine the matrix **P** of transition probabilities associated with interval Δt.

(b) Determine the steady state, state probabilities.

(c) Write the system of differential equations that present the transient or time-dependent relationships for transition.

16-3. A gambler bets $1 on each hand of blackjack. The probability of winning on any hand is p, and the probability of losing is $1 - p = q$. The gambler will continue to play until either $\$Y$ have been accumulated, or he has no money left. Let X_t denote the accumulated winnings on hand t. Note that $X_{t+1} = X_t + 1$, with probability p, and $X_{t+1} = X_t - 1$, with probability q, and that $X_{t+1} = X_t$ if $X_t = 0$ or $X_t = Y$. The stochastic process X_t is a Markov chain.

(a) Find the one-step transition matrix **P**.

(b) For $Y = 4$ and $p = .3$, find the absorption probabilities b_{10}, b_{14}, b_{30}, and b_{34}.

16-4. An object moves between four points on a circle, which are labeled 1, 2, 3, and 4. The probability of moving one unit to the right is p, and the probability of moving one unit to the left is $1 - p = q$. Assume that the object starts at 1, and let X_n denote the location on the circle after n steps.

(*a*) Find the one-step transition matrix **P**.

(*b*) Find an expression for the steady state probabilities p_j.

(*c*) Evaluate the probabilities p_j for $p = .5$ and $p = .8$.

16-5. For the single-server, queueing model presented in Section 16-7, sketch the graphs of the following quantities as a function of $\rho = \lambda/\mu$, for $0 < \rho < 1$.

(*a*) Probability of no units in the system

(*b*) Mean time in the system

(*c*) Mean time in the queue

16-6. Interarrival times at a telephone booth are exponential, with an average time of 10 minutes. The length of a phone call is assumed to be exponentially distributed, with mean 3 minutes.

(*a*) What is the probability that a person arriving at the booth will have to wait?

(*b*) What is the average queue length?

(*c*) The telephone company will install a second booth when an arrival would expect to have to wait 3 minutes or more for the phone. By how much must the rate of arrivals be increased in order to justify a second booth?

(*d*) What is the probability that an arrival will have to wait more than 10 minutes for the phone?

(*e*) What is the probability that it will take a person more than 10 minutes altogether, for the phone and to complete the call?

(*f*) Estimate the fraction of a day that the phone will be in use.

16-7. Automobiles arrive at a service station in a random manner at a mean rate of 15 per hour. This station has only one service position, with a mean servicing rate of 27 customers per hour. Service times are exponentially distributed. There is space for only the automobile being served and 2 waiting. If all three spaces are filled, the automobile will go on to another station.

(*a*) What is the average number of units in the station?

(*b*) What fraction of customers will be lost?

(*c*) Why is $L_q \neq L - 1$?

16-8. An engineering school has three secretaries in its general office. Professors with jobs for the secretaries arrive at random, at an average rate of 20 per 8-hour day. The amount of time that a secretary spends on a job has an exponential distribution with a mean of 40 minutes.

(*a*) What fraction of the time are the secretaries busy?

(*b*) How much time does it take, on the average, for a professor to get his or her jobs completed?

(*c*) If an economy drive reduced the secretarial force to two secretaries, what will be the new answers to (*a*) and (*b*)?

16-9. The mean frequency of arrivals at an airport is 18 planes per hour, and the mean time that a runway is tied up with an arrival is 2 minutes. How many runways will have to be provided so that the probability of a plane having to wait is .20? Ignore finite population effects and make the assumption of exponential interarrival and service times.

16-10. A hotel reservations facility uses inward WATS lines to service customer requests. The mean number of calls that arrive per hour is 50, and the mean service time for a call is 3 minutes. Assume that interarrival and service times are exponentially distributed. Calls that arrive when all lines are busy obtain a busy signal and are lost from the system.

(a) Find the steady state, state equations for this system.

(b) How many WATS lines must be provided to ensure that the probability of a customer obtaining a busy signal is .05?

(c) What fraction of the time are all WATS lines busy?

(d) Suppose that during the evening hours, call arrivals occur at a mean rate of 10 per hour. How does this affect the WATS line utilization?

(e) Suppose the estimated service time (3 minutes) is in error, and the true service time is really 5 minutes. What effect will this have on the probability of a customer finding all lines busy if the number of lines in (a) are used?

Chapter 17

Statistical Decision Theory

Statistical inference deals with drawing conclusions or making decisions based on a random sample of information about a phenomena. Statistical decision theory is concerned with the methodology for decision making in problems where uncertainty is present. Thus, many of the statistical inference techniques that we have studied in previous chapters can be thought of as techniques of statistical decision theory. In this chapter we discuss some of these techniques, such as parameter estimation and hypothesis testing, from a decision theory viewpoint. The development of these techniques in previous chapters has been from a *classical* viewpoint. In addition to presenting these procedures in a decision theory framework, we also present the Bayesian approach to these problems. A comparison between the Bayesian and classical approach to statistical analysis will also be given.

17-1 The Structure and Concepts of Decisions

Suppose that a decision maker must select an action or decision from among several courses of action that are available. Furthermore, the course of action will be the result of sampling a random variable with density $f(x)$, which is characterized by the unknown parameter θ. If θ were known, the density function would be completely specified, and the appropriate action would be known.

The decision maker selects a random sample, say $X_1, X_2, \ldots, X_n$. Next the set of all possible values of θ, called the parameter space, which is denoted by Ω, must be defined. In a similar fashion, the set of all possible decisions D, called the decision space, is determined. Then, a function of the sample data is computed, say

$$a = d(X_1, X_2, \ldots, X_n) \tag{17-1}$$

where a is in the decision space. The decision maker will take the action denoted by a, where $a = d(x_1, x_2, \ldots, x_n)$, if $x_1, x_2, \ldots, x_n$, is observed. The function d is usually called a decision function or strategy.

Since there are many ways in which the decision function d could be formulated, we need some way to evaluate decision functions and to select good ones. One way in which this can be done is to evaluate the consequences of the decisions associated with d. This is usually accomplished by introducing the concept of a *loss function*, say $l(a; \theta)$. A loss function is a real-valued nonnegative function that represents the loss of making decision a when the true value of the parameter is θ. Obviously, we would define $l(a; \theta) = 0$ if a is the appropriate action for θ. For a statistical decision problem, the loss function may be written as

$$l\{d(X_1, X_2, \ldots, X_n); \theta\} \tag{17-2}$$

because our decision, a, depends on the particular sample values of $X_1, X_2, \ldots, X_n$, which we observe.

We see that the loss is a random variable and depends on the sample outcome. Therefore, we choose to define the *risk* to be the expected value of the loss function. That is, the risk, say $R(d; \theta)$, is a function of θ, d, and l such that

$$R(d; \theta) = E\{l[d(X_1, X_2, \ldots, X_n); \theta]\}$$
$$= \int_{-\infty}^{\infty} \int_{-\infty}^{\infty} \cdots \int_{-\infty}^{\infty} l\{d(x_1, x_2, \ldots, x_n); \theta\} f(x_1) f(x_2)$$
$$\ldots f(x_n) \, dx_1 \, dx_2 \ldots dx_n \tag{17-3}$$

Obviously, a good decision function would be one that minimizes the risk $R(d; \theta)$ for all values of θ in the parameter space Ω.

In many applied problems, the use of decision theory is complicated because of the difficulty of specifying a realistic loss function. For example, it may be quite difficult to specify the loss realized by incorrectly estimating the parameter of a probability distribution, or making the wrong decision when testing a hypothesis. Since the losses are random variables, it may also be questionable to talk about a risk function, which is the expected value of the losses, when the decision problem is only encountered once. In practice, a precise loss function is not really necessary, because good decision procedures are insensitive to small errors in estimating the form of the loss function. Also, if the losses are measured in terms of a *utility function*, we can measure the random losses by dealing with the risk function.

We will concentrate on the two problems of statistical inference that have been treated in the previous chapters: estimation of a parameter θ, which characterizes the probability density function $f(x)$, and testing hypotheses about θ.

Consider first the problem of determining a point estimate of θ. If the decision consists of acting as if $\hat{\theta}$ were the true value of θ, then our loss function $l(\hat{\theta}; \theta) = 0$ if and only if $\hat{\theta} = \theta$. Our decision function is

$$\hat{\theta} = d(X_1, X_2, \ldots, X_n) \tag{17-4}$$

and $\hat{\theta}$ is usually called an estimator of θ. The choice of a loss function for such a problem is of considerable importance. In many problems, an appropriate loss function is

$$l(\hat{\theta}; \theta) = h(\theta)(\hat{\theta} - \theta)^2 \tag{17-5}$$

where $h(\theta) > 0$ for all θ. Because $h(\theta)$ plays a relatively minor role in determining the relative merits of two decision functions, it is often reasonable to set $h(\theta) = 1$. If $h(\theta) = 1$ in Equation (17-5), the loss function is called a squared-error loss function.

For a point estimation problem, we usually call the decision function an estimator, and the decision an estimate. Thus, $\hat{\theta}$ is an estimate or decision. Frequently, $\hat{\theta}$ is also called an estimator, in which case we mean the decision function d defined in Equation (15-4). The central problem of point estimation is to find a decision function d that will minimize the risk function

$$R(d; \theta) = h(\theta)E\{(\hat{\theta} - \theta)^2\}$$

or, if $h(\theta) = 1$,

$$R(d; \theta) = E\{(\hat{\theta} - \theta)^2\}$$

Therefore, for $l(\hat{\theta}; \theta) = (\hat{\theta} - \theta)^2$, the problem of finding an estimator with minimum risk is equivalent to finding an estimator that minimizes the mean square error $E\{(\hat{\theta} - \theta)^2\}$. Unfortunately, for most densities $f(x)$, there does not exist an estimator that minimizes the mean square error for all possible values of θ. That is, one estimator may produce a minimum mean square error for some values of θ, while another estimator may produce a minimum mean square error for other values of θ. Since the value of θ is unknown, this limits the usefulness of mean square error as a criterion for the selection of estimators. However, we can use mean square error as a guide. For example, if $\hat{\theta}_1$ and $\hat{\theta}_2$ are two estimators such that

$$\hat{\theta}_1 = d_1(X_1, X_2, \ldots, X_n)$$

and

$$\hat{\theta}_2 = d_2(X_1, X_2, \ldots, X_n)$$

we can compare the risk $R(d_1; \theta)$ with $R(d_2; \theta)$ for specified values of θ. If $R(d_1; \theta) < R(d_2; \theta)$ for a particular θ, then $\hat{\theta}_1$ is a "better" estimator than $\hat{\theta}_2$.

We may also discuss hypothesis testing from the point of view of statistical decision theory. Suppose there are two possible decisions, say a_1 and a_2, such that the appropriate decision depends on the value of an unknown parameter θ. The true value of the parameter θ is often called the state of nature, and it will be an element of the parameter space Ω. We may form two subsets of Ω, say ω_1 and $\omega_2 = \Omega - \omega_1$. Decision a_1 is to be preferred if θ is in ω_1 and decision a_2 is to be preferred if θ is in ω_2. The loss associated with decision a and the state of nature θ is $l(a; \theta)$, where

$$l(a_1; \theta) = 0 \qquad \theta \in \omega_1$$
$$l(a_2; \theta) = 0 \qquad \theta \in \omega_2 \tag{17-6}$$

and $l(a; \theta) \geq 0$ for all θ.

Now let $\mathbf{X} = X_1, X_2, \ldots, X_n$ be a random sample from the density $f(x)$ and $S_{\mathbf{X}}$ be the n-dimensional sample space for $\mathbf{X}$. We may partition $S_{\mathbf{X}}$ into the sets $S_{\mathbf{X}_1}$ and $S_{\mathbf{X}_2}$, where $S_{\mathbf{X}_2} = S_{\mathbf{X}} - S_{\mathbf{X}_1}$, such that decision a_1 is made if the sample point $\mathbf{X}$ falls in $S_{\mathbf{X}_1}$ and decision a_2 is made if the sample point $\mathbf{X}$ falls in $S_{\mathbf{X}_2}$. The risk associated with the decision function d is, from Equation (17-3),

$$
\begin{aligned}
R(d; \theta) &= \int\!\!\int \cdots \int_{S_{\mathbf{X}}} l\{d(x_1, x_2, \ldots, x_n); \theta\} f(x_1) \\
&\quad \times f(x_2) \ldots f(x_n)\, dx_1\, dx_2 \ldots dx_n \\
&= \int\!\!\int \cdots \int_{S_{\mathbf{X}_1}} l\{d(x_1, x_2, \ldots, x_n); \theta\} f(x_1) \\
&\quad \times f(x_2) \ldots f(x_n)\, dx_1\, dx_2 \ldots dx_n \\
&\quad + \int\!\!\int \cdots \int_{S_{\mathbf{X}_2}} l\{d(x_1, x_2, \ldots, x_n); \theta\} f(x_1)f(x_2) \ldots f(x_n) \\
&\quad \times dx_1\, dx_2 \ldots dx_n
\end{aligned}
$$

The decision function is such that we choose a_1 if $\mathbf{X} \in S_{\mathbf{X}_1}$, or we choose a_2 if $\mathbf{X} \in S_{\mathbf{X}_2}$, so

$$
\begin{aligned}
R(d; \theta) &= \int\!\!\int \cdots \int_{S_{\mathbf{X}_1}} l(a_1; \theta)f(x_1)f(x_2) \ldots f(x_n)\, dx_1\, dx_2 \ldots dx_n \\
&\quad + \int\!\!\int \cdots \int_{S_{\mathbf{X}_2}} l(a_2; \theta)f(x_1)f(x_2) \ldots f(x_n)\, dx_1\, dx_2 \ldots dx_n \\
&= l(a_1; \theta) \int\!\!\int \cdots \int_{S_{\mathbf{X}_1}} f(x_1)f(x_2) \ldots f(x_n)\, dx_1\, dx_2 \ldots dx_n \\
&\quad + l(a_2; \theta) \int\!\!\int \cdots \int_{S_{\mathbf{X}_2}} f(x_1)f(x_2) \ldots f(x_n)\, dx_1\, dx_2 \ldots dx_n
\end{aligned}
$$

or

$$R(d; \theta) = l(a_1; \theta)P(\mathbf{X} \in S_{\mathbf{X}_1}|\theta) + l(a_2; \theta)P(\mathbf{X} \in S_{\mathbf{X}_2}|\theta) \tag{17-7}$$

We see that $P(\mathbf{X} \in S_{\mathbf{x}_1} | \theta)$ is the probability that the sample point $\mathbf{X}$ falls in $S_{\mathbf{x}_1}$ given that the true state of nature is θ, and $P(\mathbf{X} \in S_{\mathbf{x}_2} | \theta)$ is the probability that the sample point falls in $S_{\mathbf{x}_2}$ given that the state of nature is θ.

We can easily evaluate the risk in Equation (17-7) if the parameter $\theta \in \omega_1$ as

$$R(d; \theta \in \omega_1) = l(a_1; \theta \in \omega_1)P(\mathbf{X} \in S_{\mathbf{x}_1} | \theta \in \omega_1)$$
$$+ l(a_2; \theta \in \omega_1)P(\mathbf{X} \in S_{\mathbf{x}_2} | \theta \in \omega_1) \qquad (17\text{-}8)$$

But by Equation (17-6), the first term in Equation (17-8) is zero, so

$$R(d; \theta \in \omega_1) = l(a_2; \theta \in \omega_1)P(\mathbf{X} \in S_{\mathbf{x}_2} | \theta \in \omega_1) \qquad (17\text{-}9)$$

If $\theta \in \omega_2$, a similar approach will yield

$$R(d; \theta \in \omega_2) = l(a_1; \theta \in \omega_2)P(\mathbf{X} \in S_{\mathbf{x}_1} | \theta \in \omega_2) \qquad (17\text{-}10)$$

We may combine Equations (17-9) and (17-10) as

$$R(d; \theta) = l(\theta)\xi(d; \theta) \qquad (17\text{-}11)$$

where $l(\theta)$ is the loss associated with making a particular decision if θ is the true state of nature, and $\xi(d; \theta)$ are the probabilities of making the wrong decisions. Obviously, we see that

$$l(\theta) = \begin{cases} l(a_1; \theta) & \text{if } \theta \in \omega_2 \\ l(a_2; \theta) & \text{if } \theta \in \omega_1 \end{cases} \qquad (17\text{-}12)$$

and

$$\xi(d; \theta) = \begin{cases} P(\mathbf{X} \in S_{\mathbf{x}_2} | \theta \in \omega_1) \\ P(\mathbf{X} \in S_{\mathbf{x}_1} | \theta \in \omega_2) \end{cases} \qquad (17\text{-}13)$$

The probabilities $\xi(d; \theta)$ are often called *error probabilities*, since they are the probability of making decision a_2 if θ is in ω_1 and the probability of making decision a_1 if θ is in ω_2, respectively.

From the above discussion, we may use the sets ω_1 and ω_2 to form a statement or hypothesis $H_0: \theta \in \omega_1$ and an alternative hypothesis $H_1: \theta \in \omega_2$. Thus, the decision a_1 is accepting the hypothesis H_0 and the decision a_2 is rejecting the hypothesis H_0. The decision function d, which we apply to the sample $x_1, x_2, \ldots, x_n$, and which leads to accepting or rejecting the hypothesis, is called a *test of the hypothesis*. Obviously, we should like to find a decision function, or test, such that the risk is minimized for every value of θ in Ω. It is not generally possible to do this, because for certain values of θ one decision function may be best while for other values of θ a different decision function may be best. As θ is unknown, there is no straightforward approach to determine d. Also, the appropriate form of the loss function may not be known.

A possible solution to these difficulties would be to select a decision function, or test, that minimizes the error probabilities. However, even this is not generally possible. The classical procedure is to select a probability, say α, somewhere in the interval $.01 \leq \alpha \leq .20$ and find the set of decision functions such that

$$P(\mathbf{X} \in S_{\mathbf{x}_2} | \theta \in \omega_1) \leq \alpha \qquad (17\text{-}14)$$

Of course, finding these decision functions is equivalent to finding the sets $S_{\mathbf{x}_2}$ such that Equation (17-14) holds. Then from the restricted set of decision functions defined by Equation (17-14) select the one decision function such that

$$P(\mathbf{X} \in S_{\mathbf{x}_1} | \theta \in \omega_2) \qquad (17\text{-}15)$$

is minimized. We see that $P(\mathbf{X} \in S_{\mathbf{x}_2} | \theta \in \omega_1)$ is the probability of rejecting H_0 given that it is true, or the probability of type I error, and $P(\mathbf{X} \in S_{\mathbf{x}_1} | \theta \in \omega_2)$ is the probability of accepting H_0 given that it is false, or the probability of type II error. That is,

$$P(\text{type I error}) = P(\mathbf{X} \in S_{\mathbf{x}_2} | \theta \in \omega_1)$$
$$P(\text{type II error}) = P(\mathbf{X} \in S_{\mathbf{x}_1} | \theta \in \omega_2)$$

Hence $S_{\mathbf{x}_2}$ would be the critical region or rejection region and $S_{\mathbf{x}_1}$ would be the acceptance region, in the language of previous chapters. Obviously,

$$P(\mathbf{X} \in S_{\mathbf{x}_2} | \theta \in \omega_2) = 1 - P(\mathbf{X} \in S_{\mathbf{x}_1} | \theta \in \omega_2) \qquad (17\text{-}16)$$

is the power of the test. Therefore, the "best" test of size α is the one for which Equation (17-16) is a maximum.

It may seem that the formulation of the hypothesis-testing problem in the last paragraph completely ignores the loss function $l(\theta)$. In fact, it is exactly the approach outlined previously. Actually, we have not ignored the loss function $l(\theta)$ totally; experimenters must carefully consider the consequences of type I and type II errors when they determine a value for α and the sample size. Admittedly, however, this procedure is less than completely satisfactory, and the failure to consider explicitly the loss resulting from wrong decisions is perhaps the greatest drawback of the classical approach to hypothesis testing.

From the above discussion we see that there are often many possible decision rules for a statistical decision problem, and some criterion for characterizing a particular rule as "good" or "bad" is often needed. Obviously, a decision rule d which minimized $R(d; \theta)$ is preferred, but generally such a rule cannot be found for all $\theta \in \Omega$. Furthermore, each decision rule will have *decision probabilities* associated with it. For example, in the hypothesis-testing problem, $P(\mathbf{X} \in S_{\mathbf{x}_1} | \theta)$ and $P(\mathbf{X} \in S_{\mathbf{x}_2} | \theta)$ are the decision probabilities.

Suppose we are comparing two different decision rules d_1 and d_2. If $R(d_1; \theta) \leq R(d_2; \theta)$ for all values of θ in Ω, and $R(d_1; \theta) < R(d_2; \theta)$ for at least one θ, we say that d_1 is a better decision rule than d_2. In general, a decision rule d is said to be *admissible* if there is no other decision rule $d*$ such that

$$R(d*; \theta) \leq R(d; \theta) \qquad \text{for all } \theta \in \Omega$$

and

$$R(d*; \theta) < R(d; \theta) \qquad \text{for some } \theta \in \Omega$$

Since no decision function generally gives a minimum risk for all values of θ in Ω, it seems reasonable to find the class of admissible decision functions and select one from this class.

17-2 Bayesian Inference

In the preceding chapters we have made an extensive study of the use of probability. Until now, we have interpreted these probabilities in the frequency sense; that is, they refer to an experiment that can be repeated an indefinite number of times, and if the probability of occurrence of an event A is .6, then we would expect A to occur in about 60 percent of the experimental trials. This frequency interpretation of probability is often called the objectivist or classical viewpoint.

Bayesian inference requires a different interpretation of probability, called the subjective viewpoint. We often encounter subjective probabilistic statements, such as "There is a 30 percent chance of rain today." Subjective statements measure a person's "degree of belief" concerning some event, rather than a frequency interpretation. Bayesian inference requires us to make use of subjective probability to measure our degree of belief about a state of nature. That is, we must specify a probability distribution to describe our degree of belief about an unknown parameter. This procedure is totally unlike anything we have discussed previously. Until now, parameters have been treated as unknown constants. Bayesian inference requires us to think of parameters as *random variables.*

Suppose we let $f(\theta)$ be the probability distribution of the parameter or state of nature θ. The distribution $f(\theta)$ summarizes our objective information about θ prior to obtaining sample information. Obviously, if we are reasonably certain about the value of θ, we will choose $f(\theta)$ with a small variance, while if we are less certain about θ, $f(\theta)$ will be chosen with a larger variance. We call $f(\theta)$ the *prior distribution* of θ.

Now consider the distribution of the random variable X. We denote the distribution of X by $f(x|\theta)$, to indicate that the distribution depends on the unknown parameter θ. Suppose we take a random sample from X, say $X_1, X_2, \ldots, X_n$. The joint density or *likelihood* of the sample is

$$f(x_1, x_2, \ldots, x_n | \theta) = f(x_1|\theta)f(x_2|\theta) \ldots f(x_n|\theta)$$

We define the *posterior distribution* of θ as the conditional distribution of θ, given the sample results. This is just

$$f(\theta \mid x_1, x_2, \ldots, x_n) = \frac{f(x_1, x_2, \ldots, x_n; \theta)}{f(x_1, x_2, \ldots, x_n)} \qquad (17\text{-}17)$$

The joint distribution of the sample and θ in the numerator of Equation (17-17) is the product of the prior distribution of θ and the likelihood, or

$$f(x_1, x_2, \ldots, x_n; \theta) = f(\theta) \cdot f(x_1, x_2, \ldots, x_n \mid \theta)$$

The denominator of Equation (17-17), which is the marginal distribution of the sample, is just a normalizing constant obtained by

$$f(x_1, x_2, \ldots, x_n) = \begin{cases} \displaystyle\int_{-\infty}^{\infty} f(\theta)f(x_1, x_2, \ldots, x_n \mid \theta)\, d\theta & x \text{ continuous} \\[2mm] \displaystyle\sum_{\theta} f(\theta)f(x_1, x_2, \ldots, x_n \mid \theta), & x \text{ discrete} \end{cases} \qquad (17\text{-}18)$$

Consequently, we may write the posterior distribution of θ as

$$f(\theta \mid x_1, x_2, \ldots, x_n) = \frac{f(\theta)f(x_1, x_2, \ldots, x_n \mid \theta)}{f(x_1, x_2, \ldots, x_n)} \qquad (17\text{-}19)$$

We note that Bayes' theorem has been used to transform or update the prior distribution to the posterior distribution. The posterior distribution reflects our degree of belief about θ given the sample information. Furthermore, the posterior distribution is proportional to the product of the prior distribution and the likelihood, the constant of proportionality being the normalizing constant $f(x_1, x_2, \ldots, x_n)$.

Thus, the posterior density for θ expresses our degree of belief about the value of θ given the results of the sample.

● **Example 17-1.** The time to failure of a transistor is known to be exponentially distributed with parameter λ. For a random sample of n transistors, the joint density of the sample elements, given λ, is

$$f(x_1, x_2, \ldots, x_n \mid \lambda) = \lambda^n e^{-\lambda \sum x_i}$$

Suppose we feel that a suitable prior distribution for λ is

$$f(\lambda) = ke^{-k\lambda} \qquad \lambda > 0$$
$$= 0 \qquad \text{otherwise}$$

where k would be chosen depending on the exact knowledge or degree of belief we have about the value of λ. Notice that the joint density of the sample and λ is

$$f(x_1, x_2, \ldots, x_n; \lambda) = k\lambda^n e^{-\lambda(\sum x_i + k)}$$

and the marginal density of the sample is

$$f(x_1, x_2, \ldots, x_n) = \int_0^\infty k\lambda^n e^{-\lambda(\Sigma x_i + k)} \, d\lambda$$

$$= \frac{k\Gamma(n+1)}{(\Sigma \, x_i + k)^{n+1}}$$

Therefore, the posterior density for λ, by Equation (17-19), is

$$f(\lambda \mid x_1, x_2, \ldots, x_n) = \frac{1}{\Gamma(n+1)} (\Sigma \, x_i + k)^{n+1} \lambda^n e^{-\lambda(\Sigma x_i + k)}$$

and we see that the posterior density for λ is a gamma distribution with parameters $n + 1$ and $\Sigma \, X_i + k$.

17-3   Applications to Estimation

In this section we discuss the application of Bayesian inference to the problem of estimating an unknown parameter of a probability distribution. Let $X_1, X_2, \ldots, X_n$ be a random sample of the random variable X having density $f(x \mid \theta)$. We want to obtain a point estimate of θ. Let $f(\theta)$ be the prior distribution for θ and $l(\hat\theta; \theta)$ be the loss function. As before, the risk is $E[l(\hat\theta; \theta)] = R(d; \theta)$. Since θ is considered to be a random variable, the risk is a random variable. We would like to find the function d that minimizes the *expected* risk. We write the expected risk as

$$B(d) = E[R(d; \theta)] = \int_{-\infty}^\infty R(d; \theta)f(\theta) \, d\theta$$

$$= \int_{-\infty}^\infty \left\{ \int_{-\infty}^\infty \cdots \int_{-\infty}^\infty l\{d(x_1, x_2, \ldots, x_n); \theta\} f(x_1, x_2, \ldots, x_n \mid \theta) \right.$$

$$\left. \times \, dx_1 \, dx_2 \ldots dx_n \right\} f(\theta) \, d\theta \qquad (17\text{-}20)$$

We define the *Bayes estimator* of the parameter θ to be the function d of the sample $X_1, X_2, \ldots, X_n$, that minimizes the expected risk. Upon interchanging the order of integration in Equation (17-20) we obtain

$$B(d) = \int_{-\infty}^\infty \cdots \int_{-\infty}^\infty \left\{ \int_{-\infty}^\infty l\{d(x_1, x_2, \ldots, x_n; \theta)\} f(x_1, x_2, \ldots, x_n \mid \theta) f(\theta) \, d\theta \right\}$$

$$\times \, dx_1 \, dx_2 \ldots dx_n \qquad (17\text{-}21)$$

The function B will be minimized if we can find a function d that minimizes the quantity within the large braces in Equation (17-21) for every set of the X values. That is, the Bayes estimator of θ is a function d of the X_i that minimizes

$$\int_{-\infty}^{\infty} l\{d(x_1, x_2, \ldots, x_n); \theta\} f(x_1, x_2, \ldots, x_n \mid \theta) f(\theta) \, d\theta$$

$$= \int_{-\infty}^{\infty} l(\hat{\theta}; \theta) f(x_1, x_2, \ldots, x_n; \theta) \, d\theta$$

$$= f(x_1, x_2, \ldots, x_n) \int_{-\infty}^{\infty} l(\hat{\theta}; \theta) f(\theta \mid x_1, x_2, \ldots, x_n) \, d\theta \qquad (17\text{-}22)$$

Thus, the Bayes estimator of θ is the value $\hat{\theta}$ that minimizes

$$Z = \int_{-\infty}^{\infty} l(\hat{\theta}; \theta) f(\theta \mid x_1, x_2, \ldots, x_n) \, d\theta \qquad (17\text{-}23)$$

If the loss function $l(\hat{\theta}; \theta)$ is the squared-error loss $(\hat{\theta} - \theta)^2$, then we may show that the Bayes estimator of θ, say $\hat{\theta}$, is the mean of the posterior density for θ.

● **Example 17-2.** Consider the situation in Example 17-1, where it was shown that if the random variable X is exponentially distributed with parameter λ, and if the prior distribution for λ is exponential with parameter k, then the posterior distribution for λ is a gamma distribution, with parameters $n + 1$ and $\Sigma_{i=1}^{n} X_i + k$. Therefore, if a squared-error loss function is assumed, the Bayes estimator for λ is

$$\hat{\lambda} = \frac{n + 1}{\sum_{i=1}^{n} X_i + k}$$

Suppose that in the time to failure problem in Example 17-1, a reasonable exponential prior distribution for λ has parameter $k = 140$. A random sample of size $n = 10$ yields $\Sigma_{i=1}^{10} x_i = 1500$. The Bayes estimate of λ is

$$\hat{\lambda} = \frac{n + 1}{\sum_{i=1}^{10} x_i + k} = \frac{10 + 1}{1500 + 140} = .06707$$

We may compare this with the results that would have been obtained by classical methods. The maximum likelihood estimator of the parameter λ in an exponential distribution is

$$\lambda^* = \frac{n}{\sum_{i=1}^{n} X_i}$$

Consequently, the maximum likelihood estimate of λ, based on the above sample data, is

$$\lambda^* = \frac{n}{\sum_{i=1}^{n} x_i} = \frac{10}{1500} = .06667$$

Note that the results produced by the two methods differ somewhat.

● **Example 17-3.** Let $X_1, X_2, \ldots, X_n$ be a random sample from the normal density with mean μ and variance 1, where μ is unknown. Assume that the prior density for μ is normal with mean 0 and variance 1; that is,

$$f(\mu) = \frac{1}{\sqrt{2\pi}} e^{-(1/2)\mu^2} \qquad -\infty < \mu < \infty$$

The joint conditional density of the sample given μ is

$$f(x_1, x_2, \ldots, x_n \mid \mu) = \frac{1}{(2\pi)^{n/2}} e^{-(1/2)\sum(x_i - \mu)^2}$$

$$= \frac{1}{(2\pi)^{n/2}} e^{-(1/2)(\sum x_i^2 - 2\mu\sum x_i + n\mu^2)}$$

Thus, the joint density of the sample and μ is

$$f(x_1, x_2, \ldots, x_n; \mu) = \frac{1}{(2\pi)^{(n+1)/2}} \exp\left\{-\frac{1}{2}\left[\sum x_i^2 + (n+1)\mu^2 - 2\mu n\bar{x}\right]\right\}$$

The marginal density of the sample is

$$f(x_1, x_2, \ldots, x_n) = \frac{1}{(2\pi)^{(n+1)/2}} \exp\left\{-\frac{1}{2}\sum x_i^2\right\} \int_{-\infty}^{\infty} \exp\left\{-\frac{1}{2}[(n+1)\mu^2 - 2\mu n\bar{x}]\right\} d\mu$$

By completing the square in the exponent under the integral, we obtain

$$f(x_1, x_2, \ldots, x_n) = \frac{1}{(2\pi)^{n/2}} \exp\left[-\frac{1}{2}\left(\sum x_i^2 - \frac{n^2\bar{x}^2}{n+1}\right)\right]$$

$$\times \left[\frac{1}{(2\pi)^{1/2}} \int_{-\infty}^{\infty} \exp\left[-\frac{1}{2}(n+1)\left(\mu - \frac{n\bar{x}}{n+1}\right)^2\right] d\mu\right]$$

$$= \frac{1}{(n+1)^{1/2}(2\pi)^{n/2}} \exp\left[-\frac{1}{2}\left(\sum x_i^2 - \frac{n^2\bar{x}^2}{n+1}\right)\right]$$

using the fact that the integral is $(2\pi)^{1/2}/(n+1)^{1/2}$. Now the posterior density for μ is

$$f(\mu \mid x_1, x_2, \ldots, x_n) = \frac{(2\pi)^{-(n+1)/2} \exp\left\{-\frac{1}{2}\left[\sum x_i^2 + (n+1)\mu^2 - 2n\bar{x}\mu\right]\right\}}{(2\pi)^{-n/2}(n+1)^{-1/2} \exp\left\{-\frac{1}{2}\left(\sum x_i^2 - \frac{n^2\bar{x}^2}{n+1}\right)\right\}}$$

$$= \frac{(n+1)^{1/2}}{(2\pi)^{1/2}} \exp\left\{-\frac{1}{2}(n+1)\left[\mu^2 - \frac{2n\bar{x}\mu}{n+1} + \frac{n^2\bar{x}^2}{(n+1)^2}\right]\right\}$$

$$= \frac{(n+1)^{1/2}}{(2\pi)^{1/2}} \exp\left\{-\frac{1}{2}(n+1)\left[\mu - \frac{n\bar{x}}{n+1}\right]^2\right\}$$

Therefore, the posterior density for μ is a normal density with mean $n\bar{X}/(n+1)$ and variance $(n+1)^{-1}$. If we let the loss function $l(\hat{\mu}; \mu)$ be a squared error, the Bayes estimator of μ is

$$\hat{\mu} = \frac{n\bar{X}}{n+1} = \frac{\sum_{i=1}^{n} X_i}{n+1}$$

There is a relationship between the Bayes estimator for a parameter and the maximum likelihood estimator of the same parameter. For large sample sizes the two are nearly equivalent. In general, the difference between the two estimators is small compared to $1/\sqrt{n}$. In practical problems, a moderate sample size will produce approximately the same estimate by either the Bayes or maximum likelihood method, if the sample results are consistent with the assumed prior information. If the sample results are inconsistent with the prior assumptions, then the Bayes estimate may differ considerably from the maximum likelihood estimate. In these circumstances, if the sample results are accepted as being correct, the prior information must be incorrect. The maximum likelihood estimate would then be the better estimate to use.

If the sample results do not agree with the prior information, the Bayes estimator will always tend to produce an estimate that is between the maximum likelihood estimate and the prior assumptions. If there is more inconsistency between the prior information and the sample, there will be more difference between the two estimates.

We may use Bayesian methods to construct interval estimates of parameters that are similar to confidence intervals. If the posterior density for θ has been obtained, we can construct an interval, usually centered at the posterior mean, that contains $100(1-\alpha)$ percent of the posterior probability. Such an interval is called the $100(1-\alpha)$ percent Bayes interval for the unknown parameter θ.

While in many cases the Bayes interval estimate for θ will be quite similar to a classical confidence interval with the same confidence coefficient, the interpretation of the two is very different. A confidence interval is an interval that, before the sample is taken, will include the unknown θ with probability $1 - \alpha$. That is, the classical confidence interval relates to the relative frequency of an interval including θ. On the other hand, a Bayes interval is an interval that contains $100(1 - \alpha)$ percent of the posterior probability for θ. Since the posterior probability density measures a degree of belief about θ given the sample results, the Bayes interval provides a subjective degree of belief about θ rather than a frequency interpretation. The Bayes interval estimate of θ is affected by the sample results, but not completely determined by them.

● **Example 17-4.** Suppose that the random variable X is normally distributed with mean μ and variance 4. The value of μ is unknown, but a reasonable prior density would be normal with mean 2 and variance 1. That is,

$$f(x_1, x_2, \ldots, x_n \mid \mu) = \frac{1}{(8\pi)^{n/2}} e^{-(1/2) \Sigma (x_i - \mu)^2}$$

and

$$f(\mu) = \frac{1}{\sqrt{2\pi}} e^{-(1/2)(\mu - 2)^2}$$

We can show that the posterior density for μ is

$$f(\mu \mid x_1, x_2, \ldots, x_n) = \left(\frac{n}{4} + 1\right)^{1/2} \exp\left\{\frac{-\left(\frac{n}{4} + 1\right)}{2}\left[\mu - \frac{1}{\left(\frac{n}{4} + 1\right)}\left(\frac{n\bar{x}}{4} + 2\right)\right]\right\}$$

using the methods of this section. Thus, the posterior distribution for μ is normal with mean $(n/4 + 1)^{-1}(n\bar{X}/4 + 2)$ and variance $(n/4 + 1)^{-1}$. A 95 percent Bayes interval for μ, which is symmetric about the posterior mean, would be

$$\left[\frac{n}{4} + 1\right]^{-1}\left[\frac{n\bar{X}}{4} + 2\right] - Z_{.025}\left[\frac{n}{4} + 1\right]^{-1/2}$$

$$\leq \mu \leq \left[\frac{n}{4} + 1\right]^{-1}\left[\frac{n\bar{X}}{4} + 2\right] + Z_{.025}\left[\frac{n}{4} + 1\right]^{-1/2} \qquad (17\text{-}24)$$

If a random sample of size 16 is taken and we find that $\bar{x} = 2.5$, Equation (17-24) reduces to

$$1.52 \leq \mu \leq 3.28$$

If we ignore the prior information, the classical confidence interval for μ is

$$1.52 \leq \mu \leq 3.48$$

We see that the Bayes interval is slightly shorter than the classical confidence interval, because the prior information is equivalent to a slight increase in the sample size if no prior knowledge was assumed.

17-4 Applications to Hypothesis Testing

The Bayesian approach may also be applied to hypothesis-testing problems. The usual approach consists of determining the posterior distribution and then calculating probabilities of the hypotheses of interest from this distribution. For example, suppose that X is normally distributed, with unknown mean μ and known variance $\sigma^2 = 4$. We wish to test the hypotheses

$$H_0: \mu \geq 2$$

$$H_1: \mu < 2$$

A normal prior distribution is assumed for μ, with mean 2 and variance 1. A random sample of size 16 is taken and the sample mean is $\bar{x} = 2.5$. In Example 17-4, we showed that the posterior distribution in this case is normal with mean 2.40 and variance .20.

Now let us consider the posterior probabilities of the null and alternative hypotheses. For the null hypothesis,

$$P\{H_0\} = P\{\mu \geq 2.0\} = 1 - \Phi\left(\frac{2.0 - 2.40}{\sqrt{.20}}\right) = 1 - \Phi(-.89)$$

or

$$P\{H_0\} = .8133$$

while for the alternative hypothesis,

$$P\{H_1\} = P\{\mu < 2.0\} = \Phi\left(\frac{2.0 - 2.40}{\sqrt{.20}}\right) = \Phi(-.89)$$

or

$$P\{H_1\} = .1867$$

It seems that the alternative hypothesis is relatively unlikely. We may form the *odds ratio* of H_0 to H_1 as

$$\frac{P(H_0)}{P(H_1)} = \frac{.8133}{.1867} = 4.36$$

That is, the odds are 4.36 to 1 in favor of $H_0: \mu \geq 2.0$, as opposed to $H_1: \mu < 2.0$.

In general, the results obtained from the Bayesian approach to hypothesis testing will differ numerically and in interpretation from the classical approach. However, if the prior distribution chosen by the analyst is "informationless" (often called a *diffuse* prior), then the results obtained from the Bayesian and classical procedures will agree numerically. In this case, the posterior probability $P(H_0)$ is exactly equal to the significance level in the classical procedure, where H_0 would be rejected. Since a small significance level implies that H_0 is relatively unlikely and should be rejected, and a small posterior probability $P(H_0)$ implies that H_0 is unlikely, we see that there is a Bayesian interpretation of the one-tailed classical hypothesis-testing procedure.

There is no Bayesian interpretation of the classical two-tailed procedure. For example, suppose that we wish to test

$$H_0: \mu = 2.0$$
$$H_1: \mu \neq 2.0$$

Specifying a continuous prior distribution for μ implies that the posterior distribution for μ will also be continuous. Therefore, since the null hypothesis H_0 consists of a single value of μ, the posterior probability $P(H_0) = 0$. However, by modifying the null hypothesis appropriately, the Bayesian approach can still be applied. Since even in the classical framework it is unlikely that the statement $H_0: \mu = 2.0$ means that $\mu = 2.0$ *exactly*, but rather that μ is very close to 2.0, we could modify the hypothesis to, say,

$$H_0: 1.9 \leq \mu \leq 2.1$$
$$H_1: \mu > 2.1 \text{ or } \mu < 1.9$$

Now inferences can be meaningfully made from the posterior distribution of μ.

17-5 Summary

This chapter has introduced statistical decision theory and has briefly discussed the relationships among decision theory, Bayesian inference, and classical statistics. An important aspect of decision theory is the incorporation of a loss function into the analysis. This allows the decision maker to choose an action that is optimal with respect to a specified criterion. The concepts of prior and posterior distributions were also introduced. We can make several interesting comparisons between the classical and Bayesian approaches to statistical inference. Classical inference methods use sample information only, while Bayesian methods use a blend of subjective and sample information through the updating procedure that transforms the prior distribution into the posterior distribution. In effect, the Bayesian approach allows the analyst a methodology for formally introducing any available prior information into the problem, while classical procedures either ignore prior information or treat it informally. Classical procedures require that the analyst select values for the type I and type II error probabilities. If these probabilities are selected carefully after due consideration of the consequences of the two types of errors, then the classical framework is a reasonable one. If, however, they are chosen arbitrarily, then there is a serious failure of the application of the classical approach. This is to some extent avoided by the Bayesian approach.

17-6   Exercises

17-1. Let X be a normally distributed random variable with mean μ and variance σ^2. Assume that σ^2 is known and μ unknown. The prior density for μ is assumed to be normal with mean μ_0 and σ_0^2. Determine the posterior density for μ, given a random sample of size n from X.

17-2. Suppose that X is normally distributed with mean μ and known variance σ^2. The prior density for μ is uniform on the interval (a, b). Determine the posterior density for μ, given a random sample of size n from X.

17-3. Let X be normally distributed with known mean μ and unknown variance σ^2. Assume that the prior density for $1/\sigma^2$ is a gamma distribution with parameters $m + 1$ and $m\sigma_0^2$. Determine the posterior density for $1/\sigma^2$, given a random sample of size n from X.

17-4. Let X be a geometric random variable with parameter p. Suppose we assume a beta distribution with parameters a and b as the prior density for p. Determine the posterior density for p, given a random sample of size n from X.

17-5. Let X be a Bernoulli random variable with parameter p. If the prior density for p is a beta distribution with parameters a and b, determine the posterior density for p, given a random sample of size n from X.

17-6. Let X be a Poisson random variable with parameter λ. The prior density for λ is a gamma distribution with parameters $m + 1$ and $(m + 1)/\lambda_0$. Determine the posterior density for λ, given a random sample of size n from X.

17-7. For the situation described in Exercise 17-1, determine the Bayes estimator of μ, assuming a squared-error loss function.

17-8. For the situation described in Exercise 17-2, determine the Bayes estimator of μ, assuming a squared-error loss function.

17-9. For the situation described in Exercise 17-3, determine the Bayes estimator of $1/\sigma^2$, assuming a squared-error loss function.

17-10. For the situation described in Exercise 17-4, determine the Bayes estimator of p, assuming a squared-error loss function.

17-11. For the situation described in Exercise 17-5, determine the Bayes estimator of p, assuming a squared-error loss function.

17-12. For the situation described in Exercise 17-6, determine the Bayes estimator of λ, assuming a squared-error loss function.

17-13. Suppose that $X \sim N(\mu, 40)$, and let the prior density for μ be $N(4, 8)$. For a random sample of size 25 the value $\bar{x} = 4.85$ is obtained. What is the Bayes estimate of μ, assuming a squared-error loss?

17-14. A process manufactures bolts, whose length is a random variable $X \sim N(\mu, .01)$. The prior density for μ is uniform between 1.98 and 2.20 inches. A random sample of size 4 produces the value $\bar{x} = 2.01$. Assuming a squared-error loss, determine the Bayes estimate of μ.

17-15. The time between failures of a certain type of machine is exponentially distributed with parameter λ. Suppose we assume an exponential prior on λ with mean of 1000 hours. Two machines are observed and the average time between failures is $\bar{x} = 1145$ hours. Assuming a squared-error loss, determine the Bayes estimate of λ.

17-16. The weight of boxes of soap powder filled by machine is normally distributed with mean μ and variance $\frac{1}{10}$. It is reasonable to assume a prior density for μ that is normal with mean 10 pounds and variance $\frac{1}{25}$. Determine the Bayes estimate of μ given that a sample of size 16 produces $\bar{x} = 9.95$ pounds. If boxes which weigh less than 9.75 pounds are defective, what is the probability that defective boxes will be produced?

17-17. The number of automobile accidents on a certain highway during a weekend is known to be a Poisson random variable with parameter λ. Assume that the prior density for λ is exponential with parameter .25. A total of 45 accidents were observed over a 10-weekend period. Set up an integral which defines a 95 percent Bayes interval for λ. What difficulties would you encounter in evaluating this integral?

17-18. A machine produces bolts, whose diameter is a random variable known to be normally distributed with mean μ and variance $\sigma^2 = .0001$. A normal prior is assumed for μ, with mean .25 and variance .000001. A random sample of 25 bolts yields $\bar{x} = .249$. Construct a 90 percent Bayes interval for μ. If bolts that are greater than .255 or less than .245 in diameter are defective, about what fraction defective does this process produce?

17-19. The random variable X has density function

$$f(x|\theta) = \frac{2x}{\theta^2} \qquad 0 < x < \theta$$

and the prior density for θ is

$$f(\theta) = 1 \qquad 0 < \theta < 1$$

(a) Find the posterior density for θ assuming $n = 1$.
(b) Find the Bayes estimator for θ assuming the loss function $l(\hat{\theta}; \theta) = \theta^2(\theta - \hat{\theta})^2$ and $n = 1$.

17-20. Let X follow the Bernoulli distribution with parameter p. Assume a reasonable prior density for p to be

$$f(p) = 6p(1 - p) \qquad 0 \le p \le 1$$
$$= 0 \qquad \qquad \text{otherwise}$$

If the loss function is squared error, find the Bayes estimator of p if one observation is available. If the loss function is

$$l(\hat{p}; p) = 2(\hat{p} - p)^2$$

find the Bayes estimator of p for $n = 1$.

17-21. A random variable X is normally distributed with mean μ and variance $\sigma^2 = 10$. The prior density for μ is uniform between 6 and 12. A random sample of size 9 yields $\bar{x} = 10$. Construct a 90 percent Bayes interval for μ. Could you reasonably accept the hypothesis $H: \mu = 9$?

17-22. Let X be a normally distributed random variable with mean $\mu = 4$ and unknown variance σ^2. The prior density for $1/\sigma^2$ is a gamma distribution with parameters $r = 3$ and $\lambda = 1.0$. Determine the posterior density for $1/\sigma^2$. If a random sample of size 10 yields $\Sigma(x_i - 4)^2 = 4.92$, determine the Bayes estimate of $1/\sigma^2$ assuming a squared-error loss. Set up an integral that defines a 90 percent Bayes interval for $1/\sigma^2$.

17-23. Using the data in Exercise 17-16, test the hypotheses

$$H_0: \mu \geq 10$$
$$H_1: \mu < 10$$

17-24. Using the data in Exercise 17-18, test the hypotheses

$$H_0: .245 \leq \mu \leq .255$$
$$H_1: \mu < .245 \text{ or } \mu > .255$$

17-25. Prove that if a squared-error loss function is used, the Bayes estimator of θ is the mean of the posterior distribution for θ.

References

1. Anderson, V. L., and R. A. McLean (1974), *Design of Experiments: A Realistic Approach*, Marcel Dekker, Inc., New York.
2. Berrettoni, J. M. (1964), "Practical Applications of the Weibull Distribution," *Industrial Quality Control*, Vol. 21, No. 2, pp. 71–79.
3. Bartlett, M. S. (1947), "The Use of Transformations," *Biometrics*, Vol. 3, pp. 39–52.
4. Box, G. E. P., and D. R. Cox (1964), "An Analysis of Transformations," *Journal of the Royal Statistical Society*, B, Vol. 26, pp. 211–252.
5. Cochran, W. G. (1947), "Some Consequences When the Assumptions for the Analysis of Variance are not Satisfied," *Biometrics*, Vol. 3, pp. 22–38.
6. Cochran, W. G., and G. M. Cox (1957), *Experimental Designs*, John Wiley & Sons, New York.
7. *Cumulative Probability Distribution* (1955), Harvard University Press, Cambridge, Mass.
8. Daniel, C., and F. S. Wood (1971), *Fitting Equations to Data*, John Wiley & Sons, New York.
9. Davenport, W. B., and W. L. Root (1958), *An Introduction to the Theory of Random Signals and Noise*, McGraw-Hill, New York.
10. Draper, N. R., and W. G. Hunter (1969), "Transformations: Some Examples Revisited," *Technometrics*, Vol. 11, pp. 23–40.
11. Draper, N. R., and H. Smith (1966), *Applied Regression Analysis*, John Wiley & Sons, New York.
12. Duncan, A. J. (1974), *Quality Control and Industrial Statistics*, 4th edition, Richard D. Irwin, Homewood, Ill.
13. Duncan, D. B. (1955), "Multiple Range and Multiple F Tests," *Biometrics*, Vol. 11, pp. 1–42.
14. Epstein, B. (1960), "Estimation from Life Test Data," *IRE Transactions on Reliability*, Vol. RQC-9.
15. Feller, W. (1968), *An Introduction to Probability Theory and Its Applications*, 3rd edition, John Wiley & Sons, New York.
16. Furnival, G. M., and R. W. Wilson, Jr. (1974), "Regression by Leaps and Bounds," *Technometrics*, Vol. 16, pp. 499–512.
17. Hald, A. (1952), *Statistical Theory with Engineering Applications*, John Wiley & Sons, New York.
18. Hocking, R. R. (1976), "The Analysis and Selection of Variables in Linear Regression," *Biometrics*, Vol. 32, pp. 1–49.
19. Hocking, R. R., F. M. Speed, and M. J. Lynn (1976), "A Class of Biased Estimators in Linear Regression," *Technometrics*, Vol. 18, pp. 425–437.
20. Hoerl, A. E., and R. W. Kennard (1970a), "Ridge Regression: Biased Estimation for Non-Orthogonal Problems," *Technometrics*, Vol. 12, pp. 55–67.
21. Hoerl, A. E., and R. W. Kennard (1970b), "Ridge Regression: Application to Non-Orthogonal Problems," *Technometrics*, Vol. 12, pp. 69–82.

22. Kendall, M. G., and A. Stuart (1963), *The Advanced Theory of Statistics*, Hafner Publishing Company, New York.
23. Keuls, M. (1952), "The Use of the Studentized Range in Connection with an Analysis of Variance," *Euphytica*, Vol. 1, p. 112.
24. Lloyd, D. K., and M. Lipow (1972), *Reliability: Management, Methods, and Mathematics*, Prentice-Hall, Englewood Cliffs, N.J.
25. Marquardt, D. W., and R. D. Snee (1975), "Ridge Regression in Practice," *The American Statistician*, Vol. 29, pp. 3–20.
26. Molina, E. C. (1942), *Poisson's Exponential Binomial Limit*, Van Nostrand Reinhold, New York.
27. Montgomery, D. C. (1976), *Design and Analysis of Experiments*, John Wiley & Sons, New York.
28. Mood, A. M., F. A. Graybill, and D. C. Boes (1974), *Introduction to the Theory of Statistics*, 3rd edition, McGraw-Hill, New York.
29. Neter, J., and W. Wasserman (1974), *Applied Linear Statistical Models*, Richard D. Irwin, Homewood, Ill.
30. Newman, D. (1939), "The Distribution of the Range in Samples from a Normal Population Expressed in Terms of an Independent Estimate of Standard Deviation," *Biometrika*, Vol. 31, p. 20.
31. Owen, D. B. (1962), *Handbook of Statistical Tables*, Addison–Wesley Publishing Company, Reading, Mass.
32. Romig, H. G. (1953), 50–100 *Binomial Tables*, John Wiley & Sons, New York.
33. Scheffe, H. (1953), "A Method for Judging All Contrasts in the Analysis of Variance," *Biometrika*, Vol. 40, pp. 87–104.
34. Snee, R. D. (1977), "Validation of Regression Models: Methods and Examples," *Technometrics*, Vol. 19, No. 4, pp. 415–428.
35. Tucker, H. G. (1962), *An Introduction to Probability and Mathematical Statistics*, Academic Press, New York.
36. Tukey, J. W. (1953), "The Problem of Multiple Comparisons," unpublished notes, Princeton University.
37. United States Department of Defense (1957), *Military Standard Sampling Procedures and Tables for Inspection by Variables for Percent Defective* (MIL-STD-414), Government Printing Office, Washington, D.C.
38. United States Department of Defense (1963), *Military Standard Sampling Procedures and Tables for Inspection by Attributes* (MIL-STD-105D), Government Printing Office, Washington, D.C.
39. United States Department of Defense (1965), *Life Testing Sampling Procedures for Established Levels of Reliability and Confidence in Electronic Parts Specification* (MIL-STD-690A), Government Printing Office, Washington, D.C.
40. Weibull, W. (1951), "Statistical Distribution Function of Wide Application," *Journal of Applied Mechanics*, Vol. 18, p. 293.
41. White, J. A., J. W. Schmidt, G. K. Bennett (1975), *Analysis of Queueing Systems*, Academic Press, New York.

APPENDIX

TABLE I **Cumulative Poisson Distribution**^a

x	.01	.05	.10	$\alpha = \lambda t$ .20	.30	.40	.50	.60
0	.990	.951	.904	.818	.740	.670	.606	.548
1	.999	.998	.995	.982	.963	.938	.909	.878
2		.999	.999	.998	.996	.992	.985	.976
3				.999	.999	.999	.998	.996
4					.999	.999	.999	.999
5							.999	.999

x	.70	.80	.90	$\alpha = \lambda t$ 1.00	1.10	1.20	1.30	1.40
0	.496	.449	.406	.367	.332	.301	.272	.246
1	.844	.808	.772	.735	.699	.662	.626	.591
2	.965	.952	.937	.919	.900	.879	.857	.833
3	.994	.990	.986	.981	.974	.966	.956	.946
4	.999	.998	.997	.996	.994	.992	.989	.985
5	.999	.999	.999	.999	.999	.998	.997	.996
6		.999	.999	.999	.999	.999	.999	.999
7				.999	.999	.999	.999	.999
8							.999	.999

x	1.50	1.60	1.70	$\alpha = \lambda t$ 1.80	1.90	2.00	2.10	2.20
0	.223	.201	.182	.165	.149	.135	.122	.110
1	.557	.524	.493	.462	.433	.406	.379	.354
2	.808	.783	.757	.730	.703	.676	.649	.622
3	.934	.921	.906	.891	.874	.857	.838	.819
4	.981	.976	.970	.963	.955	.947	.937	.927
5	.995	.993	.992	.989	.986	.983	.979	.975
6	.999	.998	.998	.997	.996	.995	.994	.992
7	.999	.999	.999	.999	.999	.998	.998	.998
8	.999	.999	.999	.999	.999	.999	.999	.999
9			.999	.999	.999	.999	.999	.999
10							.999	.999

TABLE I **Cumulative Poisson Distribution (*continued*)**

x	2.30	2.40	2.50	$\alpha = \lambda t$ 2.60	2.70	2.80	2.90	3.00
0	.100	.090	.082	.074	.067	.060	.055	.049
1	.330	.308	.287	.267	.248	.231	.214	.199
2	.596	.569	.543	.518	.493	.469	.445	.423
3	.799	.778	.757	.736	.714	.691	.669	.647
4	.916	.904	.891	.877	.862	.847	.831	.815
5	.970	.964	.957	.950	.943	.934	.925	.916
6	.990	.988	.985	.982	.979	.975	.971	.966
7	.997	.996	.995	.994	.993	.991	.990	.988
8	.999	.999	.998	.998	.998	.997	.996	.996
9	.999	.999	.999	.999	.999	.999	.999	.998
10	.999	.999	.999	.999	.999	.999	.999	.999
11			.999	.999	.999	.999	.999	.999
12							.999	.999

x	3.50	4.00	4.50	$\alpha = \lambda t$ 5.00	5.50	6.00	6.50	7.00
0	.030	.018	.011	.006	.004	.002	.001	.000
1	.135	.091	.061	.040	.026	.017	.011	.007
2	.320	.238	.173	.124	.088	.061	.043	.029
3	.536	.433	.342	.265	.201	.151	.111	.081
4	.725	.628	.532	.440	.357	.285	.223	.172
5	.857	.785	.702	.615	.528	.445	.369	.300
6	.934	.889	.831	.762	.686	.606	.526	.449
7	.973	.948	.913	.866	.809	.743	.672	.598
8	.990	.978	.959	.931	.894	.847	.791	.729
9	.996	.991	.982	.968	.946	.916	.877	.830
10	.998	.997	.993	.986	.974	.957	.933	.901
11	.999	.999	.997	.994	.989	.979	.966	.946
12	.999	.999	.999	.997	.995	.991	.983	.973
13	.999	.999	.999	.999	.998	.996	.992	.987
14		.999	.999	.999	.999	.998	.997	.994
15			.999	.999	.999	.999	.998	.997
16				.999	.999	.999	.999	.999
17					.999	.999	.999	.999
18						.999	.999	.999
19							.999	.999
20								.999

TABLE I   **Cumulative Poisson Distribution (*continued*)**

x	7.50	8.00	8.50	$\alpha = \lambda t$ 9.00	9.50	10.0	15.0	20.0
0	.000	.000	.000	.000	.000	.000	.000	.000
1	.004	.003	.001	.001	.000	.000	.000	.000
2	.020	.013	.009	.006	.004	.002	.000	.000
3	.059	.042	.030	.021	.014	.010	.000	.000
4	.132	.099	.074	.054	.040	.029	.000	.000
5	.241	.191	.149	.115	.088	.067	.002	.000
6	.378	.313	.256	.206	.164	.130	.007	.000
7	.524	.452	.385	.323	.268	.220	.018	.000
8	.661	.592	.523	.455	.391	.332	.037	.002
9	.776	.716	.652	.587	.521	.457	.069	.005
10	.862	.815	.763	.705	.645	.583	.118	.010
11	.920	.888	.848	.803	.751	.696	.184	.021
12	.957	.936	.909	.875	.836	.791	.267	.039
13	.978	.965	.948	.926	.898	.864	.363	.066
14	.989	.982	.972	.958	.940	.916	.465	.104
15	.995	.991	.986	.977	.966	.951	.568	.156
16	.998	.996	.993	.988	.982	.972	.664	.221
17	.999	.998	.997	.994	.991	.985	.748	.297
18	.999	.999	.998	.997	.995	.992	.819	.381
19	.999	.999	.999	.998	.998	.996	.875	.470
20	.999	.999	.999	.999	.999	.998	.917	.559
21	.999	.999	.999	.999	.999	.999	.946	.643
22		.999	.999	.999	.999	.999	.967	.720
23			.999	.999	.999	.999	.980	.787
24					.999	.999	.988	.843
25						.999	.993	.887
26							.996	.922
27							.998	.947
28							.999	.965
29							.999	.978
30							.999	.986
31							.999	.991
32							.999	.995
33							.999	.997
34								.998

[a]Entries in the table are values of $F(x) = P(C \leq x) = \sum_{c=0}^{x} e_\alpha^{-\alpha c}/c!$. Blank spaces below the last entry in any column may be read as 1.0; blank spaces above the first entry in any column may be read as 0.0.

TABLE II   Cumulative Standard Normal Distribution

$$\Phi(z) = \int_{-\infty}^{z} \frac{1}{\sqrt{2\pi}} e^{-u^2/2}\, du$$

z	.00	.01	.02	.03	.04	z
.0	.500 00	.503 99	.507 98	.511 97	.515 95	.0
.1	.539 83	.543 79	.547 76	.551.72	.555 67	.1
.2	.579 26	.583 17	.587 06	.590 95	.594 83	.2
.3	.617 91	.621 72	.625 51	.629 30	.633 07	.3
.4	.655 42	.659 10	.662 76	.666 40	.670 03	.4
.5	.691 46	.694 97	.698 47	.701 94	.705 40	.5
.6	.725 75	.729 07	.732 37	.735 65	.738 91	.6
.7	.758 03	.761 15	.764 24	.767 30	.770 35	.7
.8	.788 14	.791 03	.793 89	.796 73	.799 54	.8
.9	.815 94	.818 59	.821 21	.823 81	.826 39	.9
1.0	.841 34	.843 75	.846 13	.848 49	.850 83	1.0
1.1	.864 33	.866 50	.868 64	.870 76	.872 85	1.1
1.2	.884 93	.886 86	.888 77	.890 65	.892 51	1.2
1.3	.903 20	.904 90	.906 58	.908 24	.909 88	1.3
1.4	.919 24	.920 73	.922 19	.923 64	.925 06	1.4
1.5	.933 19	.934 48	.935 74	.936 99	.938 22	1.5
1.6	.945 20	.946 30	.947.38	.948 45	.949 50	1.6
1.7	.955 43	.956 37	.957 28	.958 18	.959 07	1.7
1.8	.964 07	.964 85	.965 62	.966 37	.967 11	1.8
1.9	.971 28	.971 93	.972 57	.973 20	.973 81	1.9
2.0	.977 25	.977 78	.978 31	.978 82	.979 32	2.0
2.1	.982 14	.982 57	.983 00	.983 41	.983 82	2.1
2.2	.986 10	.986 45	.986 79	.987 13	.987 45	2.2
2.3	.989 28	.989 56	.989 83	.990 10	.990 36	2.3
2.4	.991 80	.992 02	.992 24	.992 45	.992 66	2.4
2.5	993 79	.993 96	.994 13	.994 30	.994 46	2.5
2.6	.995 34	.995 47	.995 60	.995 73	.995 85	2.6
2.7	.996 53	.996 64	.996 74	.996 83	.996 93	2.7
2.8	.997 44	.997 52	.997 60	.997 67	.997 74	2.8
2.9	.998 13	.998 19	.998 25	.998 31	.998 36	2.9
3.0	.998 65	.998 69	.998 74	.998 78	.998 82	3.0
3.1	.999 03	.999 06	.999 10	.999 13	.999 16	3.1
3.2	.999 31	.999 34	.999 36	.999 38	.999 40	3.2
3.3	.999 52	.999 53	.999 55	.999 57	.999 58	3.3
3.4	.999 66	.999 68	.999 69	.999 70	.999 71	3.4
3.5	.999 77	.999 78	.999 78	.999 79	.999 80	3.5
3.6	.999 84	.999 85	.999 85	.999 86	.999 86	3.6
3.7	.999 89	.999 90	.999 90	.999 90	.999 91	3.7
3.8	.999 93	.999 93	.999 93	.999 94	.999 94	3.8
3.9	.999 95	.999 95	.999 96	.999 96	.999 96	3.9

TABLE II   Cumulative Standard Normal Distribution (*continued*)

$$\Phi(z) = \int_{-\infty}^{z} \frac{1}{\sqrt{2\pi}} \, e^{-u^2/2} \, du$$

z	.05	.06	.07	.08	.09	z
.0	.519 94	.523 92	.527 90	.531 88	.535 86	.0
.1	.559 62	.563 56	.567 49	.571 42	.575 34	.1
.2	.598 71	.602 57	.606 42	.610 26	.614 09	.2
.3	.636 83	.640 58	.644 31	.648 03	.651 73	.3
.4	.673 64	.677 24	.680 82	.684 38	.687 93	.4
.5	.708 84	.712 26	.715 66	.719 04	.722 40	.5
.6	.742 15	.745 37	.748 57	.751 75	.754 90	.6
.7	.773 37	.776 37	.779 35	.782 30	.785 23	.7
.8	.802 34	.805 10	.807 85	.810 57	.813 27	.8
.9	.828 94	.831 47	.833 97	.836 46	.838 91	.9
1.0	.853 14	.855 43	.857 69	.859 93	.862 14	1.0
1.1	.874 93	.876 97	.879 00	.881 00	.882 97	1.1
1.2	.894 35	.896 16	.897 96	.899 73	.901 47	1.2
1.3	.911 49	.913 08	.914 65	.916 21	.917 73	1.3
1.4	.926 47	.927 85	.929 22	.930 56	.931 89	1.4
1.5	.939 43	.940 62	.941 79	.942 95	.944 08	1.5
1.6	.950 53	.951 54	.952 54	.953 52	.954 48	1.6
1.7	.959 94	.960 80	.961 64	.962 46	.963 27	1.7
1.8	.967 84	.968 56	.969 26	.969 95	.970 62	1.8
1.9	.974 41	.975 00	.975 58	.976 15	.976 70	1.9
2.0	.979 82	.980 30	.980 77	.981 24	.981 69	2.0
2.1	.984 22	.984 61	.985 00	.985 37	.985 74	2.1
2.2	.987 78	.988 09	.988 40	.988 70	.988 99	2.2
2.3	.990 61	.990 86	.991 11	.991 34	.991 58	2.3
2.4	.992 86	.993 05	.993 24	.993 43	.993 61	2.4
2.5	.994 61	.994 77	.994 92	.995 06	.995 20	2.5
2.6	.995 98	.996 09	.996 21	.996 32	.996 43	2.6
2.7	.997 02	.997 11	.997 20	.997 28	.997 36	2.7
2.8	.997 81	.997 88	.997 95	.998 01	.998 07	2.8
2.9	.998 41	.998 46	.998 51	.998 56	.998 61	2.9
3.0	.998 86	.998 89	.998 93	.998 97	.999 00	3.0
3.1	.999 18	.999 21	.999 24	.999 26	.999 29	3.1
3.2	.999 42	.999 44	.999 46	.999 48	.999 50	3.2
3.3	.999 60	.999 61	.999 62	.999 64	.999 65	3.3
3.4	.999 72	.999 73	.999 74	.999 75	.999 76	3.4
3.5	.999 81	.999 81	.999 82	.999 83	.999 83	3.5
3.6	.999 87	.999 87	.999 88	.999 88	.999 89	3.6
3.7	.999 91	.999 92	.999 92	.999 92	.999 92	3.7
3.8	.999 94	.999 94	.999 95	.999 95	.999 95	3.8
3.9	.999 96	.999 96	.999 96	.999 97	.999 97	3.9

TABLE III Percentage Points of the χ^2 Distribution[a]

ν \ α	.995	.990	.975	.950	.900	.500	.100	.050	.025	.010	.005
1	.00+	.00+	.00+	.00+	.02	.45	2.71	3.84	5.02	6.63	7.88
2	.01	.02	.05	.10	.21	1.39	4.61	5.99	7.38	9.21	10.60
3	.07	.11	.22	.35	.58	2.37	6.25	7.81	9.35	11.34	12.84
4	.21	.30	.48	.71	1.06	3.36	7.78	9.49	11.14	13.28	14.86
5	.41	.55	.83	1.15	1.61	4.35	9.24	11.07	12.83	15.09	16.75
6	.68	.87	1.24	1.64	2.20	5.35	10.65	12.59	14.45	16.81	18.55
7	.99	1.24	1.69	2.17	2.83	6.35	12.02	14.07	16.01	18.48	20.28
8	1.34	1.65	2.18	2.73	3.49	7.34	13.36	15.51	17.53	20.09	21.96
9	1.73	2.09	2.70	3.33	4.17	8.34	14.68	16.92	19.02	21.67	23.59
10	2.16	2.56	3.25	3.94	4.87	9.34	15.99	18.31	20.48	23.21	25.19
11	2.60	3.05	3.82	4.57	5.58	10.34	17.28	19.68	21.92	24.72	26.76
12	3.07	3.57	4.40	5.23	6.30	11.34	18.55	21.03	23.34	26.22	28.30
13	3.57	4.11	5.01	5.89	7.04	12.34	19.81	22.36	24.74	27.69	29.82
14	4.07	4.66	5.63	6.57	7.79	13.34	21.06	23.68	26.12	29.14	31.32
15	4.60	5.23	6.27	7.26	8.55	14.34	22.31	25.00	27.49	30.58	32.80
16	5.14	5.81	6.91	7.96	9.31	15.34	23.54	26.30	28.85	32.00	34.27
17	5.70	6.41	7.56	8.67	10.09	16.34	24.77	27.59	30.19	33.41	35.72

ν[a]											
18	6.26	7.01	8.23	9.39	10.87	17.34	25.99	28.87	31.53	34.81	37.16
19	6.84	7.63	8.91	10.12	11.65	18.34	27.20	30.14	32.85	36.19	38.58
20	7.43	8.26	9.59	10.85	12.44	19.34	28.41	31.41	34.17	37.57	40.00
21	8.03	8.90	10.28	11.59	13.24	20.34	29.62	32.67	35.48	38.93	41.40
22	8.64	9.54	10.98	12.34	14.04	21.34	30.81	33.92	36.78	40.29	42.80
23	9.26	10.20	11.69	13.09	14.85	22.34	32.01	35.17	38.08	41.64	44.18
24	9.89	10.86	12.40	13.85	15.66	23.34	33.20	36.42	39.36	42.98	45.56
25	10.52	11.52	13.12	14.61	16.47	24.34	34.28	37.65	40.65	44.31	46.93
26	11.16	12.20	13.84	15.38	17.29	25.34	35.56	38.89	41.92	45.64	48.29
27	11.81	12.88	14.57	16.15	18.11	26.34	36.74	40.11	43.19	46.96	49.65
28	12.46	13.57	15.31	16.93	18.94	27.34	37.92	41.34	44.46	48.28	50.99
29	13.12	14.26	16.05	17.71	19.77	28.34	39.09	42.56	45.72	49.59	52.34
30	13.79	14.95	16.79	18.49	20.60	29.34	40.26	43.77	46.98	50.89	53.67
40	20.71	22.16	24.43	26.51	29.05	39.34	51.81	55.76	59.34	63.69	66.77
50	27.99	29.71	32.36	34.76	37.69	49.33	63.17	67.50	71.42	76.15	79.49
60	35.53	37.48	40.48	43.19	46.46	59.33	74.40	79.08	83.30	88.38	91.95
70	43.28	45.44	48.76	51.74	55.33	69.33	85.53	90.53	95.02	100.42	104.22
80	51.17	53.54	57.15	60.39	64.28	79.33	96.58	101.88	106.63	112.33	116.32
90	59.20	61.75	65.65	69.13	73.29	89.33	107.57	113.14	118.14	124.12	128.30
100	67.33	70.06	74.22	77.93	82.36	99.33	118.50	124.34	129.56	135.81	140.17

[a] ν = degrees of freedom.

TABLE IV **Percentage Points of the *t* Distribution**

ν \ α	.40	.25	.10	.05	.025	.01	.005	.0025	.001	.0005
1	.325	1.000	3.078	6.314	12.706	31.821	63.657	127.32	318.31	636.62
2	.289	.816	1.886	2.920	4.303	6.965	9.925	14.089	23.326	31.598
3	.277	.765	1.638	2.353	3.182	4.541	5.841	7.453	10.213	12.924
4	.271	.741	1.533	2.132	2.776	3.747	4.604	5.598	7.173	8.610
5	.267	.727	1.476	2.015	2.571	3.365	4.032	4.773	5.893	6.869
6	.265	.718	1.440	1.943	2.447	3.143	3.707	4.317	5.208	5.959
7	.263	.711	1.415	1.895	2.365	2.998	3.499	4.029	4.785	5.408
8	.262	.706	1.397	1.860	2.306	2.896	3.355	3.833	4.501	5.041
9	.261	.703	1.383	1.833	2.262	2.821	3.250	3.690	4.297	4.781
10	.260	.700	1.372	1.812	2.228	2.764	3.169	3.581	4.144	4.587
11	.260	.697	1.363	1.796	2.201	2.718	3.106	3.497	4.025	4.437
12	.259	.695	1.356	1.782	2.179	2.681	3.055	3.428	3.930	4.318
13	.259	.694	1.350	1.771	2.160	2.650	3.012	3.372	3.852	4.221
14	.258	.692	1.345	1.761	2.145	2.624	2.977	3.326	3.787	4.140
15	.258	.691	1.341	1.753	2.131	2.602	2.947	3.286	3.733	4.073
16	.258	.690	1.337	1.746	2.120	2.583	2.921	3.252	3.686	4.015
17	.257	.689	1.333	1.740	2.110	2.567	2.898	3.222	3.646	3.965
18	.257	.688	1.330	1.734	2.101	2.552	2.878	3.197	3.610	3.922
19	.257	.688	1.328	1.729	2.093	2.539	2.861	3.174	3.579	3.883
20	.257	.687	1.325	1.725	2.086	2.528	2.845	3.153	3.552	3.850
21	.257	.686	1.323	1.721	2.080	2.518	2.831	3.135	3.527	3.819
22	.256	.686	1.321	1.717	2.074	2.508	2.819	3.119	3.505	3.792
23	.256	.685	1.319	1.714	2.069	2.500	2.807	3.104	3.485	3.767
24	.256	.685	1.318	1.711	2.064	2.492	2.797	3.091	3.467	3.745
25	.256	.684	1.316	1.708	2.060	2.485	2.787	3.078	3.450	3.725
26	.256	.684	1.315	1.706	2.056	2.479	2.779	3.067	3.435	3.707
27	.256	.684	1.314	1.703	2.052	2.473	2.771	3.057	3.421	3.690
28	.256	.683	1.313	1.701	2.048	2.467	2.763	3.047	3.408	3.674
29	.256	.683	1.311	1.699	2.045	2.462	2.756	3.038	3.396	3.659
30	.256	.683	1.310	1.697	2.042	2.457	2.750	3.030	3.385	3.646
40	.255	.681	1.303	1.684	2.021	2.423	2.704	2.971	3.307	3.551
60	.254	.679	1.296	1.671	2.000	2.390	2.660	2.915	3.232	3.460
120	.254	.677	1.289	1.658	1.980	2.358	2.617	2.860	3.160	3.373
∞	.253	.674	1.282	1.645	1.960	2.326	2 576	2.807	3.090	3.291

Source: This table is adapted from *Biometrika Tables for Statisticians*, Vol. 1, 3rd edition, 1966, by permission of the Biometrika Trustees.

TABLE V Percentage Points of the *F* Distribution

$$F_{.25,\,\nu_1,\,\nu_2}$$

$\nu_2 \backslash \nu_1$	1	2	3	4	5	6	7	8	9	10	12	15	20	24	30	40	60	120	∞
1	5.83	7.50	8.20	8.58	8.82	8.98	9.10	9.19	9.26	9.32	9.41	9.49	9.58	9.63	9.67	9.71	9.76	9.80	9.85
2	2.57	3.00	3.15	3.23	3.28	3.31	3.34	3.35	3.37	3.38	3.39	3.41	3.43	3.43	3.44	3.45	3.46	3.47	3.48
3	2.02	2.28	2.36	2.39	2.41	2.42	2.43	2.44	2.44	2.44	2.45	2.46	2.46	2.46	2.47	2.47	2.47	2.47	2.47
4	1.81	2.00	2.05	2.06	2.07	2.08	2.08	2.08	2.08	2.08	2.08	2.08	2.08	2.08	2.08	2.08	2.08	2.08	2.08
5	1.69	1.85	1.88	1.89	1.89	1.89	1.89	1.89	1.89	1.89	1.89	1.89	1.88	1.88	1.88	1.88	1.87	1.87	1.87
6	1.62	1.76	1.78	1.79	1.79	1.78	1.78	1.78	1.77	1.77	1.77	1.76	1.76	1.75	1.75	1.75	1.74	1.74	1.74
7	1.57	1.70	1.72	1.72	1.71	1.71	1.70	1.70	1.70	1.69	1.68	1.68	1.67	1.67	1.66	1.66	1.65	1.65	1.65
8	1.54	1.66	1.67	1.66	1.66	1.65	1.64	1.64	1.63	1.63	1.62	1.62	1.61	1.60	1.60	1.59	1.59	1.58	1.58
9	1.51	1.62	1.63	1.63	1.62	1.61	1.60	1.60	1.59	1.59	1.58	1.57	1.56	1.56	1.55	1.54	1.54	1.53	1.53
10	1.49	1.60	1.60	1.59	1.59	1.58	1.57	1.56	1.56	1.55	1.54	1.53	1.52	1.52	1.51	1.51	1.50	1.49	1.48
11	1.47	1.58	1.58	1.57	1.56	1.55	1.54	1.53	1.53	1.52	1.51	1.50	1.49	1.49	1.48	1.47	1.47	1.46	1.45
12	1.46	1.56	1.56	1.55	1.54	1.53	1.52	1.51	1.51	1.50	1.49	1.48	1.47	1.46	1.45	1.45	1.44	1.43	1.42
13	1.45	1.55	1.55	1.53	1.52	1.51	1.50	1.49	1.49	1.48	1.47	1.46	1.45	1.44	1.43	1.42	1.42	1.41	1.40
14	1.44	1.53	1.53	1.52	1.51	1.50	1.49	1.48	1.47	1.46	1.45	1.44	1.43	1.42	1.41	1.41	1.40	1.39	1.38
15	1.43	1.52	1.52	1.51	1.49	1.48	1.47	1.46	1.46	1.45	1.44	1.43	1.41	1.41	1.40	1.39	1.38	1.37	1.36
16	1.42	1.51	1.51	1.50	1.48	1.47	1.46	1.45	1.44	1.44	1.43	1.41	1.40	1.39	1.38	1.37	1.36	1.35	1.34
17	1.42	1.51	1.50	1.49	1.47	1.46	1.45	1.44	1.43	1.43	1.41	1.40	1.39	1.38	1.37	1.36	1.35	1.34	1.33
18	1.41	1.50	1.49	1.48	1.46	1.45	1.44	1.43	1.42	1.42	1.40	1.39	1.38	1.37	1.36	1.35	1.34	1.33	1.32
19	1.41	1.49	1.49	1.47	1.46	1.44	1.43	1.42	1.41	1.41	1.40	1.38	1.37	1.36	1.35	1.34	1.33	1.32	1.30
20	1.40	1.49	1.48	1.47	1.45	1.44	1.43	1.42	1.41	1.40	1.39	1.37	1.36	1.35	1.34	1.33	1.32	1.31	1.29
21	1.40	1.48	1.48	1.46	1.44	1.43	1.42	1.41	1.40	1.39	1.38	1.37	1.35	1.34	1.33	1.32	1.31	1.30	1.28
22	1.40	1.48	1.47	1.45	1.44	1.42	1.41	1.40	1.39	1.39	1.37	1.36	1.34	1.33	1.32	1.31	1.30	1.29	1.28
23	1.39	1.47	1.47	1.45	1.43	1.42	1.41	1.40	1.39	1.38	1.37	1.35	1.34	1.33	1.32	1.31	1.30	1.28	1.27
24	1.39	1.47	1.46	1.44	1.43	1.41	1.40	1.39	1.38	1.38	1.36	1.35	1.33	1.32	1.31	1.30	1.29	1.28	1.26
25	1.39	1.47	1.46	1.44	1.42	1.41	1.40	1.39	1.38	1.37	1.36	1.34	1.33	1.32	1.31	1.29	1.28	1.27	1.25
26	1.38	1.46	1.45	1.44	1.42	1.41	1.39	1.38	1.37	1.37	1.35	1.34	1.32	1.31	1.30	1.29	1.28	1.26	1.25
27	1.38	1.46	1.45	1.43	1.42	1.40	1.39	1.38	1.37	1.36	1.35	1.33	1.32	1.31	1.30	1.28	1.27	1.26	1.24
28	1.38	1.46	1.45	1.43	1.41	1.40	1.39	1.38	1.37	1.36	1.34	1.33	1.31	1.30	1.29	1.28	1.27	1.25	1.24
29	1.38	1.45	1.45	1.43	1.41	1.40	1.38	1.37	1.36	1.35	1.34	1.32	1.31	1.30	1.29	1.27	1.26	1.25	1.23
30	1.38	1.45	1.44	1.42	1.41	1.39	1.38	1.37	1.36	1.35	1.34	1.32	1.30	1.29	1.28	1.27	1.26	1.24	1.23
40	1.36	1.44	1.42	1.40	1.39	1.37	1.36	1.35	1.34	1.33	1.31	1.30	1.28	1.26	1.25	1.24	1.22	1.21	1.19
60	1.35	1.42	1.41	1.38	1.37	1.35	1.33	1.32	1.31	1.30	1.29	1.27	1.25	1.24	1.22	1.21	1.19	1.17	1.15
120	1.34	1.40	1.39	1.37	1.35	1.33	1.31	1.30	1.29	1.28	1.26	1.24	1.22	1.21	1.19	1.18	1.16	1.13	1.10
∞	1.32	1.39	1.37	1.35	1.33	1.31	1.29	1.28	1.27	1.25	1.24	1.22	1.19	1.18	1.16	1.14	1.12	1.08	1.00

Degrees of freedom for the numerator (ν_1)

Degrees of freedom for the denominator (ν_2)

Source: Adapted with permission from *Biometrika Tables for Statisticians*, Vol. 1, 3rd edition, by E. S. Pearson and H. O. Hartley, Cambridge University Press, Cambridge, 1966.

TABLE V Percentage Points of the F Distribution (*continued*)

$F_{.10, \nu_1, \nu_2}$

ν_2 \ ν_1	1	2	3	4	5	6	7	8	9	10	12	15	20	24	30	40	60	120	∞
1	39.86	49.50	53.59	55.83	57.24	58.20	58.91	59.44	59.86	60.19	60.71	61.22	61.74	62.00	62.26	62.53	62.79	63.06	63.33
2	8.53	9.00	9.16	9.24	9.29	9.33	9.35	9.37	9.38	9.39	9.41	9.42	9.44	9.45	9.46	9.47	9.47	9.48	9.49
3	5.54	5.46	5.39	5.34	5.31	5.28	5.27	5.25	5.24	5.23	5.22	5.20	5.18	5.18	5.17	5.16	5.15	5.14	5.13
4	4.54	4.32	4.19	4.11	4.05	4.01	3.98	3.95	3.94	3.92	3.90	3.87	3.84	3.83	3.82	3.80	3.79	3.78	3.76
5	4.06	3.78	3.62	3.52	3.45	3.40	3.37	3.34	3.32	3.30	3.27	3.24	3.21	3.19	3.17	3.16	3.14	3.12	3.10
6	3.78	3.46	3.29	3.18	3.11	3.05	3.01	2.98	2.96	2.94	2.90	2.87	2.84	2.82	2.80	2.78	2.76	2.74	2.72
7	3.59	3.26	3.07	2.96	2.88	2.83	2.78	2.75	2.72	2.70	2.67	2.63	2.59	2.58	2.56	2.54	2.51	2.49	2.47
8	3.46	3.11	2.92	2.81	2.73	2.67	2.62	2.59	2.56	2.54	2.50	2.46	2.42	2.40	2.38	2.36	2.34	2.32	2.29
9	3.36	3.01	2.81	2.69	2.61	2.55	2.51	2.47	2.44	2.42	2.38	2.34	2.30	2.28	2.25	2.23	2.21	2.18	2.16
10	3.29	2.92	2.73	2.61	2.52	2.46	2.41	2.38	2.35	2.32	2.28	2.24	2.20	2.18	2.16	2.13	2.11	2.08	2.06
11	3.23	2.86	2.66	2.54	2.45	2.39	2.34	2.30	2.27	2.25	2.21	2.17	2.12	2.10	2.08	2.05	2.03	2.00	1.97
12	3.18	2.81	2.61	2.48	2.39	2.33	2.28	2.24	2.21	2.19	2.15	2.10	2.06	2.04	2.01	1.99	1.96	1.93	1.90
13	3.14	2.76	2.56	2.43	2.35	2.28	2.23	2.20	2.16	2.14	2.10	2.05	2.01	1.98	1.96	1.93	1.90	1.88	1.85
14	3.10	2.73	2.52	2.39	2.31	2.24	2.19	2.15	2.12	2.10	2.05	2.01	1.96	1.94	1.91	1.89	1.86	1.83	1.80
15	3.07	2.70	2.49	2.36	2.27	2.21	2.16	2.12	2.09	2.06	2.02	1.97	1.92	1.90	1.87	1.85	1.82	1.79	1.76
16	3.05	2.67	2.46	2.33	2.24	2.18	2.13	2.09	2.06	2.03	1.99	1.94	1.89	1.87	1.84	1.81	1.78	1.75	1.72
17	3.03	2.64	2.44	2.31	2.22	2.15	2.10	2.06	2.03	2.00	1.96	1.91	1.86	1.84	1.81	1.78	1.75	1.72	1.69
18	3.01	2.62	2.42	2.29	2.20	2.13	2.08	2.04	2.00	1.98	1.93	1.89	1.84	1.81	1.78	1.75	1.72	1.69	1.66
19	2.99	2.61	2.40	2.27	2.18	2.11	2.06	2.02	1.98	1.96	1.91	1.86	1.81	1.79	1.76	1.73	1.70	1.67	1.63
20	2.97	2.59	2.38	2.25	2.16	2.09	2.04	2.00	1.96	1.94	1.89	1.84	1.79	1.77	1.74	1.71	1.68	1.64	1.61
21	2.96	2.57	2.36	2.23	2.14	2.08	2.02	1.98	1.95	1.92	1.87	1.83	1.78	1.75	1.72	1.69	1.66	1.62	1.59
22	2.95	2.56	2.35	2.22	2.13	2.06	2.01	1.97	1.93	1.90	1.86	1.81	1.76	1.73	1.70	1.67	1.64	1.60	1.57
23	2.94	2.55	2.34	2.21	2.11	2.05	1.99	1.95	1.92	1.89	1.84	1.80	1.74	1.72	1.69	1.66	1.62	1.59	1.55
24	2.93	2.54	2.33	2.19	2.10	2.04	1.98	1.94	1.91	1.88	1.83	1.78	1.73	1.70	1.67	1.64	1.61	1.57	1.53
25	2.92	2.53	2.32	2.18	2.09	2.02	1.97	1.93	1.89	1.87	1.82	1.77	1.72	1.69	1.66	1.63	1.59	1.56	1.52
26	2.91	2.52	2.31	2.17	2.08	2.01	1.96	1.92	1.88	1.86	1.81	1.76	1.71	1.68	1.65	1.61	1.58	1.54	1.50
27	2.90	2.51	2.30	2.17	2.07	2.00	1.95	1.91	1.87	1.85	1.80	1.75	1.70	1.67	1.64	1.60	1.57	1.53	1.49
28	2.89	2.50	2.29	2.16	2.06	2.00	1.94	1.90	1.87	1.84	1.79	1.74	1.69	1.66	1.63	1.59	1.56	1.52	1.48
29	2.89	2.50	2.28	2.15	2.06	1.99	1.93	1.89	1.86	1.83	1.78	1.73	1.68	1.65	1.62	1.58	1.55	1.51	1.47
30	2.88	2.49	2.28	2.14	2.03	1.98	1.93	1.88	1.85	1.82	1.77	1.72	1.67	1.64	1.61	1.57	1.54	1.50	1.46
40	2.84	2.44	2.23	2.09	2.00	1.93	1.87	1.83	1.79	1.76	1.71	1.66	1.61	1.57	1.54	1.51	1.47	1.42	1.38
60	2.79	2.39	2.18	2.04	1.95	1.87	1.82	1.77	1.74	1.71	1.66	1.60	1.54	1.51	1.48	1.44	1.40	1.35	1.29
120	2.75	2.35	2.13	1.99	1.90	1.82	1.77	1.72	1.68	1.65	1.60	1.55	1.48	1.45	1.41	1.37	1.32	1.26	1.19
∞	2.71	2.30	2.08	1.94	1.85	1.77	1.72	1.67	1.63	1.60	1.55	1.49	1.42	1.38	1.34	1.30	1.24	1.17	1.00

Degrees of freedom for the numerator (ν_1)

Degrees of freedom for the denominator (ν_2)

TABLE V Percentage Points of the F Distribution (continued)

$$F_{.05,\,\nu_1,\,\nu_2}$$

Degrees of freedom for the numerator (ν_1)

ν_2	1	2	3	4	5	6	7	8	9	10	12	15	20	24	30	40	60	120	∞
1	161.4	199.5	215.7	224.6	230.2	234.0	236.8	238.9	240.5	241.9	243.9	245.9	248.0	249.1	250.1	251.1	252.2	253.3	254.3
2	18.51	19.00	19.16	19.25	19.30	19.33	19.35	19.37	19.38	19.40	19.41	19.43	19.45	19.45	19.46	19.47	19.48	19.49	19.50
3	10.13	9.55	9.28	9.12	9.01	8.94	8.89	8.85	8.81	8.79	8.74	8.70	8.66	8.64	8.62	8.59	8.57	8.55	8.53
4	7.71	6.94	6.59	6.39	6.26	6.16	6.09	6.04	6.00	5.96	5.91	5.86	5.80	5.77	5.75	5.72	5.69	5.66	5.63
5	6.61	5.79	5.41	5.19	5.05	4.95	4.88	4.82	4.77	4.74	4.68	4.62	4.56	4.53	4.50	4.46	4.43	4.40	4.36
6	5.99	5.14	4.76	4.53	4.39	4.28	4.21	4.15	4.10	4.06	4.00	3.94	3.87	3.84	3.81	3.77	3.74	3.70	3.67
7	5.59	4.74	4.35	4.12	3.97	3.87	3.79	3.73	3.68	3.64	3.57	3.51	3.44	3.41	3.38	3.34	3.30	3.27	3.23
8	5.32	4.46	4.07	3.84	3.69	3.58	3.50	3.44	3.39	3.35	3.28	3.22	3.15	3.12	3.08	3.04	3.01	2.97	2.93
9	5.12	4.26	3.86	3.63	3.48	3.37	3.29	3.23	3.18	3.14	3.07	3.01	2.94	2.90	2.86	2.83	2.79	2.75	2.71
10	4.96	4.10	3.71	3.48	3.33	3.22	3.14	3.07	3.02	2.98	2.91	2.85	2.77	2.74	2.70	2.66	2.62	2.58	2.54
11	4.84	3.98	3.59	3.36	3.20	3.09	3.01	2.95	2.90	2.85	2.79	2.72	2.65	2.61	2.57	2.53	2.49	2.45	2.40
12	4.75	3.89	3.49	3.26	3.11	3.00	2.91	2.85	2.80	2.75	2.69	2.62	2.54	2.51	2.47	2.43	2.38	2.34	2.30
13	4.67	3.81	3.41	3.18	3.03	2.92	2.83	2.77	2.71	2.67	2.60	2.53	2.46	2.42	2.38	2.34	2.30	2.25	2.21
14	4.60	3.74	3.34	3.11	2.96	2.85	2.76	2.70	2.65	2.60	2.53	2.46	2.39	2.35	2.31	2.27	2.22	2.18	2.13
15	4.54	3.68	3.29	3.06	2.90	2.79	2.71	2.64	2.59	2.54	2.48	2.40	2.33	2.29	2.25	2.20	2.16	2.11	2.07
16	4.49	3.63	3.24	3.01	2.85	2.74	2.66	2.59	2.54	2.49	2.42	2.35	2.28	2.24	2.19	2.15	2.11	2.06	2.01
17	4.45	3.59	3.20	2.96	2.81	2.70	2.61	2.55	2.49	2.45	2.38	2.31	2.23	2.19	2.15	2.10	2.06	2.01	1.96
18	4.41	3.55	3.16	2.93	2.77	2.66	2.58	2.51	2.46	2.41	2.34	2.27	2.19	2.15	2.11	2.06	2.02	1.97	1.92
19	4.38	3.52	3.13	2.90	2.74	2.63	2.54	2.48	2.42	2.38	2.31	2.23	2.16	2.11	2.07	2.03	1.98	1.93	1.88
20	4.35	3.49	3.10	2.87	2.71	2.60	2.51	2.45	2.39	2.35	2.28	2.20	2.12	2.08	2.04	1.99	1.95	1.90	1.84
21	4.32	3.47	3.07	2.84	2.68	2.57	2.49	2.42	2.37	2.32	2.25	2.18	2.10	2.05	2.01	1.96	1.92	1.87	1.81
22	4.30	3.44	3.05	2.82	2.66	2.55	2.46	2.40	2.34	2.30	2.23	2.15	2.07	2.03	1.98	1.94	1.89	1.84	1.78
23	4.28	3.42	3.03	2.80	2.64	2.53	2.44	2.37	2.32	2.27	2.20	2.13	2.05	2.01	1.96	1.91	1.86	1.81	1.76
24	4.26	3.40	3.01	2.78	2.62	2.51	2.42	2.36	2.30	2.25	2.18	2.11	2.03	1.98	1.94	1.89	1.84	1.79	1.73
25	4.24	3.39	2.99	2.76	2.60	2.49	2.40	2.34	2.28	2.24	2.16	2.09	2.01	1.96	1.92	1.87	1.82	1.77	1.71
26	4.23	3.37	2.98	2.74	2.59	2.47	2.39	2.32	2.27	2.22	2.15	2.07	1.99	1.95	1.90	1.85	1.80	1.75	1.69
27	4.21	3.35	2.96	2.73	2.57	2.46	2.37	2.31	2.25	2.20	2.13	2.06	1.97	1.93	1.88	1.84	1.79	1.73	1.67
28	4.20	3.34	2.95	2.71	2.56	2.45	2.36	2.29	2.24	2.19	2.12	2.04	1.96	1.91	1.87	1.82	1.77	1.71	1.65
29	4.18	3.33	2.93	2.70	2.55	2.43	2.35	2.28	2.22	2.18	2.10	2.03	1.94	1.90	1.85	1.81	1.75	1.70	1.64
30	4.17	3.32	2.92	2.69	2.53	2.42	2.33	2.27	2.21	2.16	2.09	2.01	1.93	1.89	1.84	1.79	1.74	1.68	1.62
40	4.08	3.23	2.84	2.61	2.45	2.34	2.25	2.18	2.12	2.08	2.00	1.92	1.84	1.79	1.74	1.69	1.64	1.58	1.51
60	4.00	3.15	2.76	2.53	2.37	2.25	2.17	2.10	2.04	1.99	1.92	1.84	1.75	1.70	1.65	1.59	1.53	1.47	1.39
120	3.92	3.07	2.68	2.45	2.29	2.17	2.09	2.02	1.96	1.91	1.83	1.75	1.66	1.61	1.55	1.50	1.43	1.35	1.25
∞	3.84	3.00	2.60	2.37	2.21	2.10	2.01	1.94	1.88	1.83	1.75	1.67	1.57	1.52	1.46	1.39	1.32	1.22	1.00

Degrees of freedom for the denominator (ν_2)

TABLE V Percentage Points of the *F* Distribution (*continued*)

$F_{.025, \nu_1, \nu_2}$

ν_2 \ ν_1	1	2	3	4	5	6	7	8	9	10	12	15	20	24	30	40	60	120	∞
1	647.8	799.5	864.2	899.6	921.8	937.1	948.2	956.7	963.3	968.6	976.7	984.9	993.1	997.2	1001	1006	1010	1014	1018
2	38.51	39.00	39.17	39.25	39.30	39.33	39.36	39.37	39.39	39.40	39.41	39.43	39.45	39.46	39.46	39.47	39.48	39.49	39.50
3	17.44	16.04	15.44	15.10	14.88	14.73	14.62	14.54	14.47	14.42	14.34	14.25	14.17	14.12	14.08	14.04	13.99	13.95	13.90
4	12.22	10.65	9.98	9.60	9.36	9.20	9.07	8.98	8.90	8.84	8.75	8.66	8.56	8.51	8.46	8.41	8.36	8.31	8.26
5	10.01	8.43	7.76	7.39	7.15	6.98	6.85	6.76	6.68	6.62	6.52	6.43	6.33	6.28	6.23	6.18	6.12	6.07	6.02
6	8.81	7.26	6.60	6.23	5.99	5.82	5.70	5.60	5.52	5.46	5.37	5.27	5.17	5.12	5.07	5.01	4.96	4.90	4.85
7	8.07	6.54	5.89	5.52	5.29	5.12	4.99	4.90	4.82	4.76	4.67	4.57	4.47	4.42	4.36	4.31	4.25	4.20	4.14
8	7.57	6.06	5.42	5.05	4.82	4.65	4.53	4.43	4.36	4.30	4.20	4.10	4.00	3.95	3.89	3.84	3.78	3.73	3.67
9	7.21	5.71	5.08	4.72	4.48	4.32	4.20	4.10	4.03	3.96	3.87	3.77	3.67	3.61	3.56	3.51	3.45	3.39	3.33
10	6.94	5.46	4.83	4.47	4.24	4.07	3.95	3.85	3.78	3.72	3.62	3.52	3.42	3.37	3.31	3.26	3.20	3.14	3.08
11	6.72	5.26	4.63	4.28	4.04	3.88	3.76	3.66	3.59	3.53	3.43	3.33	3.23	3.17	3.12	3.06	3.00	2.94	2.88
12	6.55	5.10	4.47	4.12	3.89	3.73	3.61	3.51	3.44	3.37	3.28	3.18	3.07	3.02	2.96	2.91	2.85	2.79	2.72
13	6.41	4.97	4.35	4.00	3.77	3.60	3.48	3.39	3.31	3.25	3.15	3.05	2.95	2.89	2.84	2.78	2.72	2.66	2.60
14	6.30	4.86	4.24	3.89	3.66	3.50	3.38	3.29	3.21	3.15	3.05	2.95	2.84	2.79	2.73	2.67	2.61	2.55	2.49
15	6.20	4.77	4.15	3.80	3.58	3.41	3.29	3.20	3.12	3.06	2.96	2.86	2.76	2.70	2.64	2.59	2.52	2.46	2.40
16	6.12	4.69	4.08	3.73	3.50	3.34	3.22	3.12	3.05	2.99	2.89	2.79	2.68	2.63	2.57	2.51	2.45	2.38	2.32
17	6.04	4.62	4.01	3.66	3.44	3.28	3.16	3.06	2.98	2.92	2.82	2.72	2.62	2.56	2.50	2.44	2.38	2.32	2.25
18	5.98	4.56	3.95	3.61	3.38	3.22	3.10	3.01	2.93	2.87	2.77	2.67	2.56	2.50	2.44	2.38	2.32	2.26	2.19
19	5.92	4.51	3.90	3.56	3.33	3.17	3.05	2.96	2.88	2.82	2.72	2.62	2.51	2.45	2.39	2.33	2.27	2.20	2.13
20	5.87	4.46	3.86	3.51	3.29	3.13	3.01	2.91	2.84	2.77	2.68	2.57	2.46	2.41	2.35	2.29	2.22	2.16	2.09
21	5.83	4.42	3.82	3.48	3.25	3.09	2.97	2.87	2.80	2.73	2.64	2.53	2.42	2.37	2.31	2.25	2.18	2.11	2.04
22	5.79	4.38	3.78	3.44	3.22	3.05	2.93	2.84	2.76	2.70	2.60	2.50	2.39	2.33	2.27	2.21	2.14	2.08	2.00
23	5.75	4.35	3.75	3.41	3.18	3.02	2.90	2.81	2.73	2.67	2.57	2.47	2.36	2.30	2.24	2.18	2.11	2.04	1.97
24	5.72	4.32	3.72	3.38	3.15	2.99	2.87	2.78	2.70	2.64	2.54	2.44	2.33	2.27	2.21	2.15	2.08	2.01	1.94
25	5.69	4.29	3.69	3.35	3.13	2.97	2.85	2.75	2.68	2.61	2.51	2.41	2.30	2.24	2.18	2.12	2.05	1.98	1.91
26	5.66	4.27	3.67	3.33	3.10	2.94	2.82	2.73	2.65	2.59	2.49	2.39	2.28	2.22	2.16	2.09	2.03	1.95	1.88
27	5.63	4.24	3.65	3.31	3.08	2.92	2.80	2.71	2.63	2.57	2.47	2.36	2.25	2.19	2.13	2.07	2.00	1.93	1.85
28	5.61	4.22	3.63	3.29	3.06	2.90	2.78	2.69	2.61	2.55	2.45	2.34	2.23	2.17	2.11	2.05	1.98	1.91	1.83
29	5.59	4.20	3.61	3.27	3.04	2.88	2.76	2.67	2.59	2.53	2.43	2.32	2.21	2.15	2.09	2.03	1.96	1.89	1.81
30	5.57	4.18	3.59	3.25	3.03	2.87	2.75	2.65	2.57	2.51	2.41	2.31	2.20	2.14	2.07	2.01	1.94	1.87	1.79
40	5.42	4.05	3.46	3.13	2.90	2.74	2.62	2.53	2.45	2.39	2.29	2.18	2.07	2.01	1.94	1.88	1.80	1.72	1.64
60	5.29	3.93	3.34	3.01	2.79	2.63	2.51	2.41	2.33	2.27	2.17	2.06	1.94	1.88	1.82	1.74	1.67	1.58	1.48
120	5.15	3.80	3.23	2.89	2.67	2.52	2.39	2.30	2.22	2.16	2.05	1.94	1.82	1.76	1.69	1.61	1.53	1.43	1.31
∞	5.02	3.69	3.12	2.79	2.57	2.41	2.29	2.19	2.11	2.05	1.94	1.83	1.71	1.64	1.57	1.48	1.39	1.27	1.00

Degrees of freedom for the numerator (ν_1)

Degrees of freedom for the denominator (ν_2)

TABLE V Percentage Points of the *F* Distribution (continued)

$$F_{.01,\ \nu_1,\ \nu_2}$$

ν_2 \ ν_1	1	2	3	4	5	6	7	8	9	10	12	15	20	24	30	40	60	120	∞
1	4052	4999.5	5403	5625	5764	5859	5928	5982	6022	6056	6106	6157	6209	6235	6261	6287	6313	6339	6366
2	98.50	99.00	99.17	99.25	99.30	99.33	99.36	99.37	99.39	99.40	99.42	99.43	99.45	99.46	99.47	99.47	99.48	99.49	99.50
3	34.12	30.82	29.46	28.71	28.24	27.91	27.67	27.49	27.35	27.23	27.05	26.87	26.69	26.60	26.50	26.41	26.32	26.22	26.13
4	21.20	18.00	16.69	15.98	15.52	15.21	14.98	14.80	14.66	14.55	14.37	14.20	14.02	13.93	13.84	13.75	13.65	13.56	13.46
5	16.26	13.27	12.06	11.39	10.97	10.67	10.46	10.29	10.16	10.05	9.89	9.72	9.55	9.47	9.38	9.29	9.20	9.11	9.02
6	13.75	10.92	9.78	9.15	8.75	8.47	8.26	8.10	7.98	7.87	7.72	7.56	7.40	7.31	7.23	7.14	7.06	6.97	6.88
7	12.25	9.55	8.45	7.85	7.46	7.19	6.99	6.84	6.72	6.62	6.47	6.31	6.16	6.07	5.99	5.91	5.82	5.74	5.65
8	11.26	8.65	7.59	7.01	6.63	6.37	6.18	6.03	5.91	5.81	5.67	5.52	5.36	5.28	5.20	5.12	5.03	4.95	4.86
9	10.56	8.02	6.99	6.42	6.06	5.80	5.61	5.47	5.35	5.26	5.11	4.96	4.81	4.73	4.65	4.57	4.48	4.40	4.31
10	10.04	7.56	6.55	5.99	5.64	5.39	5.20	5.06	4.94	4.85	4.71	4.56	4.41	4.33	4.25	4.17	4.08	4.00	3.91
11	9.65	7.21	6.22	5.67	5.32	5.07	4.89	4.74	4.63	4.54	4.40	4.25	4.10	4.02	3.94	3.86	3.78	3.69	3.60
12	9.33	6.93	5.95	5.41	5.06	4.82	4.64	4.50	4.39	4.30	4.16	4.01	3.86	3.78	3.70	3.62	3.54	3.45	3.36
13	9.07	6.70	5.74	5.21	4.86	4.62	4.44	4.30	4.19	4.10	3.96	3.82	3.66	3.59	3.51	3.43	3.34	3.25	3.17
14	8.86	6.51	5.56	5.04	4.69	4.46	4.28	4.14	4.03	3.94	3.80	3.66	3.51	3.43	3.35	3.27	3.18	3.09	3.00
15	8.68	6.36	5.42	4.89	4.56	4.32	4.14	4.00	3.89	3.80	3.67	3.52	3.37	3.29	3.21	3.13	3.05	2.96	2.87
16	8.53	6.23	5.29	4.77	4.44	4.20	4.03	3.89	3.78	3.69	3.55	3.41	3.26	3.18	3.10	3.02	2.93	2.84	2.75
17	8.40	6.11	5.18	4.67	4.34	4.10	3.93	3.79	3.68	3.59	3.46	3.31	3.16	3.08	3.00	2.92	2.83	2.75	2.65
18	8.29	6.01	5.09	4.58	4.25	4.01	3.84	3.71	3.60	3.51	3.37	3.23	3.08	3.00	2.92	2.84	2.75	2.66	2.57
19	8.18	5.93	5.01	4.50	4.17	3.94	3.77	3.63	3.52	3.43	3.30	3.15	3.00	2.92	2.84	2.76	2.67	2.58	2.49
20	8.10	5.85	4.94	4.43	4.10	3.87	3.70	3.56	3.46	3.37	3.23	3.09	2.94	2.86	2.78	2.69	2.61	2.52	2.42
21	8.02	5.78	4.87	4.37	4.04	3.81	3.64	3.51	3.40	3.31	3.17	3.03	2.88	2.80	2.72	2.64	2.55	2.46	2.36
22	7.95	5.72	4.82	4.31	3.99	3.76	3.59	3.45	3.35	3.26	3.12	2.98	2.83	2.75	2.67	2.58	2.50	2.40	2.31
23	7.88	5.66	4.76	4.26	3.94	3.71	3.54	3.41	3.30	3.21	3.07	2.93	2.78	2.70	2.62	2.54	2.45	2.35	2.26
24	7.82	5.61	4.72	4.22	3.90	3.67	3.50	3.36	3.26	3.17	3.03	2.89	2.74	2.66	2.58	2.49	2.40	2.31	2.21
25	7.77	5.57	4.68	4.18	3.85	3.63	3.46	3.32	3.22	3.13	2.99	2.85	2.70	2.62	2.54	2.45	2.36	2.27	2.17
26	7.72	5.53	4.64	4.14	3.82	3.59	3.42	3.29	3.18	3.09	2.96	2.81	2.66	2.58	2.50	2.42	2.33	2.23	2.13
27	7.68	5.49	4.60	4.11	3.78	3.56	3.39	3.26	3.15	3.06	2.93	2.78	2.63	2.55	2.47	2.38	2.29	2.20	2.10
28	7.64	5.45	4.57	4.07	3.75	3.53	3.36	3.23	3.12	3.03	2.90	2.75	2.60	2.52	2.44	2.35	2.26	2.17	2.06
29	7.60	5.42	4.54	4.04	3.73	3.50	3.33	3.20	3.09	3.00	2.87	2.73	2.57	2.49	2.41	2.33	2.23	2.14	2.03
30	7.56	5.39	4.51	4.02	3.70	3.47	3.30	3.17	3.07	2.98	2.84	2.70	2.55	2.47	2.39	2.30	2.21	2.11	2.01
40	7.31	5.18	4.31	3.83	3.51	3.29	3.12	2.99	2.89	2.80	2.66	2.52	2.37	2.29	2.20	2.11	2.02	1.92	1.80
60	7.08	4.98	4.13	3.65	3.34	3.12	2.95	2.82	2.72	2.63	2.50	2.35	2.20	2.12	2.03	1.94	1.84	1.73	1.60
120	6.85	4.79	3.95	3.48	3.17	2.96	2.79	2.66	2.56	2.47	2.34	2.19	2.03	1.95	1.86	1.76	1.66	1.53	1.38
∞	6.63	4.61	3.78	3.32	3.02	2.80	2.64	2.51	2.41	2.32	2.18	2.04	1.88	1.79	1.70	1.59	1.47	1.32	1.00

Degrees of freedom for the numerator (ν_1)

Degrees of freedom for the denominator (ν_2)

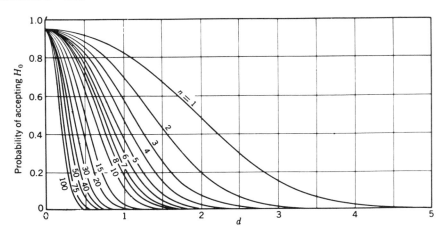

CHART VI **Operating Characteristic Curves**

(a) OC curves for different values of *n* for the two-sided normal test for a level of significance $\alpha = .05$.

(b) OC curves for different values of *n* for the two-sided normal test for a level of significance $\alpha = .01$.

Source: Charts VI*a*, *e*, *f*, *k*, *m*, and *q* are reproduced with permission from "Operating Characteristics for the Common Statistical Tests of Significance," by C. L. Ferris, F. E. Grubbs, and C. L. Weaver, *Annals of Mathematical Statistics*, June 1946.

Charts VI*b*, *c*, *d*, *g*, *h*, *i*, *j*, *l*, *n*, *o*, *p*, and *r* are reproduced with permission from *Engineering Statistics*, 2nd Edition, by A. H. Bowker and G. J. Lieberman, Prentice-Hall, 1972.

CHART VI **Operating Characteristic Curves (*continued*)**

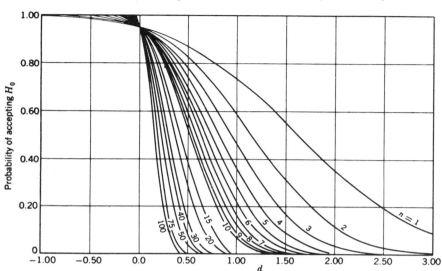

(*c*) OC curves for different values of *n* for the one-sided normal test for a level of significance $\alpha = .05$.

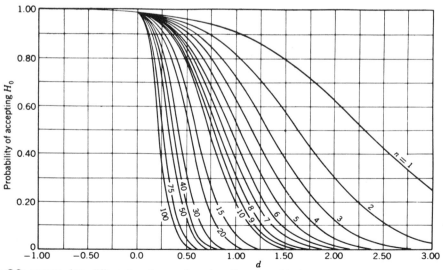

(*d*) OC curves for different values of *n* for the one-sided normal test for a level of significance $\alpha = .01$.

CHART VI Operating Characteristic Curves (*continued*)

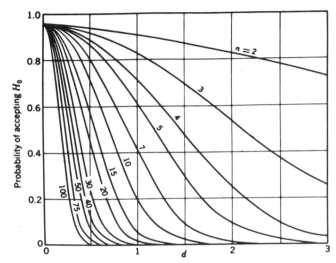

(e) OC curves for different values of n for the two-sided t test for a level of significance $\alpha = .05$.

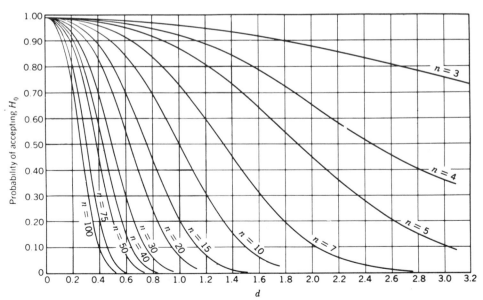

(f) OC curves for different values of n for the two-sided t test for a level of significance $\alpha = .01$.

CHART VI   **Operating Characteristic Curves (*continued*)**

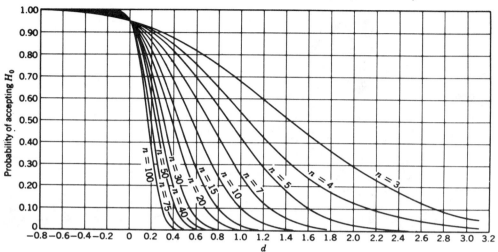

(*g*) OC curves for different values of *n* for the one-sided *t* test for a level of significance $\alpha = .05$.

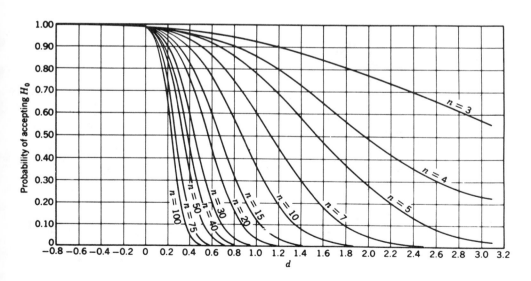

(*h*) OC curves for different values of *n* for the one-sided *t* test for a level of significance $\alpha = .01$.

CHART VI Operating Characteristic Curves (*continued*)

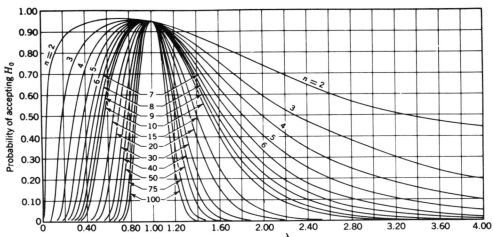

(*i*) OC curves for different values of *n* for the two-sided chi-square test for a level of significance $\alpha = .05$.

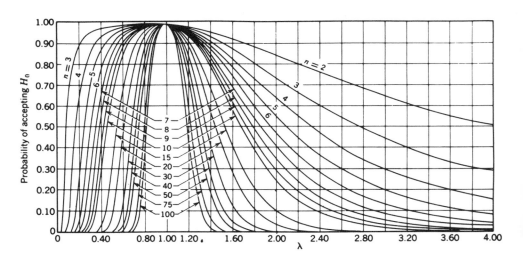

(*j*) OC curves for different values of *n* for the two-sided chi-square test for a level of significance $\alpha = .01$.

CHART VI   **Operating Characteristic Curves (*continued*)**

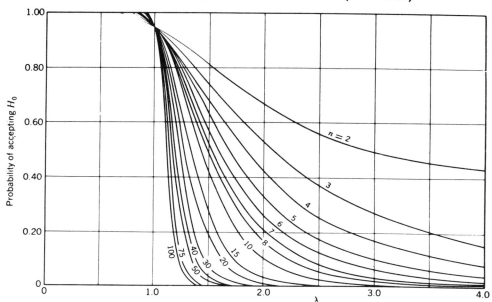

(*k*) OC curves for different values of *n* for the one-sided (upper tail) chi-square test for a level of significance $\alpha = .05$.

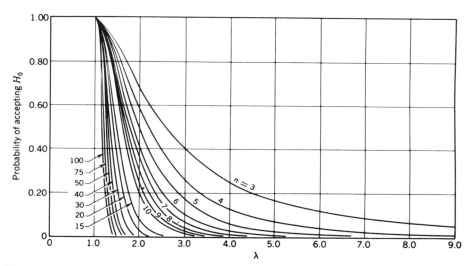

(*l*) OC curves for different values of *n* for the one-sided (upper tail) chi-square test for a level of significance $\alpha = .01$.

CHART VI **Operating Characteristic Curves (*continued*)**

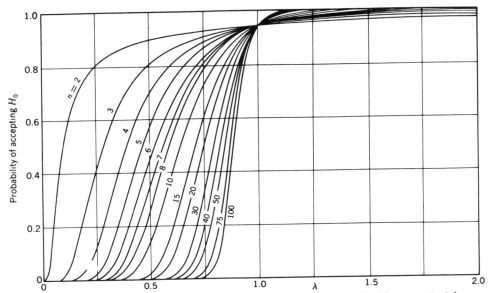

(*m*) OC curves for different values of *n* for the one-sided (lower tail) chi-square test for a level of significance $\alpha = .05$.

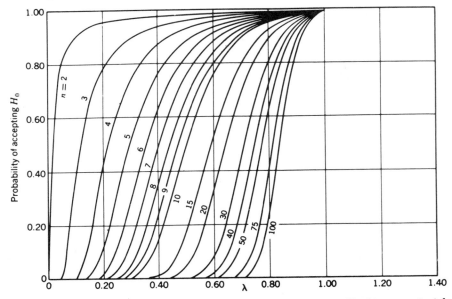

(*n*) OC curves for different values of *n* for the one-sided (lower tail) chi-square test for a level of significance $\alpha = .01$.

CHART VI **Operating Characteristic Curves (*continued*)**

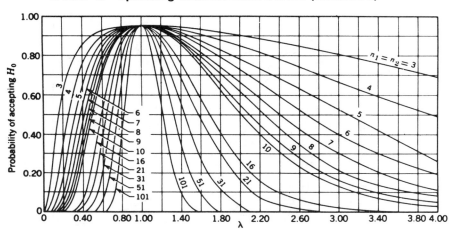

(*o*) OC curves for different values of *n* for the two-sided *F* test for a level of significance $\alpha = .05$.

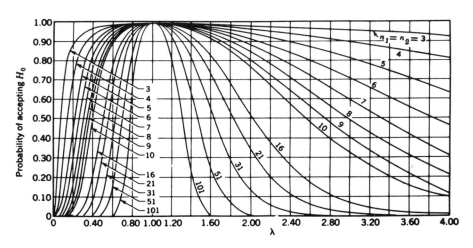

(*p*) OC curves for different values of *n* for the two-sided *F* test for a level of significance $\alpha = .01$.

CHART VI **Operating Characteristic Curves (*continued*)**

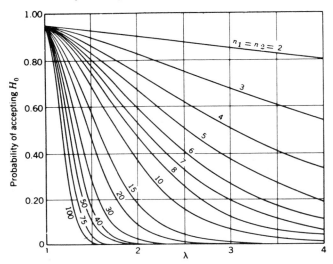

(*q*) OC curves for different values of *n* for the one-sided *F* test for a level of significance $\alpha = .05$.

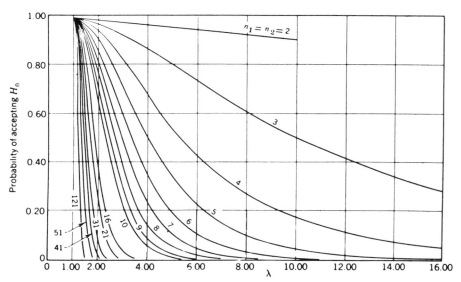

(*r*) OC curves for different values of *n* for the one-sided *F* test for a level of significance $\alpha = .01$.

CHART VII **Operating Characteristic Curves for the Fixed Effects Model Analysis of Variance.**

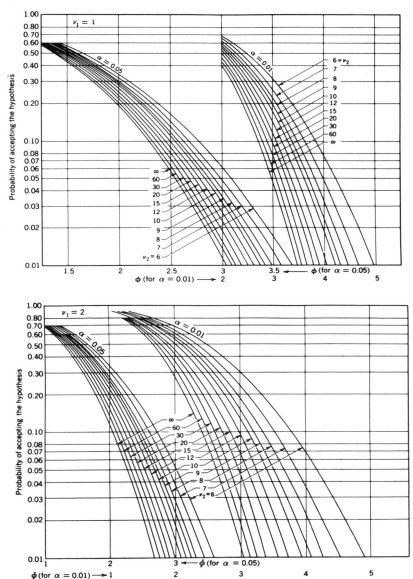

ν_1 = numerator degrees of freedom, ν_2 = denominator degrees of freedom

Source: Chart VII is adapted with permission from *Biometrika Tables for Statisticians*, Vol. 2, by E. S. Pearson and H. O. Hartley, Cambridge University Press, Cambridge, 1972.

CHART VII **Operating Characteristic Curves for the Fixed Effects Model Analysis of Variance (*continued*)**

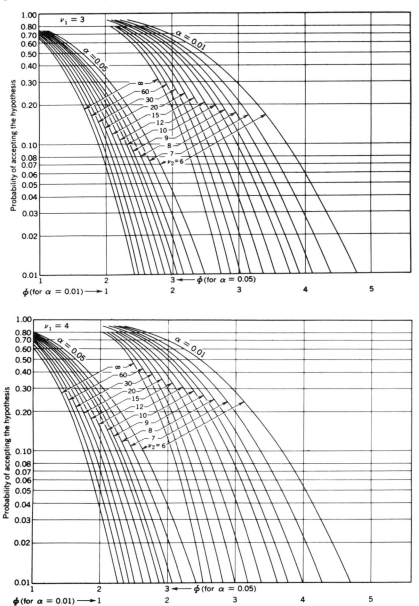

CHART VII **Operating Characteristic Curves for the Fixed Effects Model Analysis of Variance (continued)**

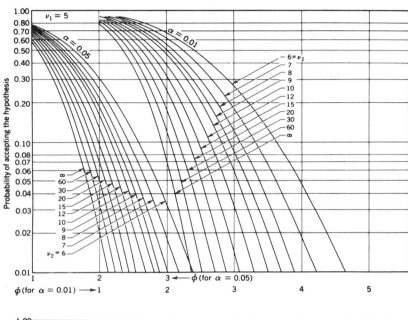

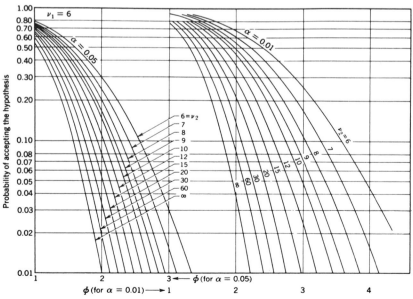

CHART VII **Operating Characteristic Curves for the Fixed Effects Model Analysis of Variance (*continued*)**

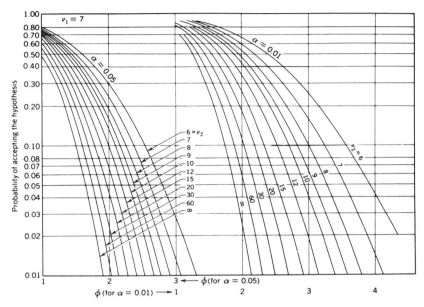

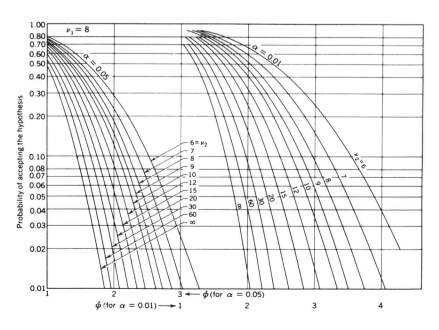

CHART VIII **Operating Characteristic Curves for the Random Effects Model Analysis of Variance**

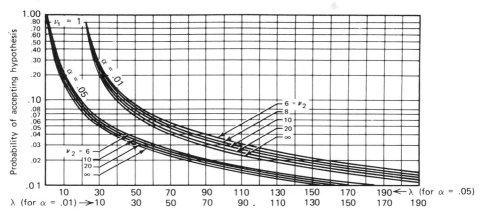

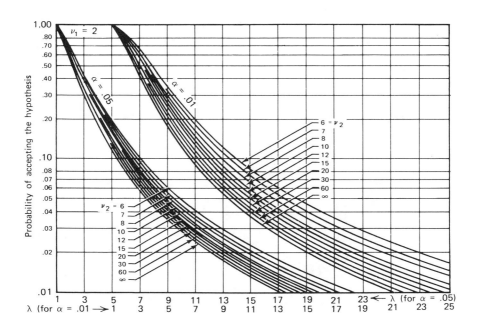

Source: Reproduced with permission from *Engineering Statistics*, 2nd Edition, by A. H. Bowker and G. J. Lieberman, Prentice-Hall, Englewood Cliffs, N.J., 1972.

CHART VIII **Operating Characteristic Curves for the Random Effects Model Analysis of Variance (*continued*)**

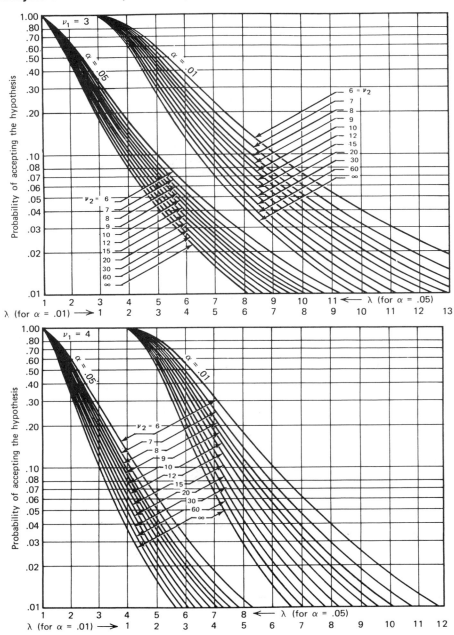

CHART VIII **Operating Characteristic Curves for the Random Effects Model Analysis of Variance (*continued*)**

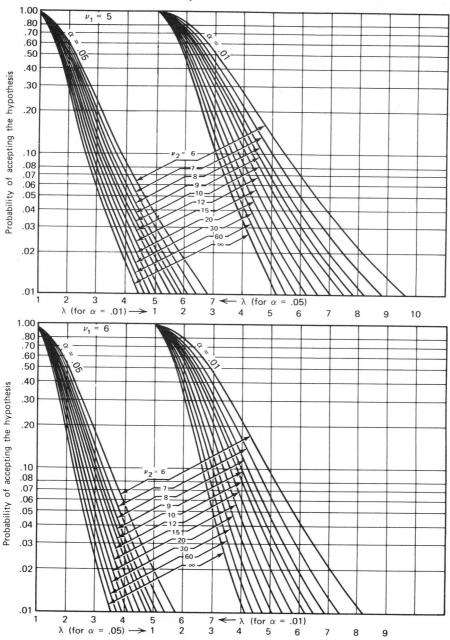

CHART VIII **Operating Characteristic Curves for the Random Effects Model Analysis of Variance (*continued*)**

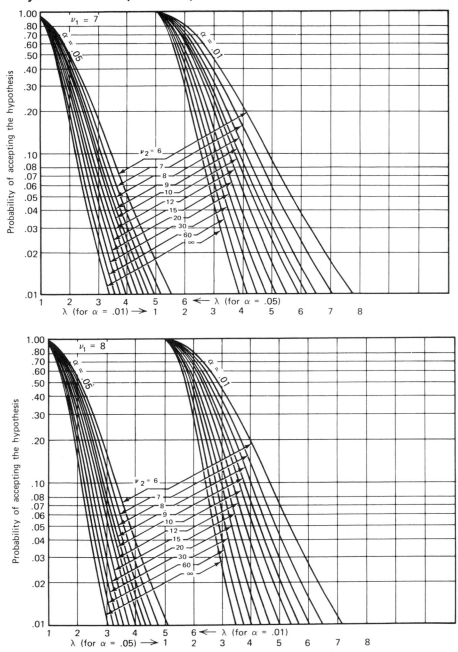

TABLE IX   Critical Values for the Wilcoxon Two-Sample Test[a]

$$R^*_{.05}$$

$n_2 \backslash n_1$	2	3	4	5	6	7	8	9	10	11	12	13	14	15
4			10											
5		6	11	17										
6		7	12	18	26									
7		7	13	20	27	36								
8	3	8	14	21	29	38	49							
9	3	8	15	22	31	40	51	63						
10	3	9	15	23	32	42	53	65	78					
11	4	9	16	24	34	44	55	68	81	96				
12	4	10	17	26	35	46	58	71	85	99	115			
13	4	10	18	27	37	48	60	73	88	103	119	137		
14	4	11	19	28	38	50	63	76	91	106	123	141	160	
15	4	11	20	29	40	52	65	79	94	110	127	145	164	185
16	4	12	21	31	42	54	67	82	97	114	131	150	169	
17	5	12	21	32	43	56	70	84	100	117	135	154		
18	5	13	22	33	45	58	72	87	103	121	139			
19	5	13	23	34	46	60	74	90	107	124				
20	5	14	24	35	48	62	77	93	110					
21	6	14	25	37	50	64	79	95						
22	6	15	26	38	51	66	82							
23	6	15	27	39	53	68								
24	6	16	28	40	55									
25	6	16	28	42										
26	7	17	29											
27	7	17												
28	7													

Source: Reproduced with permission from "The Use of Ranks in a Test of Significance for Comparing Two Treatments," by C. White, *Biometrics*, 1952, Vol. 8, p. 37.

[a]For large n_1 and n_2, R is approximately normally distributed with mean $\frac{1}{2}n_1(n_1 + n_2 + 1)$ and variance $\frac{1}{12}n_1 n_2(n_1 + n_2 + 1)$.

TABLE IX **Critical Values for the Wilcoxon Two-Sample Test (*continued*)**

$$R^*_{.01}$$

n_2 \ n_1	2	3	4	5	6	7	8	9	10	11	12	13	14	15
5				15										
6			10	16	23									
7			10	17	24	32								
8			11	17	25	34	43							
9		6	11	18	26	35	45	56						
10		6	12	19	27	37	47	58	71					
11		6	12	20	28	38	49	61	74	87				
12		7	13	21	30	40	51	63	76	90	106			
13		7	14	22	31	41	53	65	79	93	109	125		
14		7	14	22	32	43	54	67	81	96	112	129	147	
15		8	15	23	33	44	56	70	84	99	115	133	151	171
16		8	15	24	34	46	58	72	86	102	119	137	155	
17		8	16	25	36	47	60	74	89	105	122	140		
18		8	16	26	37	49	62	76	92	108	125			
19	3	9	17	27	38	50	64	78	94	111				
20	3	9	18	28	39	52	66	81	97					
21	3	9	18	29	40	53	68	83						
22	3	10	19	29	42	55	70							
23	3	10	19	30	43	57								
24	3	10	20	31	44									
25	3	11	20	32										
26	3	11	21											
27	4	11												
28	4													

TABLE X   **Critical Values for the Sign Test**[a]

$$R_\alpha^*$$

n \ α	.10	.05	.01
5	0		
6	0	0	
7	0	0	
8	1	0	0
9	1	1	0
10	1	1	0
11	2	1	0
12	2	2	1
13	3	2	1
14	3	2	1
15	3	3	2
16	4	3	2
17	4	4	2
18	5	4	3
19	5	4	3
20	5	5	3
21	6	5	4
22	6	5	4
23	7	6	4
24	7	6	5
25	7	7	5
26	8	7	6
27	8	7	6
28	9	8	6
29	9	8	7
30	10	9	7
31	10	9	7
32	10	9	8
33	11	10	8
34	11	10	9
35	12	11	9
36	12	11	9
37	13	12	10
38	13	12	10
39	13	12	11
40	14	13	11

[a]For $n > 40$, R is approximately normally distributed with mean $n/2$ and variance $n/4$.

TABLE XI **Critical Values for the Wilcoxon Signed-Rank Test**[a]

n \ α	.10	.05	.02	.01
4				
5	0			
6	2	0		
7	3	2	0	
8	5	3	1	0
9	8	5	3	1
10	10	8	5	3
11	13	10	7	5
12	17	13	9	7
13	21	17	12	9
14	25	21	15	12
15	30	25	19	15
16	35	29	23	19
17	41	34	27	23
18	47	40	32	27
19	53	46	37	32
20	60	52	43	37
21	67	58	49	42
22	75	65	55	48
23	83	73	62	54
24	91	81	69	61
25	100	89	76	68
26	110	98	84	75
27	119	107	92	83
28	130	116	101	91
29	140	126	110	100
30	151	137	120	109
31	163	147	130	118
32	175	159	140	128
33	187	170	151	138
34	200	182	162	148
35	213	195	173	159
36	227	208	185	171
37	241	221	198	182
38	256	235	211	194
39	271	249	224	207
40	286	264	238	220
41	302	279	252	233
42	319	294	266	247
43	336	310	281	261
44	353	327	296	276
45	371	343	312	291
46	389	361	328	307
47	407	378	345	322
48	426	396	362	339
49	446	415	379	355
50	466	434	397	373

Source: Adapted with permission from "Extended Tables of the Wilcoxon Matched Pair Signed Rank Statistic" by Robert L. McCornack, *Journal of the American Statistical Association*, Vol. 60, September, 1965.
[a]If $n > 50$, R is approximately normally distributed with mean $n(n + 1)/4$ and variance $n(n + 1)(2n + 1)/24$.

TABLE XII Significant Ranges for Duncan's Multiple Range Test

$$r_{.01}(p, f)$$

f^a	2	3	4	5	6	7	8	9	10	20	50	100
1	90.0	90.0	90.0	90.0	90.0	90.0	90.0	90.0	90.0	90.0	90.0	90.0
2	14.0	14.0	14.0	14.0	14.0	14.0	14.0	14.0	14.0	14.0	14.0	14.0
3	8.26	8.5	8.6	8.7	8.8	8.9	8.9	9.0	9.0	9.3	9.3	9.3
4	6.51	6.8	6.9	7.0	7.1	7.1	7.2	7.2	7.3	7.5	7.5	7.5
5	5.70	5.96	6.11	6.18	6.26	6.33	6.40	6.44	6.5	6.8	6.8	6.8
6	5.24	5.51	5.65	5.73	5.81	5.88	5.95	6.00	6.0	6.3	6.3	6.3
7	4.95	5.22	5.37	5.45	5.53	5.61	5.69	5.73	5.8	6.0	6.0	6.0
8	4.74	5.00	5.14	5.23	5.32	5.40	5.47	5.51	5.5	5.8	5.8	5.8
9	4.60	4.86	4.99	5.08	5.17	5.25	5.32	5.36	5.4	5.7	5.7	5.7
10	4.48	4.73	4.88	4.96	5.06	5.13	5.20	5.24	5.28	5.55	5.55	5.55
11	4.39	4.63	4.77	4.86	4.94	5.01	5.06	5.12	5.15	5.39	5.39	5.39
12	4.32	4.55	4.68	4.76	4.84	4.92	4.96	5.02	5.07	5.26	5.26	5.26
13	4.26	4.48	4.62	4.69	4.74	4.84	4.88	4.94	4.98	5.15	5.15	5.15
14	4.21	4.42	4.55	4.63	4.70	4.78	4.83	4.87	4.91	5.07	5.07	5.07
15	4.17	4.37	4.50	4.58	4.64	4.72	4.77	4.81	4.84	5.00	5.00	5.00
16	4.13	4.34	4.45	4.54	4.60	4.67	4.72	4.76	4.79	4.94	4.94	4.94
17	4.10	4.30	4.41	4.50	4.56	4.63	4.68	4.73	4.75	4.89	4.89	4.89
18	4.07	4.27	4.38	4.46	4.53	4.59	4.64	4.68	4.71	4.85	4.85	4.85
19	4.05	4.24	4.35	4.43	4.50	4.56	4.61	4.64	4.67	4.82	4.82	4.82
20	4.02	4.22	4.33	4.40	4.47	4.53	4.58	4.61	4.65	4.79	4.79	4.79
30	3.89	4.06	4.16	4.22	4.32	4.36	4.41	4.45	4.48	4.65	4.71	4.71
40	3.82	3.99	4.10	4.17	4.24	4.30	4.34	4.37	4.41	4.59	4.69	4.69
60	3.76	3.92	4.03	4.12	4.17	4.23	4.27	4.31	4.34	4.53	4.66	4.66
100	3.71	3.86	3.98	4.06	4.11	4.17	4.21	4.25	4.29	4.48	4.64	4.65
∞	3.64	3.80	3.90	3.98	4.04	4.09	4.14	4.17	4.20	4.41	4.60	4.68

p

Source: Reproduced with permission from "Multiple Range and Multiple F Tests," by D. B. Duncan, *Biometrics*, Vol. 11, No. 1, pp. 1–42, 1955.

$^a f$ = degrees of freedom.

TABLE XII Significant Ranges for Duncan's Multiple Range Test (continued)

$$r_{.05}(p, f)$$

f[a]	2	3	4	5	6	7	8	9	10	20	50	100
1	18.0	18.0	18.0	18.0	18.0	18.0	18.0	18.0	18.0	18.0	18.0	18.0
2	6.09	6.09	6.09	6.09	6.09	6.09	6.09	6.09	6.09	6.09	6.09	6.09
3	4.50	4.50	4.50	4.50	4.50	4.50	4.50	4.50	4.50	4.50	4.50	4.50
4	3.93	4.01	4.02	4.02	4.02	4.02	4.02	4.02	4.02	4.02	4.02	4.02
5	3.64	3.74	3.79	3.83	3.83	3.83	3.83	3.83	3.83	3.83	3.83	3.83
6	3.46	3.58	3.64	3.68	3.68	3.68	3.68	3.68	3.68	3.68	3.68	3.68
7	3.35	3.47	3.54	3.58	3.60	3.61	3.61	3.61	3.61	3.61	3.61	3.61
8	3.26	3.39	3.47	3.52	3.55	3.56	3.56	3.56	3.56	3.56	3.56	3.56
9	3.20	3.34	3.41	3.47	3.50	3.52	3.52	3.52	3.52	3.52	3.52	3.52
10	3.15	3.30	3.37	3.43	3.46	3.47	3.47	3.47	3.47	3.48	3.48	3.48
11	3.11	3.27	3.35	3.39	3.43	3.44	3.45	3.46	3.46	3.48	3.48	3.48
12	3.08	3.23	3.33	3.36	3.40	3.42	3.44	3.44	3.46	3.48	3.48	3.48
13	3.06	3.21	3.30	3.35	3.38	3.41	3.42	3.44	3.45	3.47	3.47	3.47
14	3.03	3.18	3.27	3.33	3.37	3.39	3.41	3.42	3.44	3.47	3.47	3.47
15	3.01	3.16	3.25	3.31	3.36	3.38	3.40	3.42	3.43	3.47	3.47	3.47
16	3.00	3.15	3.23	3.30	3.34	3.37	3.39	3.41	3.43	3.47	3.47	3.47
17	2.98	3.13	3.22	3.28	3.33	3.36	3.38	3.40	3.42	3.47	3.47	3.47
18	2.97	3.12	3.21	3.27	3.32	3.35	3.37	3.39	3.41	3.47	3.47	3.47
19	2.96	3.11	3.19	3.26	3.31	3.35	3.37	3.39	3.41	3.47	3.47	3.47
20	2.95	3.10	3.18	3.25	3.30	3.34	3.36	3.38	3.40	3.47	3.47	3.47
30	2.89	3.04	3.12	3.20	3.25	3.29	3.32	3.35	3.37	3.47	3.47	3.47
40	2.86	3.01	3.10	3.17	3.22	3.27	3.30	3.33	3.35	3.47	3.47	3.47
60	2.83	2.98	3.08	3.14	3.20	3.24	3.28	3.31	3.33	3.47	3.48	3.48
100	2.80	2.95	3.05	3.12	3.18	3.22	3.26	3.29	3.32	3.47	3.53	3.53
∞	2.77	2.92	3.02	3.09	3.15	3.19	3.23	3.26	3.29	3.47	3.61	3.67

p

[a]f = degrees of freedom.

TABLE XIII **Factors for Quality Control Charts**

	$\bar{X}$ Chart		R Chart			
	Factors for Control Limits		Factors for Central Line	Factors for Control Limits		
n^a	A_1	A_2	d_2	D_3	D_4	n
2	3.760	1.880	1.128	0	3.267	2
3	2.394	1.023	1.693	0	2.575	3
4	1.880	.729	2.059	0	2.282	4
5	1.596	.577	2.326	0	2.115	5
6	1.410	.483	2.534	0	2.004	6
7	1.277	.419	2.704	.076	1.924	7
8	1.175	.373	2.847	.136	1.864	8
9	1.094	.337	2.970	.184	1.816	9
10	1.028	.308	3.078	.223	1.777	10
11	.973	.285	3.173	.256	1.744	11
12	.925	.266	3.258	.284	1.716	12
13	.884	.249	3.336	.308	1.692	13
14	.848	.235	3.407	.329	1.671	14
15	.816	.223	3.472	.348	1.652	15
16	.788	.212	3.532	.364	1.636	16
17	.762	.203	3.588	.379	1.621	17
18	.738	.194	3.640	.392	1.608	18
19	.717	.187	3.689	.404	1.596	19
20	.697	.180	3.735	.414	1.586	20
21	.679	.173	3.778	.425	1.575	21
22	.662	.167	3.819	.434	1.566	22
23	.647	.162	3.858	.443	1.557	23
24	.632	.157	3.895	.452	1.548	24
25	.619	.153	3.931	.459	1.541	25

[a]$n > 25$: $A_1 = 3/\sqrt{n}$. n = number of observations in sample.

TABLE XIV **Factors for Two-Sided Tolerance Limits**

n	90% Confidence that percentage of population between limits is			95% Confidence that percentage of population between limits is			99% Confidence that percentage of population between limits is		
	90%	95%	99%	90%	95%	99%	90%	95%	99%
2	15.98	18.80	24.17	32.02	37.67	48.43	160.2	188.5	242.3
3	5.847	6.919	8.974	8.380	9.916	12.86	18.93	22.40	29.06
4	4.166	4.943	6.440	5.369	6.370	8.299	9.398	11.15	14.53
5	3.494	4.152	5.423	4.275	5.079	6.634	6.612	7.855	10.26
6	3.131	3.723	4.870	3.712	4.414	5.775	5.337	6.345	8.301
7	2.902	3.452	4.521	3.369	4.007	5.248	4.613	5.488	7.187
8	2.743	3.264	4.278	3.136	3.732	4.891	4.147	4.936	6.468
9	2.626	3.125	4.098	2.967	3.532	4.631	3.822	4.550	5.966
10	2.535	3.018	3.959	2.839	3.379	4.433	3.582	4.265	5.594
11	2.463	2.933	3.849	2.737	3.259	4.277	3.397	4.045	5.308
12	2.404	2.863	3.758	2.655	3.162	4.150	3.250	3.870	5.079
13	2.355	2.805	3.682	2.587	3.081	4.044	3.130	3.727	4.893
14	2.314	2.756	3.618	2.529	3.012	3.955	3.029	3.608	4.737
15	2.278	2.713	3.562	2.480	2.954	3.878	2.945	3.507	4.605
16	2.246	2.676	3.514	2.437	2.903	3.812	2.872	3.421	4.492
17	2.219	2.643	3.471	2.400	2.858	3.754	2.808	3.345	4.393

df									
18	2.194	2.614	3.433	2.366	2.819	3.702	2.753	3.279	4.307
19	2.172	2.588	3.399	2.337	2.784	3.656	2.703	3.221	4.230
20	2.152	2.564	3.368	2.310	2.752	3.615	2.659	3.168	4.161
21	2.135	2.543	3.340	2.286	2.723	3.577	2.620	3.121	4.100
22	2.118	2.524	3.315	2.264	2.697	3.543	2.584	3.078	4.044
23	2.103	2.506	3.292	2.244	2.673	3.512	2.551	3.040	3.993
24	2.089	2.489	3.270	2.225	2.651	3.483	2.522	3.004	3.947
25	2.077	2.474	3.251	2.208	2.631	3.457	2.494	2.972	3.904
26	2.065	2.460	3.232	2.193	2.612	3.432	2.469	2.941	3.865
27	2.054	2.447	3.215	2.178	2.595	3.409	2.446	2.914	3.828
28	2.044	2.435	3.199	2.164	2.579	3.388	2.424	2.888	3.794
29	2.034	2.424	3.184	2.152	2.554	3.368	2.404	2.864	3.763
30	2.025	2.413	3.170	2.140	2.549	3.350	2.385	2.841	3.733
35	1.988	2.368	3.112	2.090	2.490	3.272	2.306	2.748	3.611
40	1.959	2.334	3.066	2.052	2.445	3.213	2.247	2.677	3.518
50	1.916	2.284	3.001	1.996	2.379	3.126	2.162	2.576	3.385
60	1.887	2.248	2.955	1.958	2.333	3.066	2.103	2.506	3.293
80	1.848	2.202	2.894	1.907	2.272	2.986	2.026	2.414	3.173
100	1.822	2.172	2.854	1.874	2.233	2.934	1.977	2.355	3.096
200	1.764	2.102	2.762	1.798	2.143	2.816	1.865	2.222	2.921
500	1.717	2.046	2.689	1.737	2.070	2.721	1.777	2.117	2.783
1000	1.695	2.019	2.654	1.709	2.036	2.676	1.736	2.068	2.718
∞	1.645	1.960	2.576	1.645	1.960	2.576	1.645	1.960	2.576

TABLE XV Random Numbers

10480	15011	01536	02011	81647	91646	69179	14194	62590
22368	46573	25595	85393	30995	89198	27982	53402	93965
24130	48360	22527	97265	76393	64809	15179	24830	49340
42167	93093	06243	61680	07856	16376	39440	53537	71341
37570	39975	81837	16656	06121	91782	60468	81305	49684
77921	06907	11008	42751	27756	53498	18602	70659	90655
99562	72905	56420	69994	98872	31016	71194	18738	44013
96301	91977	05463	07972	18876	20922	94595	56869	69014
89579	14342	63661	10281	17453	18103	57740	84378	25331
85475	36857	53342	53988	53060	59533	38867	62300	08158
28918	69578	88231	33276	70997	79936	56865	05859	90106
63553	40961	48235	03427	49626	69445	18663	72695	52180
09429	93969	52636	92737	88974	33488	36320	17617	30015
10365	61129	87529	85689	48237	52267	67689	93394	01511
07119	97336	71048	08178	77233	13916	47564	81056	97735
51085	12765	51821	51259	77452	16308	60756	92144	49442
02368	21382	52404	60268	89368	19885	55322	44819	01188
01011	54092	33362	94904	31273	04146	18594	29852	71585
52162	53916	46369	58586	23216	14513	83149	98736	23495
07056	97628	33787	09998	42698	06691	76988	13602	51851
48663	91245	85828	14346	09172	30168	90229	04734	59193
54164	58492	22421	74103	47070	25306	76468	26384	58151
32639	32363	05597	24200	13363	38005	94342	28728	35806
29334	27001	87637	87308	58731	00256	45834	15398	46557
02488	33062	28834	07351	19731	92420	60952	61280	50001
81525	72295	04839	96423	24878	82651	66566	14778	76797
29676	20591	68086	26432	46901	20849	89768	81536	86645
00742	57392	39064	66432	84673	40027	32832	61362	98947
05366	04213	25669	26422	44407	44048	37937	63904	45766
91921	26418	64117	94305	26766	25940	39972	22209	71500
00582	04711	87917	77341	42206	35126	74087	99547	81817
00725	69884	62797	56170	86324	88072	76222	36086	84637
69011	65795	95876	55293	18988	27354	26575	08625	40801
25976	57948	29888	88604	67917	48708	18912	82271	65424
09763	83473	73577	12908	30883	18317	28290	35797	05998
91567	42595	27958	30134	04024	86385	29880	99730	55536
17955	56349	90999	49127	20044	59931	06115	20542	18059
46503	18584	18845	49618	02304	51038	20655	58727	28168
92157	89634	94824	78171	84610	82834	09922	25417	44137
14577	62765	35605	81263	39667	47358	56873	56307	61607
98427	07523	33362	64270	01638	92477	66969	98420	04880
34914	63976	88720	82765	34476	17032	87589	40836	32427
70060	28277	39475	46473	23219	53416	94970	25832	69975
53976	54914	06990	67245	68350	82948	11398	42878	80287
76072	29515	40980	07391	58745	25774	22987	80059	39911
90725	52210	83974	29992	65831	38857	50490	83765	55657
64364	67412	33339	31926	14883	24413	59744	92351	97473
08962	00358	31662	25388	61642	34072	81249	35648	56891
95012	68379	93526	70765	10592	04542	76463	54328	02349
15664	10493	20492	38391	91132	21999	59516	81652	27195

Index

SOCIAL SCIENCE LIBRARY

Manor Road Building
Manor Road
Oxford OX1 3UQ
Tel: (2)71093 (enquiries and renewals)
http://www.ssl.ox.ac.uk

This is a NORMAL LOAN item.

We will email you a reminder before this item is due.

Please see http://www.ssl.ox.ac.uk/lending.html
for details on:

- loan policies; these are also displayed on the notice boards and in our library guide.

- how to check when your books are due back.

- how to renew your books, including information on the maximum number of renewals.
Items may be renewed if not reserved by another reader. Items must be renewed before the library closes on the due date.

- level of fines; fines are charged on overdue books.

Please note that this item may be recalled during Term.